CHEMISTRY IN CONTEXT

A PROJECT OF THE
AMERICAN CHEMICAL SOCIETY

CHEMISTRY IN CONTEXT

APPLYING CHEMISTRY TO SOCIETY

A. TRUMAN SCHWARTZ
Macalester College

DIANE M. BUNCE
The Catholic University of America

ROBERT G. SILBERMAN
SUNY College at Cortland

CONRAD L. STANITSKI
University of Central Arkansas

WILMER J. STRATTON
Earlham College

ARDEN P. ZIPP
SUNY College at Cortland

Book Team

Editor *Megan Johnson*
Production Editor *Daniel Rapp*
Designer *K. Wayne Harms*
Art Editor *Mary E. Powers*
Photo Editor *Carrie Burger*
Permissions Coordinator *Mavis M. Oeth*

Wm. C. Brown Publishers

Vice President and General Manager *Beverly Kolz*
Vice President, Publisher *Earl McPeek*
Vice President, Director of Sales and Marketing *Virginia S. Moffat*
National Sales Manager *Douglas J. DiNardo*
Marketing Manager *Christopher T. Johnson*
Advertising Manager *Janelle Keeffer*
Director of Production *Colleen A. Yonda*
Publishing Services Manager *Karen J. Slaght*
Permissions/Records Manager *Connie Allendorf*

Wm. C. Brown Communications, Inc.

President and Chief Executive Officer *G. Franklin Lewis*
Corporate Senior Vice President, President of WCB Manufacturing *Roger Meyer*
Corporate Senior Vice President and Chief Financial Officer *Robert Chesterman*

Cover photo: © Hans Reinhard/Okapia/Photo Researchers, Inc.

Copyedited by Diane Beausoleil

Brief Contents

Contents

■ Chapter 10
The World of Plastics and Polymers 267

■ Chapter 11
Designing Drugs and Manipulating Molecules 295

■ Chapter 12
Nutrition: Food for Thought 325

■ Chapter 13
The Chemistry of Tomorrow 367

■ *Foreword*

The American Chemical Society has been very active in fostering innovative chemical education at all levels. The professionals of the Education Division in Washington, directed by Sylvia Ware, guided by the Society Committee on Education, and encouraged by the pro-active member Division of Chemical Education, have developed the resources needed for such worthy projects through the persistent solicitation of grants and contributions.

With this book the American Chemical Society presents a unique approach to college-level chemistry for non-science majors. Through the use of decision-making activities related to real-world societal issues that have important chemistry components, chemistry knowledge is built on a need-to-know basis that should also provide appropriate critical thinking skills for considering other issues that will be of importance in the twenty-first century.

The writing team that prepared the text has done an excellent job in integrating the societal issues with the science by including hands-on small-scale laboratory exercises, library work, and expository activities. These features should make the text as successful as the highly acclaimed *Chemistry in the Community* (*ChemCom*), developed by the American Chemical Society for non-science oriented high school students. Naturally, *Chemistry in Context* seeks to nurture more global and sophisticated points of view consistent with the level of the students, but the aim of informing and empowering the reader is common to both.

Ronald D. Archer
Advisory Committee Chair

■ Preface

*C*hemistry in Context: Applying Chemistry to Society is an innovative educational venture of the American Chemical Society (ACS). A brief history of the development of the book might help set it in its own context. In many ways, *Chemistry in Context* is the philosophical and intellectual offspring of *Chemistry in the Community (ChemCom)*, and its midwife and godmother is Sylvia Ware, Director of the Education Division of the ACS. *ChemCom*, first published in 1988 by Kendall-Hunt, is a high school text, intended primarily for college-bound students not planning a career in science. It differs from other secondary school books in that the chemistry is embedded in an exploration of broader social issues.

The success and wide acceptance of *Chemistry in the Community* prompted a decision to apply a similar pedagogical approach to college chemistry courses for non-science majors. Such courses represent the last opportunity in the formal educational system to promote chemical literacy. Yet, in most colleges and universities, chemistry has not been a popular elective or a favored means of satisfying science distribution requirements. Thus, there is a need to attract more potential poets, painters, philosophers, and politicians to the beauty and utility of chemistry. This book is an attempt to meet that need.

In the spring of 1989, the author of this preface was named as editor-in-chief and principal investigator of the project that culminated in *Chemistry in Context.* A writing team and an advisory committee were established, and work began almost immediately. The first preliminary edition was completed in less than a year and published in-house by the ACS. This version of the text was class-tested during the 1989–90 academic year by three members of the writing team with a total of about 200 students. On the basis of this experience and the comments of external reviewers, the book was revised and expanded for a second year of testing, this time with 1500 students in 18 colleges and universities. A few more test sites were added for the 1992–93 academic year, during which time there was yet another round of revisions and additions. All of this activity was supported by generous educational grants to the ACS Campaign for Chemistry. Wm. C. Brown Publishers was selected from a field of publishers who had submitted proposals to join ACS in this new venture.

The title of this text accurately describes the approach it embodies. Relevant issues are used to introduce the chemistry, and the science is set in its political, economic, social, and international context. The chemical content is considerable, but phenomena, methodology, and theory are presented as needed to help inform understanding of the issues. We thus seek to motivate readers; equip them to locate information; and develop their analytical skills, critical judgment, and the ability to assess risks and benefits. Above all, we hope to empower our readers to respond with reasoned and informed intelligence to the complexities of our modern technical age.

The pedagogical devices with which we hope to attain these ambitious goals include three types of activities. Those identified as "Your Turn" are fairly standard and straightforward drills and exercises—often involving simple calculations. The "Consider This" activities are open-ended and usually without a single "right" answer. Students may be asked to engage in risk-benefit analysis, to evaluate opposing viewpoints, to speculate on the consequences of a particular path of action, or to formulate and defend a position. Many of these assignments lend themselves to a variety of activities including library research, writing, role-playing, cooperative and small-group learning, debate, and/or discussion. Finally, in "The Sceptical Chymist," students are challenged to exercise their knowledge of chemistry (and other disciplines) and their capacity for critical thinking to check the accuracy and plausibility of assertions, especially those reported in the press.

Class testing has suggested that *Chemistry in Context* is surprisingly robust and compatible with a wide variety of courses, class sizes, students, institutions, and instructional styles. However, the authors believe that the unique character of the text can be best exploited if instructors are willing to modify their teaching practices to accommodate these innovations. Because such adaptation is often difficult, we have prepared an unusually complete and, we hope, helpful Instructors' Resource Guide to accompany the book. Furthermore, all the authors share the conviction that because chemistry is an experimental science, all students of chemistry should have the opportunity to experience chemical phenomena firsthand. To that end, a laboratory manual of over twenty tested experiments has been developed.

Chemistry in Context has been a collaborative effort involving the members of the writing team, the advisory board, our test site directors and instructors, reviewers and consultants, evaluators, and the able professional staff at the American Chemical Society and Wm. C. Brown Publishers. The manifold contributions of those whose names follow are gratefully acknowledged. The Society, its officers, members, and staff provided great financial, intellectual, and moral support to the authors, but never a hint of censorship or pressure. For that my colleagues and I are especially thankful.

A. Truman Schwartz
Editor-in-Chief and Senior Author
September 1993

■ Acknowledgments

■ Advisory Board

Ronald D. Archer, *University of Massachusetts, Chair*
William Beranek, Jr., *Indianapolis Center for Advanced Research*
Glenn A. Crosby, *Washington State University*
Alice J. Cunningham, *Agnes Scott College*
Joseph N. Gayles, *Morehouse School of Medicine*
Ned D. Heindel, *Lehigh University*
Glenn L. Taylor, *Shell Development Company*

■ Test Site Directors/Instructors

Joseph F. Bieron, *Canisius College*
Diane M. Bunce, *The Catholic University of America*
Colleen Byron, *Ripon College*
Timothy D. Champion, *Johnson C. Smith University*
Charles Eaker, *University of Dallas*
Gordon J. Ewing, *New Mexico State University, Las Cruces*
William Hendrickson, *University of Dallas*
George Gilbert, *Denison University*
Judith Kelley, *University of Lowell*
H. Graden Kirksey, *Memphis State University*
Werner Kolln, *Simpson College*
Ken Latham, *Lakewood Community College*
Joan Lebsack, *Fullerton College*
Marsha I. Lester, *University of Pennsylvania*
R. Bruce Martin, *University of Virginia*
Clifford Meints, *Simpson College*
Alan Pribula, *Towson State University*
Jeanne Robinson, *Seminole Community College*
Richard G. Scamehorn, *Ripon College*
Keith Schray, *Lehigh University*
A. Truman Schwartz, *Macalester College*
Robert G. Silberman, *State University of New York College at Cortland*
Charles Spink, *State University of New York College at Cortland*
Conrad L. Stanitski, *Mount Union College and University of Central Arkansas*

Wilmer J. Stratton, *Earlham College*
Dean Van Galen, *Northeast Missouri State University*
Arden P. Zipp, *State University of New York College at Cortland*
William Zuber, *Memphis State University*

■ Reviewers and Consultants

William F. Coleman, *Wellesley College*
Wendall H. Cross, *Georgia Institute of Technology*
Steven W. Effler, *Upstate Freshwater Institute*
John E. Frederick, *University of Chicago*
Mary L. Good, *Allied-Signal, Inc.*
B. Welling Hall, *Earlham College*
Shirley Holden Helberg, *Baltimore, MD*
A. H. Johnstone, *University of Glasgow*
Stephen Lindberg, *Oak Ridge National Laboratory*
Steven Longenecker, *Earlham College*
Kevin W. O'Connor, *Office of Technology Assessment*
Sarah Penhale, *Earlham College*
W. Ross Stevens, III, *E. I. du Pont de Nemours*
Sheila Tobias, *University of Arizona*
Chad A. Tolman, *E. I. du Pont de Nemours*
Karen Warren, *Macalester College*

■ Evaluators

Mary B. Nakhleh, *Purdue University*
Jack E. Rossmann, *Macalester College*
Timm Thorsen, *Alma College*

■ American Chemical Society Staff

Sylvia A. Ware, *Director, Education Division*
Janet M. Boese, *Head, Academic Programs Department*
T. L. Nally, *Chemistry in Context Project Director*
Alan I. Kahan, *Art Director*
Terrance Russell, *Manager, Professional Relations*
Jiwon P. Kim, *Program Assistant*
Robin Lindsey, *Program Assistant*

■ *To the Reader*

We live in a world that is shaped by science and technology. There are few aspects of our lives that are not, in some way or other, influenced by chemistry and its sister sciences. Many crucial social problems and political decisions are inextricably linked with science and it applications.

It is the thesis of this text that participants in the political processes should be informed by an understanding of the issues involved—both scientific and nonscientific. The areas where chemistry has an impact on society are far too important to be left to the chemists alone, or to the politicians, for that matter. In order to vote intelligently or even to read articles in the popular press, citizens must know something about the content and methods of chemistry. Therefore, it is the goal of this book to present chemistry, as it may be needed, to enable you, the reader, to understand a few carefully selected case studies. In every instance, the chemistry is set in its social, economic, and political context. The authors obviously cannot include all the current or potential social problems that involve chemistry. But it is our hope that the cases selected, the facts presented, and the habits of mind developed in our readers will empower them to live responsibly in the uncertain future.

As you study this text, we strongly encourage your participation in experimental laboratory investigations and in a variety of special activities that challenge you to demonstrate your skills and knowledge, to consider new issues and evidence, to find additional information, to speculate, to exercise your skepticism, and to weigh and debate controversial issues. The problems we invite you to consider are among the most important and most difficult that currently face the population of our planet. There are no easy answers, indeed, perhaps no "right" answers to many of the questions we pose. But the future depends on the information, knowledge, and wisdom that your generation can bring to our complex world.

1

The Air We
Breathe

Finally it shrank to the size of a marble, the most beautiful marble anyone can imagine.

Astronaut James Erwin, May 1969

Only a few men and women have actually observed what James Erwin saw in May 1969. But most of us have seen the spectacular photographs of the Earth taken from outer space. From that vantage point, our planet looks magnificent—a blue and white ball compounded of water, earth, air, and fire. It is home to thousands upon thousands of species of plants and animals, all interrelated in a global community. Over five billion of us belong to one particular species with a special responsibility to protect our beautiful marble.

A closer view from a satellite reveals more detail. The landforms visible in computer-enhanced photographs remind us of the great geological diversity of our planet and the many biological changes that have occurred in adaptation to these varied environments. We see mountains, forests, deserts, prairies, glaciers, jungles, lakes, and rivers. In some cases, these natural features are boundaries that, for better or worse, separate us into national communities.

But the communities that we know best are those closest to home—the cities, towns, ranches, and farms where we live, work, play, and sleep. Our neighbors, friends, and families are here. The people, customs, habits, and laws that create and constitute these regional environments shape each one of us.

As individuals, we simultaneously inhabit these concentric communities. Our personal lives are imbedded in our immediate surroundings, the countries we live in, and the entire globe. Changes in any of these environments affect us, and we, in turn, have obligations at each level of community. This book is about some of those responsibilities and the ways in which a knowledge of chemistry can help us meet them with intelligence, understanding, and wisdom.

■ Take a Breath

We begin by asking you to do something you do automatically and unconsciously thousands of times each day—take a breath. You certainly don't need textbook authors to tell you to breathe! A doctor may have encouraged your first breath with a well-placed slap, but from then on nature took over. Your childish threats to hold your breath forever probably did not worry your parents. They knew that within a minute or two you would involuntarily gasp a lungful of that invisible stuff we call air. Indeed, you could not survive more than 5–10 minutes without a fresh supply of air.

1.1   **Consider This**

One breath doesn't use much air, but what total volume of air do you inhale and exhale in a typical day? One approach to a question of this sort is simply to guess, and that might be an interesting way to begin. But educated guesses are more reliable than wild ones. By performing some simple experiments and making some estimates you can probably come up with a reasonably accurate answer. (Scientists do estimating all the time.) All you need to know is the approximate volume of one of your typical breaths and the number of breaths you take in some convenient time interval. Both of these will depend on your size, your physical condition, and your level of activity at the moment. Once you have made an estimate of the volume of your daily air intake, try to relate it to some familiar object such as a gallon jug, a home water heater, a small room, or a large room. Bring your answer to the next class session so that you can compare it with other members of the class.

You undoubtedly know why you need frequent breaths. Oxygen is the crucial substance that is necessary for life. But if you live in Los Angeles, New York, Mexico City or any of dozens of other urban centers, that lungful of air will probably contain other substances that are not good for your health. The health threat can even be so serious that laws are passed to curtail your normal ways of doing things in an effort to limit pollution, as the following newspaper headlines from this hemisphere attest:

CALIFORNIA SETS POLLUTION LIMITS ON LAWN EQUIPMENT. *Wall Street Journal,* December 17, 1990
MEXICAN CAPITAL RESTRICTS DRIVING: EFFORT TO IMPROVE AIR QUALITY KEEPS EVERY VEHICLE IDLE FOR ONE DAY A WEEK. *New York Times,* December 17, 1989
PERSONAL CARE SPRAYS CURBED IN SMOG FIGHT. *Los Angeles Times,* November 9, 1989

The bad news about polluted air is that you can't easily get away from it. If water is unsafe to drink it is possible to make it safe—or even to transport water from elsewhere. (It is an unfortunate truth, however, that people in some parts of the world do not have the luxury of resources and money to obtain safe drinking water.) If certain foods are unsafe, you can generally avoid them. Indeed for almost any common hazard in your environment, you can take steps to protect yourself. *But not for the air you breathe every day.* You need large quantities of air, you need a fresh supply every few seconds, and you need it 24 hours a day, regardless of where you are or what you are doing.

■ *Chapter Overview*

To be an informed (and healthy) citizen, you should know about the air you breathe. Specifically, you should know what pollutants are present, how they got there, why they are harmful to your health, and what can be done to make the air safer. In this chapter we will explore each of these issues. But understanding them from a scientific viewpoint requires some familiarity with certain basic concepts. Therefore, we will begin by looking at the chemical composition of air and the structure of the atmosphere. During the process you will encounter elements and compounds and the symbols and formulas used to represent these substances. In addition, you will encounter atoms and molecules, which provide a submicroscopic view of matter. Chemical reactions are at the very heart of chemistry, and we will consider a few of the more important reactions occurring in the atmosphere. That, in turn, will provide an opportunity to explore the powerful shorthand of the chemist—chemical equations. Thus armed, we return to a consideration of some of the most important air pollutants and their health effects. We will look at the sources of these pollutants and attempt to discover whether the extent of air pollution is increasing or decreasing. The chapter also includes a brief discussion of the difficult challenge of risk assessment—a theme that will frequently recur in this text. Chapter 1 ends by revisiting the breath that initiated it.

■ *What's in a Breath? The Composition of Air*

The air you breathe is a mixture of several substances. For the moment we will focus on only five: nitrogen, oxygen, carbon dioxide, argon, and water. The first four normally exist as gases. Although we usually think of water as a liquid, it can also be a gas, in which case we often call it "water vapor." Humid air contains more water vapor than does "dry" air. The normal composition of dry air is 78% nitrogen, 21% oxygen and 1% other gases. Figure 1.1 displays this information in the form of both a pie chart and a bar graph. Both of these are important, widely-used methods for displaying numerical information and we will use each at various places in this text. The

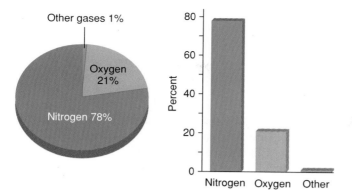

Figure 1.1
The composition of air.

pie chart emphasizes the fractions of the total, while the bar graph emphasizes the relative sizes of each. Regardless of how we present the data, notice that 99% of the total is made up of only two substances, nitrogen and oxygen. It is the 21% oxygen, of course, which is most immediately essential for sustaining life and without which we cannot survive more than a few minutes. Oxygen is absorbed into our bodies through the lungs and reacts with the foods we eat to release the energy needed for all life processes within our bodies. All life on Earth bears the stamp of oxygen. Indeed, it would be difficult to conceive of life on any planet without this remarkable element.

1.2   *Consider This*

We have become accustomed to living in an atmosphere with 21% oxygen. In such a world, cigarettes burn in about 5 minutes, a 3-inch diameter log in a fireplace burns in about 20 minutes, and you have already determined how many times a minute you inhale air. Burning, rusting, and all metabolic processes in humans, plants, and animals are dependent on oxygen. Speculate about what life on Earth would be like if the oxygen content in the atmosphere were doubled to about 40%.

Nitrogen is the most abundant substance in the air and constitutes over three-fourths of the air we inhale. However, it is much less reactive than oxygen, and it is exhaled from our lungs unchanged. Although nitrogen is essential for life and is a part of all living things, most plants and animals obtain the nitrogen they require from sources other than atmospheric nitrogen.

The remaining 1% of air is mostly argon, a substance so unreactive that it is said to be "chemically inert." There are also small amounts of carbon dioxide and water. The concentration of water vapor varies widely; it may be close to 0% in very dry desert air or 5–6% in a tropical rain forest. The amount of carbon dioxide (0.035%) is so small that it is common to describe its concentration as 350 **parts per million** (abbreviated as **ppm**). The fact that 0.035% and 350 ppm represent the same concentration can be demonstrated by setting up the corresponding ratios. Because percent means parts per hundred, we can write 0.035% as 0.035/100. Doing the division (or moving the decimal point two places to the left, which is the same thing) yields the following:

$$\frac{0.035}{100} = \frac{0.00035}{1}$$

Similarly, a concentration of 350 ppm means that out of every 1,000,000 particles in the air, 350 of them will be carbon dioxide particles. Again, we write out the ratio and perform the division.

$$\frac{350}{1,000,000} = \frac{0.00035}{1}$$

It follows that 0.035% = 350 ppm.

The relative concentrations of the major components of the atmosphere are nearly constant at all altitudes. However, you know from reading or from the experience of hiking in the mountains or flying in a jet plane that the air gets "thinner" with increasing altitude. As you climb up into the blanket of air, there is less of it, or, fewer particles in a given volume. Therefore, the **atmospheric pressure,** the force with which the atmosphere presses down on a given surface area, decreases with increasing altitude. Thus, in Boston at sea level the pressure is 14.7 pounds per square inch. A pressure of this magnitude is defined as 1 atmosphere (1 atm). In Denver, the "mile high city," the pressure is 12.0 pounds per square inch, or about 0.8 atmosphere. For every 5 kilometer increase in altitude (1 km = 0.62 mi, 5 km = 3.1 mi), the pressure drops by half. Somewhere above 100 km, the atmosphere simply fades into the almost perfect vacuum of empty outer space.

Note that Figure 1.2 includes the names given to regions of the atmosphere and a few reference points to help with orientation. The phenomena and the problems we will study in the first three chapters of this book will take us to several of these regions. The most familiar to us is the **troposphere,** the part of the atmosphere that lies directly

Figure 1.2

The pressure of the atmosphere at various altitudes.

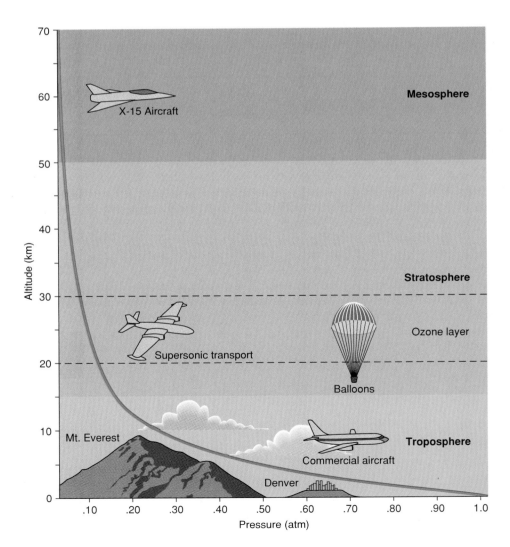

■ Table 1.1	*Typical Composition of Inhaled and Exhaled Air*	
Substance	**Inhaled air (%)**	**Exhaled air (%)**
Nitrogen	78	75
Oxygen	21	16
Argon	0.9	0.9
Carbon dioxide	0.03	4
Water	0–4	4

on the surface of the Earth. Our weather develops in the troposphere, it's where the most familiar kinds of air pollution occur, and it's where we live and breathe.

Table 1.1 gives information about the composition of the tropospheric air that we inhale and exhale. Clearly some changes have taken place that use up oxygen and give off both carbon dioxide and water. Not surprisingly, chemistry is involved. In the biological process of metabolism, oxygen reacts with foods to yield carbon dioxide and water. However, most of the water in exhaled air is simply the result of evaporation from the moist surfaces within the lungs.

These are not the only substances in air. At least 10 others are present in normal air in very small amounts. Furthermore, we know from common experience that there are differences in the air in different locations. We think of the air being different in such places as a pine forest, a bakery, an Italian restaurant, a locker room, a barnyard, or outside some chemical factories. We can't see or feel anything different, but we can *smell* a difference. The substances in Table 1.1 have no odor (to our noses), but many other chemical compounds do have pronounced odors. In fact, the human nose is an extremely sensitive detector. In some cases, only a minute trace of a compound is needed to affect the olfactory receptors in the nose. Thus, extremely small amounts of substances can have a powerful effect on our noses, as well as on our emotions.

The substances mentioned thus far are not hazardous to human health at their normal concentrations in air. When we talk about air pollutants, we are talking about other substances such as ozone, sulfur dioxide, and nitrogen oxides. These substances *are* hazardous to our health even at concentrations well below 1 ppm, and it is to some of these substances that we now turn our attention.

■ Minor Components—Major Pollutants

Some of the trace substances we can smell are pleasant and quite harmless; others can be quite dangerous. Our noses can warn us to avoid certain places because of the odors. But some of the most dangerous air pollutants have no odor at all, and others are dangerous at sufficiently low concentrations that they cannot be detected by smell. As a result, it is often necessary to rely on specialized scientific equipment to measure the presence and amounts of such substances in the air.

Table 1.2 lists the four most important gaseous air pollutants in the United States. Also included are their measured concentrations in the air above selected cities around the country. Notice that the concentrations are all expressed as parts per million, not as percentages (parts per hundred). We will first consider briefly the behavior and damaging effects of these pollutants. Then we will examine the data in the table.

The first substance, carbon monoxide, enters the bloodstream and disrupts the delivery of oxygen throughout the body. In extreme cases (as in long exposure to auto exhaust) it can lead to death. The health threat from carbon monoxide is especially serious for individuals suffering from cardiovascular disease, but healthy individuals are also affected. Visual perception, manual dexterity, and learning ability can all suffer.

Ozone is a special form of oxygen. It has a characteristic sharp odor which is frequently detected around electric motors and transformers. Unlike normal oxygen, ozone is very toxic. It affects the respiratory system and even very low concentrations

■ *Table 1.2*	*Major Gaseous Air Pollutants for Selected Cities in the United States, 1991*			
City	Carbon monoxide*	Ozone**	Sulfur oxides***	Nitrogen oxides***
(Permissible Limits)	9 ppm	0.12 ppm	0.030 ppm	0.053 ppm
New York City	10 ppm	0.18 ppm	0.018 ppm	0.047 ppm
Atlanta	7	0.13	0.008	0.025
Boston	4	0.13	0.012	0.035
Chicago	6	0.13	0.019	0.032
Los Angeles	16	0.31	0.005	0.055
Pittsburgh	6	0.12	0.024	0.031
St. Louis	7	0.12	0.016	0.026
San Francisco	8	0.07	0.002	0.031
Detroit	8	0.13	0.012	0.022
Houston	7	0.20	0.007	0.028
New Orleans	4	0.11	0.005	0.019
Indianapolis	6	0.11	0.012	0.018

*Second highest 8-hour average
**Second highest 1-hour average
***Yearly average
Source: Data from the United States Environmental Protection Agency.

will produce reduced lung function in normal, healthy people during periods of exercise. Symptoms include chest pain, coughing, sneezing, and pulmonary congestion. (You will soon see that tropospheric ozone has a number of other nasty properties, but that it is very important at high altitudes.)

Sulfur oxides and nitrogen oxides are respiratory irritants that can affect breathing and lower resistance to respiratory infections. People most affected include the elderly, young children, asthmatics, and individuals with emphysema. These gases also lead to acid rain, a subject to be explored in considerable depth in Chapter 5.

The numbers in the top row of Table 1.2 are the air quality standards established by the U.S. Environmental Protection Agency (EPA). Based on scientific studies, these are the maximum concentrations considered safe for the general population. Notice that the numbers are quite different—9 ppm for carbon monoxide, but only 0.12 ppm for ozone, and even less for the sulfur and nitrogen oxides. According to these numbers, ozone is almost 100 times more hazardous to breathe than carbon monoxide, and sulfur oxides are four times as hazardous as ozone. You should also be aware that the measured values reported in Table 1.2 are not the same sorts of averages

1.3 ■ Consider This

The top row in Table 1.2 gives the EPA limits for major air pollutants that pose a risk to human health. Answer the following questions for carbon monoxide.

  a. Which cities have carbon monoxide levels *above* the EPA limit?
  b. Of these cities, which exceed the EPA limit by a large amount? Express this in terms of a percent above the limit.
  c. Suggest some possible reasons why these particular cities would exceed the limit.

Do a similar analysis for the other three pollutants. You may wish to organize your observations into some kind of a table.

Finally, putting it all together, which of these cities would you characterize as having a serious air pollution problem (based on the EPA limits and actual levels for these four substances only)? Which city has the cleanest air?

for all four gases. For carbon monoxide and ozone, the data represent high values measured over relatively brief periods of time. Annual averages are reported for sulfur and nitrogen oxides.

■ *Classifying Matter: Elements, Compounds, and Mixtures*

In the past few pages, we have used a good deal of chemical terminology. Therefore, before proceeding, some clarification is probably necessary. For one thing, we need some understanding of the way chemists describe the composition of different types of matter. A breath of air is a **mixture**—a physical combination of two or more substances that may be present in variable amounts. Pure air and polluted air differ in composition, and we have seen that exhaled air differs from inhaled air. Much of the matter we encounter in everyday life is in the form of mixtures. The fuels we burn, the foods we eat, our bodies themselves are all complex mixtures of individual substances. By proper design of experiments, these individual substances can be isolated and shown to have reproducible characteristic properties and fixed composition.

The two most plentiful components of air are nitrogen and oxygen. These are examples of **chemical elements**—substances that cannot be broken down into simpler stuff by any chemical means. There are over 100 of these fundamental building blocks and all common forms of matter are composed of them. About 90 elements occur naturally on planet Earth and, as far as we know, in the universe. The remainder have been created from other elements through artificially induced nuclear reactions. (Plutonium is probably the best known of the artificially produced elements, although it does occur in very low concentrations in nature.) In some cases, the total amount of a newly-created form of matter is so small that there is some uncertainty (perhaps even some controversy) over just how many elements have been identified. The best estimate at the time this book went to press was 109.

An alphabetical list of the known elements and their **symbols** appears in the inside cover of this text. These symbols have been established by international agreement and are used throughout the world. The origins of some of the symbols are quite obvious to people who speak English. For example, oxygen is O, nitrogen is N, carbon is C, and sulfur is S. Others, such as Fe for iron and Pb for lead, are less apparent, unless one knows that the Latin names for iron and lead are *ferrum* and *plumbum*.

In addition to mixtures and elements, there is a third, very important class of matter: **chemical compounds.** A compound is a pure substance made up of two or more elements in a fixed, characteristic chemical combination and composition. The chemical name of a compound usually reveals the elements that make it up—at least to a chemist. Thus, the name "magnesium oxide" indicates that the compound consists of magnesium and oxygen. Similarly, "sodium chloride" implies a compound composed of the elements sodium and chlorine. Two important compounds in the atmosphere have already been mentioned: carbon dioxide and water. All pure samples of carbon dioxide contain 27% carbon and 73% oxygen by **weight** (or **mass**). The composition of carbon dioxide, no matter what its source, is always precisely the same. Moreover, its physical properties (such as boiling point) and its chemical reactivity are also constant and characteristic of the compound. However, carbon dioxide, like any compound, can be broken down by chemical reactions into its constituent elements. Thus, 100 grams of carbon dioxide will yield 27 grams of carbon and 73 grams of oxygen.

Although there are only about one hundred elements, over ten million chemical compounds have been isolated, identified, and characterized. Among these are some of the most familiar naturally occurring substances, such as water, salt, and sugar. But most of the known compounds do not exist in nature; they have been synthesized by chemists. The motivation for making new forms of matter is almost as varied as the compounds—to create a new plastic, to find a cure for AIDS, or just for the intellectual fun of it. In later chapters we will look at some examples of how chemists synthesize new compounds.

1.4 **Your Turn**

Classify each of the following as element, compound, or mixture

a. copper **b.** sulfur dioxide **c.** Coca-Cola
d. aspirin **e.** an aspirin tablet **f.** Bill Clinton

Ans. **a.** element **b.** compound **c.** mixture

■ Atoms, Molecules, and Formulas

The definitions of elements and compounds given above are valid without making any assumptions about the physical structure of matter. But it is now well established that matter is made up of **atoms.** An atom is the smallest unit of an element that can exist as a stable, independent entity (Figure 1.3). The word "atom" comes from the Greek for "uncuttable." Today we know that atoms consist of smaller particles, but atoms cannot be "split" by chemical or mechanical means. Atoms are extremely small, many billions of times smaller than anything we can see with our eyes, but they can be visualized through modern technology (Figure 1.3). Because of their small size, it is also true that there must be huge numbers of atoms in any sample of matter we can see or touch or weigh.

We can now refine our definitions of elements and compounds. Each element has a different kind of atom, but within a sample of an element, all atoms are chemically the same. By contrast, compounds are made up of the atoms of two or more elements. For example, the compound carbon dioxide has a ratio of two oxygen atoms for every carbon atom and is given the symbol CO_2. In a similar manner, water contains one oxygen atom for every two hydrogen atoms and is represented as H_2O. CO_2 and H_2O are examples of **chemical formulas,** which are symbolic representations of the elementary composition of chemical compounds.

Carbon dioxide, water, and millions of other compounds exist in molecules. A **molecule** is a combination of a fixed number of atoms, held together by chemical bonds in a certain geometric arrangement. Thus, one molecule of carbon dioxide consists of one carbon bonded to two oxygen atoms. Like the compound itself, the molecule is represented by the formula CO_2. Similarly, a molecule of water, H_2O, contains precisely two hydrogen atoms and one oxygen atom, never any other combination. The constant ratio of atoms that characterizes each compound explains why compounds always have fixed composition. Moreover, the atomic ratio of the elements in a compound is often reflected in its chemical name. This explains why CO_2 is called carbon *di*oxide, and why the chemical name for H_2O is *di*hydrogen oxide. Other prefixes used in the naming of compounds are listed in Table 1.3. Elements can exist either as molecules or as single atoms. Thus, the nitrogen and oxygen of the atmosphere are made up of diatomic N_2 and O_2 molecules, whereas argon consists of individual Ar atoms.

In summary, air is a mixture, which means that its composition can vary. Some of its components, such as nitrogen, oxygen, and argon, are individual elements; others, notably carbon dioxide and water, are compounds. All of the compounds and some of

Figure 1.3
A scanning tunnelling micrograph of gold and carbon (graphite) atoms. In this computer-enhanced image, the carbon atoms are the green spheres and the gold atoms deposited on the graphite surface are shown in yellow, red, and brown.

■ **Table 1.3**	**Prefixes Used in Naming Compounds**		
Prefix	**Meaning**	**Prefix**	**Meaning**
mono-	one	hexa-	six
di- or bi-	two	hepta-	seven
tri-	three	octa-	eight
tetra-	four	nona-	nine
penta-	five	deca-	ten

1.5 Your Turn

What information do these formulas convey about the following substances that are important atmospheric components? Give their correct chemical names.

a. CO   **b.** O_3 **c.** NO_2 **d.** SO_3

Ans. **a.** A molecule of the compound CO (carbon monoxide) consists of one atom of the element carbon combined with one atom of the element oxygen. **b.** The seldom used name of ozone is "trioxygen."

■ Table 1.4   Classification of Matter

	Observable Properties	Atomic Theory
Element	Cannot be broken down into simpler substances	Only one kind of atom
Compound	Fixed composition, but capable of being broken down into elements	Fixed combination of atoms
Mixture	Variable composition of elements and/or compounds	Variable assortment of atoms and/or molecules

the elements are present as molecules (for example, O_2 and CO_2) but some elements exist as atoms (for example, Ar). Furthermore, we can now give a more precise definition of the concentrations of gases in air. When we say that air is 78% nitrogen, we mean that in every 100 molecules of air, 78 of them will be N_2 molecules. Similarly, 350 ppm carbon dioxide means that there will be 350 CO_2 molecules in 1,000,000 molecules of air.

Table 1.4 summarizes the different kinds of matter in two ways: the behavior we can observe experimentally, and the theory or model scientists use to explain what is happening at the incredibly small atomic level.

■ Chemical Change: Reactions and Equations

The first pollutant mentioned in Table 1.2 is carbon *mon*oxide, CO, whereas unpolluted air contains carbon *di*oxide, CO_2. The formulas and names indicate that these two unique compounds are composed of the same two elements, but in different atomic ratios. It is not uncommon for two elements to form more than one compound. For example, hydrogen and oxygen form two compounds: water, H_2O, and hydrogen peroxide, H_2O_2. There are hundreds of different compounds composed of only carbon and hydrogen. Carbon monoxide and carbon dioxide both arise largely from the same major source, combustion. **Combustion** (or burning) is the rapid combination of oxygen with another material. For example, combustion reactions produce sulfur dioxide (SO_2) and water (H_2O) by the burning of sulfur and hydrogen, respectively.

Combustion is a major type of **chemical reaction,** a process whereby substances described as **reactants** are transformed into different substances called **products.** The process can be represented by a generalized expression called a **chemical equation.** At its most fundamental level, a chemical equation is very simple indeed.

$$\text{Reactant(s)} \rightarrow \text{Product(s)}$$

By convention, the reactants are always written on the left and the products on the right. The arrow represents a chemical transformation and is read as "is converted to." Thus, reactants are converted to products in the sense that the reaction creates products whose properties are different than those of the starting materials, the reactants.

Figure 1.4
The burning of charcoal in air.

The combustion of carbon to produce carbon dioxide (for example, the burning of charcoal in air in Figure 1.4) can be represented in several ways. One way is by a "word equation."

$$\text{Carbon} + \text{Oxygen} \rightarrow \text{Carbon dioxide}$$

It is much more common to use chemical symbols and formulas for the elements and compounds involved.

$$C + O_2 \rightarrow CO_2 \qquad (1.1)$$

This compact symbolic statement conveys a good deal of information to a chemist. A translation of equation 1.1 into words might read something like this: "One atom of the element carbon reacts with one molecule of the element oxygen (consisting of two oxygen atoms bonded to each other) to yield one molecule of the compound carbon dioxide (consisting of one carbon atom bonded to two oxygen atoms)."

It is possible to pack even more information into an equation by specifying the physical states of the reactants and products. A solid is designated by an "*s*" in parentheses following the symbol or formula; a liquid is designated by "*l*"; and a gas is indicated "*g*." Because carbon is a solid and oxygen and carbon dioxide are gases at ordinary temperatures and pressures, equation 1.1 becomes:

$$C(s) + O_2(g) \rightarrow CO_2(g) \qquad (1.2)$$

In this text we will designate the physical states of the substances participating in a reaction when that information is particularly important, but in many cases we will omit it for simplicity.

You will note that 1.1 is a true equation—what is on the left equals what is on the right. One carbon atom and two oxygen atoms are on the left side of the arrow and one carbon atom and two oxygen atoms are on the right. This is the test of a correctly balanced equation. The number and elementary identity of the atoms on the reactant side of the arrow must equal the number and identity of atoms on the product side. Atoms are neither created nor destroyed in a chemical reaction and their elementary identity does not change. Of course, atoms are rearranged during a chemical reaction. That is what a chemical change is all about, an alteration in chemical identity. Therefore, there is no requirement that the number of molecules must be the same on both sides of the arrow. In fact, the number of molecules changes during many reactions.

Equation 1.1 describes the combustion of pure carbon in an ample supply of oxygen. However, if the oxygen supply is limited, the product is carbon monoxide, CO, not carbon dioxide. Like any reaction, this one can be expressed in equation form. First, we write down the symbols and formulas of the reactants and products.

$$C + O_2 \rightarrow CO$$

A quick count of atoms on either side of the arrow reveals that the expression does not balance. There are two oxygen atoms on the left and one on the right. We cannot balance the equation by arbitrarily adding an additional oxygen atom to the product side. That would imply a different reaction. All we can do is to place whole-number coefficients before the various symbols and formulas. In simple cases like this, the coefficients can be found quite easily by inspection or simple experimentation. If we place a 2 in front of the symbol CO, it signifies two molecules of carbon monoxide. This corresponds to a total of two carbon atoms and two oxygen atoms. Because there are also two oxygen atoms on the left side of the arrow, the oxygen atoms have been equalized.

$$C + O_2 \rightarrow 2\,CO$$

But now the carbon atoms are out of balance. There are two on the right and only one on the left. Fortunately, this is easily corrected by placing a 2 in front of the C.

$$2\,C + O_2 \rightarrow 2\,CO \qquad (1.3)$$

The expression is now balanced. Note that it includes quantitative as well as qualitative information. It tells us how many carbon atoms and how many oxygen molecules react to form carbon monoxide. It is evident from comparing equations 1.1 and 1.3 that, relatively speaking, more oxygen is required to form CO_2 from carbon than is needed to form CO.

■ *Fire and Fuel: Burning Hydrocarbons*

Many fuels, including those obtained from petroleum, are **hydrocarbons,** compounds of hydrogen and carbon. The simplest of these is methane, CH_4, the primary component of natural gas. When a hydrocarbon burns completely, all of the carbon combines with oxygen to form carbon dioxide and all of the hydrogen combines with oxygen to form water. We can use this reaction to again illustrate the process of balancing equations. First we write formulas that qualitatively represent the combustion of methane.

$$CH_4 + O_2 \rightarrow CO_2 + H_2O$$

The expression is already balanced with respect to carbon, but not with respect to hydrogen and oxygen. It is easier to start with hydrogen because that element is present in only one substance on each side of the arrow: CH_4 on the left and H_2O on the right. Oxygen, on the other hand, is incorporated into both CO_2 and H_2O. To bring the hydrogen atoms into balance, we place the number 2 in front of H_2O.

$$CH_4 + O_2 \rightarrow CO_2 + 2 H_2O$$

The coefficient 2 multiplies H_2O and thus signifies 4 H atoms and 2 O atoms. Because a CO_2 molecule contains 2 O atoms, there are now a total of 4 O atoms on the right of the equation and 2 O atoms on the left. We equalize the number of atoms by placing a 2 before O_2.

$$CH_4 + 2 O_2 \rightarrow CO_2 + 2 H_2O \tag{1.4}$$

The nice thing about writing balanced equations is that you can always tell if you are correct by counting and comparing atoms on either side of the arrow.

Carbon:	Left:	1 CH_4 molecule × 1 C atom/CH_4 molecule	= 1 C atom
	Right:	1 CO_2 molecule × 1 C atom/CO_2 molecule	= 1 C atom
Hydrogen:	Left:	1 CH_4 molecule × 4 H atoms/CH_4 molecule	= 4 H atoms
	Right:	2 H_2O molecules × 2 H atoms/H_2O molecule	= 4 H atoms
Oxygen:	Left:	2 O_2 molecules × 2 O atoms/O_2 molecule	= 4 O atoms
	Right:	1 CO_2 molecule × 2 O atoms/CO_2 molecule	= 2 O atoms
		+ 2 H_2O molecules × 1 O atom/H_2O molecule	= 2 O atoms

1.6 | **Your Turn**

a. "Bottled gas" or liquid petroleum gas (LPG) is mostly propane, C_3H_8. Write a balanced equation for the combustion of propane to form carbon dioxide and water.

Ans. $C_3H_8 + 5 O_2 \rightarrow 3 CO_2 + 4 H_2O$

b. Cigarette lighters burn butane, C_4H_{10}. Write a balanced equation for this combustion reaction.

One of the most widely used hydrocarbon fuels in automobiles is gasoline, a mixture of dozens of individual compounds. One of the compounds is octane, C_8H_{18}. If a sufficient supply of oxygen is delivered to the auto engine when the gasoline burns, only carbon dioxide and water are formed.

$$2 C_8H_{18} + 25 O_2 \rightarrow 16 CO_2 + 18 H_2O \tag{1.5}$$

In practice, however, the amount of oxygen present and amount of time available for reaction (before the materials are ejected in the exhaust) means that some CO is formed. An extreme situation is represented by equation 1.6.

$$2\ C_8H_{18} + 17\ O_2 \rightarrow 16\ CO + 18\ H_2O \tag{1.6}$$

Most of the carbon released in automobile exhaust is in the form of CO_2, although there is also some CO. The relative amounts of these two gases indicate how efficiently the car burns the fuel, which is evidence of how well tuned the engine is. States that monitor auto emissions check for this by sampling exhaust emissions using a probe that detects CO. The measured CO concentrations are then compared to established standards (1.20% in the state of Minnesota, for example). A car with CO emissions that exceed the standard must be tuned so that it complies. Catalytic converters are also used to convert CO to CO_2, via equation 1.7.

$$2\ CO + O_2 \rightarrow 2\ CO_2 \tag{1.7}$$

1.7 *Your Turn*

Demonstrate that equations 1.5 and 1.6 are balanced by counting the number of atoms of each element on either side of the arrow.

1.8 *Your Turn*

Earlier in the chapter we mentioned that exhaled air contains carbon dioxide and water from the metabolism of foods. Carbohydrates, which are important food components, can be represented by glucose, $C_6H_{12}O_6$. Write a balanced chemical equation for the reaction of glucose with oxygen to form carbon dioxide and water.

■ Air Quality: Is it Getting Better or Worse?

Some information that may help us answer this question is presented in Figure 1.5, which was prepared from data gathered by the United States Environmental Protection Agency. This bar graph shows how the average concentrations of some important air pollutants have changed from 1975 to 1991. Included in the figure are the four gaseous pollutants listed in Table 1.2: sulfur dioxide, nitrogen oxides, carbon monoxide, and

Figure 1.5

Changes in average concentrations of air pollutants in the United States, 1975 to 1991. (Source: United States Environmental Protection Agency, *National Air Quality and Emissions Trends Report,* 1981, 1991.)

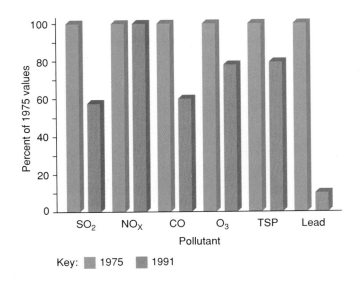

■ Table 1.5	Changes in Average Concentrations of Air Pollutants in the United States, 1975–1991		
Pollutant	**1975**	**1991**	**Change**
Sulfur Dioxide	0.0132 ppm	0.0075 ppm	43% decrease
Nitrogen Oxides	0.021 ppm	0.021 ppm	1% decrease
Carbon Monoxide	10 ppm	6 ppm	40% decrease
Ozone	0.147 ppm	0.115 ppm	22% decrease
Particulates	63 $\mu g/m^3$	47 $\mu g/m^3$	22% decrease
Lead	0.68 $\mu g/m^3$	0.048 $\mu g/m^3$	93% decrease

Source: United States Environmental Protection Agency, *National Air Quality and Emissions Trends Reports,* 1981 and 1991.

ozone. Because nitrogen forms several polluting compounds with oxygen, these oxides are often referred to collectively as NO_x, where x can represent different values. In addition, Figure 1.5 includes total solid particulates (TSP) and lead. The same information is also presented in Table 1.5.

Looking at the graph and the table, it is easy to see that the concentrations of most of these pollutants have indeed decreased. The most dramatic changes have been in lead, sulfur dioxide, and carbon monoxide. The concentrations of ozone and particulates have decreased by 22%, but nitrogen oxides have remained essentially constant. To understand why and how these changes occurred, we need to know something about the two primary sources of these pollutants: electric power generation (in coal-fired steam plants) and the automobile.

■ *The Case of the Deadly Fog*

In December 1952, a deadly fog settled over the city of London. It lasted for five days and resulted in approximately 4000 deaths from respiratory illness. A similar incident occurred in 1948 in Donora, Pennsylvania, a coal-mining community in the western part of the state. It caused illness in 40% of the population and 20 deaths. Fog is not normally toxic. Why were these cases different? The answer is that in both instances a thermal inversion had occurred, producing a layer of stagnant air that trapped smoke from coal combustion close to the ground. The combination of smoke with fog (or "smog") can produce an extremely toxic mixture. To understand what happened, we need to know something about the composition of coal and its combustion products. Although coal is mostly carbon and hydrogen, it is a complex mixture of variable composition. Most coals contain rock-like minerals and 1–3% sulfur. When coal is burned, the sulfur is converted into gaseous sulfur dioxide, and the minerals are converted into fine ash particles. If they are not intercepted, the particles and the sulfur dioxide gas go up the smoke stack.

Sulfur dioxide can react with more oxygen to form sulfur trioxide, SO_3.

$$2\ SO_2 + O_2 \rightarrow 2\ SO_3 \tag{1.8}$$

This reaction is normally quite slow, but it is much faster in the presence of the small ash particles. The particles are said to **catalyze** the reaction, that is, they increase the rate at which it occurs. These particles also aid another process: if there is high humidity in the air, the ash particles will promote the conversion of the water vapor into an aerosol of tiny water droplets which we call fog. An **aerosol** is a form of liquid in which the droplets are so small that they stay suspended in the air rather than settling out. Once sulfur trioxide is formed, it will dissolve readily in the water droplets to form an aerosol of sulfuric acid.

$$H_2O + SO_3 \rightarrow H_2SO_4 \tag{1.9}$$

When inhaled, the sulfuric acid aerosol droplets are small enough to be trapped in the lung tissue where they cause severe damage.

It is interesting to note that this so-called "London smog" requires three components simultaneously: water vapor, sulfur dioxide, and ash particles. The combination of the three is far worse than the sum of each of them acting separately. This interactive phenomenon is called **synergism.** Fortunately, such conditions are rare, as are "killer fogs." Nevertheless, sulfur dioxide is a major factor in acid precipitation, the topic of an entire chapter in this book.

Atmospheric levels of SO_2 have been decreasing slowly as a result of the Clean Air Act of 1970, which mandated reductions in power plant emissions. More stringent regulations were established in the Clean Air Act Amendments of 1990. In spite of this progress, SO_2 emissions in many parts of the United States still exceed the Federal standards, and progress will not come easily or cheaply. More information about the strategies and technologies available to reduce atmospheric SO_2 and their economic and political costs are discussed in Chapter 6.

■ *A Chemical Reactor Called a Car*

The automobile is currently the source of the nation's most serious air pollutants, nitrogen oxides and ozone. On the other hand, the greatest successes in reducing air pollution (by lowering concentrations of carbon monoxide and lead) have also involved the ubiquitous motor car. The United States has more automobiles per capita than any other nation. In 1987 we had 139 million cars, slightly more than one for every two Americans. In some cities, such as Denver, Houston, and Los Angeles, 90% of the working population commutes to work by car, often with only one person per vehicle. The vast majority of cars are powered by internal combustion engines, which run on gasoline.

The combustion reaction of octane (a major component of gasoline) has already been discussed. Ideally, from an energy and environmental standpoint, the only combustion products would be CO_2 and H_2O (see equation 1.5). But a modern high-performance automobile, capable of operating at high speeds and with fast acceleration, is a source of carbon monoxide and some incompletely burned fragments of gasoline molecules. In other words, the combustion reaction is not complete. There is either insufficient oxygen present or insufficient time in the cylinders for all the hydrocarbon to be burned to carbon dioxide and water.

The problem of carbon monoxide emissions from automobiles has been partially solved by the installation of catalytic converters that convert CO in the exhaust stream to CO_2 via reaction 1.7. These converters were introduced in 1975 and became required on all new cars soon thereafter. The bars in Figure 1.5 show the results. Despite a doubling of the number of automobiles between 1970 and 1990, the atmospheric concentration of carbon monoxide has *decreased* by nearly 40%.

There is even more good news. For many years, a compound named tetraethyl lead was added to gasoline to make it burn more smoothly by eliminating premature explosion or "knocking." It worked beautifully, but unfortunately the lead was released to the atmosphere through the tailpipe. Lead is a highly toxic element, a cumulative poison that can cause a wide variety of neurological problems, especially in children. Moreover, lead can also "poison" catalysts. With the advent of catalytic converters it became necessary to formulate gasoline without lead because that element destroys the effectiveness of the catalyst in the converter. Since 1976, all new cars and trucks sold in the United States have been designed to use only unleaded fuel. The result has been a dramatic decrease in lead emissions—from over 200,000 tons in 1970 to less than 20,000 tons in 1989.

1.9	*Consider This*
	Devise an advertising campaign, geared to first-time car buyers, for cars that produce less pollution but have poorer acceleration and a top speed of 50 mph.

Figure 1.6
Smog over Los Angeles.

▪ *Ozone, Nitrogen Oxides, and Photochemical Smog*

By far the most serious air pollution problem in urban areas today is **photochemical smog,** which results primarily from automobile use. Under the high-temperature and high-pressure conditions in an auto engine, nitrogen and oxygen react to form nitric oxide, NO, and nitrogen dioxide, NO_2. These nitrogen oxides are extremely toxic by themselves but they also react with unburned fragments of gasoline in the auto exhaust to form ozone (O_3) and several related substances that are severe respiratory irritants. There is a fascinating aspect to these reactions: they do not occur at night. In essence, solar energy provides the necessary "kick" to make the substances react. Because of the role of light, these chemical changes are called photochemical reactions and the result of these reactions is known as photochemical smog. This is the stuff of the brown Los Angeles haze and the source of most air pollution alerts in major cities (see Figure 1.6). Note that photochemical smog is very different from the London smog we discussed earlier, but it is no less serious.

The bad news is that the quantity of nitrogen oxides released to the atmosphere increased steadily up to 1980 and has only slightly decreased since then. The good news is that the situation would have been much worse (given the increased number of autos) if emission controls had not been mandated starting in 1970. The Clean Air Act of 1970 set tailpipe emission standards, to go into effect in 1975, and these were subsequently revised. Table 1.6 summarizes the changes in the federal standards since 1975, as well as the more restrictive California standards. Despite early claims from

■ *Table 1.6*	**National Tailpipe Emission Standards**	
Year	**Hydrocarbons (g/mile)**	**Nitrogen oxides (g/mile)**
1975	1.5 (0.9)	3.1 (2.0)
1980	0.41 (0.41)	2.0 (1.0)
1985	0.41 (0.41)	1.0 (0.4)
1990	0.41 (0.41)	1.0 (0.4)
1995	0.25	0.4
2003	0.125 if necessary*	0.2 if necessary*

*if 12 of 27 seriously polluted cities remain smoggy
(California standards in parentheses)

the auto industry that it would be impossible, or too costly, to meet the standards, the industry has, in fact, achieved these goals using improved catalytic converters, engine designs, and gasoline formulations. But the industry did not act until forced by Federal legislation to do so. It is also noteworthy that there were no significant changes in the standards during the decade of the 1980s, a situation not unrelated to a national administration that advocated less governmental regulation of private industry. The Clean Air Act of 1990, signed in November 1990, was the first major new clean air legislation in 20 years, but it did not come about easily. There was a great deal of political infighting as Congress and the administration struggled to find acceptable compromises.

■ *Air Quality at Home and Abroad*

Air pollution is primarily an urban problem, and more than fifty percent of all Americans live in cities with populations over 500,000. The major sources are the burning of coal and gasoline. Over the past 20 years there have been dramatic improvements. But we still have serious problems with several pollutants, notably nitrogen oxides and ozone. More than half of all Americans live in areas that do not now meet the national air quality standards designed to protect human health. Children and adults with chronic respiratory problems or heart disease are most at risk from exposure to these air pollutants, especially during vigorous physical activity. Furthermore there is evidence that the present air quality standards provide little margin of safety in protecting public health. The American Lung Association estimates that $50 billion in health benefits could be realized annually in the United States if air quality standards were met.

We face difficult political and economic choices. Are we willing to spend what would be needed to really clean up the air? What would happen if the political or economic situation changed and regulations were dropped or relaxed? Our environment is a fragile system, which could quickly revert to a status of severe air pollution.

International comparisons indicate that air pollution is no respecter of nations. It has the potential to be a major problem wherever there is industrial development and many automobiles. Although most industrially developed nations have been making progress in improving air quality, many countries have few or no controls on pollutant emissions. In some cases this is because of the political system. Often it is because of the enormous pressures to put economic development before environmental quality. Germany has ten times as many automobiles per square mile as the United States, yet it has fewer emission controls on cars (or electric power plants). As a result, parts of Germany face much more serious air pollution problems than we have in this country. The situation in Eastern Europe is especially bad because heavy industrialization has occurred without the constraints of pollution controls. Portions of the region have become nearly uninhabitable. A very low grade of coal, called "brown coal" is widely used and large lead smelters release huge amounts of lead into the atmosphere. In

1.10

Consider This

The Clean Air Act Amendments were signed into law by President Bush in 1990. This legislation gives the federal government extended powers to impose costly technological pollution controls on industry to help restore a higher level of air quality. The legislation is not without controversy, however. Senator Steve Symms (R-Idaho) believes the legislation is "so costly that it probably will do more damage to the economy than the good it may do for the air." Senate Majority Leader George Mitchell (D-Maine) takes the opposite position: "There are great costs to do nothing. . . . No single factor is as critical to our future as the health of the American people. And no single factor potentially undermines the health of every single American more than the quality of the air we breathe."

There are costs associated with implementing this legislation and costs associated with not implementing it. Identify some of these costs. Which are you willing to pay and which are you not willing to pay?

China, 28 northern cities have SO_2 and particulate concentrations that are 3–8 times higher than the limits set by the World Health Organization (WHO). In Mexico City, its 20 million residents breathe ozone levels that are more than 50% above WHO guidelines for most of the year. The ozone concentration in Mexico City has been measured at 0.37 ppm and that of sulfur dioxide at 0.039 ppm. (See Table 1.2 for comparison with cities in the United States.)

■ Taking and Assessing Risks

This chapter provides our first look at the subject of risk. This is an important topic, and one to which we will return repeatedly because it is implicit in many of the topics in this text. Indeed, it is an issue that is central to life itself, because everything we do carries a certain level of risk, although the levels vary greatly. We are often presented with warnings about certain activities because they are believed to carry high risk. For example, the law requires all cigarette packages to carry the message "WARNING: Smoking cigarettes may be dangerous to your health." Still other practices have been declared illegal because the level of risk is judged unacceptable to society. On the other hand, there are many other activities that carry no warning, presumably because the degree of risk is quite low or because the risks are unavoidable.

One of the characteristics of such a warning (because it is a characteristic of risk itself) is that it does not say that a specific individual *will* be harmed by a particular activity. It only indicates the statistical probability or chance that an individual will be affected. For example, if the odds of contracting a certain kind of cancer were reported to be one in 500, this means that, on average one person out of every 500 people would get the disease. Such predictions are not simply guesses, but are the result of evaluating scientific data and making predictions about the probabilities in an organized manner. Such studies are referred to as **risk assessment.**

For air pollutants, the assessment of risk requires knowledge of two factors: *toxicity* and *exposure*. In other words, it is necessary to consider the intrinsic hazard of a substance and the amount of the substance encountered. **Exposure** is the easier one to evaluate, because it depends simply on the concentration of the substance in the air, the length of time to which a person is exposed, and the amount of air inhaled into the lungs in a given time. As you already know, the latter depends on lung size and breathing rate. Concentrations in air are usually expressed either as parts per million (ppm), which is the number of pollutant molecules per one million air molecules, or as micrograms per cubic meter ($\mu g/m^3$). (A cubic meter of air is 1000 liters.)

Toxicity, on the other hand, is more difficult to determine accurately, in part because it is considered unethical to do controlled experiments with human subjects. This leaves scientists with three choices: human population studies, animal studies, and bacterial studies. Population studies involve collecting data on affected groups of people. For example, a researcher may determine what percentage of people who smoke one pack of cigarettes per day get lung cancer. Such studies are necessarily limited and may require many years of observation in order to obtain results that are statistically significant and reflect accurately the long-term risk. For this reason, animal studies have been a widely-used substitute. Animals are given controlled doses of the substance being tested and are observed for harmful effects. Aside from questions of animal rights, the problem is that we do not know with certainty whether specific animal species respond the same as humans. There is a growing awareness among scientists that animal studies must be interpreted with great caution. A more recent area of toxicity measurements relies on studies with bacteria. One important advantage of using bacteria is that they grow and reproduce very rapidly, thus many studies can be done quickly and inexpensively.

Even if data are available to calculate the risks from a given pollutant, we still have to ask what level of risk is acceptable and for what groups of people. Various government agencies are charged with establishing safe limits of exposure for the major air pollutants. Table 1.7 gives current air-quality standards established by the United States Environmental Protection Agency for the pollutants discussed in this chapter. Some states, including California and Oregon, have their own stricter standards.

■ *Table 1.7*	*National Ambient Air-Quality Standards, 1989*
Pollutant	**Limit**
Carbon monoxide	9 ppm over an 8-hour period, not to be exceeded more than once a year; 35 ppm for a 1-hour period, not to be exceeded more than once per year
Ozone	0.12 ppm for a 1-hour period, not to be exceeded more than once a year
Sulfur dioxide	0.03 ppm annual average; 0.14 ppm for a 24-hour period, not to be exceeded more than once per year
Nitrogen oxides	0.05 ppm annual average

1.11 ■ Consider This

There are important scientific and public policy questions involved in setting air-quality standards such as those in Table 1.7.

a. Why do some pollutants (carbon monoxide and sulfur dioxide) have several different standards in the above table?

b. Why do you think there is the stipulation "not to be exceeded more than once a year"?

c. Do you think there should be different standards for different population groups (the general public, industrial workers, the elderly, infants)?

d. Do you think it is appropriate for different states to set different standards?

■ Back to the Breath—at the Atomic Level

The maximum concentrations of pollutants specified in Table 1.7 seem very small, and they are. Nine CO molecules out of one million molecules of the mixture called air is a tiny fraction. But, as we will soon calculate, a breath of air contains a staggering number of CO molecules. This apparent contradiction is a consequence of the minuscule size of molecules and the immense numbers of them. Recall 1.1 Consider This. If you are an average sized adult in good physical condition, the capacity of your two lungs is approximately one liter (1 L), or about one quart. Determining the number of molecules in this volume of air is no easy task, but it can be done. As a result of experiments (as well as theories) we know that a typical breath contains more than 20,000,000,000,000,000,000,000 particles—molecules such as N_2 and individual atoms like Ar. The number is so huge that we will write it in **scientific notation,** to avoid turning the text into strings of zeroes. In scientific notation, this particular number is written as 2×10^{22}. The easy way to make this conversion is to simply count the number of zeroes to the right of the initial 2. There are 22 of them, and 22 becomes the exponent of 10. The number 2 is then multiplied by 10^{22} to obtain the number of gas particles in a breath.

Why $20,000,000,000,000,000,000,000 = 2 \times 10^{22}$ will take a bit more explaining. Remember that 10^{22} means 10 multiplied by itself 22 times. This is simply another case in the following series:

$$10^1 = 10$$
$$10^2 = 10 \times 10 = 100$$
$$10^3 = 10 \times 10 \times 10 = 1000$$

Note that 10^1 is 1 followed by 1 zero, 10^2 is 1 followed by 2 zeroes, and 10^3 is 1 followed by 3 zeroes. Continuing this pattern, 10^{22} is 1 followed by 22 zeroes. Therefore, 2×10^{22} must equal $2 \times 10,000,000,000,000,000,000,000$ or $20,000,000,000,000,000,000,000$.

The power of exponents can be illustrated by calculating the number of CO molecules in the breath you just inhaled. We will assume the breath contained 2×10^{22} molecules, and that the CO concentration in the air was the national ambient air-quality standard of 9 ppm. This means that out of every million (1×10^6) molecules of air, 9 will be CO molecules. To compute the number of CO molecules in the breath (let's call it n) we multiply the total number of air molecules by the fraction of carbon monoxide molecules.

$$n = 2 \times 10^{22} \text{ air molecules} \times \frac{9 \text{ CO molecules}}{1 \times 10^6 \text{ air molecules}}$$
$$= 18 \times \frac{10^{22}}{10^6} = 18 \times 10^{(22-6)} = 18 \times$$
$$10^{16} = 1.8 \times 10^{17} \text{ CO molecules}$$

In doing the calculation we multiplied 2×9 to get 18, and divided 10^{22} by 10^6 to get $10^{(22-6)}$ or 10^{16}. An equivalent way of writing 18×10^{16} is 1.8×10^{17}. (If all of this is coming at you a little too fast, please consult Appendix 2.)

It may sound surprising, but it would be more accurate to round off the answer and report it as 2×10^{17} CO molecules. Certainly 1.8×10^{17} looks more accurate, but the data that went into our calculation was not very exact. The breath contains *about* 2×10^{22} molecules, but it might be 1.6×10^{22} or 2.3×10^{22} or some other number. The jargon is that 2×10^{22} expresses a physically-based property to "one **significant figure.**" Only one digit, the initial 2, is used. That means that the number of molecules in the breath is closer to 2×10^{22} than to 1×10^{22} or to 3×10^{22}, but we can't say much beyond that. Similarly, unless the analytical data is very good, the concentration of carbon monoxide is also known to only one significant figure, 9 ppm. The product 2×9 equals 18. That is certainly correct mathematically, but this problem is based on physical data. The answer 18×10^{16} or 1.8×10^{17} CO molecules includes two significant figures, the 1 and the 8. It implies a level of knowledge that is not justified. The rule is that you cannot improve the accuracy of experimental measurements by ordinary mathematical manipulations like multiplying and dividing. The accuracy of a calculation is limited by the *least accurate* piece of data that goes into it. Therefore, the number of digits used to report a result should be the same as the number of digits (significant figures) in the least accurate piece of data. In this case, both the number of molecules in the breath and the concentration of CO were known to one significant figure, and therefore, the answer must also contain only one significant figure, hence 2×10^{17}.

You may well question the significance of all of this talk about significant figures, but it is very important in interpreting numbers. It has been observed that "figures don't lie, but liars can figure." Numbers often lend an air of authenticity to newspaper or television stories, so the popular press is full of numbers. Some are meaningful and some are not, and the informed citizen must struggle to discriminate between the two types. For example, the assertion that the concentration of carbon dioxide in the atmosphere is 348.553791 parts per million should be taken with a rather large grain of salt; the estimate of 350 ppm is reasonable.

1.12   *Your Turn*

To better help you comprehend the magnitude of the 2×10^{17} CO molecules in just one of your breaths, assume that they were equally distributed among the 5 billion (5×10^9) human inhabitants of the Earth. Calculate each person's share of the 2×10^{17} CO molecules you just inhaled.

Ans. 4×10^7 or 40,000,000 CO molecules per person.

There are other ways in which numbers sometimes introduce ambiguity. You have just encountered some conflicting information. The concentration of CO in air is very small, 9 parts per million. Nevertheless, the number of CO molecules in a breath is almost unimaginably large, 2×10^{17}. Both statements are true. The significance of these numbers is that it is impossible to completely remove pollutant molecules from the air. "Zero pollutants" is an unattainable goal; you could not even determine whether it had been achieved. At present, our most sensitive and sophisticated methods of chemical analysis are capable of detecting one target molecule out of a trillion (10^{12}). One part per trillion corresponds to the width of a human hair in the distance around the earth, a single second in 320 centuries, or a pinch of salt in 10,000 tons of potato chips. And yet, a chemical could be undetectable at this level, and a breath might still include 2×10^{10} or 20,000,000,000 molecules of the substance.

A breath of air typically contains molecules of hundreds—perhaps thousands—of different compounds, most in minuscule concentrations. For almost all of these substances, it is impossible to say whether the origin is natural or artificial. Indeed, many trace components, including the oxides of sulfur and nitrogen, come from both natural sources and those related to human activity. And, as with all chemicals, "natural" is not necessarily good and "human-made" is not necessarily bad. As you learned a few paragraphs ago, what matters is toxicity, exposure, and the assessment of risk.

In addition to being extremely small, the particles in your breath possess other remarkable characteristics. In the first place, they are in constant motion. At room temperature and pressure, a nitrogen molecule travels at about 1000 feet per second and experiences approximately 400 billion collisions with other molecules in that time interval. Relatively speaking, though, the molecules are quite far apart. The actual volume of the molecules making up the air is only about 1/1000th of the total volume of the gas. If the particles in your one liter breath were all squeezed together, their volume would be about 1 milliliter (1 mL)—about one third of a teaspoon. Sometimes people mistakenly think that air is empty space. It's 99.9% empty space, but the matter that's in it is literally a matter of life and death!

Moreover, it is matter that we continuously exchange with other living things. The carbon dioxide that we exhale is used by plants to make the food we eat, and the oxygen that plants release is essential for our existence. Our lives are linked together by the elusive medium of air. With every breath we exchange millions of molecules with each other. As you read this, your lungs contain 4×10^{19} molecules that have been previously breathed by other human beings, and 6×10^8 molecules that have been breathed by some *particular* person—say Julius Caesar, Marie Curie, or Martin Luther King, Jr. Pick your favorite hero—your body almost certainly contains atoms that were once in his or her body. In fact, the odds are very good that right now your lungs contain one molecule that was in Caesar's *last* breath. The consequences are breathtaking!

1.13 *The Sceptical Chymist*

In 1661, the Honorable Robert Boyle (1627–1691), an early investigator of the properties of air, published a book that proved to be very influential in the development of the infant science of chemistry. It had a wonderful title: *The Sceptical Chymist: or Chymico-Physical Doubts & Paradoxes. . . .* In addition to stressing the importance of experimentation, Boyle observed that scientific truth would be more solidly established "if men would more carefully distinguish those things that they know from those that they ignore or do but think." The popular press is full of statements and stories that seem to confuse what is known and what is thought. Modern "chymists" and students of "chymistry" would be well advised to develop the critical habit of mind that leads them to doubt and investigate these seeming paradoxes. At various points throughout the text, we will introduce such passages and invite your skepticism and participation.

Our first example comes from the text itself. We just claimed that your lungs currently contain one molecule that was in Caesar's last breath. That assertion is based on some assumptions and a calculation. We are not asking you to reproduce the calculation, but rather to identify some of the assumptions and arguments that we might have used. Does our estimate seem reasonable? If not, why not?

1.14 *The Sceptical Chymist*

The following passage comes from a newspaper article about a woman who is very sensitive to chemicals and has written several books on the subject. She is quoted as saying: "I had to go through the whole process of unlayering all my reactions until I got down to not reacting anymore, and when I got down to not reacting, everything in my house was natural. There were no chemicals." Analyze and, if appropriate, criticize this statement, both with respect to its accuracy of expression and to the correctness of the chemistry.

■ *Conclusion*

The air we breathe has a personal and immediate effect on our health. Our very existence depends on having a large supply of relatively pure, unpolluted air with its essential elements, oxygen and nitrogen, and two compounds, water and carbon dioxide, that are also necessary for life. But air is often polluted with toxic substances such as carbon monoxide, ozone, sulfur oxides, and nitrogen oxides. This is true especially in the urban environments of our large cities—the very places where the majority of Americans live. The major pollutants are, for the most part, relatively simple chemical substances. Carbon monoxide and the oxides of sulfur and nitrogen are compounds that exist as molecules made from atoms of their constituent elements. These compounds are formed by chemical reactions, often as unavoidable consequences of our dependence on fossil fuels for energy production in power plants and in internal combustion engines. Over the past twenty years, governmental regulations and modern technology have resulted in large reductions in many pollutants. But it is not possible to reduce pollutant concentrations to zero because of the minuscule size of atoms and molecules and their immense numbers. Rather we must ask what the risk is from a given level of pollutant and then what level of risk is acceptable for various population groups.

The air we breathe, with its life-sustaining oxygen is, of course, very close to the surface of the earth. But the Earth's atmosphere extends upward for considerable distance and contains other substances which are also essential for life on this planet. In the next two chapters we will consider two of these substances and how they may be changing as a result of human activities.

■ *References and Resources*

Cortese, A.D. "Clearing the Air." *Environmental Science & Technology* **24** (1990): 442–48.

Ember, L.R. "President's Clean Air Bill Gets Mixed Reviews." *Chemical & Engineering News*, Aug. 7, 1989: 26–27.

———. "Clean Air Debates Will Pivot on Costs of Compliance." *Chemical & Engineering News*, Jan. 22, 1990: 15–16.

———. "Conference Committee Tackles Clean Air Legislation." *Chemical & Engineering News*, July 23, 1990: 17–18.

Graedel, T.E. and Crutzen, P.J. "The Changing Atmosphere." *Scientific American* **261**, Sept. 1989: 58–68.

Lemonick, M.D. "Forecast: Clearer Skies." *Time*, Nov. 5, 1990: 33.

National Air Quality and Emissions Report, 1989. Washington: U.S. Environmental Protection Agency, 1991.

Schneider, K. "Ambitious Air Pollution Bill Sent to White House." *New York Times*, Oct. 28, 1990: A28.

U.S. Congress, Office of Technology Assessment. *Catching Our Breath: Next Steps for Reducing Urban Ozone* (Summary). Washington: U.S. Government Printing Office, 1989.

■ *Experiments and Investigations*

Note: The experiments listed at the end of chapters are particularly relevent to the topics considered. Instructions for these experiments are given in the laboratory manual that accompanies this text.

1. Preparation and Properties of Atmospheric Gases I

2. Preparation and Properties of Atmoshperic Gases II

■ *Exercises*

Note: Answers are provided in Appendix 4 for exercises printed in color. Exercises marked with an asterisk are particularly challenging; some require extensions of the concepts presented in the text.

1. A mixture of gases is prepared by combining 1.2 liters of oxygen, 1.8 liters of nitrogen and 0.6 liters of carbon dioxide for photosynthesis experiments. Determine the percentage of each gas in this mixture (by volume).

2. The percentages of gases in a mixture can be calculated on the basis of the relative numbers of molecules (or volumes) or on the basis of the relative masses. The concentration of oxygen in air is 21% based on numbers of molecules and 23% based on mass. What does this imply?

3. Express the 0.9% argon content of air in ppm (parts per million).

4. A sample of very dry air contains 82 ppm of water. Express this quantity of water as a percentage.

5. Determine the pressure of the atmosphere at the bottom of the ozone layer (altitude = 20 km) according to Figure 1.2.

6. Find the pressure of the atmosphere at the summit of Mt. Everest (8.8 km above sea level) from Figure 1.2. If the percentage of oxygen in the air at that altitude is the same as it is at sea level, why do most people who climb Mt. Everest require oxygen masks to survive?

7. Name the compounds with the following formulas.

    a. CS_2 (a very smelly liquid)

    b. KBr (somewhat similar to salt)

    c. NI_3 (an explosively unstable compound)

8. Write formulas for the following compounds.

    a. sodium fluoride (an anti-cavity ingredient in toothpaste)

    b. carbon tetrachloride (formally used as a dry cleaning solvent, but now banned)

    c. dinitrogen oxide (the anesthetic nitrous acid or "laughing gas")

9. According to Table 1.1, the percentage of nitrogen in inhaled air is 78% while that in exhaled air is 75%. Account for this difference even though nitrogen is not removed from air during breathing.

10. According to the information in the text what percentage of the molecules in a typical breath have been breathed previously by

    a. another human being?

    b. a particular human being?

11. Write balanced chemical equations for the following reactions.

    a. Nitrogen (N_2) reacts with oxygen (O_2) in an automobile engine to form

        i. nitric oxide (NO)

        ii. nitrogen dioxide (NO_2)

    b. Hydrogen (H_2) burns in oxygen (O_2) to form water (H_2O)

    c. "Gasohol" (C_2H_5OH) reacts with oxygen to form carbon dioxide (CO_2) and water (H_2O)

    d. Glucose ($C_6H_{12}O_6$) reacts with limited oxygen (O_2) to form ethanol (C_2H_5OH) and carbon dioxide (CO_2)

12. There are about 2×10^{22} molecules in one liter of air.

    a. How many molecules will there be in one cubic meter (1.0×10^3 liter)?

    b. How many molecules will there be in one milliliter (1.0×10^{-3} liter)?

    c. Assuming that air is 21% oxygen, how many molecules in (a) and (b) will be oxygen molecules?

13. The text reports that at 25°C and a pressure of 1 atmosphere, a nitrogen molecule travels 1000 feet per second. Convert this speed to miles per hour. (1 mile = 5280 feet)

14. Write the following numbers in scientific notation.

    a. 100,000

    b. 237.5

    c. 0.0058

15. Write the following numbers in conventional notation.

    a. 1×10^{-5}

    b. 7.49×10^3

    c. 3.8×10^{-2}

16. Some of the air-quality standards in Table 1.7 cover a long period of time (for example one year for nitrogen oxides). Others (such as carbon monoxide and ozone) are based on much shorter time intervals. Speculate on the reasons for these differences.

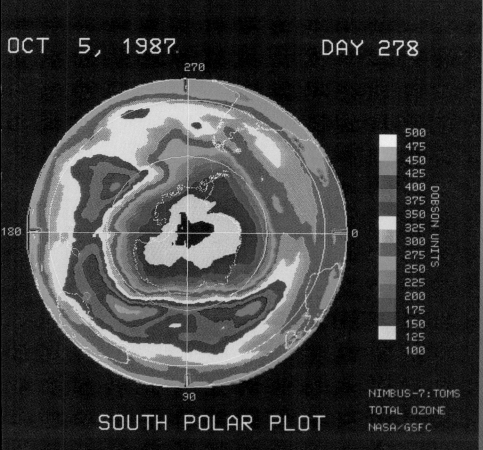

OCT 5, 1987.   DAY 278

270

180   0

90

SOUTH POLAR PLOT

500
475
450
425
400
375
350
325
300
275
250
225
200
175
150
125
100

DOBSON UNITS

NIMBUS-7:TOMS
TOTAL OZONE
NASA/GSFC

2

Protecting the Ozone Layer

∎

October 31, 1992

Ozone Hole Over Antarctica is Widening

San Francisco Chronicle

The ozone hole over Antarctica covered a record area this year, stretching over 9 million square miles, about three times the size of the continental United States, according to the World Meteorological Organization.

The hole, about 25 percent larger than in past years, spread over the southern part of Tierra del Fuego, a populated island across the Strait of Magellan from the South American mainland.

The levels "are by far the lowest ozone values ever observed at these inhabited latitudes," said a bulletin by the organization.

Researchers attributed the increased severity in part to weather but mostly to higher levels of man-made chemicals called chlorofluorocarbons, or CFCs, in the atmosphere. Though being phased out throughout much of the world, CFCs are long-lived.

"What we are looking at is ozone depletion caused by CFCs that were released back in 1987 and 1988," said F. Sherwood Rowland, a chemistry professor at the University of California at Irvine. "The expectation is that it will probably continue to get worse in the stratosphere for another decade or so."

The stratospheric ozone layer protects Earth from dangerous ultraviolet radia-

tion, and the loss of ozone is expected to increase skin cancers, possibly interfere with the human immune system and reduce crop yields. The Antarctic ozone hole was discovered in 1985.

CFCs, previously used as aerosol propellants, continue to be employed as coolants and in the making of plastics.

Fortunately for populations living in areas where the ozone layer was sharply reduced in 1992, the thinning occurred during seasons when ultraviolet radiation is normally low, not in summer.

In the Northern Hemisphere, stratospheric ozone declined sharply over parts of Western Europe this year, Rowland said.

("Ozone Hole Over Antarctic is Widening," by Maura Dolan. Copyright © 1992 *Los Angeles Times*. Reprinted by permission.)

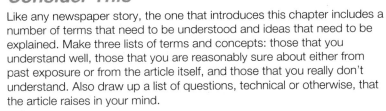

2.1   Consider This

Like any newspaper story, the one that introduces this chapter includes a number of terms that need to be understood and ideas that need to be explained. Make three lists of terms and concepts: those that you understand well, those that you are reasonably sure about either from past exposure or from the article itself, and those that you really don't understand. Also draw up a list of questions, technical or otherwise, that the article raises in your mind.

The article that introduces this chapter was published in the *San Francisco Chronicle* Saturday, October 31, 1992. It is typical of dozens of articles that appear daily in American newspapers—articles about issues that relate everyday life and the chemical sciences. The story describes a problem, but it leaves much unsaid. Some of the words used are part of the specialized vocabulary of science—ozone, ultraviolet radiation, chlorofluorocarbons. What exactly do these terms mean? By what mechanism does the stratospheric ozone layer protect Earth from ultraviolet radiation? Why is loss of ozone expected to increase skin cancers? What is the role of chlorofluorocarbons in ozone depletion? What is the evidence for the involvement of CFCs and how do these substances interact with ozone?

Such chains of interconnected questions make chemistry fascinating (and sometimes frustrating) to those who practice it. The technical issues concerning ozone depletion can involve hundreds of scientists for years of rewarding and illuminating research. But for the average citizen (and that includes the chemist, too) many of the crucial and complex issues lie in the grey area where society and chemistry intersect. The *Chronicle* article raises questions that transcend the strictly scientific: How serious is the problem of ozone depletion and increased exposure to ultraviolet radiation? Will it have a significant global impact and will it affect me personally? What can be done about the problem? Will the proposed measures work and how much will they cost?

■ Chapter Overview

In this chapter, we attempt to address the questions identified above by considering both scientific and societal issues. We begin by investigating the properties of ozone, and we soon find that an understanding of how it acts in the stratosphere to filter the Sun's harmful radiation requires some knowledge of its molecular structure and the nature of light. This leads to a section describing some fundamental properties of atomic structure and another in which these ideas are used to predict the molecular structures of a number of substances, including ozone. We next turn our attention to sunlight in particular and radiation in general. We find that light is strangely schizophrenic—it can behave both like waves and like little particles of energy. The particulate properties are especially useful in describing how ozone and oxygen molecules absorb ultraviolet radiation and how radiation damages biological materials. Two sections deal with the formation and fate of ozone and its distribution in the atmosphere. We next turn to an analysis of the various mechanisms for the depletion of stratospheric ozone, some involving naturally occurring chemicals and others involving synthetic compounds. Among the latter are the chlorofluorocarbons, whose properties and uses form the subject of another section. After a discussion of the Antarctic ozone hole, the chapter concludes with a consideration of the social and technical problems associated with reducing chlorofluorocarbon emissions and developing substitutes.

■ Ozone: What Is It?

The central substance in this chapter is ozone. If you have ever been near a sparking electric motor or an arc welding machine or in a severe lightning storm, you have probably smelled it. The odor is unmistakable, but hard to describe. One can smell concentrations as low as 10 parts per billion (ppb)—10 molecules out of one billion. Appropriately enough, the name "ozone" comes from a Greek word meaning "to smell."

Ozone is oxygen that has undergone rearrangement from the normal diatomic molecule, O_2, to a triatomic form, O_3. A simple chemical equation summarizes the reaction.

$$\text{Energy} + 3\,O_2 \rightarrow 2\,O_3 \tag{2.1}$$

We have inserted a reminder that energy must be absorbed in order for this reaction to occur, which accounts for the fact that ozone forms when oxygen is subjected to electrical discharge.

Ozone is called an **allotrope** or **allotropic form** of oxygen. Allotropes are two forms of the same element that differ in their molecular or crystal structure, and hence in their properties. The familiar allotropes of carbon, diamond and graphite, have different crystal structures (see Chapter 10). Common diatomic oxygen, O_2, and triatomic ozone, O_3, obviously differ in molecular structure. This variance is responsible for slight differences in the physical and chemical properties of the two allotropes. For example, ordinary oxygen is odorless. It condenses and changes from a colorless gas to a light blue liquid at $-183°C$ and a pressure of one atmosphere. Ozone is more easily liquified, changing its physical state from gas to a dark blue liquid at $-112°C$. Because ozone is chemically more reactive than oxygen, O_3 is used in the purification of water and the bleaching of paper pulp and fabrics. At one time it was even advocated as a deodorant for air in crowded interiors.

You learned in Chapter 1 that, at ground level, ozone contributes to photochemical smog and other forms of air pollution. But what is detrimental in one region of the atmosphere may be essential in another. In the stratosphere, at an altitude of 20 to 30 km where its concentration is the highest, ozone performs most of its filtering function on ultraviolet light. That process involves the interaction of matter and radiant energy, and to understand it requires knowledge about both of these fundamental topics. We turn first to a submicroscopic view of matter.

■ *An Aside on Atoms and Electrons*

The chemical and physical properties of oxygen and ozone and the interaction of these allotropes with sunlight are intimately related to the structure of the O_2 and O_3 molecules. Before we can speak about molecular structure, we must consider the atoms from which molecules are formed. You will recall from Chapter 1, or from your previous study, that each element consists of its own distinctive, characteristic atoms. During the twentieth century, chemists and other scientists have made great progress in discovering details about structure of atoms and the particles that make them up. The physicists have been almost too successful—they have found more than 200 subatomic particles. Fortunately, most chemistry can be explained with only three.

We now know that every atom has at its center a minuscule **nucleus.** This nucleus is composed of particles called **protons** and **neutrons.** Protons are positively charged and neutrons are electrically neutral, but both have almost exactly the same mass. This mass is about equal to that of a hydrogen atom. Indeed, the protons and neutrons in the nucleus account for almost all of an atom's mass. Well beyond the nucleus are the **electrons** that define the outer boundary of the atom. An electron has a mass equal to only about 1/2000th the mass of a proton or neutron. Moreover, an electron has a negative electrical charge that is equal in magnitude to that of a proton, but opposite in sign. The charge and mass properties of these particles are summarized in Table 2.1.

■ **Table 2.1**	**Properties of Subatomic Particles**	
	Relative Mass	**Relative Charge**
Proton	1	+1
Neutron	1	0
Electron	1/1838	−1

In any electrically neutral atom, the number of electrons equals the number of protons. This number is called the **atomic number.** The atomic number is very important because it determines the elementary identity of the atom. Each element has its own characteristic atomic number. The simplest atom is that of hydrogen. Each hydrogen atom contains one electron and one proton. The atomic number of hydrogen is thus 1. The elements are often arranged in order of increasing atomic number. The names, symbols, and atomic numbers of the first ten appear below.

Hydrogen							Helium
H							He
1							2

Lithium	Beryllium	Boron	Carbon	Nitrogen	Oxygen	Fluorine	Neon
Li	Be	B	C	N	O	F	Ne
3	4	5	6	7	8	9	10

These elements are arranged in a manner characteristic of the **periodic table** reproduced inside the cover of the book. Although variants of the periodic table were proposed by a number of scientists, the arrangement is most closely associated with Dmitri Mendeleev (1834–1907), a Russian chemist who first published his ideas in 1869. Mendeleev organized the elements according to their atomic masses and their chemical and physical properties, placing similar substances in the same columns or families. He worked before the discovery of subatomic particles, and without any knowledge of atomic structure. Indeed, atomic number does not appear on Mendeleev's original periodic tables. Nevertheless, his familiarity with the properties of the elements was so great that the fundamental form of the table has remained unchanged for well over a century. Over the past 75 years, our growing understanding of atomic structure has provided an explanation of why the properties of the elements are regularly repeated when they are arranged in order of increasing atomic number.

2.2   *Your Turn*

Using the periodic table as a guide, specify the number of electrons and protons in each atom of the following elements.

a. calcium (Ca)   **c.** gold (Au)
b. copper (Cu)   **d.** uranium (U)

Ans. **a.** 20 electrons, 20 protons **b.** 29 electrons, 29 protons

---

Today we know that the periodicity of properties is chiefly the consequence of the number and arrangement of electrons in the atoms of the elements. It can be demonstrated by experiment and calculation that the electrons are arranged in levels or shells about the nucleus. The electrons in the innermost shell are the most strongly attracted by the oppositely charged nucleus. The greater the distance between an electron and the nucleus, the weaker the attraction. We say that the more distant electron is in a higher energy level, which means that the electron itself possesses more energy.

An important feature of these energy levels is the fact that they have maximum electron capacities and are particularly stable when they are fully filled. The inner level, corresponding to the smallest electron orbit, can hold only two electrons. The second level has a maximum capacity of eight, and the higher levels are also particularly stable when they contain eight electrons.

Figure 2.1 provides schematic representations of a hydrogen atom and a sodium atom (atomic number 11). The electrons are pictured as moving in orbits about the nucleus. Reality is a good deal more complicated and abstract. For one thing, the relative size of the nucleus and the atom are very much out of proportion. If the nucleus of a hydrogen atom were the size of the central dot in Figure 2.1, the orbit of the electron would be about 30 feet in diameter. An atom is thus mostly empty space. Another oversimplification in the figure is that electrons do not really follow specific circular orbits. Rather, the distribution of electrons in an atom is best represented by probability and statistics.

Because it is difficult to accurately picture an atom, Figure 2.2 simply lists the numbers of electrons in the atoms of the first 18 elements. Note that the figure specifies the numbers of electrons in the various energy levels. The identity of the number of *outer* electrons in each elementary family is particularly important in accounting for the similarity in chemical properties.

The periodic table is a useful guide to the number of outer electrons in any particular element. In the families marked "A" the number that heads the column indicates the number of outer electrons. Thus, sodium (Na), potassium (K), rubidium (Rb), and cesium (Cs) are all in group 1A and all have one outer electron. Boron (B), aluminum (Al), and all the other members of family 3A have 3 outer electrons. Similarly, the elements of group 7A, including fluorine (F), chlorine (Cl), bromine (Br), and iodine (I) have 7 outer electrons.

Figure 2.1

Schematic representation of hydrogen and sodium atoms.

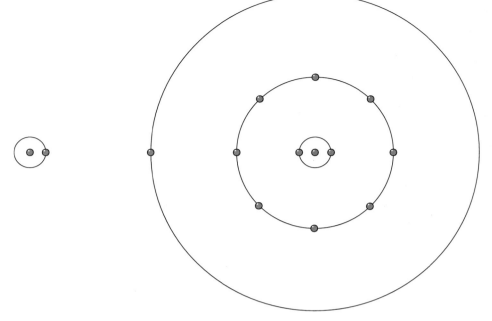

Hydrogen (atomic number 1)
  1 proton
  1 electron
  (1 outer electron)

Sodium (atomic number 11)
  11 protons
  11 electrons
  (1 outer electron)

Key: ● Nuclei   ● Electrons

Figure 2.2

Electronic arrangements in atoms of the first 18 elements.

Group	1A	2A	3A	4A	5A	6A	7A	Rare Gases
	1							2
	H							He
	1							2
	3	4	5	6	7	8	9	10
	Li	Be	B	C	N	O	F	Ne
(2) + 1		2	3	4	5	6	7	8
	11	12	13	14	15	16	17	18
	Na	Mg	Al	Si	P	S	Cl	Ar
(2) + (8) + 1		2	3	4	5	6	7	8

Number of electrons in atom (atomic number)
Number of outer electrons
(   ) indicates fully filled electron energy levels

Note that the family designation, as it appears above and in some periodic tables, corresponds to the number of outer electrons.

■ *A Matter of Mass*

In addition to atomic numbers, the periodic table specifies the **atomic mass** or **atomic weight** of each element. The mass of an atom is almost all associated with the protons and neutrons that make up the nucleus. The mass of a hydrogen atom is thus very nearly the mass of the single proton in its nucleus, 1 on the relative scale of Table 2.1. But not all hydrogen atoms are identical. To be sure, they each contain one electron and one proton—that's what makes the substance hydrogen. However, one hydrogen atom out of 6700 also has a neutron in its nucleus. Because both the proton and the neutron have relative masses of 1, the relative mass of the atom equals 2. This "heavy hydrogen" is also called deuterium. It is an example of a naturally occurring isotope of hydrogen. **Isotopes** are two (or more) forms of the same element whose atoms differ in number of neutrons and hence in atomic mass. Isotopes are identified by their **mass numbers**—the sum of the number of protons and the number of neutrons in an atom. The mass number typically follows the name or symbol of the element. Thus, deuterium is sometimes designated as H-2 or "hydrogen-2." There is also a third isotope of hydrogen, called tritium, whose atoms consist of two neutrons in addition to one proton and one electron. Tritium, a radioactive isotope that does not occur in nature, thus has a mass number of 3 (1 proton + 2 neutrons).

2.3 ■ ***Your Turn***

Specify the number of electrons, protons, and neutrons in atoms of the following isotopes.

 a. oxygen-16 (O-16)
 b. argon-40 (Ar-40)
 c. strontium-90 (Sr-90)
 d. uranium-235 (U-235)

Ans. a. no. protons = no. electrons = atomic number = 8
no. neutrons = mass number − atomic number = 16 − 8 = 8
c. no. protons = no. electrons = atomic number = 38
no. neutrons = mass number − atomic number = 90 − 38 = 52

■ *Molecules and Models*

After this excursion into the atom, we come to our primary motivation—molecular structure. The stability of fully filled electron shells can be invoked to explain why atoms bond to each other to form molecules. The simplest case is the H_2 molecule. A hydrogen atom has only one electron, but if two atoms come together, the two electrons become common property. Each atom effectively has a share in both electrons. The resulting H_2 molecule has a lower energy than two individual H atoms, and consequently the molecule is more stable. The two shared electrons constitute what is called a **single bond** or a **covalent bond.** Appropriately, the name "covalent" implies "shared strength." If we represent each atom by its symbol and each electron by a dot, the H_2 molecule can be written as follows.

H : H

This is called a dot or **Lewis structure,** after Gilbert Newton Lewis (1875–1946), an American chemist who pioneered its use. Lewis structures can be predicted for any molecule by following a few simple steps, which we will illustrate with the water molecule.

1. **Starting with the chemical formula of the compound, note the number of outer electrons contributed by each of the atoms.**

   1 O atom × 6 outer electrons per atom = 6 outer electrons
   2 H atoms × 1 outer electron per atom = 2 outer electrons

2. **Add the electrons contributed by the individual atoms to obtain the total number of outer electrons available.**

$$6 + 2 = 8 \text{ outer electrons}$$

3. **Arrange the outer electrons to maximize stability by giving each atom a share in enough electrons to fully fill its outer shell—2 electrons in the case of hydrogen, 8 electrons for most other atoms.**

$$H : \overset{..}{\underset{..}{O}} : H$$

The fact that in many molecules electrons are arranged so that every atom (except hydrogen) shares in eight electrons is called the **octet rule.** This generalization is a useful guide for predicting Lewis structures. Also note that in most molecules where there is only one atom of one element and two or more atoms of another element (or elements), the single atom goes in the center.

2.4   *Your Turn*

Use the above procedure to draw Lewis or dot structures for
a. chlorine (Cl_2), **b.** ammonia (NH_3), and **c.** methane (CH_4).

Ans. **a.** Chlorine, atomic number 17, has 17 electrons. The number of outer electrons equals $17 - 2 - 8$ or 7 (see Figure 2.2). Therefore, there are a total of 2×7 or 14 outer electrons available from the two Cl atoms. Each of the atoms can participate in an octet of electrons if two electrons are shared. This corresponds to a single covalent bond and the following Lewis structure.

$$: \overset{..}{\underset{..}{Cl}} : \overset{..}{\underset{..}{Cl}} :$$

The Lewis structure for water reveals two sorts of electrons. Four of them (two pairs) are involved in covalent bonds connecting the oxygen and hydrogen atoms. The remaining four electrons, again in two pairs, are localized on the oxygen atom and are said to be nonbonding electrons or lone pairs. Each pair of bonding electrons is often represented by a single straight line. Using this symbolism, the H_2O molecule is written as follows.

$$H - \overset{..}{\underset{..}{O}} - H$$

These Lewis representations provide more information than the simple formula H_2O, because they indicate how the atoms are connected to each other. On the other hand, Lewis structures do not directly reveal the shape of a molecule. From the structure given here, it might appear that the atoms of the water molecule all fall in a straight line. In fact, the molecule is bent. Chapter 3 describes how the Lewis structure can lead to the prediction of that geometry.

The octet rule often leads to the correct conclusions, but not always. It turns out that the O_2 molecule is a case where the rule is slightly misleading. Here we have 12 outer electrons to distribute, six from each of the atoms. There are not enough electrons to give each of the atoms a share in eight electrons if only one pair is held in common. However, the octet rule can be satisfied if the two atoms share four electrons. A four-electron covalent bond is called a **double bond** and it is represented by four dots or two lines.

$$\overset{..}{\underset{..}{O}} : : \overset{..}{\underset{..}{O}} \quad \text{or} \quad \overset{..}{\underset{..}{O}} = \overset{..}{\underset{..}{O}}$$

Double bonds are shorter, stronger, and harder to break than single bonds, and the length and strength of the bond in the O_2 molecule does correspond to a double bond. However, oxygen has a peculiar property that is not fully consistent with the Lewis structure drawn above. When liquid oxygen is poured between the poles of a strong magnet, it sticks there like iron filings. Such behavior implies that the electrons are not as neatly paired as the octet rule would suggest. But this little discrepancy is hardly a reason to discard the generalization. After all, simple scientific models seldom if ever explain all phenomena, but they can be useful approximations.

The ozone molecule introduces another structural feature. We again start with the octet rule. Each of the three oxygen atoms contributes six outer electrons for a total of 18. These 18 electrons can be arranged in three ways, all of which give each atom a share in eight outer electrons.

There are experimental techniques for determining molecular structure, and it turns out that none of the above versions is exactly correct. The nice, neat equilateral triangle on the right definitely does not conform to experiment. The other two structures both predict that the molecule should contain a single bond and a double bond. But in fact the two bonds are identical in length and strength, lying somewhere between a single bond and a double bond. Moreover, the O_3 molecule is bent, not linear. You will soon see how the structures of O_2 and O_3 molecules influence their interaction with sunlight, but first we must focus on that light.

2.5 *Your Turn*

Use the octet rule to draw a Lewis or dot structure for **a.** nitrogen (N_2), **b.** carbon monoxide (CO), and **c.** sulfur dioxide (SO_2), all of which contain multiple bonds.

$$: N : : : N :$$

Ans. **a.** Each nitrogen atom has five outer electrons, so there are ten outer electrons to distribute in N_2. The octet rule will be satisfied if the two atoms share three pairs of electrons to form a **triple bond.** Each atom will also have one nonbonding lone pair to complete the octet.

Hint: **b.** Note that CO and N_2 both have the same number of outer electrons. **c.** Note that sulfur and oxygen are in the same periodic family.

▪ *Waves of Light*

Every second, five million tons of the Sun's matter are converted into energy which is radiated into space. The fact that we can detect color indicates that the radiation that reaches us is not all identical. Prisms and raindrops break sunlight into a spectrum of colors. Each of these colors can be identified by the numerical value of its **wavelength.** The word correctly suggests that light behaves rather like a wave in the ocean. The wavelength is the distance between successive peaks (Figure 2.3). It is expressed in units of length and symbolized by the Greek letter lambda (λ).

Our eyes are sensitive to light with wavelengths between about 700×10^{-9} and 400×10^{-9} meters (m). These lengths are very short, so we typically express them in nanometers (nm). Because 1 nm is defined as 1×10^{-9} m, we can state that the visible spectrum extends from 400 nm (violet) to 700 nm (red).

Figure 2.3
Wave motion.

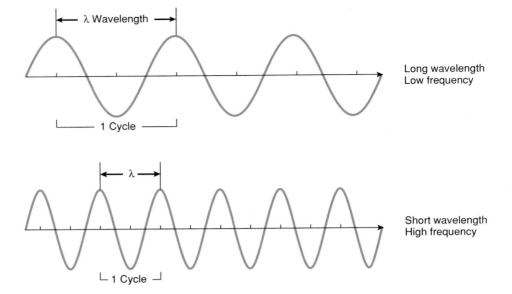

Another way of quantifying color is to express it in terms of **frequency.** If you were watching waves on the surface of a lake or the ocean, you could measure the distance between successive crests (the wavelength). But you could also time how often the crests passed your point of observation by counting the number in a particular time interval. That would give you the frequency of the waves. The same idea applies to radiation. The frequency of light is the number of waves passing a fixed point in one second. The longer the wavelength the lower the frequency, that is, the smaller the number of waves that pass the observer in one second. We have symbolized this relationship in the margin. The arrows signify that frequency increases as wavelength decreases. Instead of reporting frequency as "waves per second," the units are shortened to "per seconds" and written as 1/s or s^{-1}. This unit is also referred to as hertz (Hz), a term that may be familiar to you from radio station frequencies. Frequency is symbolized by the Greek letter nu (ν).

The relationship between frequency and wavelength that we have just described in words can be summarized in a simple equation.

$$\text{Frequency} = \nu = \frac{c}{\lambda} \qquad (2.2)$$

The letter c represents the constant speed with which the radiation travels—3.00×10^8 m/s. The smaller the value for λ, the greater the value for ν. Thus, red light, which has a wavelength of 700 nm, has a frequency of 4.3×10^{14} s^{-1}. Violet light has a shorter wavelength (400 nm) and a higher frequency, 7.5×10^{14} s^{-1}.

The light that we can see directly represents only a narrow window in the entire **electromagnetic spectrum.** Figure 2.4 indicates that the spectrum extends in both directions from the visible region. Scientists have devised a variety of detectors that are sensitive to the radiation in various parts of this broad band. As a consequence, we can speak with confidence about the regions of the spectrum that are invisible to our eyes. At wavelengths longer than red, one first encounters **infrared (IR)** or heat rays, which we can certainly feel. The microwaves used in radar and to quickly cook food have wavelengths of about 1 centimeter (cm) (1 cm = 1×10^{-2} m). At still longer wavelengths (1 m to 1000 m) are the regions of the spectrum used to transmit AM and FM radio and television signals.

In this chapter we are most concerned with the **ultraviolet (UV)** region, which lies at wavelengths shorter than those of violet. At still shorter wavelengths are the **X-rays** used in medical diagnosis and the determination of crystal structure, **gamma rays** that are given off in certain radioactive processes, and the highly energetic **cosmic rays** that bombard the planet. Some cosmic rays can have wavelengths as short as 10^{-14} m.

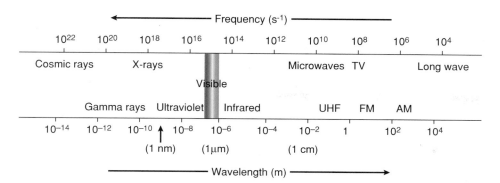

Figure 2.4
The electromagnetic spectrum.

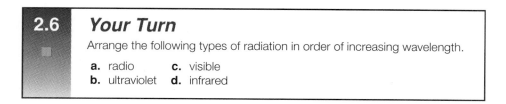

2.6 | **_Your Turn_**

Arrange the following types of radiation in order of increasing wavelength.

a. radio   **c.** visible
b. ultraviolet **d.** infrared

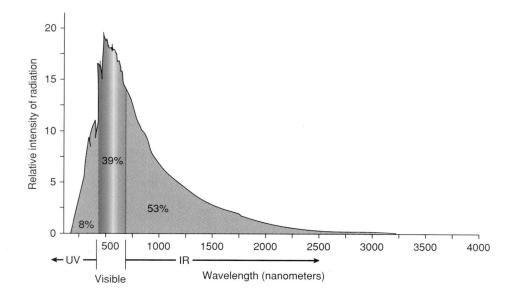

Figure 2.5
Energy distribution in solar radiation
above the Earth's atmosphere.
(Source: From Muhammad Iqbal,
An Introduction to Solar Radiation.
Copyright © 1983 Academic Press,
Inc., Orlando, FL. Reprinted by
permission.)

Our local star, the Sun, emits infrared, visible, ultraviolet, and cosmic radiation, but not all with equal intensity. This is evident from Figure 2.5, which is a plot of the relative intensity of solar radiation as a function of wavelength. The curve represents the spectrum as measured *above* the atmosphere, before there has been opportunity for interaction of radiation with the molecules of the air. The peak indicating the greatest intensity is in the visible region. However, infrared radiation is spread over a much wider wavelength range, with the result that 53% of the total energy emitted by the Sun falls in this part of the spectrum. This is the major source of heat for the planet. Approximately 39% of the energy comes to us as visible light and only about 8% as ultraviolet. (The areas under the curve give you an indication of these percentages.) But in spite of its small percentage, the Sun's UV radiation is potentially the most damaging to living things. To understand why, we must look at radiation in a different light.

■ *"Particles" of Energy*

The idea that radiation can be described in terms of wave-like character is well established and very useful. However, around the beginning of the twentieth century, scientists found a number of phenomena that seemed to contradict this model. In 1900, a German physicist named Max Planck (1858–1947) argued that the shape of the energy distribution curve pictured in Figure 2.5 could be explained only if the energy of the radiating body were the sum of many energy levels of minute but discrete size. In other words, the energy distribution is not really continuous, but consists of many individual steps. The energy is said to be **quantized.** An often-used analogy is that the energy of a radiating body is like a piano that plays specific notes, not like a violin that can produce a continuous range of pitch.

Five years later, in the work that won him his Nobel Prize, an amateur violinist went a little further. Albert Einstein (1879–1955) suggested that radiation itself should be viewed as constituted of individual bundles of energy called **photons.** One can regard these photons as "particles of light," but they are definitely not particles in the usual sense. For example, they have no mass.

The atomization of energy by quantum theory did not displace the utility of the wave model. Both are valid descriptions of radiation. This dual nature of radiant energy seems to defy common sense. How can light be two different things at the same time—both waves and particles? There is no obvious answer to that very reasonable question—that's just the way nature is. The two views are linked in a simple relationship that is one of the most important equations in modern science. It is also an equation that is very relevant to the role of ozone in the atmosphere.

$$E = h\nu = hc/\lambda \qquad (2.3)$$

Here E represents the energy of a single photon. It is proportional to ν, the frequency of radiation. The symbol h is **Planck's constant,** which has a value of 6.63×10^{-34} joule second (J s).

The higher the frequency, the greater the energy associated with a single photon. Thus, an FM radio wave with a frequency of 100 megahertz (100×10^6 s^{-1}) can be viewed as a stream of photons each possessing an energy of 6.63×10^{-26} joule.

$$E = 6.63 \times 10^{-34} \text{ J s} \times 100 \times 10^6 \text{ s}^{-1} = 6.63 \times 10^{-26} \text{ J}$$

By a similar calculation, the energy of a photon of ultraviolet light with a wavelength of 300 nm and a frequency of 1.00×10^{15} s^{-1} can be shown to have an energy of 6.63×10^{-19} joule. Both of these energies are very small. (One joule is approximately equal to the energy required to raise a 1 kilogram [2.2 lb] book 10 centimeters [4 in] against the force of gravity.) The point to remember is that the energy of a photon of the ultraviolet light is ten million times greater than that of a photon from your favorite radio station. One consequence of that fact is explained in the marginal note.

Wavelength ↓
Frequency ↑
Energy ↑

On Tanning and Listening to the Radio

You have no doubt observed that you can't get a tan from listening to the radio—unless the radio happens to be in the Sun. Whether or not your radio is turned on, you are continuously bombarded by radio waves. Your body can't detect them, but your radio can. The energy associated with each of the radio photons is very low—about 6.63×10^{-26} joule. That is not enough energy to result in a local increase in the skin pigment, melanin. The tanning process involves a quantum jump—an electronic transition that requires approximately 6.63×10^{-19} joule. Your body cannot store up the 10 million low-energy photons of radio frequency that would be necessary to equal the energy required for the tanning reaction. It's an either/or situation. Either a photon has enough energy to cause a specific chemical change or it doesn't. The photons of ultraviolet radiation of 300 nm or shorter do have sufficient energy to bring about the changes that result in tanning. (You may be interested in knowing that it was essentially this reasoning, on a different system, that won Einstein his Nobel Prize. It seems pretty simple, doesn't it?)

2.7 ■ *Your Turn*

Arrange the following colors of the visible spectrum in order of increasing energy per photon.

a. yellow   **c.** red
b. blue   **d.** orange

Ans. red, orange, yellow, blue

Arrange the following types of radiation in order of increasing energy per photon.

a. X-ray   **c.** visible
b. microwave   **d.** ultraviolet

Matter and Radiation

The recognition that the energy associated with any physical system can have only certain specific values and the discovery that radiant energy comes in small, discrete packets have led to major reinterpretations of physical reality. It has given rise to **quantum mechanics,** one of the most important scientific developments in this century. Much of our current understanding of atomic and molecular structure has come from quantum mechanics, but for the purposes of this chapter, our major concern is with the interaction of radiation and matter.

Every second, the Sun bombards the Earth with countless photons—indivisible packages of energy. Many of these photons are absorbed by the atmosphere, the surface of the planet, and its living things. Radiation in the infrared region of the spectrum warms the Earth and its oceans, causing molecules to move, rotate, and vibrate. The cells of our retinas are tuned to the wavelengths of visible light. Photons of the right frequency are absorbed and the energy is used to "excite" electrons in biological molecules. The electrons jump to higher energy levels, triggering a series of complex chemical reactions that ultimately lead to sight. Green plants capture photons in an even narrower region of the visible spectrum (corresponding to red light) and use the energy to convert carbon dioxide and water into food, fuel, and oxygen in the process of photosynthesis. As the frequency of light increases, so does the energy carried by each photon. Consequently, the interaction of radiation and matter becomes more violent. Photons in the UV region of the spectrum carry sufficient energy to eject electrons from atoms and molecules and convert them into positively charged ions. Often bonds are broken and molecules come apart. In living things, cells are disrupted, and the seeds of birth defects and cancer are sometimes planted.

It is part of the fascinating symmetry of nature that this interaction of radiation and matter explains both the potentially damaging effects of ultraviolet radiation and the mechanism that protects us from it. We turn first to the shield of oxygen and ozone.

The Oxygen/Ozone Screen

The presence of oxygen and ozone in the Earth's atmosphere means that the radiation that reaches the surface of the planet is different from that emitted by the Sun in some important respects. Much of the ultraviolet is blocked by these two allotropes of the same element. As we noted in Chapter 1, approximately 21% of the atmosphere consists of ordinary, elementary diatomic oxygen. The forms of life that inhabit our planet are absolutely dependent upon the chemical properties of this gas and on its interaction with ultraviolet radiation. The strong covalent bond holding the two oxygen atoms together in the O_2 molecule can be broken by the absorption of a photon of the proper radiant energy. The photon excites an electron to a higher energy level, causing the atoms to come apart. But the bond will be broken and the molecule will dissociate only if the photon has energy corresponding to a wavelength of 242 nm or less ($\lambda \le 242$ nm). This value is in the ultraviolet region of the spectrum.

$$O_2 + photon \rightarrow 2\ O \qquad\qquad (2.4)$$
$$\lambda \le 242 \text{ nm}$$

Because of this reaction, stratospheric oxygen shields the surface of the Earth from high-energy radiation. As green plants flourished on the young planet, they released oxygen into the atmosphere. The increasing oxygen concentration led to more effective interception of ultraviolet radiation. Consequently, forms of life evolved that were less resistant to UV radiation than they would have been otherwise.

Ordinary diatomic oxygen screens out radiation with wavelengths shorter than 242 nm. However, if O_2 were the only UV absorber in the atmosphere, the surface of the Earth and the creatures that live on it would still be subjected to potentially damaging radiation in the 242–320 nm range. It is here that ozone plays its protective role. The fact that ozone is more reactive than diatomic oxygen suggests that the O_3 molecule is more easily broken apart than O_2. Recall that the atoms in the latter molecule

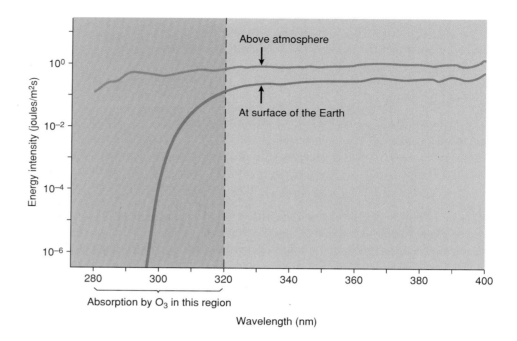

are connected with a strong double bond. The bonds in O_3 are somewhat weaker.
Therefore, photons of a lower energy should be sufficient to separate the atoms in O_3.
This is in fact the case; radiation of 320 nm or less will induce the following reaction.

$$O_3 + photon \rightarrow O_2 + O \qquad (2.5)$$
$$\lambda \leq 320 \text{ nm}$$

Because of this reaction and that represented by equation 2.4, only a relatively small
fraction of the Sun's ultraviolet radiation reaches the surface of the Earth. However,
what does arrive can do damage.

■ Biological Effects of Ultraviolet Radiation

The impact of ultraviolet radiation on living things depends on two factors: the intensity of UV radiation and the sensitivity of organisms to that radiation. The vertical
scale of Figure 2.6 indicates the solar energy falling on a surface area of one square
meter in one second. The graph shows how this energy intensity varies with wavelength. For any wavelength, the total amount of energy will be the product of the number of photons striking the surface and the energy per photon. The rather flat upper
curve reveals that the energy input above the atmosphere does not depend significantly
on wavelength. If the Earth's atmosphere did not exist, the surface of the planet and
the creatures on it would be subjected to these punishingly high levels of radiant energy. However, the lower curve indicates that the energy reaching the surface of the
Earth varies markedly with wavelength. It starts dropping at 330 nm and falls off
sharply as the wavelength decreases.

In fact, the decrease in UV radiation is a good deal more dramatic than the figure
at first suggests. The vertical scale is a logarithmic one, a method of presenting data
that permits the inclusion of a wide range of values. Every mark on the axis represents
an energy value that is one tenth of that corresponding to the mark immediately above
it. Thus, at 320 nm, where ozone starts absorbing, the energy input to the Earth's surface is 10^{-1} or 0.1 joules per square meter per second. At 300 nm, the value has
dropped to 10^{-4} or 0.0001 J/m² s.

Solar radiation at wavelengths below 300 nm is almost completely screened out
by O_2 and O_3 in the atmosphere. This is most fortunate, because radiation in this region
of the spectrum is particularly damaging to living things. This relationship is evident

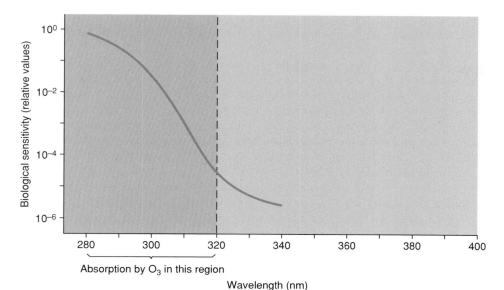

Figure 2.7
Variation of biological sensitivity of DNA with wavelength of UV radiation. (Reprinted by permission of John E. Frederick, The University of Chicago.)

from Figure 2.7, where biological sensitivity is plotted versus wavelength. As defined here, biological sensitivity is based on experiments in which the damage to deoxyribonucleic acid (DNA) is measured at various wavelengths. In the figure, sensitivity is expressed in relative units. Note that once more, the scale is logarithmic. Biological sensitivity at 320 nm is about 10^{-5} or 0.00001 unit. But at 280 nm, the sensitivity is 10^0 or 1 unit. This means that radiation at 280 nm is 10^5 or 100,000 times more damaging than radiation at 320 nm. As we have seen, this is because the energy per photon and the potential for biological damage increase as the wavelength decreases. Highly energetic photons can excite electrons and break bonds in biological molecules, rearranging them and altering their properties. (A discussion of DNA, the chemical basis of heredity, appears in Chapter 13.)

Wavelength ↓
Frequency ↑
Energy ↑
DNA Damage ↑

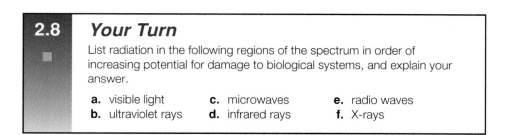

2.8 **_Your Turn_**

List radiation in the following regions of the spectrum in order of increasing potential for damage to biological systems, and explain your answer.

a. visible light   **c.** microwaves   **e.** radio waves
b. ultraviolet rays   **d.** infrared rays   **f.** X-rays

Because of the crucial role of ozone as an absorber of high-energy photons, a reduction in the concentration of ozone in the stratosphere would increase exposure to biologically damaging ultraviolet radiation. A quantitative prediction of the extent of this effect requires some knowledge of how the intensity of the radiation in this spectral region is dependent on ozone concentration. The complexity of the problem means that only estimates are possible. These estimates are, in turn, coupled with projections of the rate of global ozone depletion. The situation is further complicated by other factors. For example, changes in cloud cover and low altitude pollution can also change the amount of UV reaching the ground.

In spite of these great complexities, mathematical models have been created to predict changes in ozone concentration. Current extrapolations suggest that by the years 2020 to 2040, the ozone level at 60°N (the latitude of Oslo or Saint Petersburg) may decrease by 5%. Other calculations predict that such a change will increase the flux of biologically active UV radiation by 10%. Some researchers conclude this would result in a 10% increase in skin cancer, especially the more easily treated form, non-melanoma skin cancer. This condition is considerably more common among Caucasians than those having more heavily pigmented skin, which offers better protection against the harmful effects of radiation.

There is good evidence linking the incidence of non-melanoma skin cancer with the intensity of UV radiation. For example, the disease becomes more prevalent as one moves farther south in the Northern Hemisphere. Those who endure the long nights and short days of a Maine winter are compensated by a level of non-melanoma skin cancer that is only about half that of those who enjoy year-round Florida sunshine. In fact, the geographical effect on radiation intensity and skin cancer is, at least to date, much greater than that due to ozone depletion. A 1% decrease in ozone levels is roughly equivalent to moving 20 km closer to the equator.

That seemingly reassuring statement should not be interpreted as suggesting that the problem of ozone depletion can be ignored. Human beings are, after all, not the only creatures on the globe; our existence is inextricably linked to the entire ecosystem. Plant growth is suppressed by UV radiation, and phytoplankton seem to be UV sensitive, too. These photosynthetic microorganisms live in the oceans where they occupy a fundamental niche in the food chain. The phytoplankton ultimately supply the food for all the animal life in the oceans, and any significant decrease in their number could have a major impact. Moreover, these tiny plant-like organisms play an important role in the carbon dioxide balance of the planet by absorbing CO_2. Thus, it is possible that ozone depletion may influence another atmospheric problem—the greenhouse effect. In any case, it is essential that scientists and the general public come to understand some of the chemistry that occurs 15 miles above the surface of the Earth.

■ *Stratospheric Ozone: Its Formation and Fate*

Every day, 300 million tons of stratospheric ozone are formed, and an equal mass is destroyed. Of course, matter is not really created or destroyed. As in any chemical or physical change, matter merely undergoes changes in its chemical or physical form. In this particular case, a **steady state** is established in which the concentration of ozone remains constant. This steady state is the net result of four reactions that constitute the **Chapman cycle,** named after Sydney Chapman who first proposed it in 1929 and 1930. The cycle is summarized in Figure 2.8.

You have already encountered the reaction labeled 1 in Figure 2.8. It is the process by which oxygen molecules absorb photons of UV radiation and dissociate into individual oxygen atoms. These atoms tend to combine readily with other atoms and molecules. One such reaction is Step 2, which occurs when an oxygen atom strikes an oxygen molecule to generate an ozone molecule. As we have already seen, once an O_3 molecule is generated, it can absorb a photon of UV radiation, causing it to dissociate and regenerate O_2 and O (Step 3). It is, of course, by means of this reaction that ozone screens out UV radiation. Most of the oxygen molecules and atoms formed

Figure 2.8
The Chapman Cycle.

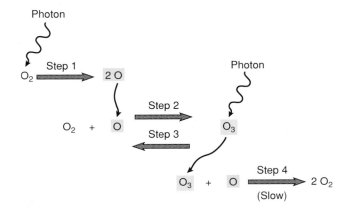

in Step 3 recombine to form ozone molecules via Step 2. Occasionally, however, an O_3 molecule collides with an atom to form two O_2 molecules (Step 4). This slow reaction removes the O and O_3 "odd oxygen" species from the cycle. A typical O_3 molecule "lives" 100–200 seconds before it is dissociated.

The four reactions of Figure 2.8 constitute a cyclic system in which the rate of O_3 formation equals the rate of O_3 destruction. Although the reactions are occurring, no net change in concentration of the reactants or products is observed. In this particular case, the steady-state concentration of ozone depends on such factors as the intensity of the UV radiation, the concentration of O_2, and the rates and efficiencies of the individual steps.

When these variables are properly evaluated and included, it becomes apparent that the Chapman cycle does not tell the whole story. It is fundamentally correct, but the real world is inevitably more complicated than idealized abstractions. The concentrations of the reacting species, the intensity of the dissociating radiation, the temperature, and consequently the rates of the individual reactions all vary with altitude. Moreover, winds transport gases vertically and horizontally, mixing them up. Efforts have been made to include these variables in mathematical models designed to mimic the atmosphere. The test of these models is how well their predictions of ozone concentration conform to measured values.

■ Distribution of Ozone in the Atmosphere

The total amount of ozone in a vertical column of air of known volume can be determined with relative ease. It is done from the surface of the Earth by measuring the amount of UV radiation reaching the detector—the lower the intensity of the radiation, the greater the amount of ozone. Such data have been collected since the late 1930s. Since the 1970s, measurements of total ozone have also been made from the top of the atmosphere. Detectors mounted on satellites record the intensity of the ultraviolet radiation that is scattered by the atmosphere. The results can be related to the amount of O_3 present.

Measuring the ozone concentration at various altitudes is a good deal more difficult. Detectors are carried aloft on airplanes, rockets, balloons, and satellites. The results of some of these measurements are summarized in Figure 2.9. This plot includes a good deal of important information and is worth some careful study. First of all, note

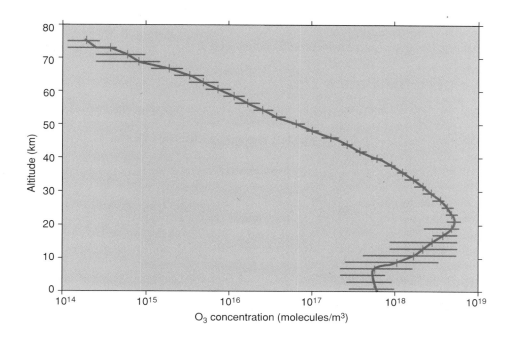

Figure 2.9

Ozone concentrations at various altitudes. (Source: Data from *United States Standard Atmosphere, 1976.*)

that the concentration scale on the horizontal axis has units of O_3 molecules per cubic meter. Now look at the numbers along this axis. Once more we are using a logarithmic scale to include a wide range of data. (If a typical linear graph were used to present these data, it would be inconveniently long.) There is a tenfold increase in concentration between two successive markers. Thus, 10^{15} is ten times larger than 10^{14}. Although the distances between successive numbers on the horizontal axis are equal, the concentration intervals are not. The difference between 10^{16} and 10^{15} is ten times the difference between 10^{15} and 10^{14}.

The horizontal lines that cross the curve are called uncertainty bars. Each bar is a composite of the individual concentration values obtained in a number of separate measurements. Not all measurements give identical results, and the length of the bar indicates the scatter or variability. In some cases, especially at low and at high altitudes, the variation is considerable. This variability may be due to limitations in instrument sensitivity, experimental error, or genuine variation in concentration. It is impossible to determine the source from the graph. In this case, most of the scatter is due to seasonal and geographic variation.

The message is that when evaluating experimental results published in a newspaper or a scientific journal, it is important to know the limitations of the data. Just how good were the measurements and how many were performed? Are the results statistically significant? How great is the uncertainty? How much confidence can you have in the conclusions? A political poll reporting that 51% ± 10% of those surveyed favored candidate X and 49% ± 10% favored candidate Y would be of little value. Healthy and informed skepticism is a useful attribute in analyzing any experimental evidence or statistical summary.

In Figure 2.9 the curve is drawn through the midpoints of the uncertainty bars. These are points that represent the averages of the measurements. Even allowing for uncertainties, it is clear that the highest concentration of ozone occurs between 10 and 30 km, with a maximum around 20 km. Roughly 91% of the Earth's ozone is found in the stratosphere, between the altitudes of 10 and 50 km.

Because this 40 km band is so broad, the concept of an ozone layer can be a little misleading. At the altitudes of the maximum ozone concentration, the atmosphere is very thin so the total amount of ozone is surprisingly small. Suppose all of the O_3 in the atmosphere could be isolated and brought to the average pressure and temperature at the surface of the Earth (1 atmosphere and 15°C). The resulting layer of gas would be 0.30 cm thick, or about 1/8 of an inch. On a global scale, this is a minuscule amount of matter. Yet it shields the surface of the Earth and its inhabitants from the harmful effects of ultraviolet radiation. Because ozone is present in a small and finite quantity, it is important that we protect and preserve it.

■ *Paths for Ozone Destruction*

Over the past 40 years, ozone concentrations have been measured at experimental stations spread over the planet. The results are clear; the concentration levels are lower than those predicted from the simple Chapman mechanism. That in itself is neither cause for alarm nor proof that the lower values are the consequence of human intervention. It merely suggests that the processes determining the steady-state concentration of ozone are more complicated than originally believed.

For one thing, the concentration of ozone is not uniform over all parts of the globe. On the average, the total O_3 concentration increases the closer one gets to either pole (with the exception of the well-known "hole" over the Antarctic). Moreover, ozone fluctuates with the seasons, reaching its maximum concentration (in the Northern Hemisphere) in March and its minimum in October. The winds blowing through the atmosphere cause these variations over time and space, some of them on a seasonal basis and some over a 28-month cycle. The Sun's emission is not constant, either. It changes over an 11- to 12-year cycle related to sunspot activity. Because the production of ozone depends on ultraviolet radiation, this variation can influence O_3

concentrations, but only by 1 to 2%. To further complicate matters, random fluctuations in composition often occur in large samples, such as the atmosphere. Finally, it is very likely that other gases, some of them artificially produced, are also involved in the reduction of the ozone concentration.

One of the causes of ozone destruction is a series of reactions involving water vapor and its breakdown products. The great majority of the H_2O molecules that evaporate from the oceans and lakes fall back to the surface of the Earth as rain or snow. A few, however, reach the stratosphere, where the H_2O concentration is about 5 ppm. There ultraviolet radiation triggers chemical reactions resulting in the dissociation of water molecules into hydrogen atoms and hydroxyl **free radicals** (equation 2.6). Free radicals are unstable species with an unpaired electron.

$$H_2O \rightarrow H + OH \qquad (2.6)$$

These species participate in many reactions, but two are particularly important because they involve ozone. Equation 2.7 represents the destruction of an ozone molecule by a hydrogen atom. The hydroxyl radical produced reacts with an oxygen atom according to equation 2.8 and regenerates another hydrogen atom. This H atom can now participate in reaction 2.7. Reactions 2.7 and 2.8 thus constitute a cycle. The net effect of this cycle becomes apparent when we add the two equations, just as we might add two mathematical equations.

$$H + O_3 \rightarrow OH + O_2 \qquad (2.7)$$
$$OH + O \rightarrow H + O_2 \qquad (2.8)$$
$$\cancel{H} + O_3 + \cancel{OH} + O \rightarrow \cancel{OH} + \cancel{H} + 2\,O_2 \qquad (2.9)$$

Note that H and OH appear on both sides of equation 2.9. We again treat this chemical equation as if it were a mathematical equation. We subtract H and OH from both sides, or, if you prefer, we cancel the Hs and the OHs.
What remains is equation 2.10.

$$O + O_3 \rightarrow 2\,O_2 \qquad (2.10)$$

Equation 2.10 should look familiar. It is the ozone-removing step of the Chapman mechanism (Step 4). Interaction with H and OH thus provides another path for the destruction of ozone. It turns out that this is the most efficient mechanism for destroying ozone at altitudes greater than 50 km. The fact that H and OH are both reactants and products in equation 2.9 is an important one. These species are both consumed and regenerated in the cycle, so there is no net change in their concentration. Such behavior is characteristic of a **catalyst**—a chemical substance that can participate in a chemical reaction and influence its speed without undergoing permanent change. Because the catalytic molecules are regenerated as fast as they are consumed, a single pair of H and OH radicals can trigger the destruction of a large number of O_3 molecules.

Water molecules and their breakdown products, hydrogen atoms and hydroxyl radicals, are not the only catalysts for ozone destruction. Another is nitric oxide, NO, which promotes a series of reactions similar to those just discussed. Most of the NO in the stratosphere is of natural origin. It is formed when nitrous oxide, N_2O, reacts with oxygen atoms. The N_2O is produced in the soil and oceans by microorganisms and gradually drifts up to the stratosphere. There is really little that can or should be done to control this process. It is part of a cycle involving compounds of nitrogen and living things.

However, not all the nitric oxide in the atmosphere is of natural origin; human activities can alter steady-state concentrations. That is why, in the 1970s, chemists became concerned about the increase in NO that would result from developing and deploying a fleet of supersonic transport (SST) airplanes. These planes were designed to fly at altitudes of 15–20 km, the region of the ozone layer. The scientists calculated that much additional NO would be generated by the direct combination of nitrogen and oxygen.

$$Energy + N_2 + O_2 \rightarrow 2\,NO \qquad (2.11)$$

The equation emphasizes that this reaction requires large amounts of energy, which can be supplied by lightning or in the high-temperature jet engines of the SSTs. Many experiments and calculations were carried out, and predictions were made about the net effect of a fleet of SSTs. The conclusion was that the potential risks outweighed the benefits, and the decision was made, partly on scientific grounds, not to build an American fleet. The Anglo-French Concorde is the only commercial plane that currently operates at this altitude.

Subsequent research has indicated that atmospheric reactions involving NO and other oxides of nitrogen are more numerous and more complicated than originally thought. NASA is currently sponsoring a project to investigate effects of another generation of high altitude supersonic aircraft. As you already know from your reading of Chapter 1, a more serious pollution problem involving the oxides of nitrogen occurs at ground level.

■ *The Case of the Absent Ozone*

The fact that the steady-state concentration of ozone in the stratosphere is lower than that predicted by the simple Chapman model can be explained by the existence of catalytic pathways involving water, nitric oxide, and other naturally occurring chemical species. However, these alternate routes cannot fully account for the decrease in ozone concentration found since 1970. Figure 2.10 summarizes some of the data. The percent change in ozone concentration per decade is plotted, for both winter and summer, as a function of latitude. There is considerable scatter in the data points, and the results for

Figure 2.10

Trends in total ozone over the Northern Hemisphere, 1970–1988. (Data from the World Meteorological Organization *Ozone Report #20.*)

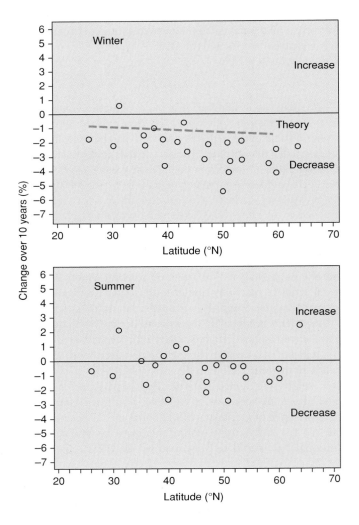

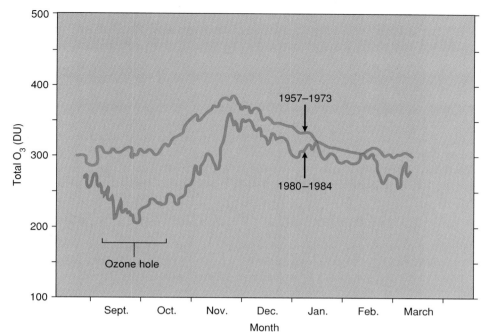

Figure 2.11
Monthly variation of total ozone over Halley Bay, Antarctica. (Source: Susan Solomon, *Reviews in Geophysics*. Vol. 26, pp. 134–148, 1988. Copyright by the American Geophysical Union.)

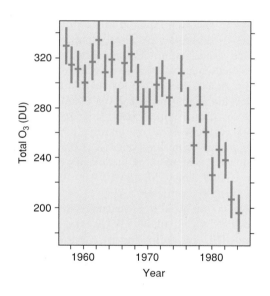

Figure 2.12
Development of the Antarctic "ozone hole," 1975–1985. Based on data gathered in October each year by the British Antarctic Survey at Halley Bay. The vertical lines represent the scatter in the data; the cross bars indicate average values. (Source: Susan Solomon, *Reviews in Geophysics*. Vol. 26, pp. 134–148, 1988. Copyright by the American Geophysical Union.)

summer are not statistically significant. One can, however, conclude from the graph that at 45°N (the latitude of Minneapolis-St. Paul), that total winter ozone levels have decreased by 2–3% per decade.

The loss of ozone over the South Pole has been far more dramatic. Indeed, it is so pronounced that when the British monitoring team at Halley Bay in Antarctica first observed it in the 1970s, they thought their instruments were malfunctioning. Some of their data are reproduced in Figures 2.11 and 2.12. Both report total ozone levels above the surface of the Earth in Dobson units (DU). For our purposes, the precise definition of a Dobson unit is less important than the fact that a value of 320 DU represents the average O_3 level over the northern United States. A value of 250 DU is typical at the equator.

Figure 2.11 compares results for 1957–1973 with those for 1980–1984. There has apparently always been a seasonal variation of ozone concentration, with a minimum in October, which is the Antarctic spring. What was unprecedented was the dramatic decrease in this minimum that has been observed over the last decade.

Figure 2.12 shows that this "ozone hole" has been getting deeper year by year. In 1987, the October ozone level over the pole dropped to less than half of its pre-1970 value. The photograph that introduces this chapter is a representation of ozone concentrations over the South Pole on October 5, 1987. The data were acquired by NASA, the National Aeronautics and Space Administration. Concentrations (in Dobson units) are represented in various colors. The black, pink, and purple regions are those where the greatest O_3 destruction has occurred.

The dramatic decrease in stratospheric ozone poses a series of intriguing questions: What is causing the decline? Is it a totally natural process or does it involve some chemical species of human creation? Why is the decrease occurring over the South Pole? Are similar changes occurring elsewhere in the atmosphere? How serious are these changes and what, if anything, can be done to halt or reverse them?

■ *Chlorofluorocarbons: Properties and Uses*

Answers to most of the questions that conclude the previous section have been found through a masterful piece of scientific sleuthing by F. Sherwood Rowland, Mario Molina, and other chemists. Vast quantities of atmospheric data have been collected and analyzed, hundreds of chemical reactions have been studied, and complicated computer programs have been written in an effort to identify the chemical culprit. As with most scientific results, uncertainties remain, but compelling evidence has been gained implicating an unlikely group of compounds—the **chlorofluorocarbons (CFCs).**

As the name implies, chlorofluorocarbons are compounds composed of the elements chlorine, fluorine, and carbon. Fluorine (symbol F) and the more familiar chlorine (Cl) are members of the same elementary family, the halogens. The other halogens are bromine (Br) and iodine (I). These elements appear in a column labeled Group 7A in the periodic table. At ordinary temperatures and pressures, fluorine and chlorine exist as gases made up of diatomic molecules, F_2 and Cl_2. Bromine and iodine also form diatomic molecules, but the former is a liquid and the latter a solid at room temperature. Fluorine is the most reactive element known. It combines with many other elements to form a wide variety of compounds, including those in "Teflon" (a trademark of the DuPont Company) and other synthetic materials. Chlorine is best known as a water purifier, but it is also a very important starting material in the chemical industry. The element itself ranks ninth among all chemicals in total production in the United States. Moreover, four chlorine-containing compounds are among the top 50 industrial chemicals.

Chlorofluorocarbons do not occur in nature; they are artificially produced. Two of the most widely used have the formulas CCl_2F_2 and CCl_3F. They are commonly known as CFC-12 and CFC-11, following a scheme developed in the 1930s by chemists at the DuPont Company. Their scientific names and Lewis structures are given in Table 2.2.

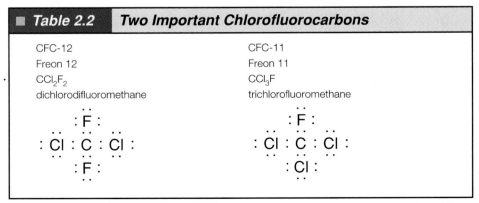

■ *Table 2.2*	*Two Important Chlorofluorocarbons*
CFC-12	CFC-11
Freon 12	Freon 11
CCl_2F_2	CCl_3F
dichlorodifluoromethane	trichlorofluoromethane

Note that the scientific names for these two compounds are based on methane, CH_4. The prefixes di- and tri- specify the number of halogen atoms that substitute for hydrogen atoms. ("Freon" is a trademark of the DuPont Company.)

The introduction of CFC-12 (Freon 12) as a refrigerant in the 1930s was rightly hailed as a great triumph of chemistry and an important advance in consumer safety. This synthetic substance replaced ammonia or sulfur dioxide, two naturally occurring toxic and corrosive compounds that made leaks in refrigeration systems extremely hazardous. In many respects, CFC-12 was (and is) an ideal substitute. It has a boiling point of $-30°C$, which is in the right range, and it is almost completely inert. It is not poisonous, it does not burn, and the CCl_2F_2 molecule is so stable that it does not react with much of anything.

The many desirable properties of CFCs soon led to other uses—as propellants in aerosol spray cans, as the gases blown into polymer mixtures to make expanded plastic foams, as solvents for oil and grease, and as sterilizers for surgical instruments. Similar compounds in which bromine replaces some of the chlorine or fluorine, have proved to be very effective fire extinguishers. These Halons are used to protect property that would be especially vulnerable to water and other conventional fire-fighting chemicals. Thus they have found applications in electronic and computer installations, chemical storerooms, aircraft, and rare book rooms.

By 1985, the combined annual international production of CFC-11 and CFC-12 was approximately 850,000 tons. Much of this was released into the atmosphere. That same year, the ground-level atmospheric concentration of CFCs was about 6 molecules out of every 10 billion (0.6 ppb), a value that has been increasing by about 4% per year. Of course, the fact that ozone levels have been decreasing while CFC levels have been increasing does not prove that the two are casually related. However, other evidence suggests that there is a connection. Ironically, the very property—inertness—that makes CFCs so ideal for so many applications may pose a threat to the environment.

■ *Interaction of CFCs with Ozone*

Chlorofluorocarbons represent a classic case where a virtue becomes a liability. Many of the uses of these compounds capitalize on their low reactivity. The carbon-chlorine and carbon-fluorine bonds in the CFCs are so strong that the molecules can remain unreacted for long periods of time. For example, it has been estimated that an average CCl_2F_2 molecule will persist in the atmosphere for 120 years before it is destroyed. In a much shorter time, typically about five years, many CFC molecules penetrate to the stratosphere with their structures intact. There, the interaction of radiation and matter again comes into play. High energy photons of UV light break carbon-chlorine bonds, releasing chlorine atoms.

$$CCl_2F_2 + \text{photon} \rightarrow CClF_2 + Cl \qquad (2.12)$$

Similar reactions occur with other CFCs.

Unlike CFCs, chlorine atoms are very reactive. A Cl atom has seven outer electrons, and hence exhibits a strong tendency to achieve a stable octet by combining and sharing electrons with another atom. When a chlorine atom encounters an ozone molecule, the following reactions occur.

$$Cl + O_3 \rightarrow ClO + O_2 \qquad (2.13)$$
$$ClO + O \rightarrow Cl + O_2 \qquad (2.14)$$
$$\cancel{Cl} + O_3 + \cancel{ClO} + O \rightarrow \cancel{ClO} + \cancel{Cl} + 2\,O_2 \qquad (2.15)$$

The chlorine atom pulls an oxygen atom away from the O_3 molecule, forming chlorine monoxide, ClO, and leaving an O_2 molecule (equation 2.13). The ClO molecule is another free radical that readily reacts with an oxygen atom (equation 2.14), yielding an O_2 molecule and a Cl atom. The sum of these two reactions is represented by equation 2.15, which corresponds to equation 2.10 and Step 4 in Figure 2.8, except that here, Cl catalyzes O_3 destruction. Again and again, chlorine atoms are regenerated and recycled to remove more ozone molecules. On the average, a single Cl atom may destroy as many as 100,000 O_3 molecules before it is carried back to the lower atmosphere by winds.

Figure 2.13

Ozone and chlorine monoxide concentrations over the Antarctic. (Reprinted by permission of Dr. James G. Anderson, Harvard University.)

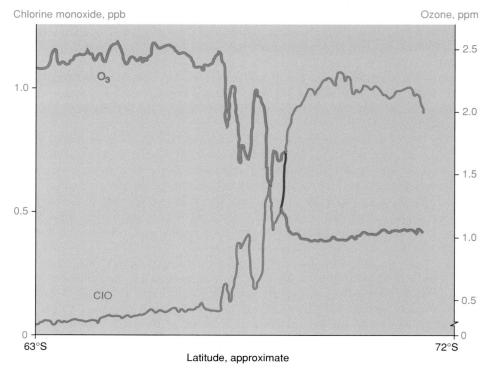

Instruments aboard NASA's ER-2 research airplane measured concentrations of chlorine monoxide and ozone simultaneously as the plane flew from Punta Arenas, Chile (53°S), to 72°S. The data shown above were collected on September 16, 1987. As the plane entered the ozone hole, concentrations of chlorine monoxide increased to about 500 times normal levels while ozone plummeted.

The Cl atoms can also become incorporated into stable compounds that do not react to destroy ozone. Hydrogen chloride, HCl, and chlorine nitrate, $ClONO_2$, are two of these "safe" compounds that are quite readily formed at altitudes below 30 km. Thus, chlorine atoms are fairly effectively removed from the region of highest ozone concentration (about 20 km). Maximum ozone destruction by chlorine atoms appears to occur at about 40 km, where the normal ozone concentration is quite low.

Perhaps the most damning evidence for the involvement of chlorine in the destruction of stratospheric ozone is presented in Figure 2.13. The same graph contains two plots: one of O_3 concentration over Antarctica and the other of the ClO concentration. Both are plotted versus the latitude of the sampling airplane when the measurement was made. The two curves mirror each other almost perfectly, the O_3 concentration decreasing as the ClO concentration increases. Because ClO and Cl are linked by equations 2.13 and 2.14, the conclusion is compelling.

While most of the attention has focused on chlorofluorocarbons, it is important to recognize that not all of the chlorine implicated in ozone destruction comes from CFCs. In 1985, there was a total of about 3.2 ppb of chlorinated carbon compounds capable of reaching the ozone layer. Of this, about 0.6 ppb came from natural sources, 1.6 ppb was due to CFCs, and the remaining 1.0 ppb was from other synthetic compounds.

Using computer simulations of atmospheric chemistry, researchers have predicted changes in the total atmospheric ozone that may occur if CFCs continue to be released at the 1980 rate. Some typical results are presented in Figure 2.14. Note that greater depletion is predicted for the Northern Hemisphere than for the equator. Partially in response to such studies, regulations have been introduced to reduce the levels of CFC emission. The use of CFCs in spray cans was banned in North America in 1978, and their use as foaming agents for plastics was discontinued in 1990. The problem, however, is a global one, and it requires international cooperation. A major step in this direction was the signing, in 1987, of the Montreal Protocol on Substances that Deplete

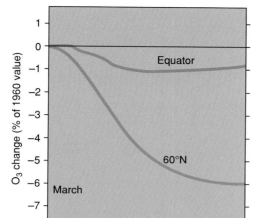

Figure 2.14
Predicted changes in total ozone in response to releases of CFCs at the 1980 rates. (Reprinted by permission of John E. Frederick, The University of Chicago.)

the Ozone Layer, in which the participating nations agreed to reduce CFC production to one-half of the 1986 levels by 1998. Based on a growing understanding of the cause of the ozone hole and the potential for global ozone depletion, more stringent limits were set in the summer of 1990, when representatives of approximately 100 nations met in London and agreed to ban the use of CFCs in 2000. Even that phase-out time is being accelerated. In February 1992, President George Bush ordered a complete halt to CFC production in the United States after December 31, 1995.

2.9   *Consider This*

■

American society has come to rely heavily on the availability of relatively inexpensive fast food. Over the past 25 years, the increase in sales has been dramatic and the competition among the fast food chains has been intensive. Initially food chains wrapped the food in paper, then they switched to styrofoam containers to take advantage of their better insulating and protective properties. The public was pleased with these packaging improvements until it became known that the styrofoam containers were formed using CFCs. As the effect of CFCs on the environment and the resistance of styrofoam to biodegradation became more apparent, the public called for a return to paper wrappers and containers. Although CFCs are no longer used to manufacture styrofoam, paper is once again widely used in the fast food industry. But the use of paper packaging does not solve all our problems. To produce paper, we must cut down trees, and if the used paper is incinerated, it produces CO_2 and H_2O, which contribute to global warming.

Analyze the risks and benefits of paper versus foam containers by considering the economic, social, ecological, and political factors involved. Devise another container that avoids some of the ecological problems associated with paper or styrofoam while at the same time meeting the public's need for a fast food container at a reasonable price.

■ *The Antarctic Ozone Hole*

The predictions of Figure 2.14 do not take into consideration the seasonal O_3 depletion over Antarctica. Evidence suggests that a special mechanism is operative in that region. This mechanism is related to the fact that the lower stratosphere over the South Pole is the coldest spot on Earth. From June to September, during the Antarctic winter, circular winds blowing around the pole prevent warmer air from entering the region. Temperatures get as low as $-90°C$. Under these conditions, the small amount of water vapor present freezes into thin stratospheric clouds of ice crystals. Chemical reactions occurring on the surface of these ice crystals convert otherwise safe molecules like $ClONO_2$ and HCl to more reactive species such as HOCl and Cl_2. When the Sun comes out in October to end the long Antarctic night, the radiation breaks down the HOCl and Cl_2, releasing Cl atoms. The destruction of ozone, which is catalyzed by these atoms, accounts for the "hole."

As the sunlight warms the stratosphere, the ice clouds evaporate, halting the chemistry that occurs on the ice crystals. Moreover, air from lower latitudes flows into the polar regions, replenishing the depleted ozone levels. Thus, by the end of November, the hole is pretty much refilled. However, annual repetitions of this process could result in a general lowering of ozone concentrations in the Antarctic region and eventually across the entire atmosphere. There is already evidence that the ozone reduction over the Southern Hemisphere is greater than one would predict solely on the basis of the mid-latitude chlorine cycle. *Time* on February 17, 1992 reported that Australian scientists believe that wheat, sorghum, and pea production has already been lowered as a result of increased ultraviolet radiation. Health officials there have also observed significant increases in skin cancers. Ultraviolet alerts have even been issued in Australia. These may be among the first manifestations of the consequences of ozone destruction over the South Pole.

Recent measurements have indicated that some of the same conditions also apply over the North Pole, where potentially destructive chemical species such as ClO have been detected. In fact, the highest stratospheric concentration of ClO ever observed (1.5 ppb) was measured in January 1992 by the Second Airborne Arctic Stratospheric Expedition. However, a large hole did not develop the following spring. The Arctic atmosphere is not as cold as that down south, and the air trapped over the North Pole generally begins to diffuse out of the region before the Sun gets bright enough to trigger much ozone destruction. Thus, the problem over the Arctic may not be as serious as that over the Antarctic. Nevertheless, scientists are giving it their close attention.

2.10	*The Sceptical Chymist*
	In spite of all you have just read, good experimental evidence indicates that the level of ultraviolet radiation reaching the surface of the Earth in urban areas has in fact decreased in the past decade or two. What reasons can you offer to explain this observation? Is the ozone depletion scenario just a scientific scam or perhaps the overreaction of some environmentalists? Write a short newspaper article explaining this apparent discrepancy.

■ Substitutes for the Future

Where do we go from here? Robert T. Watson, program manager of upper atmospheric research at the National Aeronautics and Space Administration has summed up the current situation and looked into the future. "There's about 3 ppb of chlorine in the atmosphere today. Under the Montreal protocol, chlorine will continue to increase to about 6 ppb. Even if we were to stop producing the fully halogenated CFCs today, the Antarctic ozone hole would be here for centuries to come. Time is of the essence. Every year we continue to put chlorine in adds 10 years to the time it will take to remove it."

Scientists estimate that even under the most stringent international controls on the use of ozone-depleting chemicals, the atmospheric chlorine concentration would not drop to 2 ppb until 2075. This value is significant because the Antarctic ozone hole first appeared when chlorine reached 2 ppb. We cannot scrub the atmosphere of chlorine and chlorofluorocarbons, we must wait until nature removes them—a slow process given the century-long lifetimes of many CFCs. Nor can we, as some have suggested, replenish the ozone layer by releasing O_3 from planes flying in the lower stratosphere. Sherwood Rowland, one of the pioneers in the study of ozone destruction, has estimated that "the energy that would be needed to move the ozone up [to the stratosphere] is about 2 1/2 times all of our current global power use." Even if we could temporarily replace the lost ozone, the steady-state cycle would soon reestablish itself.

2.11 *The Sceptical Chymist*

The Sceptical Chymist even questions the accuracy of cartoons. The one that follows was taken from *What's so Funny About Science?* by Sidney Harris.

"OH, FOR PETE'S SAKE, LET'S JUST GET SOME OZONE AND SEND IT BACK UP THERE!"

(© 1976 by Sidney Harris—American Scientist Magazine.)

What's so funny about this industrialist's answer to the ozone problem? To check whether it is a reasonable solution, suppose you determine the number of jumbo jet-loads of O_3 that would be required to replace 10% of the atmospheric ozone. Clearly, there are many different assumptions you might make and a variety of strategies you might use, but we offer some suggestions on how to proceed.

a. Let's focus on the region of the atmosphere between 10 and 50 km above the Earth's surface. Use Figure 2.9 to estimate the ozone concentration in this region.

2.11   *Continued*

b. You need to know the total number of O_3 molecules in this region, and therefore must calculate its volume. The region in question is a spherical shell, 40 km thick. The volume of such a shell is equal to $4\pi r^2 t$, where r is the average radius of the shell and t is its thickness. Since the shell of air surrounds the Earth, you need to include the radius of the Earth (6300 km) in your calculation. A drawing might help. If you express r and t in kilometers, the volume will be in cubic kilometers or km^3.

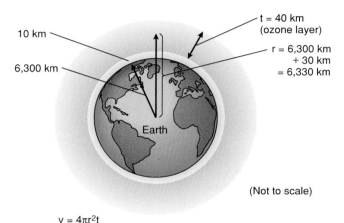

10 km

6,300 km

t = 40 km (ozone layer)

r = 6,300 km
  + 30 km
  = 6,330 km

Earth

(Not to scale)

$$v = 4\pi r^2 t$$
$$= 4(3.14)(6,330 \text{ km})^2(40 \text{ km})$$

c. Now you can use the results from a. and b. to compute the number of O_3 molecules in this ozone-rich region ($1 \text{ km}^3 = 10^9 \text{ m}^3$).

d. You also need to find out the number of O_3 molecules that can be carried in a jumbo jet. This in turn requires knowledge of the volume of a plane. Let's assume the jet can be approximated by a cylinder with a radius (r) of 5 m (15 ft) and a length (l) of 65 m (200 ft). The volume of the cylinder is $\pi r^2 l$.

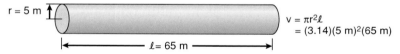

r = 5 m

$$v = \pi r^2 l$$
$$= (3.14)(5 \text{ m})^2(65 \text{ m})$$

$l = 65$ m

e. At 25°C and 1 atmosphere pressure, 25 L of gas contain about 6×10^{23} molecules. Use this information and your answer to d. to calculate the number of O_3 molecules that could be contained in a single jumbo jet under these conditions of temperature and pressure ($10^3 \text{ L} = 1 \text{ m}^3$).

f. Finally, use the results of c. and e. to find the number of jet-loads of O_3 that would be required to replenish 10% of the ozone molecules in this region.

Note: We got a final answer of 1.7×10^7 or 170 million planeloads, but you might be skeptical of that, too.

While we cannot undo what already has been done, we can stop doing it. There is already widespread agreement among atmospheric scientists, environmentalists, chemical manufacturers, and government officials that the Montreal Protocol is not sufficiently stringent. The elimination of CFCs must occur more rapidly than originally proposed. Considerable progress has already been made. The United Nations Environmental Program reports a 40% drop in CFC consumption between 1986 and 1991. However, there is still considerable disagreement over what is safe, prudent, and feasible.

Key to the process will be the success of chemists in finding replacements for CFCs. No one advocates the return to ammonia and sulfur dioxide in refrigeration apparatus. In designing replacement molecules, chemists are concentrating on compounds similar to the CFCs. The assumption is that the substitute molecules will include one or two carbon atoms, at least one hydrogen atom, and several chlorine and/or fluorine atoms. The rules of molecular structure limit the options. For example, in the molecules under consideration each carbon atom forms single bonds to four other atoms.

The chemical and physical properties of all compounds depend on elementary composition and molecular structure. In synthesizing substitutes for CFCs, chemists must weigh three potentially undesirable properties: toxicity, flammability, and extreme stability, and attempt to achieve the most suitable compromise. Compounds containing only carbon and fluorine (fluorocarbons) are neither toxic nor flammable, and they are not decomposed by ultraviolet radiation, even in the stratosphere. Consequently, they would not catalyze the destruction of ozone. This would be ideal, were it not for the fact that the undecomposed fluorocarbons would eventually build up in the atmosphere and contribute to the greenhouse effect by absorbing infrared radiation (see Chapter 3).

Introducing hydrogen atoms in place of one or more halogen atoms reduces molecular stability and promotes their destruction at low altitudes, long before they enter the ozone-rich regions of the atmosphere. However, too many hydrogen atoms increase flammability and too many chlorine atoms seem to increase toxicity. For these reasons, chloroform, $CHCl_3$, would not be a good substitute. Moreover, when a halogen atom is replaced by a much lighter hydrogen atom, the total mass of the molecule is decreased. This results in a decrease in boiling point, a trend observed in many families of similar compounds. A boiling point in the -10 to $-30°C$ range is an important property for a refrigerant. Therefore, the relationship between composition, molecular structure, boiling point, and proposed use must be considered along with toxicity, flammability, and stability.

Fortunately, chemists already know a good deal about how these variables are related, and they have used this knowledge to synthesize some promising replacements for CFCs. Some are hydrofluorocarbons (HFCs), which are compounds of hydrogen, fluorine, and carbon. HFC-134a, CF_3CH_2F, with a boiling point of $-26°C$, may well be the substitute of choice for CFC-12. It has no chlorine atoms to interact with ozone, and its two hydrogen atoms facilitate its decomposition in the lower atmosphere without making it flammable under normal conditions. HCFC-22, CHF_2Cl, is already being used in air conditioners and in the production of foamed fast food containers. Its ozone-depleting potential is about 5% that of CFC-12.

All of this has major economic consequences. The annual worldwide market for CFCs is $2 billion, but that is only the beginning. In the United States alone, chlorofluorocarbons are used in or used to produce goods valued at about $28 billion per year. Currently, $135 billion worth of equipment, including your refrigerator and automobile air conditioner, rely on CFCs. The conversion to new compounds will be very costly. Research and development efforts to find replacements are expensive, and it is possible that the new compounds will be more costly to manufacture. In addition, companies that produce refrigerators, air conditioners, insulating plastics, and other goods will need to learn how to use the new compounds. They will also be required to invest in equipment compatible with the replacements. Yet another economic and environmental consideration is the fact that most substitutes for CFC refrigerants appear to be less energy efficient, hence increasing energy consumption.

To be sure, the introduction of substitute compounds holds out some prospect of profit for chemical manufacturers, but the initial investment is significant. Should there be special incentives to encourage industry to do the necessary research and development and build the new plant capacity? Or should industry be subject to a special tax on profits resulting from the increasing prices of the dwindling supplies of CFCs?

Again, issues are complex. Stephen Anderson, an EPA official, has been quoted as saying "Business is moving faster than the laws require. They're finding they can save money and improve performance."

On a domestic level, the political dimension of CFC regulation raises many issues. What agency establishes the limits? Where is the legislation enacted? Is this a national, state, or local affair? Who will enforce the regulations? What limitations and what time constraints are reasonable, responsible, or possible? How much testing is necessary before replacement compounds can be introduced? How can the country be confident that those making the political, legal, and economic decisions are getting the best scientific advice and interpreting it correctly? There are no easy answers to these questions, maybe not even any right or wrong answers, but the activity that follows gives you an opportunity to struggle with some of them.

2.12 ■ Consider This

The Environmental Protection Agency has ruled that its federal regulations governing production of CFCs do not preempt the rights of states and cities to enact their own legislation. As a consequence, over 90 bills have been introduced in more than 20 states. Much of this proposed legislation involves restrictions on the use of CFCs in automobile air conditioners or requirements for recycling refrigerants.

Suppose a bill has been introduced into your state legislature to ban the use of CFCs in automobile air conditioners. Car dealers, manufacturers, and service personnel are opposed to the bill. Environmentalists favor it. Align yourself with one of these groups and draft a letter to your state representative, outlining your positions and your reasons for either supporting or not supporting the bill.

Developing countries face another set of economic problems and priorities. Chlorofluorocarbons have played an important part in improving the quality of life in the industrialized nations. Few would be willing to give up the convenience and health benefits of refrigeration, or the comfort of air conditioning. It is understandable that millions of people over the globe aspire to the lifestyle of the industrialized West. As an example, over the past decade, the annual production of refrigerators in China has increased from 500,000 to 8 million. But if the developing nations are banned from using the relatively cheap CFC-based technology, they may not be able to afford alternatives. "Our development strategies cannot be sacrificed for the destruction of the environment caused by the West," asserts Ashish Kothari, a member of an Indian environmental group.

Clearly, such technical, economic, and political issues make international agreements on protecting the ozone layer more difficult to achieve. On the other hand, the problem is global, and its solution requires global cooperation.

2.13 ■ Consider This

Maneka Gandhi, former Indian Minister of the Environment and delegate to the Montreal Protocol has summed up the stance of his government: "India recognizes the threat to the environment and the necessity for a global burden sharing to control it. But is it fair that the industrialized countries who are responsible for the ozone depletion should arm-twist the poorer nations into bearing the cost of their mistakes?" Discuss this pertinent question and identify and evaluate alternatives to "arm-twisting."

2.14 *Consider This*

■ Both India and China refused to sign the original Montreal Protocol because they felt that it discriminated against developing countries. In 1990, the industrially developed nations created a special $240 million fund to help developing countries phase out CFCs. China has now signed the revised protocol and India is expected to do so. The United States initially resisted contributing to this fund, but ultimately donated 25% of the total. Organize a debate in which the positions of India and the United States are accurately and forcefully articulated.

■ *Conclusion*

Chemistry is intimately entwined with the story of ozone depletion. Chemists created the chlorofluorocarbons whose near-perfect properties only recently revealed their dark side as predators of stratospheric ozone. Chemists discovered the mechanism by which CFCs destroy ozone and warned of the dangers of increasing ultraviolet radiation. And chemists will synthesize the substitutes that will soon replace CFCs. But the issues involve more than just chemistry. Philip Elmer-Dewitt said it well in the article in *Time* on February 17, 1992, that provided some of the quotations used in this chapter: "Chlorofluorocarbons have worked their way deep into the machinery of what much of the world thinks of as modern life—air-conditioned homes and offices, climate-controlled shopping malls, refrigerated grocery stores, squeaky-clean computer chips. Extricating the planet from the chemical burden of that high-tech life style—for both those who enjoy it and those who aspire to it—will require not just technical ingenuity but extraordinary diplomatic skill."

■ *References and Resources*

Elmer-Dewitt, P. "How Do You Patch a Hole in the Sky That Could Be as Big as Alaska?" *Time,* Feb. 17, 1992: 64–68.

Lemonick, M.D. "The Ozone Vanishes." *Time,* Feb. 17, 1992: 60–63.

Makhijani, A.; Bickel, A.; and Makhijani, A. "Still Working on the Ozone Hole." *Technology Review,* May–June, 1990: 53–59.

Monastersky, R. "The Two Faces of Ozone." *Science News,* Sept. 2, 1989: 154–55.

O'Sullivan, D. A. "International Gathering Plans Ways to Safeguard Atmospheric Ozone." *Chemical & Engineering News,* June 26, 1989: 33–36.

"Our Ozone Shield." *Reports to the Nation,* Fall, 1992. Boulder, Colorado: University Corporation for Atmospheric Research.

Roan, S. L. *Ozone Crisis: The 15-Year Evolution of a Sudden Global Emergency.* New York: Wiley, 1989.

Rowland, F. S. "Stratospheric Ozone in the 21st Century." *Environmental Science & Technology* **25** (1991): 622–28.

Zurer, P. S. "Producers, Users Grapple with Realities of CFC Phaseout." *Chemical & Engineering News,* July 24, 1989: 7–13.

———. "EPA Proposes Nationwide Recycling Program for Ozone-Depleting CFCs." *Chemical & Engineering News,* May 7, 1990: 44–45.

———. "Ozone-Safe Technology Fund Likely Vital to Montreal Treaty Compliance." *Chemical & Engineering News,* Aug. 6, 1990: 19–20.

———."Arctic Ozone Loss: Fact-Finding Mission Concludes Outlook is Bleak." *Chemical & Engineering News,* March 6, 1992: 29–31.

———. "Industry, Consumers Prepare for Compliance with Pending CFC Ban." *Chemical & Engineering News,* June 22, 1992: 7–13.

■ *Experiments and Investigations*

4. Molecular Models

■ Exercises

1. Use a periodic table and information from the text to complete the following table for elements with atomic numbers between 11 and 18.

Element	Number of electrons	Number of outer electrons	Element resembled
Na	___	___	___
___	13	___	___
___	___	5	___
___	___	___	Cl

2. Specify the atomic numbers and mass numbers of the following isotopes.

   a. carbon-14 or C-14 (a radioactive isotope used to date artifacts)

   b. iodine-131 or I-131 (a radioactive isotope used to treat overactive thyroid)

   c. iron-56 or Fe-56 (the most plentiful isotope of iron)

3. Indicate the number of electrons, neutrons, and protons in one atom of each of the isotopes listed in Exercise 2.

4. Draw electron dot structures for

   a. hydrogen sulfide (H_2S)

   b. carbon dioxide (CO_2)

   c. methanol (CH_3OH)

5. Convert the following wavelengths to meters and identify the spectral region where they would be found.

   a. 300 nanometers

   b. 10 micrometers

   c. 2 centimeters

6. Calculate the frequency that corresponds to each of the wavelengths in Exercise 5.

7. Calculate the energy of a photon for each wavelength in Exercise 5.

8. The bonds in O_2 and O_3 can be broken with light of 242 nm and 320 nm, respectively. Calculate the energy of a photon of light at each of these wavelengths.

9. The favorite radio station of one of the authors of this text broadcasts at a frequency of 91.3×10^6 s^{-1}. Determine the length of the waves that correspond to this frequency. Repeat the calculation for your favorite radio station.

10. It is suggested in this chapter that a 5% decrease in the ozone levels at 60°N latitude could lead to a 10% increase in skin cancer. Estimate the increase in skin cancer rates that might be expected if variations in the solar cycle led to a 2% decrease in ozone levels. State the assumptions that you made in doing this calculation.

*11. Draw electron dot structures for the species (H, OH, NO) that are especially effective at catalyzing the reaction:

$$O_3 + O \rightarrow 2\,O_2$$

   What do these structures have in common? How might this feature contribute to their catalytic effectiveness?

12. Propose a scheme involving NO as the catalyst for the destruction of O_3. Hint: equations 2.13–2.15 may serve as a useful guide. Describe the important features of a catalyst on the basis of this scheme.

13. The total quantity of energy that reaches the Earth in the infrared region is greater than that in the ultraviolet region. Despite this fact, there is less concern about IR radiation than about UV. Explain why this is the case.

14. If a typical O_3 molecule "lives" only 100 to 200 seconds before undergoing dissociation, how can O_3 offer any protection from ultraviolet radiation?

15. Speculate about the energy needed to dissociate N_2 relative to that needed to dissociate O_2 or O_3. Account for your prediction on the basis of electron dot structures.

16. Use Figure 2.9 to calculate the ratio of the concentration of O_3 at 50 km to that at: a) 10 km b) 20 km. Describe any assumptions or approximations you made to carry out this calculation.

17. Estimate the percentage decrease in the total O_3 level over Antarctica from 1960 to 1980 from the data in Figure 2.12.

18. According to Figure 2.14 and other appropriate information, what percentage increase in skin cancer would be expected at 60°N latitude by the year 2040? What assumptions did you make to carry out this calculation?

19. List three characteristics of CFCs that led to their widespread adoption. Which of these have proven to be problematic?

3

The Chemistry of Global Warming

■

In 1992, a United States Senator from Tennessee published a book in which he proposed the following as part of the role of this country in a "Global Marshall Plan."

> **That we create an Environmental Security Trust fund, with payments into the Fund based on the amount of CO_2 put into the atmosphere.** Production of gasoline, heating oil and other oil-based fuels, coal, natural gas, and electricity generated from fossil fuels would trigger incremental payments of the CO_2 tax according to the carbon content of the fuels produced. These payments would be reserved in a trust fund, which would be used to subsidize the purchase by consumers of environmentally benign technologies—such as low-energy light bulbs or high-mileage automobiles. A corresponding reduction in the amount of taxes paid on incomes and payrolls in the same year would ensure that the trust fund plan does not raise taxes but leaves them as they are—while having sufficient flexibility to ensure progressivity and to deal equitably with special hardships encountered in the transition to renewable energy sources (such as those faced by someone with no immediate alternative to the purchase of large quantities of heating oil, gasoline, or the like). I am convinced that a CO_2 tax which is completely offset by decreases in other taxes is rapidly becoming politically feasible.

The title of the book is *Earth in the Balance: Ecology and the Human Spirit;* its author is Al Gore. The book and proposals such as these suddenly took on national significance (and controversy) when, in the summer of 1992, the Democratic party nominated Gore as its vice-presidential candidate. Indeed, the CO_2 tax was one of the items of contention in the nationally televised debate between Senator Gore and then Vice-President Dan Quayle. And in some ways, the energy tax bill, introduced by the Clinton administration in early 1993, is a direct descendent of Gore's proposed carbon dioxide tax.

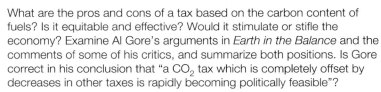

3.1 ■ **Consider This**

What are the pros and cons of a tax based on the carbon content of fuels? Is it equitable and effective? Would it stimulate or stifle the economy? Examine Al Gore's arguments in *Earth in the Balance* and the comments of some of his critics, and summarize both positions. Is Gore correct in his conclusion that "a CO_2 tax which is completely offset by decreases in other taxes is rapidly becoming politically feasible"?

To be sure, taxing chemical compounds is not new. Some societies tax salt, and ethyl alcohol carries a heavy levy in many countries. But why this fuss about an essential component of the atmosphere—a gas that all animals exhale and all green plants absorb? This chapter is an attempt to answer that question by investigating the way in which carbon dioxide and other gases, in part generated by human activity, contribute to global warming.

■ *Chapter Overview*

The two sections that immediately follow this overview provide a general description of the greenhouse effect and its relationship to the evolution of the Earth and its atmosphere. Central to the issue of global warming is the Earth's energy balance and the molecular mechanism by which carbon dioxide and other compounds absorb the infrared radiation emitted by the planet. Some knowledge of molecular structure and shape is necessary to understand this mechanism. Therefore, we develop a general method for predicting molecular geometry and then relate it to infrared-induced vibrations. The section on the carbon cycle makes it clear that most of the CO_2 in the atmosphere is of natural origin, but increased human contributions are the chief cause of the current concern about the greenhouse effect. These concerns have a significant quantitative component; we need numbers to help assess the seriousness of the situation. That need justifies several sections in which we introduce and illustrate some fundamental chemical concepts, including atomic and molecular mass and the mole concept. Thus armed, we return to a brief look at methane and several other greenhouse gases. A discussion of predictions based on computer modeling of the climate leads to an assessment of the current situation. The chapter ends with some suggested answers to the all-important question: "What can we do?"

■ *In the Greenhouse*

The brightest and most beautiful body in the night sky, after our own moon, is Venus. It is ironic that the planet named for the goddess of love is a most unlovely place. Spacecraft launched by the United States and the Soviet Union have revealed a desolate, eroded surface with an average temperature of about 450°C (840°F). The Venusian atmosphere has a pressure 90 times greater than that of the Earth, and it is 96% carbon dioxide, with clouds of sulfuric acid. It makes the worst smog-bound Southern California inversion seem like a breath of country air. The beautiful blue-green ball we inhabit has an average annual temperature of 15°C (59°F). The point of this little astronomical digression is that both Venus and Earth are warmer than one would expect solely on the basis of their distances from the Sun and the amount of solar radiation they receive. If that were the only determining factor, the temperature of Venus would average approximately 100°C, the boiling point of water. The Earth, on the other hand, would have an average temperature of −18°C (0°F), and the oceans would be frozen year-round.

3.2 ■	## *Consider This*

A number of successful writers of science fiction began their careers as science majors. Their best work reveals a sound understanding of scientific phenomena and principles. Often a good science fiction story assumes a slightly different scientific reality than the one we know. For example, *Dune* by Frank Herbert takes place on a desert planet. Here is an opportunity to exercise your imagination in a different climate. Suppose the planet had an average temperature of −18°C (0°F)? What would human life be like? Write a brief description of a day on a frozen planet. (Residents of Minnesota should have a great advantage in this exercise.)

The composition of the atmosphere is central to understanding why our planet is 33°C warmer than we would expect, considering the amount of solar energy reaching its surface. The moderating effect is primarily due to two of the minor constituents of the atmosphere: water vapor and carbon dioxide. There is a sort of wonderfully harmonious symmetry in the fact that the two compounds that keep our planet warm enough to sustain life are also among the essential ingredients of all living things.

The idea that atmospheric gases might somehow be involved in trapping some of the Sun's heat was first proposed around 1800 by the French mathematician and physicist, Jean Baptiste-Joseph Fourier (1768–1830). Fourier compared the function of the atmosphere to that of the glass in a "hothouse" (his term) or **greenhouse.** Although he did not understand the mechanism or know the identity of the gases responsible for the effect, his metaphor has persisted. Some 60 years later, John Tyndall (1820–1893) in

England experimentally demonstrated that carbon dioxide and water vapor absorb heat radiation. In addition, he calculated the warming effect that would result from the presence of these two compounds in the atmosphere.

■ *The Testimony of Time*

In the 4.5 billion years that our planet has existed, its atmosphere and its climate have varied widely. Evidence from the composition of volcanic gases suggests the concentration of carbon dioxide in the early atmosphere of the Earth was perhaps 1000 times what it is today. Much of the CO_2 dissolved in the oceans became incorporated in rocks such as limestone, which is calcium carbonate, $CaCO_3$. But the high concentration of carbon dioxide also made possible the most significant event in the history of our planet. Although the Sun's energy output was 25–30% less than it is today, the ability of CO_2 to trap heat kept the Earth sufficiently warm to permit the development of life. As early as 3 billion years ago, the oceans were filled with primitive plants such as cyanobacteria. Like their more sophisticated descendants, these simple plants were capable of **photosynthesis.** They were able to capture sunlight and use its energy to combine carbon dioxide and water to form more complex molecules such as glucose.

$$\overset{\text{chlorophyll}}{6\,CO_2 \;+\; 6\,H_2O \quad \rightarrow \quad \underset{\text{glucose}}{C_6H_{12}O_6} \;+\; 6\,O_2} \tag{3.1}$$

Photosynthesis not only dramatically reduced the CO_2 concentration of the atmosphere, it increased the amount of O_2 present. The microbiologist, Lynn Margulis, has called this "the greatest pollution crisis the Earth has ever endured." We and our kin are its beneficiaries. The increase in oxygen concentration made possible the evolution of animals. But even in the time of the dinosaurs, 100 million years ago, the average temperature is estimated to have been 10–15°C warmer than it is today and the CO_2 concentration is assumed to have been considerably higher.

Reasonably reliable evidence is available about temperature fluctuations during the past 200,000 years—only yesterday in geological terms. Deep drilling cores from the ocean floor give us a slice through time. The number and nature of the microorganisms present at any particular level indicate the temperature at which they lived. Supplementing this, the alignment of the magnetic field in particles in the sediment provides an independent measure of time.

Other relevant information comes from the analysis of ice cores. The Soviet drilling project at the Vostok Station in Antarctica has yielded over a mile of ice formed from the snows of 160 millennia. The ratio of deuterium to ordinary hydrogen in the ice can be measured and used to estimate the local temperature at the time the snow fell. In addition, the bubbles of air trapped in the ice can be analyzed for composition. Both sorts of data are incorporated in Figure 3.1. The upper curve (corresponding to the scale on the left) is a plot of parts per million of carbon dioxide in the atmosphere versus time over a span of 160,000 years. The lower plot and the right-hand scale indicate how the average global temperature has varied over the same period. For example, the figure shows that 20,000 years ago, during the last ice age, the average temperature of the Earth was about 9°C below the 1950–1980 average. At the other extreme, a maximum temperature (just over 16°C) occurred approximately 130,000 years ago.

What is particularly striking about Figure 3.1 is the fact that temperature and carbon dioxide concentration parallel each other. When the CO_2 concentration was high, the temperature was high. Other measurements show that periods of high temperature have also been characterized by high atmospheric concentrations of methane (CH_4). Admittedly, such correlations do not necessarily prove that elevated atmospheric CO_2 and CH_4 caused the temperature increases. Presumably, the converse could have taken place. But the fact is that these compounds are known to trap heat, and there is no doubt that they can and do contribute to global warming.

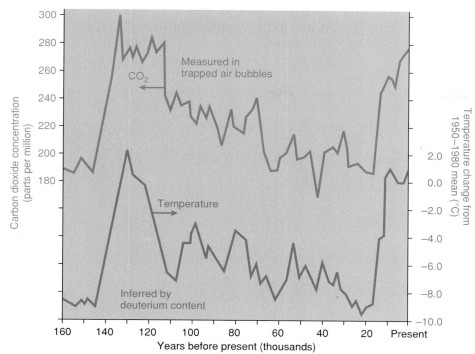

Figure 3.1

Atmospheric CO_2 concentration and average global temperature over 160,000 years (data from ice cores). (Reprinted with permission from *Chemical & Engineering News*, March 13, 1989, **67** (11), p. 36. Copyright © 1989 American Chemical Society.)

To be sure, other mechanisms are also involved in the periodic fluctuations of global temperature. Temperature maxima seem to come at roughly 100,000-year intervals, with interspersed major and minor ice ages. Over the past million years, the Earth has experienced 10 major periods of glaciation and 40 minor ones. Some of this temperature variation is probably caused by minor changes in the Earth's orbit, which affect the distance of the Earth to the Sun and the angle with which sunlight strikes the planet. However, this hypothesis cannot fully explain the observed temperature fluctuations. It is likely that the orbital effects are coupled with terrestrial events such as changes in reflectivity, cloud cover, airborne dust, and carbon dioxide and methane concentration. These factors can diminish or enhance the orbital-induced climatic changes. The feedback mechanism is complicated and not well understood. One thing is clear: the Earth is a far different place in the 1990s than it was at the time of our last temperature maximum 130,000 years ago. Our ancestors had discovered fire by then, but they had not learned to exploit it as we have.

■ *The Earth's Energy Balance*

The source of the Earth's energy is of course the Sun. About half of the radiant energy that strikes our atmosphere is either reflected or absorbed by the molecules that make up this envelope of air. You know from your study of Chapter 2 that oxygen and ozone intercept much of the ultraviolet radiation. The rays that do reach the surface of the planet are largely in the visible and infrared (heat) regions of the spectrum. This radiation is absorbed by the Earth, and as a result, the continents and oceans are warmed. The current average temperature of the planet, about 15°C, is much higher than the –270°C of outer space. Consequently, the Earth acts like a global radiator, radiating heat to its frigid surroundings.

Figure 3.2 is a schematic representation of the Earth's energy balance. The width of the arrows is intended to indicate that the rate at which energy escapes the surface of the Earth is over twice the rate at which the planet directly absorbs energy from the Sun. The Earth thus acts a little like an extravagant college student, spending money faster than he or she earns it. If this process were the only one occurring, the student

Figure 3.2

The Earth's energy balance. The width of the arrows is roughly proportional to energy flow.

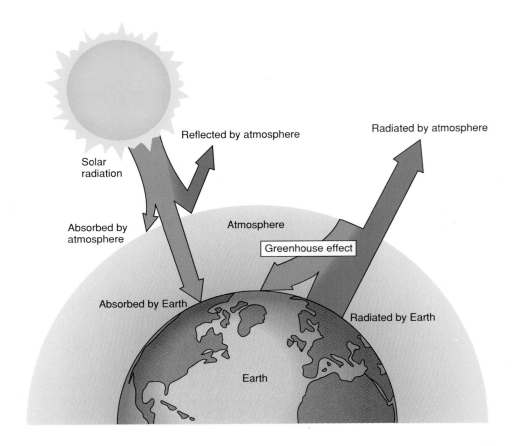

would soon be deeply in debt and the Earth would be a very cold place. Fortunately, the planet has something few students do—a sort of built-in forced savings account. Almost 84% of the heat it radiates is absorbed by the atmosphere and then reradiated back to the surface. As a result of this exchange, the books are balanced and the total energy input from the Sun balances the energy output from the Earth. A steady state is established, with more or less constant average terrestrial temperatures.

The "more or less" is of course the reason for the current concern over global warming. It is this return of 84% of the energy radiated from the surface of the Earth that has been termed the **greenhouse effect.** The "windows" of this greenhouse are made of molecules that are transparent to visible light, but absorb in the infrared region of the spectrum. They permit the radiation coming from the Sun to pass through, but trap much of the heat emitted by the Earth.

Obviously, the greenhouse effect is essential in keeping our planet habitable with the species that have evolved here. But if some CO_2 in the atmosphere is a good thing, more is not necessarily better. An increase in the concentration of this infrared absorber will very likely mean that more than 84% of the radiated energy will be returned to the Earth's surface, with an attendant increase in average temperature. Back around 1900, the Swedish chemist, Svante Arrhenius (1859–1927), estimated the extent of this effect. He calculated that doubling the concentration of CO_2 would result in an increase of 5 to 6°C in the average temperature of the planet's surface. At the turn of the century, the Industrial Revolution was already well under way in Europe and America, and it was "picking up steam" as well as generating it.

Over the past century, the concentration of CO_2 in the atmosphere has risen by about 25%, and the average temperature of the planet has increased by somewhere between 0.5 and 0.7°C. Figure 3.3 indicates the temperature change from 1880 to 1987. The values plotted here are based on five-year temperature averages. Although this smoothes out some of the year-to-year fluctuation, there is still a good deal of variability in the temperature data. Nevertheless, there seems to be a clear trend; the average temperature of the earth is increasing.

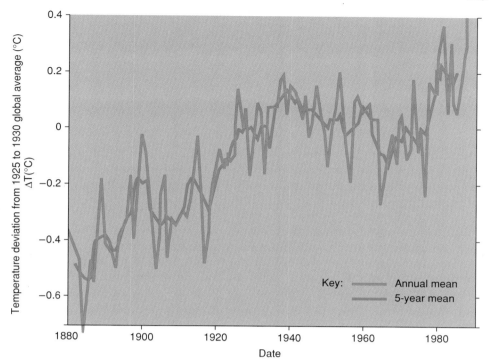

Figure 3.3
Global temperature from 1880 to 1987. (The temperature changes are expressed relative to the 1925–30 average.) Data from Hansen and Lebedeff, 1988.

When such temperature measurements are extrapolated into the future, Arrhenius' predictions must be revised downward. Current estimates are that doubling the CO_2 concentration will result in a temperature increase of between 1.5 and 4.5°C. If and when that doubling will occur depends, to a considerable extent, on the human beings who inhabit this planet. We are a long way from the out-of-control hothouse of Venus, but we face difficult decisions.

These decisions may not be made easier, but they will be better informed with an understanding of the mechanism by which greenhouse gases interact with radiation. For that we must again assume a submicroscopic view of matter.

3.3	*Consider This*

List what you have done today to increase the concentration of atmospheric carbon dioxide.

■ *Molecules: How They Shape Up*

Methane, water, and carbon dioxide are greenhouse gases; nitrogen and oxygen are not. The obvious question is "Why?" The not-so-obvious answer has to do with molecular structure and shape. When you encountered molecular structure in Chapter 2, it was at the level of Lewis structures. The octet rule provides a generally reliable method for predicting bonding in molecules. Moreover, in the case of diatomic molecules, it also predicts molecular geometry. In molecules such as O_2 and N_2, shape is unambiguous; the atoms can only be in a straight line.

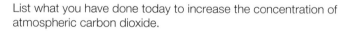

With molecules of three or more atoms, differences in molecular geometry become possible. Fortunately, knowing where the electrons are located provides insight into molecular shape. To illustrate the procedure, we look first at methane, CH_4. The carbon atom (atomic number 6) carries four outer electrons, and each of the four hydrogen atoms contributes one electron. These eight electrons are arranged around the central carbon atom in four bonding pairs, each pair connecting the carbon atom to a hydrogen atom.

$$
\begin{array}{c}
\text{H} \\
\text{H} : \overset{..}{\underset{..}{\text{C}}} : \text{H} \\
\text{H}
\end{array}
$$

The Lewis structure for CH_4 seems to imply that the molecule is flat or planar. But here we are restricted to the two dimensions of a sheet of paper. The architecture of molecules is three dimensional, and we must look into that third dimension. When we do, using special instruments, we find that the methane molecule is far from flat. The hydrogen atoms are as far from each other as they can be and still be covalently bonded to the central carbon atom. The molecule is reminiscent of the base of a folding music stand, the four bonds corresponding to the three evenly spaced legs and the vertical shaft. The angle between each pair of bonds is 109.5°. This shape is said to be **tetrahedral,** because the hydrogen atoms correspond to the corners of a **tetrahedron,** a four-cornered figure with four equal sides

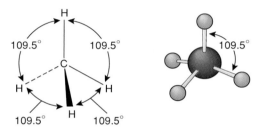

The drawing on the left is an attempt to convey this structure. The wedge-shaped line represents a bond that is coming out of the paper at an angle but generally toward the reader, the dashed line represents a bond pointing away from the reader, and the solid lines are assumed to be in the plane of the paper. This is an improvement, but the best way to visualize molecules is with wooden or plastic models, as in the drawing on the right.

The general principle behind this specific case is that the electron pairs in a molecule are usually arranged in such a way that they are as far from each other as possible, within the constraints of the bonding. In CH_4, each of the covalent bonds consists of a pair of electrons. These negative charges repel each other, and the structure of the molecule represents maximum separation of the electron pairs.

We now apply this rule to the H_2O molecule. The Lewis structure, given below, discloses eight electrons on the central atom—two pairs involved in bonding and two lone pairs.

$$
\text{H} : \overset{..}{\underset{..}{\text{O}}} : \text{H}
$$

If these four pairs of electrons are arranged so that they are as far apart as possible, the distribution will be similar to that in methane. To be sure, two pairs are bonding and two pairs are nonbonding, but we assume that will have minimal effect on their mutual repulsion. Therefore, we predict that the angle between the two O—H bonds will be approximately 109°. Experiments indicate a value of approximately 105°, suggesting that our model is reasonably reliable.

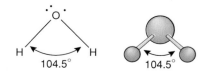

3.4

■

Your Turn

Using the strategies described above, predict and sketch the shapes of the following molecules.

a. CCl_4 (carbon tetrachloride)
b. CCl_2F_2 (Freon 12)
c. NH_3 (ammonia)

Ans. a. The first step in predicting a structure is to write the correct Lewis structure. Each of the four chlorine atoms has seven outer electrons, and the carbon atom has four. The chlorine atoms bond to the central carbon atom, forming four single bonds. Each bond is a shared pair of electrons. Thus, each atom is surrounded by eight electrons.

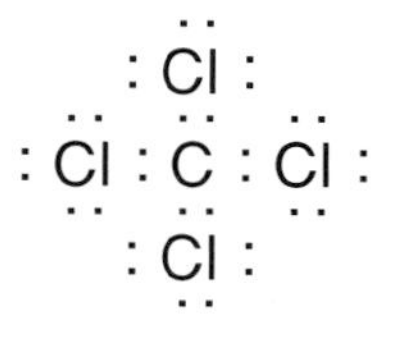

$$
\begin{array}{c}
\ddot{\,}\ddot{\,} \\
:\ddot{Cl}: \\
\ddot{\,}\ddot{\,} \\
:\ddot{Cl}:\ddot{C}:\ddot{Cl}: \\
\ddot{\,}\ddot{\,} \\
:\ddot{Cl}: \\
\ddot{\,}\ddot{\,}
\end{array}
$$

The bonding electron pairs and the attached chlorine atoms will arrange themselves so that their separation is a maximum. It follows that the shape of a carbon tetrachloride molecule is tetrahedral—the same as a methane molecule.

Carbon dioxide is our third example. A count of outer electrons reveals a total of 16: four contributed by the carbon atom and six from each of the two oxygen atoms. There are not enough electrons to provide eight electrons for each of the atoms, if only single bonds are involved. That would require 20 electrons. However, the octet rule will be obeyed if the central carbon atom shares four electrons with each of the oxygen atoms. This means that two double bonds are formed.

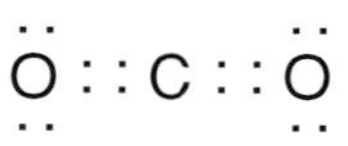

$$
\ddot{O}::C::\ddot{O}
$$

In the CO_2 molecule, there are thus two groups of four electrons associated with the central atom. These groups of electrons will again repel each other, and the most stable configuration will be that corresponding to the furthest separation of the negative charges. This will occur when the angle between them is 180°. In other words, the model predicts that all three atoms in a CO_2 molecule will be in a straight line. This is, in fact, the case. Here is an instance where the simple Lewis structure reveals the correct molecular geometry.

 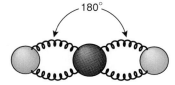

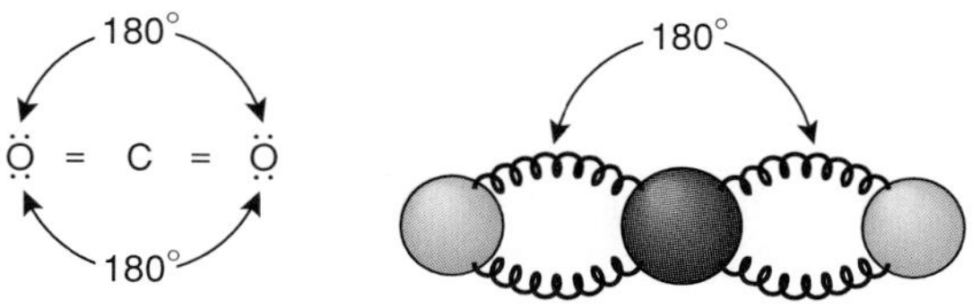

We have applied the idea of electron pair repulsion to molecules in which there are four groups of electrons (CH_4 and H_2O) and two groups of electrons (CO_2). It also applies reasonably well to molecules that include three, five, or six groups of electrons. In most molecules, the electrons and atoms are arranged to keep the separation of the electrons at a maximum.

3.5 ■ Your Turn

Predict and sketch the shapes of the following important molecules.

    a. O_3 **b.** SO_2 **c.** SO_3

Ans. **a.** First we draw the Lewis structure. Each of the three oxygen atoms contributes six outer electrons for a total of 18. As we saw in Chapter 2, these arrange themselves to form one single and one double bond.

$$:\overset{..}{\underset{..}{O}}:\overset{..}{C}::\overset{..}{\underset{..}{O}} $$

The octet rule is satisfied. Now we note that the electrons around the central oxygen atom are in three groups: the pair that constitutes the single bond, the two pairs that form the double bond, and the lone, nonbonding pair. These three groups repel each other, and the minimum energy will correspond to their furthest separation. This corresponds to an arrangement in which the electron groupings are at an angle of about 120° from each other. As a consequence, the molecule will be bent, and the angle made by the three oxygen atoms will also be about 120° (actually 117°).

    or

(The fact that the two bonds are really identical, each corresponding to about 1 1/2 bonds, does not alter the validity of this argument.)

■ Vibrating Molecules and the Greenhouse Effect

Now that we know the molecular shapes of some important greenhouse gases, we can turn to an investigation of how these molecules interact with infrared radiation. When a molecule absorbs a photon, it responds to the added energy. You have already learned that if the photon corresponds to the UV region of the spectrum, it can have sufficient energy to promote electrons to higher levels within the molecule. This can cause covalent bonds to break, as in the dissociation of O_2 and O_3.

Radiation in the infrared region of the spectrum is not sufficiently energetic to cause such molecular disruption. However, a photon of IR radiation can start a molecule vibrating. The covalent bonds holding atoms together are rather like springs, and the atoms can move back and forth. Depending on the molecular structure, only certain vibrations are permitted, and each of these vibrations has a characteristic set of permissible energy levels. The energy of the photon must correspond to the vibrational energy of the molecule in order for the photon to be absorbed. This means that different molecules absorb radiation at different wavelengths.

We illustrate these ideas with the carbon dioxide molecule, representing the atoms as balls and the bonds as springs. A CO_2 molecule can vibrate in the four ways pictured below. The arrows indicate the direction of motion. In the vibration labeled A, the central carbon atom is stationary and the oxygen atoms move back and forth in opposite directions. Alternatively, the oxygen atoms can move in the same direction and the carbon atom in the opposite direction (vibration B). Vibrations C and D look very much alike. In both cases, the molecule bends from its normal linear shape. The bending counts as two vibrations because it can occur in either of two planes.

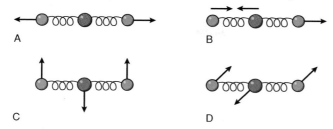

In any molecule, the amount of energy required to cause vibration will depend on the nature of the motion, the "stiffness" and strength of the bonds, and the masses of the atoms that move. If you have ever examined a spring (or played with a "Slinky") you might expect that more energy is required to stretch a CO_2 molecule than to bend it. That, in fact, is the case. It means that more energetic photons—corresponding to shorter wavelengths—are needed to excite vibrations A or B than C or D. The two bending motions (C and D) are both stimulated when the molecule absorbs IR radiation with a wavelength of 15.000 micrometers (µm). [A micrometer is equal to one millionth of a meter: $1\mu m = 1 \times 10^{-6}$ m.] Vibration B requires more energy; it will occur only if radiation of wavelength of 4.257 µm is absorbed. It turns out that vibration A cannot be triggered by the direct absorption of IR radiation because of the symmetry of the motion and the molecule. However, the other vibrations are enough to account for the greenhouse properties of carbon dioxide.

These absorption characteristics are measured with an instrument called an infrared spectrometer. Heat radiation from a glowing filament is passed through a sample of the compound to be studied, in this case gaseous carbon dioxide. A detector measures the amount of radiation, at various frequencies, that is transmitted by the sample. This information is recorded on a chart, where radiation intensity is plotted versus wavelength. The result is called the infrared spectrum of the compound. Figure 3.4 is the infrared spectrum of CO_2, obtained in just this way. There are two steep valleys where the intensity of the transmitted radiation drops almost to zero. This means that most of the radiation is absorbed by the CO_2 molecules. Note that these absorbencies occur at approximately 4.26 and 15.00 µm.

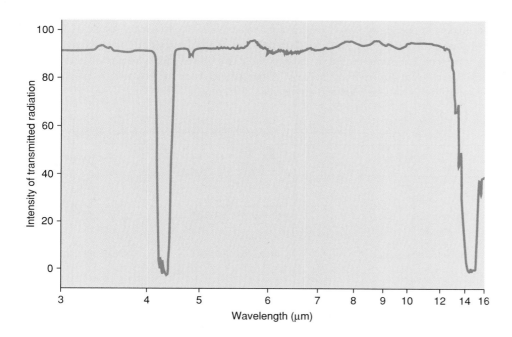

Figure 3.4
Infrared spectrum of carbon dioxide.

The same phenomenon occurs in the atmosphere. Carbon dioxide molecules absorb infrared energy at these wavelengths. They vibrate for a while and then re-emit the energy and return to their normal unexcited or "ground" state. It is by this means that carbon dioxide captures and returns the infrared radiation coming from the surface of the Earth. Significantly, this phenomenon also permits chemists to study the structures of molecules. The details of infrared spectra can yield information about bond lengths, bond angles, and bond strengths.

Any molecule that can vibrate in response to the absorption of infrared radiation is potentially a greenhouse gas. There are many such substances. Carbon dioxide and water are the most important in maintaining the temperature of the Earth. However, methane (CH_4), nitrous oxide (N_2O), ozone (O_3), and chlorofluorocarbons (such as CCl_2F_2) are among the other substances that help retain planetary heat. Diatomic N_2 and O_2 are not among them. Although molecules consisting of two identical atoms do vibrate, these vibrations cannot be triggered by the absorption of infrared radiation.

3.6 *Your Turn*

Indicate which of the following chemical species definitely do not contribute to the greenhouse effect and which might. Explain your reasoning.

a. Ar   **c.** SO_2 **e.** $CHCl_3$
b. CO   **d.** Cl_2

Ans. a. No. A single atom cannot vibrate and, hence, cannot absorb IR radiation. **b.** Yes. The CO molecule absorbs IR radiation and vibrates in response to it.

You have encountered two responses of molecules to radiation. Highly energetic photons with high frequencies and short wavelengths (such as UV radiation) can break up molecules. The less energetic photons of infrared light cause many molecules to vibrate. Both these processes are depicted in Figure 3.5, but the figure also includes another response of molecules to radiant energy that is probably a good deal more familiar to you. It happens in a microwave oven. The radiation generated in such a device is of relatively long wavelength, about a centimeter. This means that the frequency and the energy per photon are quite low. This energy is insufficient to cause a molecule to vibrate or dissociate, but it is enough to set the molecule spinning. Microwave ovens are tuned to generate radiation that causes water molecules to rotate. As the H_2O molecules absorb the photons and speed up their spinning, the resulting friction warms up the leftovers. The same region of the spectrum is used for radar. Beams of microwave radiation are sent out from a generator. When they strike an object, such as an airplane, the microwaves bounce back and are detected by a sensor.

Figure 3.5

Molecular response to radiation.

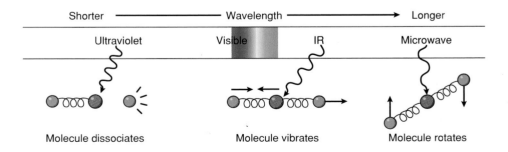

The Carbon Cycle

In a book entitled *The Periodic Table,* the late Primo Levi, chemist and author, wrote eloquently about carbon dioxide: "This gas which constitutes the raw material of life, the permanent store upon which all that grows draws, and the ultimate destiny of all flesh, is not one of the principal components of air but rather a ridiculous remnant, an 'impurity' thirty times less abundant than argon, which nobody even notices. . . . [F]rom this ever renewed impurity of the air we come, we animals and we plants, and we the human species, with our four billion discordant opinions, our millenniums of history, our wars and shames, nobility and pride."

In the essay from which this quotation is taken, Levi traces a brief portion of the life history of a carbon atom from a piece of limestone (calcium carbonate, $CaCO_3$) where it lies "congealed in an eternal present," to a CO_2 molecule, to a molecule of glucose in a leaf, and ultimately to the brain of the author. And yet, that is not the final destination. "The death of atoms, unlike our own," writes Levi, "is never irrevocable." That carbon atom, already billions of years old, will continue to persist into the unimagined future.

This marvelous continuity of matter, a consequence of its conservation, is beautifully illustrated by the carbon cycle. Even without Primo Levi's poetic gifts, the story is a fascinating one. It is summarized in Figure 3.6. Approximately 200 billion metric tons (bmt) of carbon are removed from the atmosphere in the form of CO_2 each year (1 metric ton = 1000 kg or 2200 lb). Slightly over half of it, 110 bmt, is "fixed" by photosynthesis and incorporated into plant tissue. Most of the rest dissolves in the oceans, concentrates in coral and sea shells, and ultimately finds its way into limestone and other rocks. The Earth thus serves as a vast reservoir for carbon dioxide.

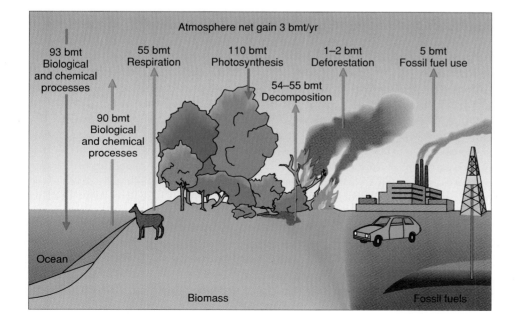

Figure 3.6

Carbon Cycle
Carbon is exchanged between the atmosphere and the land and water reservoirs on Earth. The numbers give the approximate annual transfers of carbon (in the form of CO_2) in billions of metric tons (bmt). The existing cycles remove about as much carbon from the atmosphere as they add, but human activity is currently increasing atmospheric carbon by about 3 billion metric tons per year. (These data are based on work by Bert Bolin of the University of Stockholm.)

This is a dynamic, steady-state system that returns as much CO_2 to the atmosphere as is extracted. Plants die and decay, releasing CO_2. Other plants enter the food chain where their complex molecules are broken down into CO_2, H_2O, and other simple substances. Animals exhale CO_2, carbonate rocks decompose, and carbon dioxide escapes through the vents of volcanoes. And the cycle goes on and on. Michael B. McElroy of Harvard University has estimated, "The average carbon atom has made the cycle from sediments through the more mobile compartments of the Earth back to sediments, some 20 times over the course of Earth's history."

As the carbon moves from gaseous to liquid to solid environments, from vegetable to animal to mineral, from living to dead and back again, it encompasses much of chemistry. The subdiscipline of **biochemistry** deals with the chemical processes in living things; carbon-containing compounds that typically have their origins in living things are studied in **organic chemistry;** and substances derived from the mineral realm are the subject of **inorganic chemistry. Physical chemistry** seeks to elucidate the structure of matter and discover the general principles governing its transformation. But nature knows no such boundaries. Calcium carbonate and carbon dioxide may be the provence of the inorganic chemist, but calcium carbonate derives from sea shells and carbon dioxide is the ultimate source of all organic matter.

■ Human Contributions to Atmospheric Carbon Dioxide

As members of the animal kingdom, we *homo sapiens* participate in the carbon cycle along with our fellow creatures. But we do more than our share; we do more than simply inhale and exhale, ingest and excrete, live and die. We have acquired certain skills that permit us to perturb the system. Thanks to our involvement, the cycle is out of balance. Every year, human activity releases into the atmosphere 6–7 billion tons of carbon more than has been immediately removed from it. We do so largely by spending CO_2 and sunlight that has been stored for eons in petroleum and coal. We burn fossil fuel.

The Industrial Revolution, which began in Europe in the late eighteenth and early nineteenth centuries, was fueled largely by coal. The coal was used to power steam engines in mines, factories, locomotives, ships, and later, in electrical generators. The subsequent discovery and exploitation of vast deposits of petroleum made possible the development of automobiles. To a very considerable extent, the Industrial Revolution was a revolution in energy sources and energy transfer.

As the generation of energy and the consumption of fossil fuels increased, so did the quantity of combustion products released to the atmosphere. Since 1860, the CO_2 concentration has increased from 290 ppm to 350 ppm, and the current rate of increase is about 1.5 ppm per year. At the present level, fossil fuels containing 5 billion tons of carbon are burned annually.

In addition, deforestation releases 1–2 billion tons of carbon (in the form of CO_2) to the atmosphere each year. It is estimated that in Brazil alone, a forested area the size of Pennsylvania is cut down or burned annually. As a result, a very efficient absorber of CO_2 is removed from the cycle. If the wood is burned, vast quantities of CO_2 are generated; if it is left to decay, that process also releases carbon dioxide. Even if the lumber is harvested for construction purposes and the land is replanted in cultivated crops, there is an 80% loss in CO_2 absorbing capacity.

About half of the carbon dioxide released by burning fossil fuels and by deforestation is apparently recycled into the oceans and the biosphere. The balance ends up in the atmosphere, adding 3 billion metric tons of carbon per year to the existing base of 740 bmt. Of course we are primarily concerned with the increase in carbon dioxide, because this compound is implicated in global warming. Therefore, it would be useful to know the mass of CO_2 added to the atmosphere each year. In other words, what is the mass of CO_2 that contains 3 billion tons of carbon? Chemistry can give us the answer, but it will require a rather lengthy detour before we get there.

■ *Weighing the Unweighable*

In order to solve the problem just posed, we need to know the mass fraction or mass percent of carbon in carbon dioxide. This information can be obtained experimentally by burning a weighed sample of carbon in oxygen and capturing and weighing the carbon dioxide formed. Alternatively, we could break down a known mass of CO_2 and weigh the carbon and oxygen formed. But such experiments are not easy to do. Therefore, we will calculate an answer.

The procedure requires the use of the atomic masses or atomic weights of the elements involved. You will recall from Chapter 2 that most of the mass of an atom is attributable to the neutrons and protons in the nucleus. Thus, elements differ in atomic mass because their atoms differ in composition. The internationally accepted atomic mass standard is carbon-12, the isotope that makes up 98.9% of all carbon atoms. Carbon-12 has a mass number of 12 because each atom has a nucleus consisting of six protons and six neutrons plus six orbiting electrons. The mass of one of these atoms is arbitrarily assigned a value of exactly 12 **atomic mass units (amu).** The atomic masses of all other elements are expressed relative to this standard.

You will note, however, that in the periodic table in your text, the atomic mass of carbon is reported as 12.011, not 12.000. This is not an error. Rather, it reflects the fact that carbon naturally exists in three isotopes. Although C-12 predominates, 1.1% of carbon is C-13, with six protons and *seven* neutrons per atom and an isotopic mass very close to 13. In addition, natural carbon contains a trace (1 atom in 10^{12}) of C-14, whose nuclei consist of six protons and *eight* neutrons. Carbon-14 is radioactive, and you may have heard about its use in dating objects. The tabulated atomic mass value of 12.011 is a weighted average that takes into consideration the masses of the three isotopes of carbon and their natural abundances.

Most of the atomic masses listed in the periodic table are experimentally obtained values that correspond to averages based on individual isotopic masses and the natural distribution of those isotopes. Even so, the atomic masses of many elements closely approximate whole numbers. There are two reasons for this. In the first place, the nucleus of any atom consists of a whole number of particles. These neutrons and protons each have a relative mass of very nearly one atomic mass unit. Second, for most elements one isotope dominates and is by far the most plentiful. Therefore, the experimentally determined atomic mass will be close to the sum of protons and neutrons in the most plentiful isotope. This is the case with nitrogen (N, with an atomic mass of 14.0067 for the naturally occurring isotopic mixture), argon (Ar, 39.948), uranium (U, 238.0289), and, as we have seen, carbon. When the tabulated atomic mass differs significantly from a whole number, as it does for chlorine (Cl, 35.453) or copper (Cu, 63.546) it indicates that the natural distribution involves sizeable concentrations of two or more isotopes. For some of the artificially produced heavy elements at the end of the periodic table, the atomic mass reported is the mass number of the most plentiful isotope.

Of course it is impossible to weigh a single atom. A typical laboratory balance can detect a minimum mass of 0.1 milligram (mg), and that corresponds to 5×10^{18} or 5,000,000,000,000,000,000 carbon atoms. An atomic mass unit is far too small to measure in a conventional chemistry laboratory. The gram is the chemist's mass unit of choice. Therefore, scientists use exactly 12 g of carbon-12 as the reference for the atomic masses of all the elements. The number of atoms in this mass of C-12 is called **Avogadro's number** after an Italian scientist with the impressive name of Lorenzo Romano Amadeo Carlo Avogadro di Quaregna e di Ceretto (1776–1856). (His friends called him Amadeo.) You will note that Avogadro's name consists of 52 letters; his number requires 24 numerals: 602,000,000,000,000,000,000,000. It is more compactly written in scientific notation as **6.02×10^{23}.**

Avogadro's number may be shorter than his name, but it is so large that about the only way to hope to comprehend it is through analogies. For example, an Avogadro number of regular-sized marshmallows would cover the surface of the United States to a depth of 650 miles. Or, if you are more impressed by money than marshmallows, assume 6.02×10^{23} pennies were distributed evenly among the five billion inhabitants of the Earth. Every man, woman, and child could spend one million dollars every hour, day and night, and half of the pennies would still be left unspent at death. And remember, this is the number of atoms in 12 g of carbon—a tablespoonful of soot.

Because one Avogadro number of carbon-12 atoms has a mass of 12 g, the mass of an equal number of oxygen atoms should correspond to the atomic mass of oxygen. All we need to do is count out 6.02×10^{23} atoms and weigh them. We could use some help with this assignment, so suppose we enlist all the human beings currently alive and set them counting—one atom per second per person, 24 hours a day, 365 days a year. Even at that rate, things don't look good. It would take almost four million years for all of us to complete the job. Fortunately, we don't need to count the atoms. There are fairly accurate ways of estimating the number. And besides, we can be off by 5×10^{18} and never notice! Without going into the experimental details, the result is that one Avogadro number of oxygen atoms weighs 15.9994 g. That means that the atomic mass of oxygen is 15.9994. This value is consistent with the structure of the oxygen atom. By far the most common isotope of oxygen is O-16, with atoms consisting of 8 electrons, 8 protons, and 8 neutrons.

3.8 *Your Turn*

Uranium (U) has an atomic mass of 238.0289 and an atomic number of 92.

a. Predict the number of electrons, protons, and neutrons in the most common isotope of the element.

Ans. 92 protons, 92 electrons, 146 neutrons

b. Now specify the number of electrons, protons, and neutrons in an atom of uranium-235, an isotope that undergoes splitting or fission.

3.9 *Your Turn*

Magnesium (Mg) has an atomic mass of 24.305. The isotope Mg-24 makes up 79.0% of the naturally occurring isotopic mixture. Predict whether the other common isotope (or isotopes) of magnesium will have mass numbers larger or smaller than 24. Explain your answer.

■ *Of Molecules and Moles*

An Avogadro number of anything is called a **mole.** It is a chemist's way of counting—a very large dozen. Usually the mole is used to count atoms, molecules, electrons, or other small particles. Thus, one mole of carbon consists of 6.02×10^{23} C atoms, one mole of oxygen gas is 6.02×10^{23} O_2 molecules, and one mole of carbon dioxide corresponds to 6.02×10^{23} CO_2 molecules.

Moles are fundamental to chemistry because chemistry involves the interaction of individual atoms and molecules. As you already know, chemical formulas and equations are written in terms of atoms and molecules. For example, reconsider the equation for the reaction of carbon and oxygen.

$$C + O_2 \rightarrow CO_2$$

In Chapter 1 we interpreted this expression as stating that one atom of carbon combines with one molecule of oxygen to yield one molecule of CO_2. The equation reflects the ratio in which the particles interact. Thus, it would be equally correct to say that 10 carbon atoms react with 10 oxygen molecules to form 10 carbon dioxide molecules. Or, for that matter, we could say 6.02×10^{23} C atoms combine with 6.02×10^{23} O_2 molecules to yield 6×10^{23} CO_2 molecules. The latter statement is equivalent to saying: "one *mole* of carbon plus one *mole* of oxygen gas yields one *mole* of carbon dioxide." The point is that the numbers of atoms and molecules taking part in a reaction are proportional to the numbers of moles of the same substances. The atomic or molecular ratio reflected in a chemical formula or equation is identical to the molar ratio.

In the laboratory and the factory, the quantity of matter required for a reaction is usually measured by mass or weight. The mole concept is a way to simplify matters (and matter) by relating number of particles and mass. Central to this approach is the **molar mass (MM),** defined as the mass of one Avogadro number of whatever particles are specified. Thus, the mass of a mole of carbon atoms, rounded to the nearest tenth of a gram, is 12.0 g. Similarly, a mole of oxygen atoms has a mass of 16.0 g. But we can also speak of a mole of O_2 molecules. Because there are two oxygen atoms in each oxygen molecule, there are two moles of atoms in each mole of O_2. Consequently, the molar mass of O_2 is 32.0 g—twice the molar mass of O. Some books refer to this as the **molecular mass** or **molecular weight** of O_2, emphasizing its similarity to atomic mass or atomic weight.

The same logic applies to compounds of two or more elements, which brings us, at last, to the composition of carbon dioxide. The formula, CO_2, and the molecular structure reveal that each molecule contains one carbon atom and two oxygen atoms. Scaling up by 6.02×10^{23}, we can say that each mole of CO_2 consists of one mole of C and two moles of O. But remember that we are interested in the mass composition of carbon dioxide—the number of grams of carbon per gram of CO_2. This requires the molar mass of carbon dioxide, which we obtain by adding the molar mass of carbon to twice the molar mass of oxygen.

$$MM\ CO_2 = 1 \times MM\ C + 2 \times MM\ O$$

Substituting numerical values for the molar masses of the elements gives the desired result.

$$
\begin{aligned}
1\ \text{mole C} \times 12.0\ \text{g/mole C} &= 12.0\ \text{g C} \\
+\ 2\ \text{mole O} \times 16.0\ \text{g/mole O} &= \underline{32.0\ \text{g O}} \\
\text{molar mass } CO_2 &= 44.0\ \text{g } CO_2
\end{aligned}
$$

3.10	***Your Turn***

Compute the molar masses of the following substances important in atmospheric chemistry.

a. O_3 **c.** NO_2
b. CO   **d.** SO_3

***Ans.* a.** MM O_3 = 3 mole × 16.0 g/mole O = 48.0 g O_3 **b.** 28.0 g CO

■ *Manipulating Moles and Mass with Math*

Because 44.0 g CO_2 contain 12.0 g C, we can easily find the ratio (by mass) of carbon to carbon dioxide. It is 12.0 g C/44.0 g CO_2. Out of every 44.0 g CO_2, 12.0 g are C. This mass ratio holds for all samples of carbon dioxide, and we can use it to calculate the mass of carbon in any known mass of carbon dioxide. For example, we could compute the number of grams of C in 500 g CO_2 in the following manner.

$$\text{mass C} = 500\ \text{g } CO_2 \times \frac{12.0\ \text{g C}}{44.0\ \text{g } CO_2} = 136\ \text{g C}$$

Note that carrying along the labels, "g CO_2" and "g C," helps you do the calculation correctly. In the center term of the above expression, "g CO_2" appears in the top (numerator) and the bottom (denominator). Hence, they can be canceled, and you are left with the desired label, "g C." This is a useful strategy in solving many problems.

To find the mass of carbon dioxide that contains 3 bmt of carbon, we use the same mass ratio and a similar approach. We could convert 3 bmt to grams, but it's really not necessary. As long as we use the same units for the mass of C and the mass of CO_2, the same numerical ratio holds: 12.0/44.0. But there is one important difference, this time we're solving for the mass of CO_2, not the mass of C. Therefore, we invert the mass ratio:

$$\text{mass } CO_2 = 3 \times 10^9 \text{ bmt C} \times \frac{44.0 \text{ bmt } CO_2}{12.0 \text{ bmt C}}$$
$$= 11 \times 10^9 \text{ bmt } CO_2$$

Once again the labels cancel and the answer comes out in the desired form, bmt CO_2.

Our innocent question, "What is the mass of carbon dioxide added to the atmosphere each year?" has finally been answered: 11 billion metric tons. Of course, our not-so-hidden agenda was to demonstrate the problem solving power of chemistry and to introduce five of its most important ideas: atomic mass, molecular mass, Avogadro's number, mole, and molar mass. The next few activities provide opportunity to practice your skill with these concepts and manipulations.

3.11 Your Turn

a. Calculate the mass ratio of C to CO in carbon monoxide. Also find the percent (by mass) of C in CO.

Ans. mass C/mass CO = 12.0 g C/28.0 g CO = 0.429 g C/g CO
To determine percentage (parts per hundred) multiply the ratio by 100.

percent C in CO (by mass) = 0.429 × 100 = 42.9%

b. Calculate the percent (by mass) of C in CH_4.

3.12 Your Turn

a. 60 million metric tons (mmt) of CO were released in the United States in 1989. Calculate the mass of carbon in this quantity of CO.

Ans. The simplest way to solve this problem is to use the mass ratio obtained in 3.11a.

mass C = 60 mmt × 12.0 mmt C/28.0 mmt CO = 25.7 mmt C

Alternatively, one can use the equivalent percent:

mass C = 60 mmt CO × 42.9 mmt C/100.0 mmt CO = 25.7 mmt C

b. Later in this chapter, the statement is made that 73 million metric tons of methane, CH_4 are released annually by cattle. Use your answer to 3.11 Your Turn, part b. to calculate mass of carbon (in units of mmt) present in this quantity of CH_4.

If you know how to apply these ideas, you have gained some measure of control over the media. You have been empowered with the ability to critically analyze certain statements and judge their accuracy. For example, William McKibben seeks to personalize carbon dioxide production in "The End of Nature," an article in *The New Yorker* September 11, 1989. He reports that "the average American car driven the average American distance—ten thousand miles—in an average American year releases its own weight in carbon into the atmosphere." Elsewhere in the same article, McKibben writes about the carbon emitted per gallon of gasoline consumed. One can either take such statements on faith or check their accuracy by mathematically manipulating the relevant chemical concepts, as we do in the activity that follows. Obviously, there is insufficient time to check every assertion, but we hope that readers develop questioning and critical attitudes toward all statements about chemistry and society, even those found in this book.

3.13 ■   *The Sceptical Chymist*

The following statement by William McKibben appears on page 52 of *The New Yorker*, September 11, 1989: "A clean-burning engine . . . will emit about five and a half pounds of carbon in the form of carbon dioxide for every gallon of gasoline it consumes." On what basis is this assertion made and is it correct?

This deceptively simple statement includes a lot of chemistry, but we can check it by making some reasonable assumptions and applying a few principles. In the first place, we note that the quotation implies that the gasoline is the source of the carbon that is emitted as CO_2. Chapter 1 reported that gasoline is a mixture of hydrocarbons, and identified octane, C_8H_{18}, as one of the major components. Therefore, we will use this compound to represent gasoline.

When a hydrocarbon burns "cleanly" all of the hydrogen combines with oxygen to form water and all of the carbon combines with oxygen to form carbon dioxide. It follows that if we knew the mass of C in one gallon of octane, that number would equal the mass of C released as CO_2. First, however, we need to find the mass of a gallon of C_8H_{18}. That information is not listed in the *Handbook of Chemistry and Physics*, but that same source does list the density of C_8H_{18} as 0.692 g/mL. The *Handbook* also informs us that 1 gallon has the same volume as 3790 milliliters (mL). We now know the number of milliliters in one gallon of octane and the mass per milliliter. To find the mass of this volume of octane, we multiply:

$$\text{mass } C_8H_{18} = 3790 \ \cancel{mL} \times 0.692 \ \text{g/}\cancel{mL} = 2620 \ \text{g}$$

Because 454 g = 1 lb, a gallon of gasoline weighs just under six pounds. According to McKibben, the mass of carbon that it contains is about 5.5 pounds. The estimate seems reasonable, but we can check it more precisely. The chemical formula, C_8H_{18}, enables us to calculate the mass fraction of octane that is carbon. First we find the molar mass of C_8H_{18}, using the approximate molar masses of carbon and hydrogen.

$$\text{MM } C_8H_{18} = 8 \ \text{mole C} \times 12.0 \ \text{g/mole} + 18 \ \text{mole H} \times 1.0 \ \text{g/mole}$$
$$= 96.0 \ \text{g C} + 18.0 \ \text{g H} = 114.0 \ \text{g } C_8H_{18}$$

This means that out of 114 g C_8H_{18}, 96.0 g are carbon and 18.0 g are hydrogen. It follows that the mass ratio of carbon in octane is 96.0/114.0. Multiplying the total mass of 1 gallon of octane by this fraction gives the mass of carbon in that volume.

$$\text{mass C} = 2620 \ \cancel{\text{g } C_8H_{18}} \times \frac{96.0 \ \text{g C}}{114.0 \ \cancel{\text{g } C_8H_{18}}} = 2200 \ \text{g C}$$

The only thing remaining is to change the mass in grams to pounds, making use of the fact that 1 lb = 454 g.

$$\text{mass C} = 2200 \ \text{g C} \times \frac{1 \ \text{lb}}{454 \ \text{g}} = 4.85 \ \text{lb C}$$

(You will note that in every calculation, labels were carried along with numbers to provide a check for the correctness of the expression.)

■ *Methane and Other Greenhouse Gases*

Recent estimates suggest that about half of global warming may be attributable to compounds other than carbon dioxide. Methane, CH_4, is 15 to 30 times more effective than CO_2 in its infrared trapping characteristics. The atmospheric concentration of methane is relatively low, but its current level of 1.7 ppm is more than twice that before the Industrial Revolution. Data gathered since 1979 indicate an annual increase of about 1%.

Methane comes from a wide variety of sources. Most of them are natural, but they have been magnified by human activities. For example, because methane is a major component of natural gas, some has always leaked into the atmosphere from rock fissures. But the exploitation of these deposits and the refining of petroleum has led to increased emissions. Similarly, CH_4 has always been released by decaying vegetable matter. Its early name, "marsh gas," reflects this origin. Any human activity that contributes to similar conditions leads to increased methane release. Thus, the decaying organic matter in landfills and from the residue of cleared forests generates CH_4. Methane formed in the main New York City landfill is used for residential heating, but at most landfills it simply escapes into the atmosphere. Another major source of CH_4 is cultivated rice paddies.

Agriculture has also contributed additional methane as the number of cattle and sheep has increased. The digestive systems of these ruminants contain bacteria that break down cellulose. In the process, CH_4 is formed and released through belching and flatulence—a staggering 73 million metric tons each year according to Bill McKibben. Even termites, who carry on similar chemistry in their guts, generate methane. And there is more than half a ton of termites for every man, woman, and child on Earth.

There is a possibility that global warming may exacerbate the release of methane from ocean muds, bogs, peatlands, and the permafrost of northern latitudes. In these areas, a substantial amount of CH_4 appears to be trapped in "cages" made of water molecules. As the temperature increases, the escape of CH_4 becomes more likely. Fortunately, CH_4 is quite readily converted to less harmful chemical species. It has a relatively short average lifetime of 7–10 years, compared to about 500 years for CO_2. The details of the generation and fate of methane are sufficiently complex that it is difficult to speak with a high degree of certainty about its future effect on the average temperature of the planet.

Nitrous oxide, N_2O, also known as "laughing gas," is used as an inhaled anesthetic for dental and medical purposes. In the atmosphere, it is less useful. There, a typical N_2O molecule will persist for about 150 years, absorbing and emitting infrared radiation. Over the past decade, atmospheric concentrations of the compound have shown a slow but steady rise. Major sources are artificial fertilizers and the burning of biomass. In addition to its role in the greenhouse effect, nitrous oxide contributes to stratospheric ozone depletion. Near the surface of the Earth, however, the reactions of nitrogen oxides and hydrocarbons (like methane) lead to the production of ozone. Ozone can also act like a greenhouse gas, but its efficiency depends very much upon altitude. It appears to have its maximum warming effect in the upper troposphere (around 10 km). Depletion of ozone has a cooling effect in the stratosphere and it may also promote light cooling at the surface of the Earth. Chlorofluorocarbons (CFCs), already implicated in the destruction of stratospheric ozone, also absorb infrared radiation.

■ *Climatic Modeling*

The previous paragraphs have merely hinted at the complexity of atmospheric chemistry. To accurately model global climate, one must also include a number of often poorly understood astronomical, meteorological, geological, and biological factors. Among these are variations in the intensity of the Sun's radiation as a consequence of sunspot activity, winds and air circulation patterns, cloud cover, volcanic activity, dust and soot, aerosols, ice sheets, the oceans, and the extent and nature of living things, especially human beings. Moreover, the situation is further complicated by a variety of feedback mechanisms that relate these variables.

For example, we know from the solubility properties of most gases that increasing the temperature of the oceans will decrease the solubility of CO_2, releasing more of it into the atmosphere. An increase in the temperature of the oceans may promote the growth of tiny photosynthetic plants called phytoplankton, and hence increase CO_2 absorption. But the result could be just the opposite. Water in a warmer ocean will not circulate as well as it does now, which may inhibit plankton growth and CO_2 fixing. Decreased snow and ice cover, which would attend global warming, would lower the amount of sunlight reflected from the Earth's surface. The resultant increase in absorbed radiation would promote a further increase in temperature.

A warmer Earth would presumably mean that the tree line would move north, bringing with it added CO_2 absorbing capacity. Countering this, related reductions in rainfall might turn areas that are currently covered by vegetation into deserts, thus reducing carbon dioxide absorption. Global warming would also cause more water to evaporate, increasing the average relative humidity and thus adding to the greenhouse effect. More clouds would form, but one cannot generalize their influence. It seems that high clouds contribute little to the greenhouse effect and reflect sufficient sunlight so that they have a net cooling effect on the surface of the Earth. Low clouds have a net warming effect.

In spite of such formidable problems and sometimes countervailing effects, scientists are writing computer programs to model the Earth's climate. As supercomputers have become more powerful, models have become more sophisticated. The oceans are represented as a multilayer circulating system and the model atmosphere is assumed to contain ten or more interacting layers.

Typically, the surface of the planet is divided into about 10,000 cells, not enough to provide detailed predictions, but sufficient to include general patterns of weather development. One test of these simulations of global climate is how well they predict the 0.5°C temperature increase observed over the last century when CO_2 concentrations increased by 25%. Most models estimate a temperature increase of about twice that actually measured. This suggests that certain relevant factors may have been omitted or that some variables may have been incorrectly weighted. One possibility is that the models underestimate the amount of heat absorbed by the oceans. Much of the heat radiated by the greenhouse gases may be going into the oceans, which are acting as a thermal buffer. But although the oceans are very important in moderating the temperature of the planet, there are limits to their capacity to do so.

Given this complexity of the global system, there is considerable uncertainty associated with extrapolating the climate and the weather into the future. There is no wonder, then, that experts sometimes disagree. First of all, there is the matter of projected levels of greenhouse gases. The *rate* of their emission is currently increasing about 1.5% per year. This is largely a consequence of growing global population, agricultural production, and industrialization. The population of the planet has tripled in this century and it is expected to double or triple again before reaching a plateau sometime in the next century. Industrial production is 50 times what it was 100 years ago. In the next 50 years, it will probably grow to 5 or 10 times what it is today. Most of this growth has been powered by the combustion of fossil fuels. Every year, 2–3% more energy is generated than in the previous one, and most of it comes from the burning of coal. If these rates of fuel consumption continue, the atmospheric concentration of CO_2 will be double its 1860 level sometime between the years 2030 and 2050.

All models predict that this doubling will result in an increase in the average global temperature, but the magnitude of that increase is variously estimated between 1.5 and 6°C. Many predictions fall in the 3.5–4.5°C range. Figure 3.7, based on the model developed by the National Aeronautics and Space Administration, includes observed data and extrapolations assuming three different scenarios. The worst case, which assumes a continuation of current growth in CO_2 emission, predicts a 2°C increase over a 50 year interval starting in 1980. If drastic cuts were made and the greenhouse effect leveled out after 2000, it would still be at a temperature as high as the last recorded global maximum.

Some indication of what we might expect, even under these best of conditions, can be gained by looking at the geological evidence for what the world was like 130,000 years ago. The average temperature was about 16°C, but very likely the poles

Figure 3.7

Predicted changes in global temperature based on three scenarios. *Scenario A* assumes greenhouse gas emissions will continue to increase at the current rate of 1.5% per year. *Scenario B* assumes that the rate of change in greenhouse gas emissions will decrease with time until the net annual change is constant. *Scenario C* assumes drastic cuts in greenhouse gas emissions so that there is a zero annual increase by 2000.

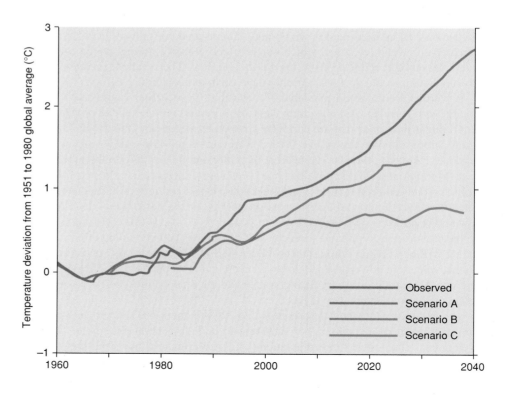

were considerably warmer than they are today. As a result, large quantities of ice melted and the oceans were approximately 5 m (15 ft) higher than they are today. Such an increase in sea level would inundate the Netherlands, many islands, and the land where half the 100 million people of Bangladesh currently live. Even if rises in the oceans were significantly smaller—say 0.5 to 1.5 m as some predict—it would endanger New York, New Orleans, Miami, Venice, Bangkok, Taipei, and many other coastal cities. Of course, it is far from certain that major increases in sea level will occur, and if they do, they will take place over many years, providing considerable time for preparation and protection.

There is even more uncertainty associated with the regional weather patterns predicted by the various models. One of the more controversial forecasters is James Hansen of NASA. Hansen has estimated that doubling the concentration of greenhouse gases would mean that New York City could expect 48 days a year with temperatures above 90°F instead of the current 15. In Dallas, the number of days per year with temperatures above 100°F would increase from 19 to 87. It is important to note that many scientists have questioned Hansen's estimates.

In a warmer world, summers are expected to be drier and winters to be wetter in the Northern Hemisphere. The regions of greatest agricultural productivity would probably change. Drought and high temperatures could reduce crop yields in the American midwest, but the growing range might extend farther into Canada. It is also possible that some of what is now desert could get sufficient rain to become arable. One nation's loss may well become another nation's gain.

3.17   Consider This

■

One example of national vs. international interests in the global warming story emerged a few years ago when M. I. Budyko, a widely respected Russian meteorologist, calculated that global warming would lengthen the growing season, increase precipitation, and extend the arable area of what was then the Soviet Union. As a result, agricultural productivity would increase. Warmer oceans could also keep northern ports ice-free all winter long. Citing his forecasts, Budyko concluded that an increase in the average temperature of the Earth would be in the best economic, social, and political interests of the Soviet Union. He therefore urged that the USSR, at the time second only to the United States in CO_2 emission, should not limit its release of the gas.

Scientific data such as the quantity of CO_2 emissions in the Russian Commonwealth and the effect this level of emission has on global warming are neutral facts. Interpretation of these facts and a specific community's response are affected by political, economic, and social factors. Cite some of the conditions in the USSR and the related social, political, and economic factors that might have influenced Budyko's position. Have subsequent events in the former Soviet Union altered the context, and if so, how?

■ Has the Greenhouse Effect Already Started?

Of course the answer to that question is "yes." Without it, we would not be here—or perhaps better, we would be very different creatures. As generally asked, that question implies alterations in climate that are a consequence of human activity, and here the answer is "probably yes." It turns out it is very difficult to separate natural and human contributions. Although there are strong differences of opinion, many experts conclude that the increase in the concentration of the greenhouse gases since the start of the Industrial Revolution is causally connected to the increase in the average temperature of the Earth over that same period.

It does not necessarily follow that the widespread North American drought of 1988 was evidence of the warming trend. Fluctuation in temperature and precipitation occurring over short periods of time are common, and they may not signal large-scale and long-range changes. On the other hand, the fact that the entire decade of the 1980s was uncommonly warm may be significant and a predictor of things to come. According to a British study, the six warmest years since 1880, when systematic meteorological measurements began, were 1988, 1987, 1983, 1981, 1980, and 1986. Furthermore, temperature measurements of the land, water, and lower atmosphere all revealed 1990 to be the warmest year on record. Evidence of changes in ocean level is inconclusive. There has been a report that sea level has been rising about 2 mm a year, but such measurements are difficult to verify.

3.18   Consider This

Given the rate at which new information about environmental effects is being generated, some parts of this book may be out of date before it is published.

Consult various appropriate sources, such as the *General Science Index* to find two articles on global warming. Attempt to find an example of an objective article and one that you believe is biased or contains other flaws such as sweeping generalizations that lead to misconceptions or misinterpretation of scientific data. Compare the two articles on the quality of their science and objectivity of their presentation. Then describe how you go about deciding if an article is trustworthy.

■ What Can We Do?

One thing is clear: if the models are reasonably accurate, we will start seeing significant climatic changes within a decade or so. But can we prudently wait that long, or is prompt action essential? Whether or not to act, and how to act are not scientific issues. "There is no single scientifically correct view of what should be done about greenhouse warming," wrote Bette Hileman in *Chemical & Engineering News,* March 13, 1989. "This is a value question involving perceptions of risk and uncertainty."

Those who dare to answer that question can be segregated (somewhat arbitrarily and unfairly) into three extreme camps. Some advocate more study, arguing that the uncertainties in our predictive powers are so great that it would be wasteful of money and effort to undertake preventative or ameliorative action at this time. Our ignorance is great, and without more knowledge we run the risk of making more mistakes. Yet, it is this attitude that Jaromir Nemec, chief of water resources for the Food and Agriculture Organization of the United Nations attacked in these words: "The risk is too large not to look for possible alternatives and means of mitigation even if the search is based on predictions with a large margin of error. Forecasting is a dangerous game. Blind and passive waiting for a possible disaster is even more dangerous."

In marked contrast to the "wait and study" school are the "do anything but do something fast" activists. For example, some have advocated dumping dust into the upper atmosphere to reflect sunlight and hence counter global warming. Given the uncertainties that have plagued arguments about the effects of a "nuclear winter," such action seems ill-advised. Presumably there are a few who would even ban the burning of fossil fuels, but the social cost of such action would be incalculable.

Still others look at the magnitude of the potential problems associated with global warming and conclude that there is nothing that can be humanly done to halt or reverse the process. The generation of energy, and with it, carbon dioxide, is an essential feature of modern industrial life. We must therefore learn to live with its consequences and begin adapting to our warm new world.

You are encouraged to debate and discuss these and other options. Clearly, your opinions and the evidence you marshall to support them are important. However, the authors of this text hope that there will be at least some advocates who develop a compromise strategy out of the best characteristics of the extreme positions just described. There is, after all, an element of truth in each position.

We take it as a given that no matter what happens, careful study is essential to improve our options. It is a response that carries little risk and the potential for great benefit. Extensive environmental monitoring is necessary to provide more reliable climatic and meteorological data. Appropriate technology must be developed and transferred to those who need it. And the public must be educated and informed. However, we also believe that complete evidence and absolute certainty will never be ours, and that intervention, based on the best available scientific and technical information, is called for. Some of this activity can and should be designed to reduce the extent of global warming; some should be designed to mitigate the temperature increases that will very likely occur, no matter how earnest our efforts.

The most obvious strategy would seem to be to reduce global reliance on fossil fuels. But such action is incredibly difficult, not only because this energy source is so important to our modern economy, but because of its international dimensions. Although the developing countries may well become the major producers of carbon dioxide and other greenhouse gases in the future, the developed countries have a prodigious lead. The annual carbon output (in the form of CO_2) of the United States is 5 tons for each of its inhabitants (Figure 3.8). On this per capita basis we are second only to the region formerly known as East Germany, which obtains almost all of its energy from coal. Comparable values for China and India are about 0.4 and 0.2 tons, respectively. Even so, the Peoples' Republic of China ranks third, behind the United States and the former Soviet Union, in total carbon dioxide emissions from fossil

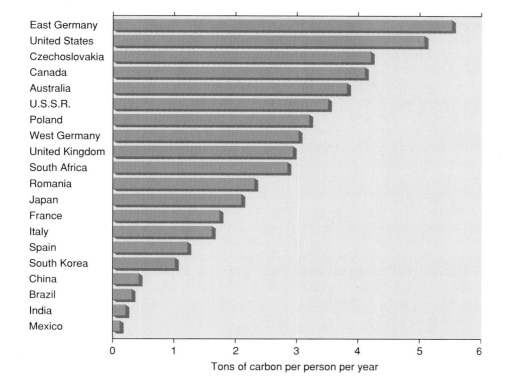

Figure 3.8

Per capita carbon dioxide emissions in 20 countries. (1986 values, measured as tons of carbon in CO_2 emissions from burning fossil fuels.) (Reprinted with permission from *Chemical & Engineering News,* March 13, 1989, **67** (11), p. 30. Copyright © 1989 American Chemical Society.)

fuels. If China were to succeed in raising its per capita gross national product to 15% of the U.S. figure, the increase in CO_2 production would approximately equal the current U.S. annual emissions from coal. It is unrealistic (and possibly unethical) for the developed countries to expect the nations of the Third World to abandon their hopes for economic growth and become good, non-polluting global citizens. In response to this dilemma, the anthropologist Margaret Mead and William W. Kellogg of the National Council on Atmospheric Research proposed as early as 1976 an international "law of the air" in which worldwide emission standards would be established and enforced by assigning "polluting" rights to each nation.

3.20 *Consider This*

In your opinion, is the proposal to assign "polluting rights" to each nation an ethical approach? Is it practical? Justify your answer. Then consider this further question: If such a plan were adopted by the global community, what basis should be used to establish relative polluting rights for each country?

Various technologies have been proposed to trap the carbon dioxide released from burning fossil fuels, but these would all be very expensive, both in terms of money and energy. One such strategy would separate the CO_2 from flue gas, liquify it, and pump it to the ocean floor. At a depth of 3000 m, the CO_2 would remain in a liquid state. It has been estimated that this approach would double the cost of power plants and increase the cost of electricity by about 75%. Extracting CO_2 from the atmosphere, liquefying it, and disposing of it in the ocean would be even more costly because of its relatively low concentration. An ideal solution might seem to be a method that would convert CO_2 to fuel by reacting it with hydrogen to yield hydrocarbons. But the energy input required to drive the process would be 50% greater than the energy content of the fuel produced. It has been estimated that the cost of the hydrocarbons produced in this manner would be well above $120 per barrel, compared to current crude oil costs of approximately $20 per barrel.

More economical approaches include shifting from coal to natural gas, which would significantly reduce CO_2 output, or to nuclear power, which generates no greenhouse gases. The latter conversion, however, is not without risk, and its costs and benefits will be analyzed in Chapter 8. Renewable energy sources and wind, solar, hydroelectric, and geothermal power have been proposed. Clearly, the development of alternate energy sources must be a high global priority. Deposits of coal, natural gas, and petroleum are not only finite, they are too valuable as raw materials to be totally consumed by combustion. Conserving energy by making power plants, furnaces, factories, and automobiles more efficient is a strategy worth pursuing under any circumstances, but there are limits imposed by laws of nature as well as by engineering skill and the properties of materials.

Humans can also intervene by rectifying their assaults on the planet. Deforested areas can be replanted, converting a CO_2 source back to a CO_2 sink. The growth of phytoplankton can be enhanced by fertilizing the oceans with phosphates and artificially improving ocean circulation. And while promoting the growth of other species, we must be aware of the fact that many of our environmental problems are related to the expanding human population.

It is unlikely that these preventative measures, though necessary, will be sufficient to avert a temperature increase. We must also be prepared to meet and mitigate future climate changes. This includes protecting arable soil, improving water management, prudently using agricultural technology and agricultural chemicals, maintaining global food reserves, and establishing an effective mechanism for disaster relief.

3.21   *Consider This*

What do you think poses the more serious problems, ozone depletion or global warming? Which can be more easily solved? Support your answers with evidence and argument.

■ *Conclusion*

> For the first time in my life I saw the horizon as a curved line. It was accentuated by a thin seam of dark blue light—our atmosphere. Obviously this was not the ocean of air I had been told it was so many times in my life. I was terrified by its fragile appearance.
>
> *Ulf Merbold*

This chapter and the two that preceded it have disclosed that our atmosphere is more robust than it appeared to the German astronaut, Ulf Merbold. Nevertheless, over the past century, the air upon which our very existence depends has been subjected to repeated assaults. The fact that most of these environmental insults were unintentional and, in some cases, the unexpected consequences of social progress, does not alter the problems we face. We have only recently recognized the potential harm that air pollution, ozone depletion, and global warming can bring to our personal, regional, national, and global communities. To reverse the damage already done and to prevent more, all of these communities must respond with intelligence, compassion, and wisdom. It is instructive that even in the absence of threats such as the greenhouse effect, much of what has been advocated in the preceding section would be sound, prudent, and responsible stewardship of our planet.

■ *References and Resources*

Boden, T. A.; Kanciruk P.; and Farrell, M. P. *Trends '90: A Compendium of Data on Global Change.* Oak Ridge, Tennessee: Carbon Dioxide Information Analysis Center, Oak Ridge National Laboratory, 1990.

"The Climate System." *Reports to the Nation,* Winter 1991. Boulder, Colorado: University Corporation for Atmospheric Research.

Goldemberg, J. "How to Stop Global Warming." *Technology Review,* Nov./Dec. 1990: 25–31.

"The Greenhouse Effect is for Real." (editorial) *New York Times,* Jan. 27, 1989: A30.

Hileman, B. "Global Warming." *Chemical & Engineering News,* March 13, 1989: 25–44.

———. "Web of Interactions Makes it Difficult to Untangle Global Warming Data." *Chemical & Engineering News,* April 27, 1992: 7–19.

Kerr, R. A. "Hansen vs. the World on the Greenhouse Threat." *Science* **244,** June 2, 1989: 1041–43.

McKibben, W. *The End of Nature.* New York: Random House, 1989. Excerpted in *The New Yorker,* Sept. 11, 1989: 47–105.

Rosswall, T. "Greenhouse Gases and Global Change: International Collaboration." *Environmental Science & Technology* **25** (1991): 567–73.

Schneider, S. H. "The Changing Climate." *Scientific American* **261,** Sept. 1989: 70–79.

———. *Global Warming: Are We Entering the Greenhouse Century?* San Francisco: Sierra Club Books, 1989.

U.S. Congress, Office of Technology Assessment. *Changing by Degrees: Steps to Reduce Greenhouse Gases (Summary).* Washington: U.S. Government Printing Office, 1991.

Zurer, P. S. "International Negotiations Inching Toward Global Climate Treaty." *Chemical & Engineering News,* March 5, 1990: 13–15.

■ Experiments and Investigations

1. Preparation and Properties of Atmospheric Gases I

3. The Weights of Equal Volumes of Gases

4. Molecular Models

7. Chemical, Not Biological Moles

■ Exercises

*1. The age of the Earth is estimated at 4.5 billion years (4.5×10^9 years). a. The so-called "agricultural revolution" began about 10,000 years ago, when humans first domesticated animals and cultivated plants. What percentage of the total age of the Earth does this 10,000-year period represent? b. Assume that the age of the Earth corresponds to a 24 hour day, with the present time set equal to midnight. At what time on this clock did the agricultural revolution begin? c. The height of the Roman Empire and the birth of Christ occurred about 2000 years ago. At what time did these events occur on this 24 hour clock? d. Finally, at what time did Fourier first propose the greenhouse effect?

2. Account for the fact that the increasing CO_2 levels of the atmosphere influence the amount of energy leaving the Earth's atmosphere but not the quantity entering it from the Sun.

3. The solar energy striking the surface of the Earth corresponds to 169 watts per square meter, but 390 watts are radiated from every square meter of the planet's surface. Given these numbers (which are correct) one would expect the earth to cool rapidly. Explain why this does not happen.

4. Write Lewis electron dot structures for the following molecules. (The order of bonding of the atoms is indicated.)

    a. H C N   b. N N O

       O   H
       | |
    c. H C H ·   d. H N H

5. Predict the geometries of the molecules in Exercise 4.

*6. The molecule BF_3 is planar with 120° angles between the B—F bonds while the NF_3 molecule is pyramidal with N—F bond angles of about 103°. Account for this difference on the basis of the electron dot structures of these two species. (Hint: BF_3 does not obey the octet rule.)

7. Carbon dioxide and water vapor absorb radiation in the infrared region of the spectrum. From your everyday experience with these compounds, offer evidence that they do not absorb visible light.

8. Arrange the different CO_2 reservoirs depicted in Figure 3.6 in order of the expected residence time for an average CO_2 molecule. Account for the order you have given.

9. The total mass of carbon on the Earth is estimated at 7.5×10^{22} g. Of this total, the greatest portion, 6.0×10^{22} g, is found in sedimentary rocks. The oceans contain 4.2×10^{19} g of carbon, and fossil fuel deposits represent 4.0×10^{18} g C. Calculate the percentages of total terrestrial carbon in these three reservoirs.

10. The total mass of carbon in living plants and animals is approximately 5.6×10^{17} g. How many parts per million of the total mass of earthly carbon does this represent? (See Exercise 9.)

11. It has been suggested that global warming could cause lower CO_2 concentrations in the oceans and, therefore, higher CO_2 concentrations in the atmosphere, thus leading to greater global warming. Describe a common experience that supports this prediction. (If you have difficulty coming up with a reasonable answer, perhaps drinking a carbonated beverage will inspire you.)

12. Give the chemical symbols, atomic numbers, and mass numbers for the elements whose atoms have the following combinations of electrons, protons, and neutrons.

	Number of electrons	Number of protons	Number of neutrons
a.	17	17	18
b.	17	17	20
c.	47	47	60
d.	88	88	136

*13. From Avogadro's number and the atomic masses of the elements given below, calculate the mass of an "average" atom of each element.

    a. carbon   12.01   b. sulfur   32.07

    c. iron   55.85   d. silver   107.9

    e. barium   137.3   f. uranium 238.0

14. What do 207.2 g Pb (lead), 44.0 g CO_2, 28.0 g N_2 and 342.3 g $C_{12}H_{22}O_{11}$ (sugar) have in common?

15. Calculate the molar masses of the following greenhouse gases.
    a. N_2O
    b. H_2O
    c. CCl_2F_2

16. Calculate the percentages (by mass) of each of the elements present in each of the compounds listed in Exercise 15.

17. Calculate the mass (in grams) of each of the elements present in 500 g of each of the compounds listed in Exercise 15.

*18. The chemical equation for the combustion of methane is given below. Fill in the blanks in the sentences that follow.

$$CH_4 + 2\,O_2 \rightarrow CO_2 + 2\,H_2O$$

The equation signifies that 1 mole of CH_4 reacts with _____ mole O_2 to yield _____ mole CO_2 and _____ mole H_2O. It follows that 5 mole CH_4 would require _____ mole O_2 to react completely. Because the molar mass of CH_4 is _____ g, the mass of 5 mole CH_4 is _____ g. The mass of O_2 required to react completely with 5 mole CH_4 is _____ g. This reaction would yield _____ g CO_2 and _____ g H_2O.

19. Determine the number of moles of the indicated substance in each of the following samples.
    a. 2.30 grams of sodium
    b. 0.127 grams of copper
    c. 454 grams of uranium
    d. 100 grams of water
    e. 50. grams of carbon dioxide
    f. 1.8 grams of octane (C_8H_{18})

20. In Europe and North America, 1816 was known as the "year without a summer" because average temperatures were significantly lower than normal. This cooling has been attributed to the vast quantity of volcanic ash and dust released into the atmosphere by the eruption of Mount Tambora, Indonesia in April 1815. Assuming this hypothesis to be correct, speculate on a combustion product that might tend to offset the global warming attributable to increased CO_2 concentrations. Also suggest a mechanism for its cooling effect.

4

Energy,
Chemistry,
and Society

Housewives in hair curlers knit sweaters at the wheels of their station wagons in the predawn blackness of Miami. Young couples in Manhattan, armed with sandwiches and hot chocolate, invite friends along for an evening of gasoline shopping. Connecticut executives regale each other with lurid tales of mile-long queues and two-hour waits at the pump. Otherwise sane citizens are in the cold grip of the nation's newest obsession: gasoline fever.

In these vivid (and sexually stereotyped) words, *Time* magazine on February 18, 1974 described the country's last energy crisis. As the lines at gasoline stations grew longer, tempers grew shorter (Figure 4.1). Verbal threats and fist fights became frequent occurrences, and a gas station owner in Gary, Indiana was shot and killed by an irate customer.

The 1974 fuel shortage was triggered by the Arab oil embargo. Imports of crude oil, the starting material for home heating oil and gasoline, dropped from 7.5 million barrels a day to 5 million when Arab nations significantly cut petroleum output. Faced with an impending shortage, the Federal Energy Office did not allow refineries to speed up their seasonal conversion from production of heating oil to gasoline. As a consequence, service stations were given monthly gasoline allotments. Some states attempted to stretch this supply by permitting motorists with license plates ending in odd numbers to purchase gasoline only on odd days of the month, even on even days. Fuel sales were severely curtailed on weekends, and many stations set dollar limits on minimum and maximum purchases. But when the allotments were used up, the stations closed until the next shipment arrived. Many state legislatures called for the development of a national gasoline rationing plan, and the federal government even went so far as to print rationing coupons, which were never used.

The oil crisis also affected other aspects of American life. Airlines cut back sharply on their flight schedules. Truckers went on strike to protest skyrocketing diesel fuel costs and limited availability. This in turn led to a shortage of food and materials, closed factories and mines, and threw at least 100,000 people temporarily out of work.

The truckers who did stay on the job were forced to travel in convoys with police or National Guard escorts for protection against robbery, sabotage, or sniper attack. Adding to this chaos was the overriding public concern that government officials were mismanaging the crisis.

For a time, some good seemed to come out of the energy crisis. A massive campaign was launched to convince the public to conserve energy by buying smaller, more fuel-efficient cars, switching off unnecessary electric lights, turning down the furnace, and turning off the air conditioner. But only a few years later there was a "gusher" in the supply of crude oil and its price fell precipitously. As the price of gasoline dropped, the horsepower of new cars began to rise. Thermostats were reset, lights came on all over America, and happy days were here again—except perhaps in the oil-producing regions of the country.

But the days of plentiful and cheap fuel are obviously limited. The planet has a finite supply. And the ominous headline of the cover story in the February 1990 issue of *Nation's Business* asks whether we will soon be contending with **"A NEW ENERGY CRISIS?"** Many Americans seem to think so. The journal *Environment* reported in 1988 that "an overwhelming majority of energy opinion leaders and the public believe the United States is likely to face a major energy shortage in the 1990s." There is even evidence that the shortage has already begun. As the article in *Nation's Business* noted, "Gas and electric companies have asked customers to lower the settings on their thermostats and have even imposed blackouts for several hours in a time when energy demand has exceeded supply." And that was before the 1991 war with Iraq over the invasion of Kuwait—a violent demonstration of the political significance of energy sources and fuel supplies.

4.1 **Consider This**

Suppose you were transported back to 1974. The oil crisis has deepened and nationwide gasoline rationing is now in effect. Each household in your area has been allotted 30 gallons of gas a month. Assume your car averages 18 miles to the gallon. Prepare a detailed budget of how you and the other members of your family would use your allocation.

■ *Chapter Overview*

The prospect of another energy shortage is a compelling reason to attend to issues related to energy. But even in the absence of such a crisis, our lives are inexorably intertwined with this familiar yet elusive concept. Energy is a common thread that runs through the first three chapters of this book. Polluting oxides of sulfur and nitrogen are released by coal-burning power plants and internal combustion engines. Stratospheric ozone shields the earth by absorbing highly energetic ultraviolet radiation, and it is being destroyed by chemicals used to transfer energy in air conditioners and refrigerators. Carbon dioxide, a product of energy production from the combustion of fossil fuels, absorbs and re-emits infrared radiation, thus warming the Earth and its atmosphere. The time has obviously come to take a closer look at energy.

To begin, we need agreement on the meanings of some fundamental concepts such as energy, work, and heat. These essential ideas are linked in the first law of thermodynamics—a generalization that describes some of the constraints governing the generation and use of energy. After a macroscopic survey of current energy sources, we focus on the molecular level as we attempt to explain the origin of the energy changes observed in chemical reactions. Most of the energy currently used in home and industry comes from fossil fuels, so the topics of coal and petroleum are explored in some depth and detail. But the sources of these fuels are limited, so the text next turns to substitutes from various sources.

Any source of energy and any mechanism for generating it are also subject to the second law of thermodynamics, another natural constraint. We introduce the second law to explain the inescapable inefficiencies of energy transformation and, along the way, we develop the concept of entropy. Finally, the chapter concludes with some observations on the importance of conserving energy and fuel.

■ Energy: Hard Work and Hot Stuff

Energy is one of those words that everyone uses, but whose precise meaning is not well understood. Unfortunately, the dictionary definition, the capacity to do work, doesn't help much because **work** is another common but poorly defined concept. To a scientist, work is done when movement occurs against a restraining force, and it is equal to the force multiplied by the distance over which the motion occurs. Thus, when you lift a book against the force of gravity, you are doing work. When you read a book without moving it, you are not, strictly speaking, working. On the other hand, you are again doing something that will not happen by itself, and that does require energy.

The source of much of the work done on our planet is another familiar form of energy—**heat.** The formal definition sounds a little strange: **heat** is that which flows from a hotter to a colder body. But a child once burned knows the meaning of heat. Our understanding of temperature is also based on experience, and we know that temperature and heat are not the same thing. But a standard definition of temperature sounds awkward and circular: **temperature** is a property that determines the direction of heat flow. When two bodies are in contact, heat always flows from the object at the higher temperature to that at the lower temperature. Heat is a consequence of motion at the molecular level. When matter, for example liquid water in a pan, absorbs heat, its molecules move more rapidly. Temperature is a statistical measure of the average speed of that motion. Hence, temperature rises as the amount of heat energy in a body increases.

Before we turn to matters of energy demand, we need a unit in which to express it. Historically, there have been many, but recently there has been an international agreement to make the common unit the **joule.** One joule (1 J) is approximately equal to the energy required to raise a 1 kg (2 lb) book 10 cm (4 in) against the force of gravity. As the name implies, one kilojoule (1 kJ) is equal to 1000 J.

Much of the published data relating to energy is reported not in joules, but in calories. The **calorie** was introduced with the metric system in the late eighteenth century as a measure of heat. Originally, the calorie was defined as the amount of heat necessary to raise the temperature of exactly one gram of water by one degree Celsius. It has been redefined as exactly 4.184 J. Calories are perhaps most familiar when used to express the energy released when foods are metabolized. The values tabulated on package labels and in diet books are, in fact, kilocalories (1 kcal = 1000 cal). When Calorie is written with a capital C, it generally means kilocalorie. Thus, the energetic equivalent of a doughnut is 425 Cal (425 kcal). For most purposes we will use joules and kilojoules in this chapter, but when it seems more appropriate or more easily understandable, we will express energy in calories or kilocalories. We will not worry about British Thermal Units (BTU), ergs, or foot-pounds, but you are warned that the world also expresses energy in these terms.

4.2 Your Turn

a. Convert the 425 kcal released when a donut is metabolized to kilojoules. Then calculate the number of books you could lift to a shelf 6 feet off the floor with that amount of energy.

Ans. The first part is easy because 1 cal = 4.184 J, 1 kcal = 4.184 kJ

$$Energy = 425 \; kcal \times \frac{4.184 \; kJ}{1 \; kcal} = 1778 \; kJ = 1778 \times 10^3 \; J$$

We next determine how many joules will be required to lift one book 6 feet. The text reports that it takes 1 J to lift one book 4 inches. It will obviously require a good deal more energy to lift the book 6 feet. We can calculate how much more by noting that 6 ft × 12 in/ft = 72 in.

$$72 \; in/4 \; in = 18$$

In other words, 6 feet is 18 times longer than 4 inches. Therefore, it must take 18 times as much energy to raise a book 6 feet as it does to raise it 4 inches. Because the latter requires 1 joule, 18 joules are needed to lift one book 6 ft. We now can calculate the number of books that can be lifted 6 ft with 1778×10^3 J of energy.

$$no. \; books = 1778 \times 10^3 \; J \times \frac{book}{18 \; J} = 99 \times 10^3 = 9.9 \times 10^4$$

In round numbers, about 100,000 books! Lots of exercise is required to work off one donut.

b. A 12 oz can of a soft drink has an energy equivalent of 92 kcal. Assume that you use this energy to lift concrete blocks that weigh 22 lb (10 kg) each. How many of these blocks could you lift to a height of 4 ft with this quantity of energy?

Ans. 3200 blocks

4.3 The Sceptical Chymist

Our solution to 4.2a makes an important, and unwarranted assumption, that leads to an incorrect answer. Try to identify the assumption and predict whether the answer obtained is too large or too small.

■ Energy Conservation and Consumption

Strictly speaking, energy is not consumed. The **first law of thermodynamics,** also called the **law of conservation of energy,** states that energy is neither created nor destroyed. Energy is often transformed as it is transferred, but the energy of the universe is constant. However, energy sources such as coal, oil, and natural gas are consumed. In the United States we burn a prodigious quantity of these fossil fuels to generate a huge amount of energy. Figure 4.2 compares our annual per capita delivered energy use with those of 16 other countries. The energy comes from many different sources, but in the bar graph it is expressed as if it were all generated from oil. Thus, the energy share of the average resident of the United States was 41 barrels of oil (about 57 million kcal) in 1988. Our neighbors to the north used even more energy per person. In stark contrast, the average Indian consumed the energetic equivalent of about 2 barrels of oil per year—a good deal of it in the form of traditional fuels such as wood, grass, or animal dung.

Figure 4.2

Annual per capita energy consumption (1988). (Adapted from page 57, "Energy for Planet Earth" by Ged R. Davis (illustrated by Andrew Christie), September 1990. Copyright © by *Scientific American, Inc.* All rights reserved.)

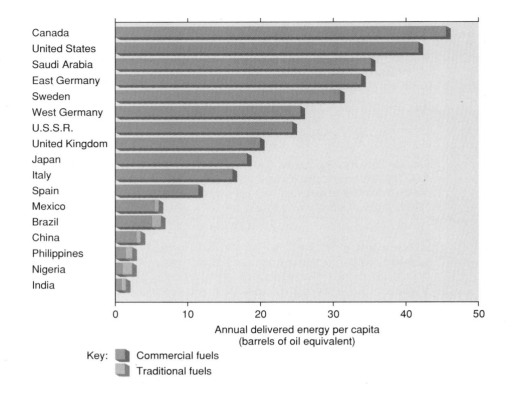

The actual fuel consumption is even greater than that suggested by Figure 4.2 because the numbers on the graph refer to "delivered energy." "Delivered" means that the energy is available to the consumer in usable form—electricity at the wall plug, gasoline at the station, natural gas in the home, and so on. These figures do not include the often very high energy costs of generating electricity, refining and distributing oil, or delivering natural gas.

It is no coincidence that the nations at the top of Figure 4.2 are industrialized and wealthy, and that those on the bottom are struggling with poverty. Energy appears to drive industrial and economic progress, and gross national product correlates well with energy consumption. The United States leads in both categories by a considerable margin.

The international burst of energy consumption is of relatively recent origin. Figure 4.3 is an attempt to represent individual daily energy use at various times throughout history. The heights of the bars along the vertical scale are proportional to energy, but the horizontal scale is completely out of proportion to time. Two million years ago, before our primitive ancestors learned to use fire, the sole source of energy available to an individual was that of his or her own body. *Homo habilis* probably consumed the equivalent of 2000 kcal per day and expended most of it finding food. This roughly corresponds to the energy output of a 100 watt light bulb. The discovery of fire and the domestication of beasts of burden increased the energy available to an individual about six times. Hence, we estimate that a farmer about the beginning of the common (Christian) era, with an ox or donkey, would have roughly 12,000 kcal at his disposal each day. The Industrial Revolution brought another five- or six-fold increase in the energy supply, most of it from coal via steam engines. The past century has seen yet another energy jump. In 1988, the total energy used in the United States (from all sources and for all purposes) corresponded to about 250,000 kcal per person per day. This translates to an annual equivalence of 62 barrels of oil or 15 tons of coal for each American. Or, to put it in human terms, the energy available to each resident of the

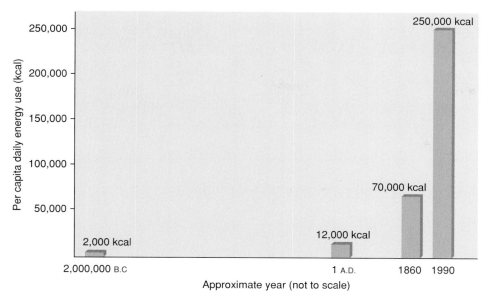

Figure 4.3
Individual daily energy consumption vs. time.

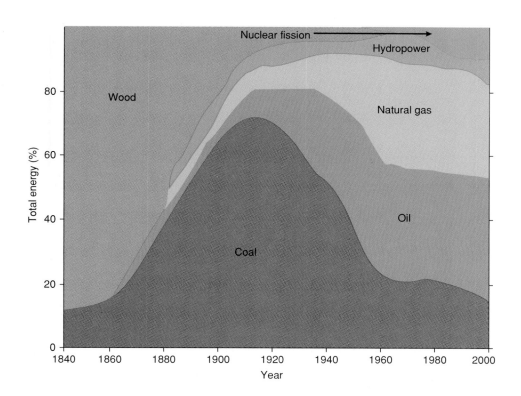

Figure 4.4
U.S. energy obtained from various sources (1840–2000). (Source: H. Landsburg and S. Schurr, *Energy in the United States: Sources, Uses, and Policy Issues,* 1963: McGraw-Hill, Inc.)

United States would require the physical labor of 125 workers. Yet, there are still people on the planet whose energy use and life style closely approximate those of 2000 years ago.

The history of increasing energy consumption is closely related to changing energy sources and the development of devices for extracting and transforming that energy. Figure 4.4 plots the percentage of energy input from various fuels for the years from 1840 with projections to 2000. Wood was originally the major energy source in the United States, and it continued to be until around 1890, when it was surpassed by coal.

Figure 4.5

Annual U.S. energy consumption from various sources (1850–2000). (Adapted from page 39, "Energy and Power" by Chauncey Starr (illustrated by Allen Beechel), September 1991. Copyright © by *Scientific American, Inc.* All rights reserved.)

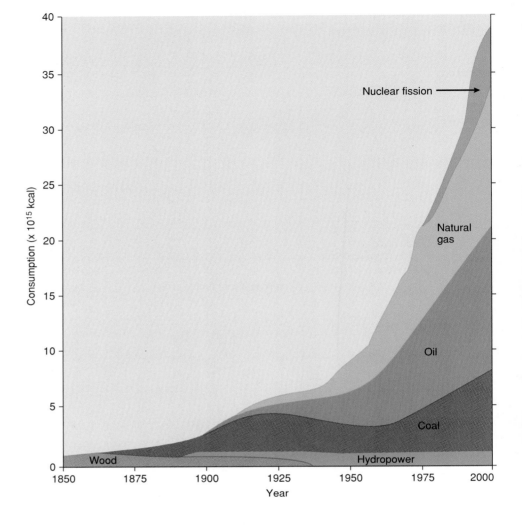

Figure 4.6

Sources of electrical energy in the United States. (Source: Data from *Energy Information Administration Annual Energy Review, 1988.*)

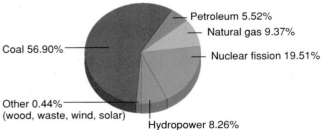

Coal provided more than 50% of the nation's energy from then until about 1940. By 1950, oil and gas were the source of more than half of the energy used in this country, and today petroleum accounts for 60%. But Figure 4.4 can be misleading because it deals only with percentages. The real magnitude of the increase is evident from Figure 4.5, where actual quantities of energy consumed in the United States are plotted against time.

Figures 4.4 and 4.5 both indicate how reliant our society is on fossil fuels—oil, coal, and gas—and how small is our utilization of energy from nuclear, hydro, geo-thermal, and solar sources. This distribution is particularly evident in Figure 4.6, a pie chart that indicates the origins of electrical energy in the United States. Some of the currently underutilized alternate sources will be considered in Chapters 8 and 9.

4.4   *Consider This*

The consumption of most fuels follows the bell-shaped curve characteristic of coal in Figure 4.4. It starts low, increases with time, peaks, and then decreases with time. Explain the technical, social, political, and economic factors contributing to this trend. Also suggest some fuel sources that might not follow this pattern and explain why.

4.5   *Consider This*

Suppose you were put in a time machine and transported 200 years into the future, You become an instant celebrity. The talk-show host of the day invites you to be interviewed. The first question is "How could the people of your century feel justified in using up so much of the world's store of non-renewable resources such as oil and coal?" What is your answer? One member of the class should be the talk-show host, the other the time traveler.

■ *Energy: Where from and How Much?*

At a time when the nation is seeking new sources of energy, it is reasonable to ask what it is that makes some substances such as coal, gas, oil, or wood usable as fuels, while many others are not. To find an answer, we must consider the properties of fuels and the means by which energy is released from them. The most common energy-generating chemical reaction is burning or combustion. **Combustion** is the combination of the fuel with oxygen to form product compounds. In such a chemical transformation, the potential energy (on a molecular scale) of the reactants is greater than that of the products. Because energy is conserved the difference in energy is given off, primarily as heat.

We illustrate the process with the combustion of methane, the principal component of natural gas.

$$CH_4(g) + 2\,O_2(g) \rightarrow CO_2(g) + 2\,H_2O(g) + \text{Energy} \qquad (4.1)$$
$$\text{methane}$$

The above reaction is said to be **exothermic** because it is accompanied by the release of heat. The quantity of energy released, the **heat of combustion,** can be measured. Not surprisingly, it depends on how much fuel and oxygen are present and reacting. A known mass of CH_4 is burned in O_2 and the heat given off in the process is absorbed by a known mass of water. As a result, the temperature of the water increases. The total amount of heat released by the burning methane can be calculated from the mass of the water, its specific heat, and its temperature increase (see Chapter 5). In this particular case, experiment shows that 802.3 kJ of heat are given off when one mole of methane reacts with two moles of oxygen to form one mole of carbon dioxide and two moles of water vapor. We can also calculate the number of kilojoules released when one gram of methane is burned. The molar mass of CH_4, calculated from the atomic masses of carbon and hydrogen, is 16.0 g/mole. Thus, the heat of combustion per gram of CH_4 is obtained as follows:

$$802.3 \text{ kJ/mole } CH_4 \times \frac{1 \text{ mole } CH_4}{16.0 \text{ g } CH_4} = 50.1 \text{ kJ/g } CH_4$$

The fact that heat is evolved signals that there is a decrease in the energy of the chemical system during the reaction. In other words, the reactants (methane and oxygen) are in a higher energy state than the products (carbon dioxide and water). This process is thus somewhat like a waterfall or a falling object. In both cases, potential energy decreases and is manifested in some other form. This decrease is signified by the negative sign that is traditionally attached to the energy change for all exothermic reactions.

Figure 4.7
Energy differences in an exothermic chemical reaction.

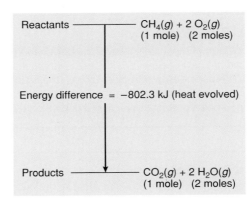

For the combustion of methane, the energy change is listed as −802.3 kJ/mole. Figure 4.7 is a schematic representation of this process.

We still have not adequately explained the origin of the energy released in an exothermic reaction. To do that, we investigate the structure of the molecules involved. We have already encountered all of the reactants and products in the combustion of methane. Therefore, we can write structures for all of the molecular species.

$$\text{H}-\underset{\underset{\text{H}}{|}}{\overset{\overset{\text{H}}{|}}{\text{C}}}-\text{H} + 2\ \ddot{\text{O}}{=}\ddot{\text{O}} \longrightarrow \ddot{\text{O}}{=}\text{C}{=}\ddot{\text{O}} + 2\ \overset{:\text{O}:}{\underset{\text{H}\quad\text{H}}{/\quad\backslash}} \qquad (4.2)$$

The reaction represented by this equation is a rearrangement of atoms. It requires the breaking and forming of chemical bonds. Our task is to calculate the total energy change in this reaction. For the purposes of this calculation we assume that all the bonds in the reactant molecules are broken and then the individual atoms are reassembled into the product molecules. Breaking bonds is an **endothermic** process requiring the absorption of energy; forming bonds is an exothermic process that liberates energy. Because we are doing energetic bookkeeping, we need to keep track of the total energy change. To do this, we assume that energy that is absorbed carries a positive sign, like a deposit to your checkbook. On the other hand, energy evolved is like money spent, it bears a negative sign. We now apply this sign convention to the above reaction, using the bond energies of Table 4.1. **Bond energy** is the amount of energy that must be absorbed to break a specific chemical bond. Notice that these energies are all given in units of kJ per mole of bonds, and the values all apply to the gaseous state.

Remember that equation 4.2 is written in terms of moles. Because there are four C—H bonds per CH_4 molecule, there are four moles of C—H bonds per mole of CH_4. Similarly, there are two moles of C=O bonds per mole of CO_2 and two moles of O—H bonds per mole of H_2O. Coupling this knowledge with the tabulated bond energies we obtain the following results.

Breaking Bonds (endothermic):

1×4 moles C—H bonds $\times$ 411 kJ/mole = 1644 kJ
2 moles O=O bonds $\times$ 494 kJ/mole = 988 kJ
Total energy absorbed = 2632 kJ

Making Bonds (exothermic):

1×2 moles C=O bonds $\times$ (−799 kJ/mole) = −1598 kJ
2×2 moles O—H bonds $\times$ (−459 kJ/mole) = −1836 kJ
Total energy evolved = −3434 kJ
Net energy change: 2632 + (−3434) = − 802 kJ

A schematic representation of this calculation is presented in Figure 4.8. Here the energy of the reactants, CH_4 and O_2, is arbitrarily set at zero. The blue arrows pointing

■ *Table 4.1*	*Bond Energies (in kJ/mole)*

Single Bonds

	H	C	N	O	S	F	Cl	Br	I
H	432								
C	411	346							
N	386	305	167						
O	459	358	201	142					
S	363	272	—	—	226				
F	565	485	283	190	284	155			
Cl	428	327	313	218	255	249	240		
Br	362	285	—	201	217	249	216	190	
I	295	213	—	201	—	278	208	175	149

Multiple Bonds

$C{=}C$	602	$C{=}N$	615	$C{=}O$		799
$C{\equiv}C$	835	$C{\equiv}N$	887	$C{\equiv}O$		1072
$N{=}N$	418	$N{=}O$	607	$S{=}O$ (in SO_2)		532
$N{\equiv}N$	942	$O{=}O$	494	$S{=}O$ (in SO_3)		469

(From *Inorganic Chemistry* by James E. Huheey. Copyright © 1993 by James E. Huheey. Data in table from Ebbing, Darrell D., *General Chemistry,* Fourth Edition. Copyright © 1993 by Houghton Mifflin Company. Used with permission of Houghton Mifflin Company and HarperCollins Publishers, Inc.)

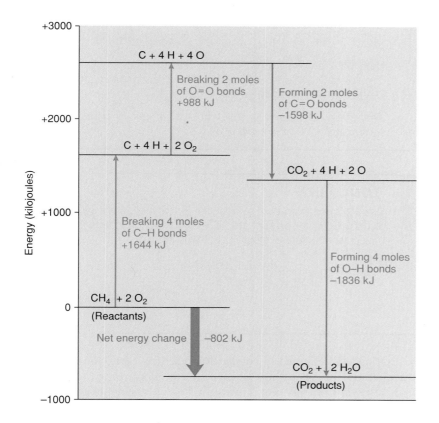

Figure 4.8

Schematic representation of the calculation of reaction energy from bond energies.

upward signify energy absorbed to break bonds. The red arrows pointing downward represent energy released as new bonds are formed. The heavy red arrow corresponds to the net energy change of –802 kJ, signifying that the overall combustion reaction is strongly exothermic. Simply stated, burning one mole of methane to carbon dioxide and water releases 802 kJ of heat energy, its heat of combustion. Note that the release of heat corresponds to an overall *decrease* in the energy of the chemical system or a *negative* energy change.

This energy change, calculated from bond energies, compares very favorably with the experimentally determined value of –802.3 kJ. In fact, the reaction does not occur by completely breaking apart the reactant molecules and reassembling them, but the agreement of the two answers supports the correctness of the central idea. Heat is evolved in a chemical reaction when the bonds in the product molecules are stronger than those in the reactant molecules. More energy is given off in forming the products than is required to break apart the reactants. The bonds in the molecules of most stable compounds are stronger than those in the parent elements.

4.6 Your Turn

Use the bond energies from Table 4.1 and the procedure illustrated above to calculate the heat of combustion of propane, C_3H_8 (LP or "bottled gas"), in kJ/mole and kJ/g. The equation for the reaction, written with structural formulas, is given below.

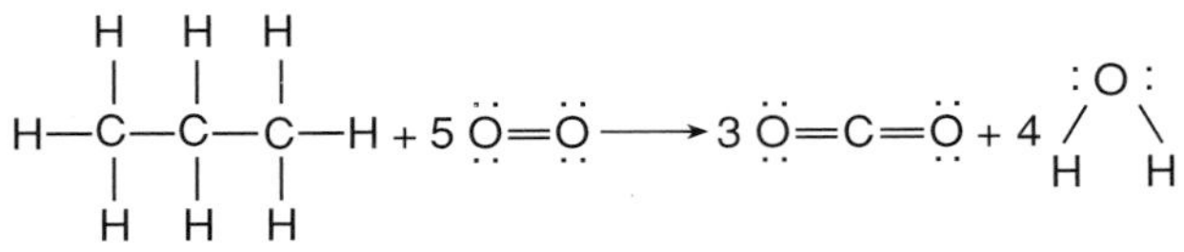

Note that there are 8 moles of C—H bonds, 2 moles of C—C bonds, 5 moles of O=O bonds, 6 moles of C=O bonds, and 8 moles of O—H bonds involved in the reaction.

Ans. Energy change = –2016 kJ/mole or –45.8 kJ/g; heat of combustion = 2016 kJ/mole or 45.8 kJ/g.

4.7 Your Turn

Use the same procedure to calculate the heat of combustion of ethyl alcohol, C_2H_5OH, present at a 10% concentration in "gasohol." The molecular structure of ethyl alcohol is as follows.

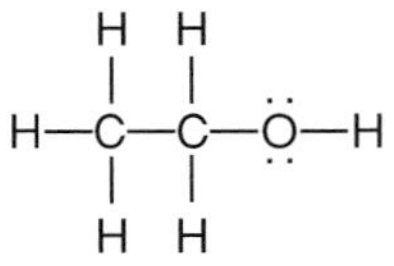

The CO_2 and H_2O formed in combustion reactions cannot be used as fuels because there are no substances into which they can be converted that have stronger bonds and are lower in energy. For this reason, Pogo's "Smogmobile," which first appeared during the 1970s as a facetious solution to air pollution, will remain a figment of the cartoonist Walt Kelly's imagination. We cannot run a car on its exhaust. We cannot expect to obtain energy without paying for it—at the very least by having the energy source lose its potential to produce additional energy.

4.8	*Consider This*
■	Look at the bond energies in Table 4.1 and identify some of the strong (high energy) and weak (low energy) bonds. What generalizations can you make about what makes bonds strong or weak?

■ *Getting Started: Activation Energy*

Just because two substances can react in an exothermic process does not mean that they will do so, even if they are in intimate contact. For example, if you turn on the gas jet at a Bunsen burner, methane and oxygen will be present in a potentially combustible mixture. But they will not react unless a spark, a flame, or some other source of energy is supplied. This turns out to be a fortunate feature of matter. Wood, paper, and many other common materials are energetically unstable and potentially capable of exothermic conversion to water, carbon dioxide, and other simple molecules, but they do not undergo spontaneous combustion.

The energy necessary to initiate a reaction is called its **activation energy.** Figure 4.9 is a schematic of the energy changes that might occur in a typical exothermic reaction. It looks a little like the cross-section of a hill. The activation energy corresponds to a peak over which a boulder must be pushed before it will roll downhill. Although energy must be expended to get the reaction (or the boulder) started, a good deal more energy is given off as the process proceeds to a lower potential energy state. Reactions that do not occur readily generally have higher activation energies than those that proceed more easily.

Activation energy is also involved in another aspect of chemical reactions that determines whether a given substance can be used as a fuel. Useful fuels react at rates that are neither too fast nor too slow. Slow reactions are of little use in producing energy because the energy is released over too long a period of time. For example, it would not be very practical to try to warm your hands over a piece of rotting wood, even though the overall reaction is similar to burning and forms CO_2 and H_2O. On the other hand, fast reactions can release energy too rapidly to be put to convenient use. In fact, such reactions often lead to explosions because the gases produced expand rapidly.

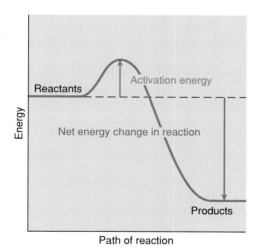

Figure 4.9

Energy-reaction pathway diagram.

One way to speed up the rate of a reaction is to divide the fuel into small particles. This principle is used in fluidized-bed power plants in which pulverized coal is burned in a blast of air. The fine coal dust is quickly heated to the kindling point and the large surface area means that oxygen reacts rapidly and completely with the fuel. The combustion actually occurs at a lower temperature than that required to ignite large pieces of coal. As a result, the generation of oxides of nitrogen is minimized. If finely divided limestone (calcium carbonate) is mixed with the powdered coal, sulfur dioxide is also removed from the effluent gas. The potential contribution to acid precipitation is thus reduced while the efficiency of coal combustion is enhanced.

4.9   Consider This

A spark is sufficient to ignite the natural gas in a Bunsen burner or a stove, but it is quite difficult to light charcoal briquets. Offer an hypothesis to account for this difference.

■ There's No Fuel Like An Old Fuel

Coal, oil, and natural gas possess many of the properties needed in a fuel. Therefore, most of the energy that drives the engines of our economy comes from these remnants of the past. In a very real sense, these fossil fuels are sunshine in the solid, liquid, and gaseous state. The sunlight was captured millions of years ago by green plants that flourished on the prehistoric planet. The reaction is the same one that is carried out by plants today.

$$\text{chlorophyll}$$
$$2800 \text{ kJ} + 6 \text{ CO}_2(g) + 6 \text{ H}_2\text{O}(l) \rightarrow \underset{\text{glucose}}{\text{C}_6\text{H}_{12}\text{O}_6(s)} + 6 \text{ O}_2(g) \qquad (4.3)$$

This conversion of carbon dioxide and water to glucose and oxygen is endothermic. It requires the absorption of 2800 kJ per mole of $C_6H_{12}O_6$ or 15.5 kJ/g. The reaction could not occur without the absorption of energy and the participation of a green pigment molecule called chlorophyll. The chlorophyll interacts with photons of visible sunlight and uses their energy to drive the photosynthetic process.

You are already aware of the essential role of photosynthesis in the initial generation of the oxygen in the Earth's atmosphere, in maintaining the planetary carbon dioxide balance, and in providing food and fuel for creatures like us. In our bodies, we run the above reaction backwards, like living internal combustion engines.

$$\text{C}_6\text{H}_{12}\text{O}_6(s) + 6 \text{ O}_2(g) \rightarrow 6 \text{ CO}_2(g) + 6 \text{ H}_2\text{O}(l) + 2800 \text{ kJ} \qquad (4.4)$$

We extract the 2800 kJ that are released per mole of glucose "burned" and use that energy to power our muscles and nerves, though we do not do it with perfect efficiency (see 4.3 Sceptical Chymist). The same overall reaction occurs when we burn wood, which is primarily composed of repeating glucose units.

When plants die and decay, they are also largely transformed into CO_2 and H_2O. However, under certain conditions, the glucose and other organic compounds that make up the plant only partially decompose and the residue still contains substantial amounts of carbon and hydrogen. Such conditions arose at various times in the prehistoric past of our planet, when vast quantities of plant life were buried beneath layers of sediment in swamps or the ocean bottom. There these remnants of vegetable matter were protected from atmospheric oxygen, and the decomposition process was halted.

■ *Table 4.2*	*Classification, Composition, and Fuel Value of Various U.S. Coals*[a]					
		Analysis, weight % before drying				
Fuel	State of origin	Moisture	Volatile matter	Carbon	Ash	Heat content (kJ/g)
Anthracite	PA	4.4	4.8	81.8	9.0	30.5
Bituminous						
Low volatile	MD	2.3	19.6	65.8	12.3	30.7
Medium volatile	AL	3.1	23.4	63.6	9.9	31.4
High volatile	OH	5.9	43.8	46.5	3.8	30.6
Subbituminous	WA	13.9	34.2	41.0	10.9	24.0
	CO	25.8	31.1	34.8	4.7	19.9
Lignite (brown coal)	ND	36.8	27.8	30.2	5.2	16.2
Peat	MS	—	—	—	—	13.
Wood[b]	—	—	—	—	—	10.4–14.1

[a]Most data from *Energy and the Future,* Table 1, A. L. Hammond, W. D. Metz, and T. H. Maugh II. © 1973, American Association for the Advancement of Science.
[b]Includes waste.

From J. W. Moore and E. A. Moore, *Environmental Chemistry,* p.94, Academic Press, 1976. Used with permission of the authors.

However, other chemical transformations did occur in Earth's high-temperature and high-pressure reactor. Over millions of years, the plants that captured the rays of a young Sun were transmuted into the fossils we call coal and petroleum.

■ *King Coal*

The great exploitation of fossil fuels began with the Industrial Revolution, about two centuries ago. The newly built steam engines consumed large quantities of fuel, but in England, where the revolution began, wood was no longer readily available. Most of the forests had already been cut down. Coal turned out to be an even better energy source than wood because it yields more heat per gram. Burning one gram of coal will release approximately 30 kJ, compared to 10–14 kJ per gram of wood.

The reason for this difference has to do with the chemical composition of the coal. Coal is not a single substance, but a complex mixture of compounds that occurs naturally in varying grades. It can be approximated by the chemical formula $C_{135}H_{96}O_9NS$. The carbon, hydrogen, oxygen, and nitrogen atoms come from the original plant material. In addition, samples of coal typically contain small amounts of silicon, sodium, calcium, aluminum, nickel, copper, zinc, arsenic, lead, and mercury. Soft lignite or brown coal is the lowest grade. The vegetable matter that makes it up has undergone the least amount of change, and its chemical composition is similar to that of wood or peat. Consequently, the heat content of lignite is only slightly greater than that of wood (see Table 4.2). The higher grades of coal, bituminous and anthracite, have been exposed to higher pressures in the earth. In the process they have lost more oxygen and moisture and have become a good deal harder—more mineral than vegetable. Anthracite has a particularly high carbon content and low concentrations of sulfur, both of which make it the most desirable grade of coal. But the deposits of anthracite are relatively small and the United States supply is almost exhausted. Therefore, we now rely most heavily on bituminous and subbituminous coal.

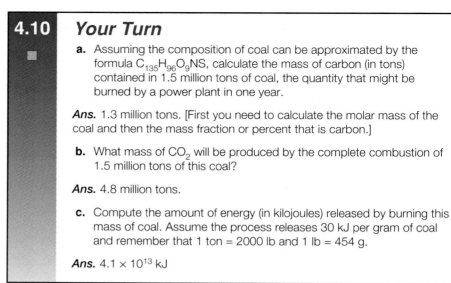

4.10 Your Turn

a. Assuming the composition of coal can be approximated by the formula $C_{135}H_{96}O_9NS$, calculate the mass of carbon (in tons) contained in 1.5 million tons of coal, the quantity that might be burned by a power plant in one year.

Ans. 1.3 million tons. [First you need to calculate the molar mass of the coal and then the mass fraction or percent that is carbon.]

b. What mass of CO_2 will be produced by the complete combustion of 1.5 million tons of this coal?

Ans. 4.8 million tons.

c. Compute the amount of energy (in kilojoules) released by burning this mass of coal. Assume the process releases 30 kJ per gram of coal and remember that 1 ton = 2000 lb and 1 lb = 454 g.

Ans. 4.1×10^{13} kJ

Figure 4.10
Schematic of energy released by burning C and CO.

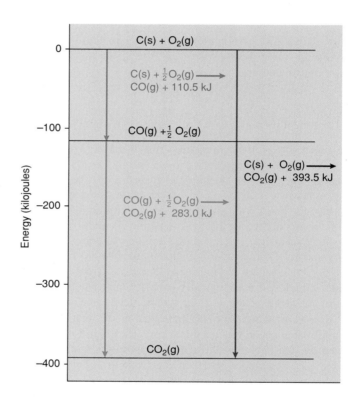

Generally speaking, the less oxygen a compound contains, the more energy per gram it will release on combustion. It is higher up on the potential energy scale. This explains why burning one mole of carbon to form carbon dioxide yields about 40% more energy than that obtained from burning one mole of carbon monoxide (see Figure 4.10). Although coal is a complex mixture, not a single compound, the same principles apply. Anthracite and bituminous coals consist primarily of carbon. Their heat of combustion is, gram for gram, about twice that of lignite.

Although the global supply of coal is large and it remains a widely used fuel, it is not without some serious drawbacks. In the first place, it is difficult to obtain. Underground mining is dangerous and expensive. In *Invention and Technology*, Summer 1992, Mary Blye Howe reports that since 1900 more than 100,000 workers have been killed in American coal mines by accidents, cave-ins, fires, explosions, and poisonous gases. Many thousands more have been injured or incapacitated by respiratory diseases. If the coal deposits lie sufficiently close to the surface, surface or strip mining can be used. In this method, the overlying soil and rock are stripped away to reveal the coal seam which is then removed by heavy machinery. However, much care is necessary to prevent serious environmental deterioration. Great holes in the earth and heaps of eroding soil dot regions of abandoned strip mines. Current regulations require the replacement of earth and top soil and the planting of trees and vegetation. Once the coal is out of the ground, its transportation is complicated by the fact that it is a solid. Unlike gas and oil, coal cannot be pumped unless it is finely divided and suspended in a water slurry.

Perhaps the most widely discussed negative characteristic of coal is the fact that it is a dirty fuel. It is, of course, physically dirty, but its dirty combustion products may be more serious. The unburned soot from countless coal fires in the nineteenth and early twentieth centuries blackened buildings and lungs in many cities. Less visible but equally damaging are the oxides of sulfur and nitrogen that are formed when certain coals burn. If these compounds are not trapped, they can contribute to the acid precipitation that forms the subject for Chapter 6. In addition, coal suffers from the same drawback of all fossil fuels: the greenhouse gas, carbon dioxide, is an inescapable product of its combustion.

In spite of these less than desirable properties, it is likely that the world's energy dependence on coal may increase rather than decrease. The recoverable world supply of coal is estimated as 20 to 40 times greater than its petroleum reserves. As the latter becomes depleted, reliance on coal will increase, unless alternative energy sources are developed. It is, however, possible that coal will not be burned in its familiar form, but rather converted to cleaner and more convenient liquid and gaseous fuels. That is a subject for a subsequent section, after we consider the properties of petroleum.

■ Black Liquid Gold

Children in the average American city or town would be hard pressed to find lumps of coal for the traditional eyes of a snow man. Indeed, they may have never seen coal, but they have undoubtedly seen gasoline. Somewhere around 1950, petroleum surpassed coal as the major energy source in the United States. The reasons are relatively easy to understand. Petroleum, like coal, is partially decomposed organic matter, but it has the distinct advantage of being liquid. It is easily pumped to the surface from its natural, underground reservoirs, transported via pipelines, and fed automatically to its point of use. Moreover, petroleum is a more concentrated energy source than coal, yielding approximately 40–60% more energy per gram. Typical figures are 48 kJ/g for petroleum and 30 kJ/g for coal.

There is, however, one property of petroleum that impeded its initial acceptance. Unlike coal, crude oil is not ready for immediate use when it is extracted from the ground. Petroleum must first be processed—a feature that has given gainful employment to many chemists and chemical engineers (and quite a few others). It has also provided an amazing array of products. Petroleum is a complex mixture of thousands of individual compounds. The great majority are hydrocarbons, their molecules consisting only of hydrogen and carbon atoms. Concentrations of sulfur and other contaminating elements are generally quite low, minimizing polluting combustion products.

■ Refining Petroleum

The oil refinery has become a symbol of the petroleum industry (Figure 4.11). In the refining process, the crude oil is separated into individual compounds or, more often, into fractions that consist of compounds with similar properties. This fractionation is accomplished by **distillation.** The petroleum is pumped into an industrial sized retort or still, and the mixture is heated. As the temperature increases, the components with

Figure 4.11

An oil refinery: Symbol of the petroleum industry.

the lowest boiling points are the first to vaporize. The gaseous molecules escape from the liquid and move up a tall distillation column or tower. There the cooled vapors recondense into the liquid state, only this time in a much purer condition. By varying the temperature of the still and the column, the engineer can regulate the boiling-point range of the fraction distilled and condensed. Higher temperatures mean higher boiling compounds.

Figure 4.12 is a schematic drawing of a distillation tower and a listing of some of the fractions obtained. They include gases such as methane, liquids such as gasoline and kerosene, waxy solids, and a tarry asphalt residue. Note that the boiling point goes up with increasing number of carbon atoms in the molecule, and hence with increasing molecular mass. Heavier, larger molecules are attracted to each other with stronger intermolecular forces than are lighter, smaller molecules. Higher temperatures are required to overcome these forces and vaporize compounds with greater molecular masses.

Because of differences in properties, the various fractions distilled from crude oil have different uses. Indeed, the great diversity of products obtained has made petroleum a particularly valuable source of matter and energy. The lowest boiling components are gases at room temperature, and are used as "bottled gas" and other fuels. Sometimes flames at the tops of refinery towers signal that the gas is being burned off in what seems to be an unnecessary waste of a valuable resource. The gasoline fraction is particularly important to our automotive civilization. Efforts at designing and manufacturing self-propelled vehicles were largely unsuccessful until petroleum provided a convenient and relatively safe liquid fuel. The kerosene fraction is somewhat higher boiling, and it finds use as a fuel in diesel engines and jet planes. Still higher boiling fractions are used to fire furnaces and as lubricating oils.

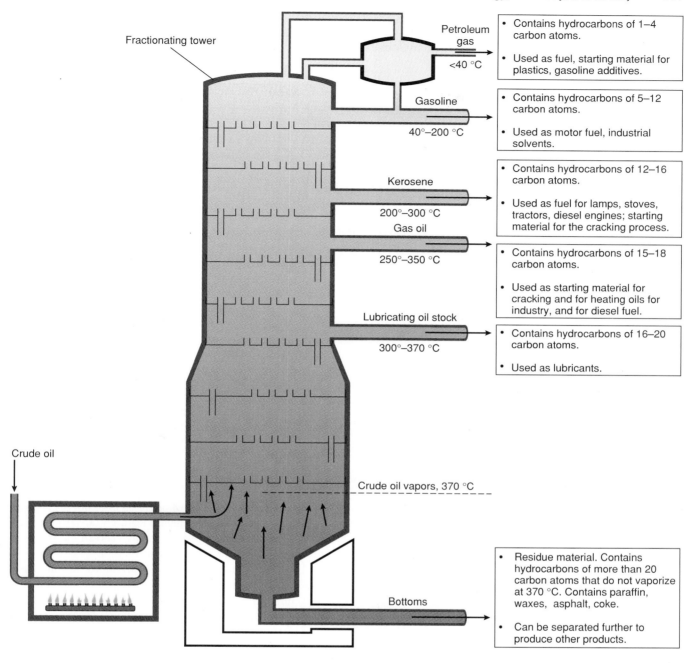

Fractionating tower

Petroleum gas

<40 °C

- Contains hydrocarbons of 1–4 carbon atoms.
- Used as fuel, starting material for plastics, gasoline additives.

Gasoline

40°–200 °C

- Contains hydrocarbons of 5–12 carbon atoms.
- Used as motor fuel, industrial solvents.

Kerosene

200°–300 °C

- Contains hydrocarbons of 12–16 carbon atoms.
- Used as fuel for lamps, stoves, tractors, diesel engines; starting material for the cracking process.

Gas oil

250°–350 °C

- Contains hydrocarbons of 15–18 carbon atoms.
- Used as starting material for cracking and for heating oils for industry, and for diesel fuel.

Lubricating oil stock

300°–370 °C

- Contains hydrocarbons of 16–20 carbon atoms.
- Used as lubricants.

Crude oil

Crude oil vapors, 370 °C

Bottoms

- Residue material. Contains hydrocarbons of more than 20 carbon atoms that do not vaporize at 370 °C. Contains paraffin, waxes, asphalt, coke.
- Can be separated further to produce other products.

■ *Manipulating Molecules*

Research has shown that many of the compounds distilled from crude oil are not ideally suited for the desired applications. Nor does the normal distribution of molecular masses correspond to the prevailing use pattern. The demand for gasoline is considerably greater than that for higher boiling fractions. Therefore, chemistry is used to rearrange matter by breaking large molecules into smaller ones. This cracking process is typified by the following reactions.

$$C_{16}H_{34} \rightarrow C_8H_{18} + C_8H_{16} \text{ or } C_5H_{12} + C_{11}H_{22} \qquad (4.5)$$

When the cracking is achieved by heating the starting materials, the process is called thermal cracking. However, valuable energy can be saved if catalysts are used to speed up the molecular breakdown at lower temperatures in an operation called catalytic or cat cracking. The catalysts employed in this process are typically synthetic aluminosilicates,

Figure 4.12

Diagram of a distillation tower and various fractions. (Reprinted from *Chemistry in the Community (ChemCom)*, copyright 1988, with permission of the American Chemical Society.)

similar to the zeolites used in water softening (see Chapter 5). If there is an excess of small molecules and a need for intermediate sized ones, the former can be catalytically combined to form the latter.

$$4\ C_2H_4 \rightarrow C_8H_{16} \tag{4.6}$$

The refining process can also rearrange the atoms within a molecule. It turns out that not all the molecules with a single chemical formula are necessarily identical. For example, octane, an important component of gasoline, has the formula C_8H_{18}. Careful analysis discloses that there are 18 different compounds with this formula. While the chemical and physical properties of these forms are similar, they are not identical. For example, the substance called normal-octane (or n-octane) has a boiling point of 125°C whereas that of the compound commonly known as iso-octane is 99°C. Different compounds with the same formula are called **isomers.** In Chapter 11 you will learn that isomers differ in molecular structure—the way in which the constituent atoms are arranged. For the present, the most important fact is that the various isomers of C_8H_{18} perform differently in internal combustion engines.

Both n- and iso-octane have essentially the same heat of combustion, but the former compound ignites much more easily. In a well-behaved car engine, gasoline vapor and air are drawn into a cylinder, compressed by a piston, and ignited by a spark. But compression alone is often enough to explode pure n-C_8H_{18}. This premature firing or preignition gives rise to a knocking sound. Moreover, knocking reduces the efficiency of the engine because the energy of the exploding and expanding gas is not applied to the pistons at the optimum time.

On the other hand, iso-C_8H_{18} is very resistant to preignition. It is the standard of excellence for rating the tendency of fuels to knock, though some compounds are even better. The performance of iso-octane in an automobile engine has been measured and arbitrarily assigned an octane rating of 100. On this scale, n-octane has an octane rating of −19. However, it is possible to rearrange or "reform" n-octane to iso-octane, thus greatly improving its performance. This is accomplished by passing the n-octane over a catalyst consisting of rare and expensive elements such as platinum (Pt), palladium (Pd), rhodium (Rh), or iridium (Ir).

Reforming isomers to improve octane rating has become particularly important because of the nationwide efforts to ban lead from gasoline. Lead does not occur naturally in petroleum, but in the 1920s, the compound tetraethyl lead, $Pb(C_2H_5)_4$, was first added to gasoline to reduce knocking and increase octane rating. The strategy proved successful, but it was not without environmental consequences. In a very short time, internal combustion engines became a major source of lead introduced into the environment. Lead is a heavy metal poison, with cumulative neurological effects that are particularly damaging to young children. Therefore, major efforts have been launched to discontinue the practice of adding lead compounds to motor fuel. Since 1976, all new cars and trucks sold in the United States have been designed to run on unleaded gasoline. The results have been dramatic. In 1970, approximately 200 thousand metric tons of lead were released into the atmosphere. Over the next ten years, lead emissions dropped to less than one tenth of that value. On balance, this decrease has been environmentally beneficial, but there have been associated costs, as the next activity suggests.

| 4.11 | *Consider This* |

Modern car engines designed to burn unleaded gasoline are somewhat less efficient than those burning leaded fuel. It can therefore be argued that the switch to unleaded gasoline has contributed to greater fuel consumption and to air pollution by unburned exhaust residues. Moreover, some of the hydrocarbons introduced into gasoline when lead was phased out have been identified as possibly causing cancer. Draw up a list of risks and benefits associated with leaded and unleaded gasoline, and indicate the technical information you would need to appropriately weigh these risks and benefits.

■ *Seeking Substitutes*

Because the world's coal supply far exceeds the available oil reserves, there is interest in converting coal into gaseous and liquid fuels that are identical with or similar to petroleum products. As a matter of fact, some of the appropriate technology is quite old. Before large supplies of natural gas were discovered and exploited, cities were lighted with water gas. This is a mixture of carbon monoxide and hydrogen, formed by blowing steam over hot coke (the impure carbon that remains after volatile components have been distilled from coal).

$$C(s) + H_2O(g) \rightarrow CO(g) + H_2(g) \qquad (4.7)$$

This same reaction is the starting point for the Fischer–Tropsch process for producing synthetic gasoline. The carbon monoxide and hydrogen are passed over an iron or cobalt catalyst, which promotes the formation of hydrocarbons. These can range from the small molecules of gases like methane, CH_4, to the medium-sized molecules found in gasoline. This process, which was developed in Germany, is only economically feasible where coal is plentiful and cheap and oil is scarce and expensive. This is the case in South Africa, where 40% of the gasoline is obtained from coal. In the future, such technology may also become competitive in other parts of the world.

Concerns about the dwindling supply of petroleum have also led to the use of renewable energy sources. This generally means **biomass**—materials produced by biological processes. One such source, which was much touted during the energy crisis, is wood. But the energy demands of our modern society cannot possibly be met by burning wood. Burning the trees would also destroy effective absorbers of carbon dioxide while adding that greenhouse gas and other pollutants to the atmosphere. In some parts of the country, the use of wood-burning stoves has been severely curtailed because of their negative effect on air quality.

4.12   *Consider This*

■   Suppose the power and light company serving Chicago decided to conduct a feasibility study to investigate switching from coal to wood as the fuel source for the city. Prepare a list of questions that would have to be answered in this study before a conversion could be considered.

Ethyl alcohol or ethanol, C_2H_5OH, is another alternative fuel produced from renewable biomass. It is formed by the fermentation of carbohydrates such as starches and sugars. Enzymes released by yeast cells catalyze the reaction that is typified by this equation.

$$C_6H_{12}O_6 \rightarrow 2\ C_2H_5OH + 2\ CO_2 \qquad (4.8)$$
$$\text{glucose}$$

The burning of ethyl alcohol in the following reaction releases 1367 kJ per mole of C_2H_5OH.

$$C_2H_5OH(l) + 3\ O_2(g) \rightarrow 2\ CO_2(g) + 3\ H_2O(l) + 1367\ kJ \qquad (4.9)$$

The energy output corresponds to 29.7 kJ/g. This value is somewhat lower than the 47.8 kJ/g produced by C_8H_{18}, because the C_2H_5OH is already partially oxidized. Nevertheless, ethyl alcohol is already being mixed with gasoline to form "gasohol." At the usual concentration of 10% ethanol, gasohol can be used without modifying standard automobile engines. Higher concentrations of alcohol would require changes in design, but these have already been made for racing engines that run on pure methyl alcohol, CH_3OH. Of the 13 million vehicles in Brazil, more than 4 million use pure ethanol, made from fermenting sugar cane juice. Most of the rest of Brazilian cars operate on a mixture of ethanol and gasoline.

Whether ethanol will make a significant contribution to energy production depends on other factors, especially agriculture. The great variety of alcoholic beverages indicates that C_2H_5OH can be prepared from almost any plant product—corn, wheat, barley, rice, sugar beets, sugar cane, grapes, apples, dandelions, and so on. But these sources also serve as food for humans or other animals. Therefore, the use of agricultural products for the production of fuel must depend on supply and demand, surpluses and shortages. Currently, the United States produces a significant surplus of corn and other grains that could be converted to ethanol. But in a recent paper, Bernard Gilland, a civil engineer, estimates that meeting only 10% of the world's current primary energy demand with alcohol would require that one-quarter of the world's cropland would have to be removed from food and feed production. Clearly, there are limits to the amount of energy we can obtain from biomass.

4.13   *Consider This*

Running automobiles on gasohol would be a way to cut down on petroleum use while providing farmers with a new market for their products. Such a fuel conversion could have a significant impact on agricultural states such as Iowa, and on oil producing states such as Texas. Imagine that a bill has been introduced into the Senate requiring that ethyl alcohol constitute 50% of gasoline by the year 2000. Identify two members of the class as senators from Iowa and Texas, and organize a debate on the issue. The remainder of the "Senate" should be prepared to vote on the bill.

Yet another potential energy source is a commodity that is always present in abundant supply and always being renewed—garbage. No one is likely to design a car that will run on orange peels and coffee grounds, but there are approximately 140 power plants in the United States that do just that. One of these, pictured in Figure 4.13, is the Hennepin County Resource Recovery Facility in Minneapolis, Minnesota. Hennepin county generates about one million tons of solid waste each year. About one third of that is burned in the waste-to-energy facility. The heat evolved is used to generate electricity via a process described in the next section. At full capacity, the

Figure 4.13

Hennepin County Resource Recovery Facility—a garbage burning power plant.

Minneapolis plant produces 37,000 kJ per second, enough energy to meet the needs of 40,000 homes. One truckload of garbage (about 27,000 lbs) will generate the same quantity of energy as 21 barrels of oil.

This resource recovery approach, as it is sometimes called, simultaneously addresses two major problems—the growing need for energy and the growing mountain of waste. The great majority of the trash is converted to carbon dioxide and water and no supplementary fuel is needed. The unburned residue is disposed of in landfills, but it represents only about 10% of the volume of the original refuse. Although some environmentalists have expressed concern about gaseous emissions from garbage incinerators, the stack effluent is carefully monitored and must be maintained within established composition limits. Both Japan and Germany are making considerably greater use of waste-to-energy technology than is the United States.

4.14	*Consider This*
■	The Hennepin County waste-burning power plant has been the subject of a good deal of controversy. Suppose you are a resident of Minneapolis, living one mile from the plant. A special information meeting has been scheduled with plant managers and engineers and representatives of the state Pollution Control Agency. What questions will you ask?

■ *Transforming Energy*

Essentially all of the fuels we have been considering in this chapter—coal, oil, alcohol, or garbage—give up their energy through combustion. They are burned to generate heat. For the most part, however, heat is not the form in which the energy is ultimately used. Heat is nice to have around on a cold winter day, but it is a cumbersome form of energy. It is dangerous if uncontrolled, difficult to transport, and hard to harness for other purposes. The industrialization of the world's economy began only with the invention of devices to convert heat to work. Chief among these was the steam engine, developed in the latter half of the eighteenth century. The heat from burning wood or coal was used to vaporize water which in turn was used to drive pistons and turbines. The resulting mechanical energy was used to power pumps, mills, looms, boats, and trains. The smoke-belching mechanical monsters of the English midlands soon replaced humans and horses as the primary source of motive power in the West.

A second energy revolution occurred early in the 1900s with the commercialization of electric power. Today, one third of the energy produced in the United States is electrical. Most of it is generated by the descendants of those early steam engines. Figure 4.14 is a schematic of a modern power plant. Heat from the burning fuel is used to boil water, usually under high pressure. The elevated pressure serves two purposes: it raises the boiling point of the water and it compresses the water vapor. The hot, high-pressure vapor is directed at the fins of a turbine. As the gas expands and cools, it gives up some of its energy to the turbine, causing it to spin like a pinwheel in the wind. The shaft of the turbine is connected to a large coil of wire that rotates within a magnetic field. The turning of this dynamo generates an electric current—a stream of electrons that represents energy in a new and particularly convenient form. Meanwhile, the water vapor leaves the turbine and continues in its closed cycle. It passes through a heat exchanger where a stream of cooling water carries away the remainder of the heat energy originally acquired from the fuel. The water condenses into its liquid state and reenters the boiler, ready to resume the energy transfer cycle.

This process of energy transformation can be summarized in the three steps diagrammed in Figure 4.15. **Potential energy** in the chemical bonds of fossil fuels is first converted to **heat energy.** The heat released in combustion is absorbed by the

Figure 4.14

Diagram of a power plant for the conversion of heat to work to electricity. (From A. Truman Schwartz, *Chemistry: Imagination and Implication,* copyright © 1973 Harcourt Brace & Company, reproduced by permission of the publisher.)

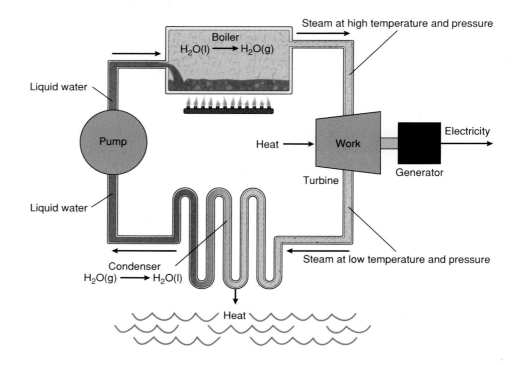

Figure 4.15

Energy transformation in a power plant.

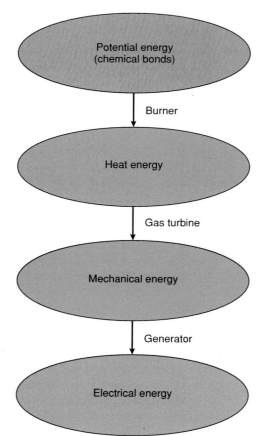

vaporizing water. This heat is then transformed into **mechanical energy** in the spinning turbine that turns the generator that changes the mechanical energy into **electrical energy.** In compliance with the first law of thermodynamics, energy is conserved throughout these transformations. To be sure, no new energy is created, but none is lost, either. We may not be able to win, but we can at least break even . . . or can we?

■ *Energy and Efficiency*

The last question is not as facetious as it might sound. In fact, we can't break even. No power plant, no matter how well designed, can completely convert heat into work. Inefficiency is inevitable, in spite of the best engineers and the most sincere environmentalists. There are, of course, energy losses due to friction and heat leakage that can be corrected, but these are not the major problems. The chief difficulty is nature, and specifically, the nature of heat and work.

The maximum efficiency with which a power plant can convert heat to work depends on the difference between the highest temperature to which the water vapor is heated (T_{hi}) and the lowest temperature to which the condensed water is cooled (T_{lo}). The mathematical expression for the efficiency is given below.

$$\text{Efficiency} = \frac{(T_{hi} - T_{lo})}{T_{hi}}$$

It is important to note that the temperatures in this equation are expressed in the **absolute** or **Kelvin scale.** Zero on the Kelvin scale is absolute zero or –273°C, the lowest temperature possible. In fact, it can't quite be attained. To convert a temperature from the Celsius scale to the Kelvin scale, one simply adds 273.

Temperature (in Kelvins) = Temperature (in degrees Celsius) + 273

A well-designed modern fossil-fuel burning power plant operates between a high temperature of about 550°C and a low temperature of 30°C. This means that, on the Kelvin scale, $T_{hi} = 550° + 273 = 823$ K, and $T_{lo} = 30° + 273 = 303$ K. It follows that

$$\text{Efficiency} = \frac{(823 \text{ K} - 303 \text{ K})}{823 \text{ K}} = \frac{520 \text{ K}}{823 \text{ K}} = 0.63$$

This value of 0.63 is the maximum theoretical efficiency of a power plant operating between these two temperatures. It means that at best only 63% of the heat that is obtained from the burning fuel is actually converted to work. The remainder is discharged to the cooling water, which consequently warms up. There are, in addition, other inefficiencies associated with friction, loss over long-distance power transmission lines, and so forth. Table 4.3 lists the efficiencies of a number of steps in the energy production. The overall efficiency is the product of the efficiencies of the individual steps. The net result is that today's most advanced power plants operate at an overall efficiency of only about 42%.

■ *Table 4.3*	*Some Typical Efficiencies in Power Production*
Maximum Theoretical Efficiency:	55–65%
Efficiency of Boiler:	90%
Mechanical Efficiency of Turbine:	75%
Efficiency of Electrical Generator:	95%
Efficiency of Power Transmission:	90%

4.15 The Sceptical Chymist

Electric heat is often advertised as being clean and efficient, but the electricity must first be generated, usually by a fossil-fuel power plant. This exercise is a critical examination of that assertion.

a. Determine the overall efficiency in the production of electric heat making the following assumptions: the energy is obtained from a methane-burning power plant with a maximum theoretical efficiency of 60%; the efficiencies of the boiler, turbine, and electrical generator, and power transmission lines are those given in Table 4.3; and the efficiency of converting electrical energy back to heat energy in the home is 98%.

Ans. Efficiencies are multiplicative. In order to obtain the overall efficiency, it is necessary to multiply the efficiencies of the individual steps, expressing the percentages in fractional form.

$$\text{Overall efficiency} = .60 \times .90 \times .75 \times .95 \times .90 \times .98 = 0.34$$

b. In a typical January, 250 million kJ (2.5×10^8 kJ) are required to heat a home in Minnesota. Assume that the home is heated with electricity generated by this power plant. Recall that the combustion of methane releases 50.1 kJ per gram of CH_4. Calculate the number of grams of methane required to heat the house for the month of January.

Ans. Because of the overall efficiency, only 34% or .34 times the total heat energy derived from the methane will actually be available to warm the house. The total quantity of heat (let's call it q) required to yield 2.5×10^8 kJ in the house is obtained as follows:

$$2.5 \times 10^8 \text{ kJ} = q \times 0.34$$
$$q = 2.5 \times 10^8 \text{ kJ}/0.34 = 7.4 \times 10^8 \text{ kJ}$$

Each gram of burning CH_4 yields 50.1 kJ, so the total mass of CH_4 that must be burned to supply the 7.4×10^8 kJ is calculated below.

$$\text{mass } CH_4 = 7.4 \times 10^8 \text{ kJ} \times 1 \text{ g } CH_4/50.1 \text{ kJ} = 1.5 \times 10^7 \text{ g}$$

c. The house could also have been heated directly with a gas furnace burning methane at an efficiency of 85%. Calculate the number of grams of methane required in January using this method of heating.

Ans. Because the only inefficiency is that of the furnace, we can do a calculation similar to that in part b, but using .85 as the efficiency. The energy required is 2.9×10^8 kJ, which corresponds to 5.8×10^6 g CH_4, 39% of the methane used to provide the heat electrically.

d. On the basis of your answers comment on the claim that electric heat is "clean" and efficient.

4.16 Your Turn

A coal-burning power plant generates electrical power at a rate of 500 megawatts, that is, 500×10^6 or 5.00×10^8 joules per second. The plant has an overall efficiency of 0.375 for the conversion of heat to electricity.

a. Calculate the total quantity of electrical energy (in joules) generated in one year of operation and the total quantity of heat energy used for that purpose.

Ans. 1.58×10^{16} J generated; 4.20×10^{16} J used

b. Assuming the power plant burns coal that releases 30 kJ per gram, calculate the mass of coal (in grams and metric tons) that will be burned in one year of operation. (1 metric ton = 1×10^3 kg = 1×10^6 g.)

Ans. 1.40×10^6 metric tons

■ *Improbable Changes and Unnatural Acts*

Although we have asserted that heat cannot be completely converted into work, we have offered little evidence and no explanation for this fact of nature. To illustrate the problem, push this book off the desk, and wait for it to come back up by itself. Be prepared to wait quite a while! The idea of a book picking itself off the ground and rising against the force of gravity is so bizarre, it is unbelievable. In fact, it is just unlikely—extremely unlikely. To understand why, let's examine the process in a little more detail.

You probably recall that a book resting on a table has potential energy by virtue of its position above the floor. When the book is dropped, the potential energy is converted to kinetic energy, the energy of motion. When the book strikes the floor, the kinetic energy is released with a bang. Some of it goes into the shock wave of moving air molecules that transmit the sound. Most of the energy goes to increase the motion of the atoms and molecules in the book and in the floor beneath it. This microscopic motion is the origin of what we call heat. A careful measurement would show that the book and the floor, and the air immediately around them are all very slightly warmer than they were before the impact. Energy has been conserved, but it has also been dissipated. **Heat** or **thermal energy** is characterized by the random motion of molecules. They chaotically move in all directions.

Now consider what would be necessary for the book to rise by itself, and thus to do work against the force of gravity. The molecules in the book would all have to move upward at the same time. At that instant, all of the molecules in the floor under the book would also have to move in an upward direction, giving the book a little shove. Needless to say, such agreement among the 10^{25} or so molecules involved is very unlikely. Yet, such a change is necessary to convert heat into work—to transform random thermal motion into uniform motion. The first law of thermodynamics may confidently assert that all forms of energy are equal, but the fact remains that some forms of energy are more equal than others! The chaotic, random motion that is heat is definitely low-grade energy.

■ *Order versus Entropy*

The inability of a power plant to convert heat into work with 100% efficiency or for dropped books to spontaneously rise are both manifestations of the same law of nature, the **second law of thermodynamics.** There are many versions of the second law, but all describe the directionality of the universe. One version states that it is impossible to completely convert heat into work without making some other changes in the universe. Another observes that heat will not of itself flow from a colder to a hotter body. The falling book provides another version of the second law. That, too, is a transformation of ordered kinetic energy into random heat energy. Like all naturally occurring changes, it involves an increase in disorder or randomness of the universe. This randomness in position or energy level is called **entropy.** The most general statement of the second law of thermodynamics is the entropy of the universe is increasing. This means that organized energy, the most useful kind for doing work, is always being transformed into chaotic motion.

A helpful way to look at the increase in universal entropy that characterizes all changes is in terms of probability. Disordered states are more probable than ordered ones, and natural change always proceeds from the less probable to the more probable. Let's suppose you define perfect order as a beautifully organized sock drawer, all the socks matched, folded, and placed in rows. This would represent a condition of zero entropy (no randomness). If you are like most people, this is probably a rather unlikely arrangement. It certainly didn't occur by itself; it took work to organize the socks. Without the continuing work of organization, it is quite possible that, over the course of a week or a month or a semester, the entropy and disorder of that sock drawer will increase. The point is that there are lots of ways in which the socks can be mixed up, and therefore disorder is more probable than order. Conversely, it is not very likely that you will open your drawer some morning and find that the socks are in perfect order and the entropy in that particular part of the universe has suddenly and spontaneously decreased. That sort of change from disorder to order is essentially what is involved in the conversion of heat to work. Henry Bent of the University of Pittsburgh has estimated that the probability of the complete conversion of one calorie of heat to work is about the same as the likelihood of a bunch of monkeys typing Shakespeare's complete works 15 quadrillion times in succession without a mistake.

4.17 ■ *Consider This*

All around us there are examples of the natural tendency for things to get messed up. Some have even been enshrined in literature: "All the king's horses and all the king's men, couldn't put Humpty together again." Cite some examples of your own.

Perhaps by now some Sceptical Chymist in the class has objected that there are also lots of earthly instances where order increases. Sock drawers do get organized; power plants convert heat to work; dropped objects get picked up; water can be decomposed into hydrogen and oxygen; refrigerators transfer heat from a colder to a hotter body; and students learn chemistry. All of these are "unnatural" events—**nonspontaneous** in the vocabulary of thermodynamics. They will not occur by themselves; they require that work be done by someone or something. An input of energy is necessary to reduce the entropy and increase the order. And in every case, the work that is done generates more entropy somewhere in the universe than it reduces in one small part of the universe. Even when entropy appears to decrease in a **spontaneous** change that occurs "by itself," for example the freezing of water at temperatures below 0°C, there are balancing increases in entropy. In this particular case, the heat given off by the freezing water adds to the disorder of its surroundings. In short, when the entire universe is considered, entropy always increases. THERE IS NO FREE LUNCH!

4.18 ■ *Consider This*

During midterm time, many students become very serious about their studying, and for hours on end will concentrate on the plays of Shakespeare or the causes of World War II. This *decrease* in intellectual entropy is often associated with an *increase* in the entropy of the student's room. Identify another process in which entropy appears to decrease but is actually coupled with an increase in entropy elsewhere in the universe.

4.19 ■ *Consider This*

People have observed that many environmental problems can be interpreted as the consequence of the second law of thermodynamics at work. Think about this way of looking at things and use it as the unifying theme for an essay on some aspect of environmental pollution.

■ *The Case for Conservation*

A fundamental feature of the universe is that energy and matter are conserved. However, the process of combustion degrades both energy and matter, converting them to less useful forms. For example, the energy stored in hydrocarbon molecules is eventually dissipated as heat when those molecules are converted to carbon dioxide and water—essential compounds, to be sure, but unusable as fuels. As residents of the universe, we have no choice; we must obey its inexorable laws. Nevertheless, there are many options within those constraints. One of the most important is to make a human contribution to the conservation of energy and matter.

The planet's store of fossil fuels is, of course, finite. A recent *Scientific American* article estimates that the amount of recoverable fossil fuels remaining is sufficient to last another 170 years at current consumption rates. But 170 years is a very short time in the context of human history and an almost undetectable instant in the age of the Earth. Moreover, world energy consumption is expected to increase by 50 or 60 percent by 2010, bringing with it increased emissions of CO_2, SO_2, and NO_x.

Another consideration is that fossil fuels are so important as feed stocks for chemical synthesis that it is a great waste to burn them. Late in the nineteenth century, Dimitri Mendeleev, the great Russian chemist who proposed the periodic table of the elements, visited the oil fields of Pennsylvania and Azerbijan. He is said to have remarked that burning petroleum as a fuel "would be akin to firing up a kitchen stove with bank notes." Mendeleev recognized that oil could be a valuable starting material for a wide variety of chemicals and the products made from them. But he would, no doubt, be amazed at the fibers, plastics, rubber, dyes, medicines, and pharmaceuticals that are currently produced from petroleum. Yet, we continue to ignore Mendeleev's warning and burn over 90% of the oil pumped from the ground.

In short, the arguments for conserving energy and the fuels that supply it are compelling. Fortunately, some promising strategies are available, and considerable savings have already been realized. Although energy production and fuel consumption have increased since the 1974 oil crisis, there has also been a significant increase in the efficiency with which the fuels are used. The production of electricity by utilities companies is the major use of energy in the United States, making up 38% of the total. The conversion of heat to work is, of course, limited by the second law of thermodynamics. But power plants currently operate well below the thermodynamic maximum efficiency. Better design will bring power plants closer to that upper limit, perhaps to overall efficiencies of 50 or 60%. A particularly appealing approach is an integrated system that uses "waste" heat from a power plant to warm buildings.

Once the electricity is generated, great savings can be realized in its use. Estimates of the technically feasible savings in electricity range from 10 to 75%. The wide range in these predictions is worrisome, but specific data are encouraging. For example, in an article in *Scientific American* in September 1990, Arnold P. Fickett, Clark W. Gellings, and Amory B. Lovins make the following statement: "If a consumer replaces a single 75-watt bulb with an 18-watt compact fluorescent lamp that lasts 10,000 hours, the consumer can save the electricity that a typical U.S. power plant would make from 770 pounds of coal. As a result, about 1600 pounds of carbon dioxide and 18 pounds of sulfur dioxide would not be released into the atmosphere." In the process, about $100 would be saved in the cost of generating electricity. Improvements in the design of electric motors and refrigeration units also hold considerable potential for increased efficiency.

4.20 *The Sceptical Chymist*

The quotation from Fickett, Gellings, and Lovins provides a marvelous opportunity for the Sceptical Chymist to apply his or her knowledge of chemistry. For example, let us check their assertion that replacement of a 75-watt (W) bulb with an 18-W fluorescent lamp will save the electricity made from 770 lb coal.

First of all, we note that the difference in the rate of energy consumption of the two bulbs is 75 W − 18 W = 57 W or 57 J/s. The total projected energy savings over the life of the bulb (10,000 hr) is obtained as follows:

$$\text{Energy savings} = 57 \text{ J/s} \times 10,000 \text{ hr} \times 60 \text{ min/hr} \times 60 \text{ s/min}$$
$$= 2.05 \times 10^9 \text{ J}$$

The energy comes from coal, and coal typically yields 30 kJ/g or 30×10^3 J/g. To determine the mass of coal that must be burned to obtain 2.05×10^9 J, the following operation is performed:

$$\text{mass coal} = 2.05 \times 10^9 \text{ J} \times \frac{1 \text{ g}}{30 \times 10^3 \text{ J}} = 6.84 \times 10^4 \text{ g}$$

Converting this to pounds yields the final answer.

$$6.84 \times 10^4 \text{ g} \times 1 \text{ lb/454 g} = 150 \text{ lb coal}$$

This is a significant discrepancy from the quoted value of 770 lb. Possibly Fickett, Gellings, and Lovins made an error, or perhaps an assumption that we neglected. Explore the latter possibility, and suggest what the assumption might have been.

4.21 *Your Turn*

Now it's your turn to exercise your skepticism and your computational skills by checking the other two claims. Calculate the mass of CO_2 and SO_2 that would *not* be released into the atmosphere if a 75-W bulb were replaced by an 18-W compact fluorescent lamp.

4.22 *The Sceptical Chymist*

Fickett, Gellings, and Lovins claim that the bulb replacement we have been discussing would save about $100 in the cost of generating electricity. What value are they assuming for the cost of electricity?

Electricity is generally priced per kilowatt-hour (kWh), so we need to know the number of kWh saved over the 10,000 hr lifetime of the bulb. First we multiply the 57 watts saved by 10,000 hr.

$$57 \text{ watts} \times 10,000 \text{ hr} = 570,000 \text{ watt hr (Wh)}$$

Then we convert the answer to kilowatt-hours, recognizing that 1 kWh = 1000 Wh.

$$570,000 \text{ Wh} \times 1 \text{ kWh/1000 Wh} = 570 \text{ kWh}$$

If 570 kWh of electricity cost $100, as the writers imply, what is the cost per kilowatt-hour? Once you have calculated the answer, find the cost of electricity to consumers in your city. Compare the results.

It is noteworthy that some utility companies have done much to promote consumer education and provide financial inducements for conserving electricity. Sophisticated economic planning, new financing arrangements, and pricing policy are all part of efforts to save energy. One particularly important concept is "payback time," the period necessary before a private consumer, an industry, or a power company recaptures in savings the initial cost of a more efficient refrigerator, manufacturing process, or power plant.

Recent advances in information technology and data processing have also made possible sizeable energy savings. "Smart" office buildings or homes feature a complicated system of sensors, computers, and controls that maintain temperature, airflow, and illumination at optimum levels for comfort and conservation. Similarly, the computerized optimization of energy flow and the automation of manufacturing processes have brought about major transformations in industry. Over the past 20 years, industrial production in the United States has increased substantially, but the associated energy consumption has actually gone down. A case in point is the low-pressure, gas-phase process developed by Union Carbide chemical engineers for making polyethylene, which is the world's most common plastic. This new process uses only one quarter of the energy required by previous high-pressure methods. Although the capital investment associated with such conversions is often substantial, consumers and manufacturers may ultimately enjoy financial savings and increased profits. Mention should also be made of the energy conservation that results from recycling materials, especially aluminum. Because of the high energy cost of extracting aluminum from its ore, recycling the metal yields an energy saving of about 70%. To put things in perspective, recycling 250,000 aluminum cans saves energy equivalent to 3.5 million gallons of gasoline. You could watch television for three hours on the energy saved by recycling just one can.

One final area where energy conservation has a direct impact on lifestyle is transportation. About one-fourth of the total energy used and one-half of the world's oil production goes to power motor vehicles. But even here, we are making progress in conservation. During the last 15 years, gasoline consumption in the United States has dropped by one-half. Much of this saving is attributable to lighter-weight vehicles, thanks to the use of new materials, and to new engine designs. Research and development continues in both areas. The Volkswagen Eco-Polo test car has achieved combined city/highway fuel consumption of 62 miles per gallon and the Volvo LCO 2000 has reached 81 mpg on the highway.

Such improvements in fuel economy are impressive, but the fact remains that the automobile is an energy-intensive means of transportation. A mass transit system is far more economical, provided it is heavily used. In Japan, 47% of travel is by public transportation, compared to only 6% in the United States. Of course, Japan is a compact country with a high population density. The great expanse of North America is not ideally suited to mass transit. And one must also reckon with the long love affair between Americans and their automobiles.

4.23   *Consider This*

■

Because oil is such a valuable resource and its supply is limited, perhaps we should curtail our principal use of oil-gasoline. If voluntary conservation methods are not working, the government could force conservation by taxing gasoline so that the average price was $3.00 per gallon—a price that is currently typical in Western Europe and Japan. Because this is more than half the minimum hourly wage, we can be reasonably sure that less gasoline would be used. Draft a letter to your representative in Congress, either supporting or protesting this hypothetical new gasoline tax. What effect, if any, would the proposed use of the new revenues have on your position?

■ *Conclusion*

To a considerable extent, taste ultimately influences what technology can do to conserve energy. As individuals and as a society, we must decide what sacrifices we are willing to make in speed, comfort, and convenience for the sake of our dwindling fuel supplies and the good of the planet. The costs might include higher taxes, more expensive gasoline and electricity, fewer and slower cars, warmer buildings in summer and cooler ones in winter, perhaps even drastically redesigned homes and cities. One thing seems to be clear: the best time to examine our options, our priorities, and our will is before we face another full-blown energy crisis.

■ *References and Resources*

Anderson, E. V. "Ethanol's Role in Reformulated Gasoline Stirs Controversy." *Chemical & Engineering News,* Nov. 2, 1992: 7–13.

"Gas Fever: Happiness Is a Full Tank." *Time,* Feb. 18, 1974: 35–36.

Gibbons, J. H.; Blair, P. D.; and Gwin, H. L. "Strategies for Energy Use." *Scientific American* **261,** Sept. 1989: 136–43.

Levine, M. D.; Meyers, S. P.; and Wilbanks, T. "Energy Efficiency and Developing Countries." *Environmental Science & Technology* **25,** April 1991: 584–89.

"No End in Sight to Gas-Pump Lines." *U.S. News & World Report,* Feb. 25, 1974: 13–14.

Roberts, P. "Energy Use in the Developing World: A Crisis of Rising Expectations." *Environmental Science & Technology* **25,** April 1991: 580–83.

Scientific American **263,** Sept. 1990. This entire issue is devoted to energy-related topics. Of particular relevance to this chapter are the following articles:

> Belviss, D. L. and Walzer, P. "Energy for Motor Vehicles." 103–109.
>
> Davis, G. R. "Energy for Planet Earth." 54–62.
>
> Fickett, A. P.; Gellings, C. W.; and Lovins, A. B. "Efficient Use of Electricity." 65–74.
>
> Fulkerson, W.; Judkins, R. R.; and Sanghvi, M. K. "Energy from Fossil Fuels." 129–35.

U.S. Congress, Office of Technology Assessment. *Energy Use and the U.S. Economy* (Background Paper). Washington: U.S. Government Printing Office, 1990.

■ *Experiments and Investigations*

5. Hot Stuff: An Investigation

6. Energy Content of Fuel

■ *Exercises*

1. Which sample contains the greater quantity of heat energy?

   a. 100 g $H_2O(l)$ at 25°C or 10 g $H_2O(l)$ at 25°C?
   b. 20 g $Fe(s)$ at 50°C or 20 g $Fe(s)$ at 100°C?
   c. 1 g $H_2O(l)$ at 100°C or 1 g $H_2O(g)$ at 100°C?

2. A friend tells you that hydrocarbons made up of larger molecules are better fuels than those consisting of smaller molecules. She supports this position by citing the following heats of combustion (in kJ/mole).

CH_4: 810	C_4H_{10}: 2859
C_6H_{14}: 4163	$C_{12}H_{22}$: 6736

   Do you agree with her conclusion and her evidence? Why or why not? Use appropriate calculations to support your position.

3. Using the bond energies of Table 4.1, calculate the energy changes associated with the following reactions. Indicate which reactions are exothermic and which are endothermic.

    a. $H_2 + Cl_2 \rightarrow 2\ HCl$
    b. $N_2 + O_2 \rightarrow 2\ NO$
    c. $H_2S + 2\ O_2 \rightarrow H_2O + SO_3$
    d. $N_2 + 3\ H_2 \rightarrow 2\ NH_3$

*4. The text reports that energy change for the following reaction is –802 kJ. In other words, 802 kJ are given off per mole of CH_4 burned.

$$CH_4(g) + 2\ O_2(g) \rightarrow CO_2(g) + 2\ H_2O(g)$$

By contrast, the energy change for the following reaction is –890 kJ per mole of CH_4 burned.

$$CH_4(g) + 2\ O_2(g) \rightarrow CO_2(g) + 2\ H_2O(l)$$

Account for this difference.

*5. Explain why the bond energy of a specific bond, for example O—H, is not equal in all compounds.

6. Calculate the heat of combustion of ethane, C_2H_6, under the following conditions. Begin by writing the appropriate equation for each of the reactions.

    a. It burns "completely" to form CO_2 (g) and H_2O (g).
    b. It burns to form CO (g) and H_2O (g).

*7. Bond energies such as those in Table 4.1 are usually found by "working backwards" from heats of reaction. A reaction is carried out and the heat absorbed or evolved is measured. From this value and known bond energies, unknown bond energies can be calculated. For example, the energy change associated with the following reaction is +81 kJ.

$$NBr_3 + 3\ H_2O \rightarrow 3\ HOBr + NH_3$$

Use this value plus Table 4.1 to calculate the energy of the N—Br bond.

8. Draw a diagram similar to Figure 4.9 to represent each of the following types of reactions.

    a. an endothermic reaction
    b. an exothermic reaction
    c. a "slow" reaction and a "fast" reaction with the same overall energy change
    d. a reaction in which there is no net energy change

9. Suggest why the headline in the check-out counter tabloid, **SLEEPING MAN BURSTS INTO FLAME,** is very likely not true.

10. Specify three reasons why petroleum was so important to the development of self-propelled vehicles.

11. The text states that catalysts are used to speed up cracking reactions in oil refining. Draw a figure similar to Figure 4.9 in which you illustrate energy changes in a reaction in the presence and the absence of a catalyst. Explain how the figure illustrates the effect of the catalyst.

12. Here's another chance to demonstrate your skill as a Sceptical Chymist. Some textbooks define a catalyst as "a substance that influences the rate of a reaction (usually by speeding it up) without participating in the reaction." Analyze that definition and formulate a more accurate version.

13. Suggest why the energy released by burning a gram of ethanol, C_2H_5OH, (29.7 kJ) is considerably less than that released by burning a gram of octane, C_8H_{18}, (47.8 kJ). *Hint:* Consider the reactants and the products in the combustion reaction. Writing equations might help.

14. Write a chemical equation for the combustion of water gas (equal moles of CO and H_2) and calculate the heat of combustion of such a mixture. Comment on the advantages and disadvantages of using water gas instead of CH_4 as a fuel for home heating.

15. One third of the one million tons of solid waste generated in Hennepin County each year is burned for energy. Recall that 27,000 pounds of this waste yields the same quantity of energy as 21 barrels of oil.

    a. Calculate the number of barrels of oil saved per year.
    b. Calculate the energy released per year. (One barrel of oil releases 6×10^6 kJ of energy.)
    c. If the steam power plant and electric generator operate at an overall efficiency of 42%, what is the total quantity of heat energy that is converted to electricity per year?

16. Suggest reasons why efficiency calculations are done using the Kelvin temperature scale rather than the Celsius scale.

17. Perform the following temperature conversions.

    a. 100°C (the boiling point of water) to K
    b. 4 K (the boiling point of helium) to °C
    c. –40°C (the temperature where the Celsius and Fahrenheit read the same value) to K
    d. 5780 K (the average surface temperature of the Sun) to °C

18. Assume that three power plants have been proposed, operating at the temperatures given below.

    Plant I T_{hi} = 1200°C, T_{lo} = 0°C

    Plant II T_{hi} = 600°C, T_{lo} = 20°C

    Plant III T_{hi} = 450°C, T_{lo} = –150°C

    a. Calculate the maximum efficiency of each plant.

    b. Identify the factors that effect the efficiency.

    c. Discuss the practical limits that govern such efficiencies. Which of the above power plants is most likely to be built?

19. There is a third law of thermodynamics which, in one of its formulations, states that absolute zero cannot be attained. What significance does this law have for the maximum efficiency of a steam engine?

*20. Over the years, hundreds of so-called "perpetual motion machines" have been designed and built. All claim to run without an external energy source and none succeed. Explain why.

21. Learning chemistry is a nonspontaneous, "unnatural" process in which the entropy of your brain is supposed to decrease. Why doesn't this contradict the second law of thermodynamics?

*22. Entropies of substances can be experimentally determined. Listed below are some typical values, all at 25°C and expressed in joules/K mole.

C (*diamond*)	2.4	CH_3OH (*l*)	127
H_2 (*g*)	131	O_2 (*g*)	205
H_2O (*l*)	70	H_2O (*g*)	189

    a. What generalizations can you deduce from these data about the factors that influence the entropy of a substance?

    b. Explain your rules on the molecular level.

    c. How would you expect the entropy of solid water to compare with the entropy values of liquid and gaseous water? Why?

    d. How would you expect the entropy of a substance to change as its temperature is increased? Why?

23. The overall cost of electrical generation (in energy and in money) is influenced by the maximum theoretical efficiency of the power plant, but there are many other expenses involved. Identify the factors that must be included in an overall accounting of the cost of electricity.

5

The Wonder
of Water

If there is magic on this planet, it is contained in water. . . . Its substance
reaches everywhere; it touches the past and prepares the future; it moves
under the poles and wanders thinly in the heights of the air. It can assume
forms of exquisite perfection in a snowflake, or strip the living
to a single shining bone
cast up by the sea.

Loren Eiseley

These poetic words of noted anthropologist and essayist Loren Eiseley capture our fas-
cination with water. Arguably, water is the most important compound on the face of
the Earth. In fact, it covers about 70% of that face, giving the planet the lovely blue
color we have seen in photographs from outer space. Water is literally everywhere.
Our bodies, too, are aqueous environments—approximately 60% water. If necessary,
humans can survive without food for about two weeks, but without water, we die
within about a week. Indeed, water is absolutely indispensable for all of the forms of
life that have evolved on Earth. Even in our flights of science fiction fantasy, it is difficult to imagine life of any sort without water.

■ *Chapter Overview*

In this chapter we will consider this remarkable compound from several perspectives. We first focus on use patterns, both personal and commercial, and find that
a flood of water is used daily for a wide range of purposes. The industrial, agricultural, and personal uses of water are a consequence of its familiar but unusual
properties, which range from super solvent to heat exchanger. Therefore, we take
a close look at these properties and at the molecular structure that accounts for
them. The emphasis then shifts to the sources of water and the means used to
purify, soften, and otherwise treat these supplies. The chapter concludes by
merging personal and public considerations by asking "Who owns the water?" It
begins by asking you to do a familiar and routine act—to take a drink of water—
and, more importantly, to think about it.

■ *Take a Drink*

Chapter 1 began with an invitation to "Take a breath of air." We now ask you to "Take
a drink of water." Nothing could be more familiar than this clear, colorless, and (usually) tasteless liquid. We drink it, cook and wash with it, swim in it, and skate, ski, and
boat on it. And yet, we generally take water for granted. We turn a faucet for a drink
or a shower and simply expect a sufficient quantity of water to come flowing out of
the tap, almost as a foregone conclusion. Unless there is a water emergency brought
on by drought or contamination of our municipal water supply, we seldom think about
where the water comes from, what it contains, how pure it is, or how long the supply
will last. Well, the time has come to take the plunge.

5.1   *Consider This*

How much water do you use daily: 5 quarts? 20 liters? 100 gallons?
In order to get an approximate answer to this question, list the ways
you used water yesterday and estimate the volume, in liters, associated
with each of the uses. One liter equals about 1 quart or more exactly,
1.00 L = 1.06 qt. Table 5.1, which gives the volume of water used by
various household fixtures, should be of help.

■ *Table 5.1*	*Typical Water Usage by Household Fixtures*
Fixture	**Water use**
Toilet	19 L/flush
Clothes washer	140 L/load
Shower	19 L/min
Faucets	12 L/min

If you are like most people, you probably underestimated your daily personal water consumption. The national average is about 300 liters used only for personal consumption and hygiene. The 3–4 liters of water we each drink each day, either as the pure compound or in some other beverage, is only about 1% of our 300-liter share. Most of it goes down the drain. Household use of water in this country has increased significantly because of the installation of indoor plumbing, showers, and electric clothes washers and dishwashers. For example, we use about eight liters (two gallons) of water daily just for drinking and cooking. This does not include that used for bathing, washing clothes, or flushing toilets. Note from Table 5.1 that a single flush of a toilet at 19 liters (water quality aside) uses enough water to furnish your cooking and drinking needs for over two days; a five minute shower could supply enough for almost two weeks. The 140 liters of water used to wash just a single load of clothes in a conventional washer is sufficient to meet your drinking and cooking water requirements for nearly a month!

5.2 ■ Consider This

According to Table 5.1, taking a shower uses about 19 liters of water per minute. How much water do you use for showering or bathing each day? Do a determination by taking a wide-mouthed container of known volume into the shower and measuring the time required to fill the container. Use this information and the length of your average shower to calculate the number of gallons of water used in your daily shower and in a year of showering (1 gal = 4 qt; 1 L = 1.06 qt).

Ans. Obviously, answers will vary according to shower heads and showering habits. One year of showering for 5 minutes a day with a shower that discharges 19 L/min uses 9200 gallons of water—enough to fill a swimming pool.

Water rationing is not uncommon during periods of drought, when car washing and lawn watering are often restricted. But several states, including California, Florida, and New York, have laws requiring new homes, apartments, and offices to install water-efficient appliances. Some of these fixtures, their water consumption, and the percent of water saved are listed in Table 5.2. Note, for example, that an air-assisted shower head uses only 2 liters of water per minute, an 89% saving over the conventional design. It has been estimated that simple, inexpensive home conservation measures, such as those in the table, could reduce domestic water use by about one-third. But if climatic changes associated with the greenhouse effect significantly reduce precipitation on a global scale, it will take considerably more than a switch to air-assisted shower heads to make the desert bloom.

Table 5.2	Potential Water Savings with More Efficient Household Fixtures		
Fixture		**Water use (L)**	**Water saved (%)**
Toilets (per flush)			
Conventional		19	—
Low flush		13	32
Air-assisted		2	89
Clothes washers (per load)			
Conventional		140	—
Wash recycle		100	29
Front-loading		80	43
Shower heads (per min.)			
Conventional		19	—
Flow-limiting		7	63
Air-assisted		2	89
Faucets (per min.)			
Conventional		12	—
Flow-limiting		6	50

Sources: Figures for low-flush fixtures from Brown and Caldwell (Inc.), *Residential Water Conservation Projects,* U.S. Department of Housing and Urban Development, Washington, D.C., 1984. All others from Robert L. Siegrist, "Minimum-Flow Plumbing Fixtures," *Journal of the American Water Works Association,* July, 1983.

From *State of the World 1986,* Lester R. Brown (Project Director) and Linda Starke (Editor).
© 1986 W. W. Norton Company, New York. Reprinted by permission.

5.3 ■ Consider This

Because of the effects of a statewide five-year drought, residents of Marin County, north of San Francisco, were limited to a daily water ration of 50 gallons per person during the winter of 1991. If you lived in Marin County during the rationing, how would you choose to use your limited supply of water?

The 300 L (71 gal) that the average American uses each day for personal purposes is only a drop in a very large bucket. In fact, the daily per capita water consumption in the United States is close to 7200 L (1900 gallons). The great majority of this flood is attributable to the indirect usage required to produce, process, and transport food and manufactured goods. Water use is hidden in just about everything you encounter—the foods you eat, the clothes you wear, the books you read, the videotapes you watch, the CDs you listen to, and almost any other manufactured item you can imagine.

Table 5.3 includes estimates of the volume of water required to manufacture various familiar products. To be sure, such figures are only approximate, because it is very difficult to decide what water usage to assign. Nevertheless, it is undoubtedly true that

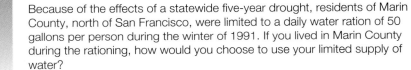

Table 5.3	Estimates of Water Required to Produce Certain Products	
	Product	**Volume of water (gal)**
	1 ton of steel	65,000
	1 ton of paper	40,000
	1 barrel of beer	500
	1 pound of aluminum	160
	1 gallon of gasoline	10

the ten-fold growth in water consumption in the United States since the beginning of the century is largely due to increases in industrial and agricultural applications. As Figure 5.1 indicates, 57% of the water used in this country goes to industry. Much of the 34% employed in agriculture goes to irrigate vast areas of the Southwest. Only 9% is used for domestic and municipal purposes.

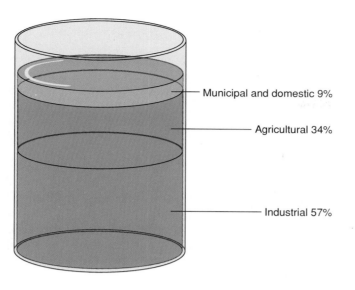

Figure 5.1

Daily water use in the United States.

Municipal and domestic 9%

Agricultural 34%

Industrial 57%

The United States, the thirstiest nation on Earth, uses almost 1.7×10^{12} L of water per day. Table 5.4 indicates that on a per capita basis, our withdrawal of water is twice that of the former Soviet Union, more than five times that of the United Kingdom, and ten times that of Indonesia. The profligate consumption of water by the United States and other industrialized nations has placed severe demands on limited supplies. The majority of the water is used industrially and agriculturally, and therefore is not generally required to meet high standards for purity and cleanliness. On the other hand, the problems of individuals in a very arid region such as Saudi Arabia should serve as continuing reminders of the daily personal requirements we have for fresh water. Only 37% of sub-Saharan Africans have clean drinking water, and the severe famine that has ravished parts of the continent is due, in part, to prolonged drought. Whether for personal, agricultural, or industrial uses, there is simply no substitute for water.

■ *Table 5.4*	*Estimated Water Use in Selected Countries—Total, Per Capita, and by Sector, 1980*					
	Daily water withdrawals			**Share withdrawn by major sectors**		
Country	**Total**	**Per capita**		**Agriculture**	**Industrial**	**Municipal**[a]
	(billion L)	(thousand L)		(percent)		
United States	1683	7.2		34	57	9
Canada	120	4.8		7	84	9
Soviet Union	967	3.6		64	30	6
Japan[b]	306	2.6		29	61	10
Mexico[b]	149	2.0		88	7	5
India[b]	1058	1.5		92	2	6
United Kingdom	78	1.4		1	85	14
Poland	46	1.3		21	62	17
China	1260	1.2		87	7	6
Indonesia[b]	115	0.7		86	3	11

[a]Along with residential use, figures may include commercial and public uses, such as watering parks and golf courses.

[b]1975 figures for Mexico; 1977 for India, Indonesia, and Japan.

Sources: U.S. data, U.S. Geological Survey; Canadian data, Harold D. Foster and W. R. Derick Sewell, *Water: The Emerging Crisis in Canada* (Toronto: James Lorimer & Company, 1981); Soviet, U.K., Polish data, U.N. Economic Commission for Europe; Japanese, Indian, Indonesian data, *Global 2000 Report;* Mexican data, U.N. Economic Commission on Latin America; Chinese data, Vaclav Smil, *The Bad Earth.*

■ *H₂O: Surprising Stuff*

The many and varied uses of water are a consequence of its properties. It is an effective solvent for a wide range of materials, and it has relatively high boiling and freezing points. Furthermore, water has a great capacity to absorb heat, and it participates in a wide variety of chemical reactions. Water is so ubiquitous that it has become the standard for many of the units of modern science, including the Celsius temperature scale, the kilogram, and the calorie. Yet, for all of that, the physical properties of water are quite peculiar, and we are very fortunate that they are. If water were a more conventional compound, we would be very different creatures.

This most common of liquids is full of surprises. For example, it is surprising that water is a liquid and not a gas at room temperature (about 25°C) and pressure (1 atmosphere). The molecular mass of water is 18.0, and almost all compounds with molecular masses that low are gases under these conditions. Consider three common atmospheric gases: nitrogen has a molecular mass of 28.0, oxygen is 32.0, and carbon dioxide is 44.0. All have molecular weights greater than that of water, yet we breathe them rather than drink them.

Moreover, not only is water a liquid under these conditions, it has an anomalously high boiling point of 100°C. This temperature is one of the reference points for the Celsius temperature scale. The other is the freezing point of water, set at 0°C. And when water freezes, it exhibits another bizarre property, it expands. Better behaved liquids contract when they solidify. These and other unusual properties all derive from water's chemical composition and its molecular structure.

The composition of the compound is known to practically everyone. Indeed, the formula for water, H₂O, is very likely the world's most widely known bit of chemical information. However, most people would probably be hard pressed to offer supporting evidence for the formula, other than the questionable authority of a textbook or a teacher. We will not provide definitive proof here, but we can at least offer experimental evidence that is consistent with the correct formula. It comes from **electrolysis**—the electrical decomposition of a compound (in this case water) into its constituent elements. A direct current is passed through a sample of water containing a little sulfuric acid or other compound added to enable it to conduct electricity. Bubbles of gas form at each of the two metal **electrodes.** If the reaction is carried out in an apparatus similar to that pictured in Figure 5.2, the two gases can be collected individually.

Figure 5.2

Electrolysis of water.

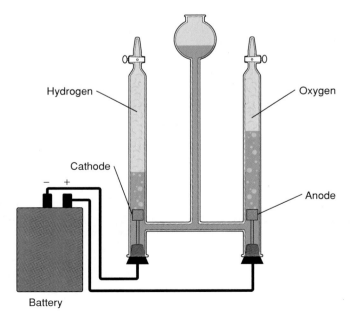

The colorless gas formed at the **cathode,** the electrode connected to the negative terminal of the battery, burns rapidly and is much less dense than air. It is hydrogen. The gas generated at the **anode,** the positive electrode in this apparatus, is also colorless, but it is slightly more dense than air. It does not burn, but allows other substances to burn in it. The gas is, of course, oxygen. Whenever the electrolysis of water is carefully carried out, the volume of hydrogen generated is always found to be twice the volume of oxygen. Experimentation with a variety of gases has shown that if temperature and pressure are maintained constant, equal volumes of gas contain equal numbers of molecules. Because the volume of hydrogen is twice the volume of oxygen, we conclude that the number of hydrogen molecules formed during electrolysis must be twice the number of oxygen molecules. This suggests (although it does not unambiguously prove) that a water molecule contains twice as many hydrogen atoms as oxygen atoms. Further support comes with the realization that oxygen and hydrogen both exist as diatomic molecules. This knowledge permits us to write an equation that agrees with the experimental results.

$$2\ H_2O(l) \rightarrow 2\ H_2(g) + O_2(g) \qquad (5.1)$$

During electrolysis, electrical energy is supplied to decompose water into hydrogen and oxygen. The reaction is thus endothermic. The reverse exothermic reaction releases an equal amount of energy. Hence, the burning of hydrogen in air or oxygen has been proposed and used as an energy source, a topic that is again considered in Chapter 9. The product, H_2O, is certainly non-polluting, but considerable caution is required when dealing with oxygen and hydrogen mixtures, which can be explosive. The destruction of the German dirigible Hindenberg in 1937 and the Challenger space shuttle explosion in 1986 are graphic evidence of the power released when hydrogen and oxygen combine.

■ *Molecular Structure and Physical Properties*

By now you know that the elementary composition of a chemical compound is only one factor in determining the chemical and physical properties of the substance. Another important (and related) factor is molecular structure. You will recall from Chapter 2 that water is a covalent compound. The oxygen atom is at the center of the bent molecule, attached to the hydrogen atoms by covalent bonds. Each of the two single bonds consists of one pair of shared electrons. It turns out that the bonding electrons are not equally shared between the oxygen and hydrogen atoms. The oxygen atom attracts the electron pair more strongly than does the hydrogen. To use the appropriate technical term, oxygen is said to have a higher **electronegativity,** or a greater attraction for the shared electron pair than hydrogen. According to Table 5.5, the electronegativity of oxygen is 3.5; that of hydrogen is 2.1. Because of this difference, the shared electrons are actually pulled toward the oxygen and away from the hydrogen. This unequal sharing gives the oxygen end of the bond a net negative charge and the

■ *Table 5.5*	**Electronegativity Values**		
Hydrogen (H)	2.1		
Lithium (Li)	1.0	Sodium (Na)	0.9
Beryllium (Be)	1.5	Magnesium (Mg)	1.2
Boron (B)	2.0	Aluminum (Al)	1.5
Carbon (C)	2.5	Silicon (Si)	1.8
Nitrogen (N)	3.0	Phosphorus (P)	2.1
Oxygen (O)	3.5	Sulfur (S)	2.5
Fluorine (F)	4.0	Chlorine (Cl)	3.0

hydrogen end a net positive charge. Because the bond has oppositely charged ends or poles, it is said to be a **polar covalent bond.** Polar bonds arise whenever atoms of differing electronegativities are covalently attached to each other. The greater the electronegativity differences of the elements involved, the more polar the bond.

5.4   *Consider This*

Examine Table 5.5 and offer some generalizations about the electronegativity values listed there. For example, you might look at the periodic table and try to relate position and electronegativity.

Many of the unique properties of water are a consequence of its molecular shape and the polarity of its bonds. Both are shown in this drawing of an H_2O molecule.

Arrows are used to indicate the direction in which the electron pairs are displaced, and the net charges are indicated by the appropriate signs. Note that the hydrogen atoms are positive and that negative charge appears to be concentrated in the two nonbonding pairs of electrons on the oxygen atom.

Now consider what happens when two water molecules approach each other. Because opposite charges attract, one of the positively charged hydrogen atoms of one molecule will be attracted to one of the regions of negative charge associated with the nonbonding electron pairs of the other molecule. This is an example of *inter*molecular attraction *between* molecules. The fact that each H_2O molecule has two hydrogen atoms and two unbonded pairs of electrons increases the opportunities for intermolecular attraction. Each of the two nonbonding electron pairs can form a loose bond to a hydrogen atom of another water molecule. Similarly, each of the two hydrogen atoms in a molecule can attract an electron pair from another molecule. Thus, a single H_2O molecule can simultaneously bond to as many as four others, as pictured in Figure 5.3.

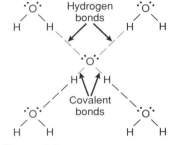

Figure 5.3
Hydrogen bonding in water.

▪ *Hydrogen Bonding*

These attractions between water molecules are called **hydrogen bonds.** They are only about one-tenth as strong as the covalent *intra*molecular bonds that connect atoms together *within* molecules. Nevertheless, as intermolecular forces go, hydrogen bonds are quite strong. Their strength is manifested in the relatively high boiling point of water (100° C) and the large amount of energy required to convert liquid water into vapor (540 cal/g). In order to boil water, the H_2O molecules must be separated from their relatively close contact in the liquid state and changed into the gaseous state where they are much farther apart. In other words, their intermolecular hydrogen bonds must be overcome. If the hydrogen bonds in water were weaker, water would have a much lower boiling temperature and require less energy to boil. If water had no hydrogen bonding at all, it would boil at about −75° C, making life very uncomfortable, if not impossible.

Although water is wet, boiling points may seem rather dry and impersonal until you reflect on the fact our bodies, at a relatively constant temperature of 37° C, are about 60% water. Because of hydrogen bonding, most of our body's water, whether in cells, blood, or other body fluids, is in the liquid state, well below the boiling point. Without hydrogen bonding, we would be a gas!

It is important to note that hydrogen bonds are not restricted to water. There is evidence for similar intermolecular attraction in many molecules that contain hydrogen atoms covalently bonded to oxygen, nitrogen, or fluorine atoms. Thus, ammonia, NH_3, and hydrogen fluoride, HF, also have unusually high boiling points, but less so than water. The net effect of hydrogen bonding in them is weaker than in water. Furthermore, hydrogen bonding is important in stabilizing the shape of large biological molecules, such as proteins and nucleic acids. In proteins, which are major components in skin, hair, and muscle, hydrogen bonding occurs between hydrogen atoms and oxygen or nitrogen atoms. The coiled, double-helical structure of DNA (deoxyribonucleic acid) is stabilized by thousands of hydrogen bonds formed between particular segments of the linked DNA strands. So in this respect, too, hydrogen bonding plays an essential role in the life process (see Chapter 13).

Hydrogen bonding also explains why ice cubes and icebergs float in water. An ice crystal is a regular array in which every H_2O molecule is hydrogen bonded

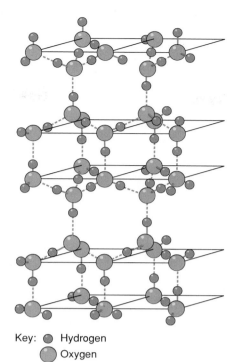

Key: Hydrogen
Oxygen

Figure 5.4

The hydrogen-bonded lattice structure of ice I, the common form of solid H_2O. Note the open channels, which contribute to the fact that ice is less dense than liquid water.

to four others. The pattern is pictured in Figure 5.4. Note that the pattern includes a good deal of empty space in the form of hexagonal channels. When ice melts, this regular array begins to break down, though a significant level of hydrogen-bonded structure is still retained. Upon melting, individual H_2O molecules can enter the open channels. As a result, the molecules in the liquid state are, on the average, more closely packed than in the solid state. A volume of one cubic centimeter (1 cm^3) of liquid H_2O contains more molecules than one cubic centimeter of ice. Consequently, liquid water has a greater mass per cubic centimeter than does ice. This is simply another way of saying that the **density** of water is greater than that of ice.

You will recall that the **density** of a substance is defined as the mass per unit volume. For most purposes, the "unit volume" used to express density is one cubic centimeter, which is identical to one milliliter (mL). A cube one centimeter on a side is about the same volume as one-fifth of a teaspoonful. At its normal freezing point of $0°C$, this volume of liquid water weighs 0.99987 g. In other words, its density is 0.99987 g/cm^3 or 0.99987 g/mL. In contrast, the density of ice at $0°C$ is 0.917 g/cm^3. It follows that ice must float in water.

5.5 *Your Turn*

A student sets out to determine the density of a sample of olive oil by measuring the mass of 24.6 cm^3 of the oil. She finds that this volume of oil weighs 22.5851 g. Calculate its density. Then predict where to find the oil layer in an olive oil and vinegar salad dressing.

Ans. Given the definition of density, we need to find the mass to volume ratio for the olive oil.

$$\text{density} = \frac{\text{mass}}{\text{volume}} = \frac{22.5851 \text{ g}}{24.6 \text{ cm}^3} = 0.918 \text{ g/cm}^3$$

Note: If you used a fancy calculator for this problem, you might be tempted to report the answer as 0.918093495935 g/cm^3. Why is 0.918 g/cm^3 a more appropriate value?

5.6 Your Turn

Toluene, C_7H_8, is a component of gasoline and has a density of 0.867 g/cm^3. Calculate the mass of one gallon of toluene in grams and pounds. (1 gal = 3790 cm^3 and 1 lb = 454 g)

Ans. 3290 g or 7.24 lb

Figure 5.5

Density of water at various temperatures.

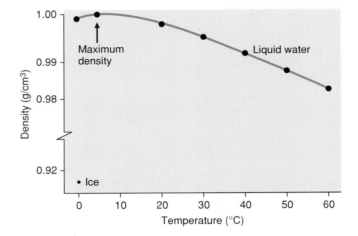

The density of liquid water increases slightly with increasing temperature between 0°C and 3.98°C, where it reaches its maximum value (Figure 5.5). This point was used to establish the fundamental unit of mass in the metric system. The kilogram (kg) was defined as the mass of 1000 cm^3 of water at its temperature of maximum density. Because 1 kg = 1000 g, this is equivalent to stating that the mass of 1000 cm^3 of water at its maximum density is 1000 g. In other words, one cubic centimeter of water has a mass of exactly one gram at 3.98°C and its density equals 1.00000 g/cm^3. Above this temperature, water is better behaved. Its density decreases as temperature increases, as is the case with most liquids. At 25°C it has a value of 0.99707 g/cm^3, and at the boiling point the density of liquid H_2O is 0.95838 g/cm^3.

For the great majority of substances, the solid state is more dense than the liquid. The fact that water shows the reverse behavior means that lakes freeze from the top down, not the bottom up. This topsy-turvy behavior is convenient for aqueous plants, fish, and ice skaters. On the other hand, the fact that water expands as it freezes is not so convenient for people whose water pipes and radiators freeze and burst.

■ Water As a Solvent

Many of the uses of water depend upon the fact that it is an excellent **solvent** for a wide variety of substances. The substances that dissolve in a solvent are called **solutes,** and the resulting mixture is, of course, a **solution.** Industrial processes are frequently carried out in aqueous solutions; and the necessity of water for life is, in part, a consequence of its solvent properties. Water dissolves and transports the minerals and nutrients necessary for plant growth, and blood and other body fluids are water solutions of biologically important solutes. Essentially all of the water we consume is in the form of solutions, including soft drinks, coffee, tea, and even tap water. If we're lucky, our tap water is a very dilute solution of harmless chemicals.

"Sugar water" and "salt water" are examples of two main classes of aqueous solutions. A significant difference between the two can be demonstrated with a conductivity meter. A simple form of such a device is pictured in Figure 5.6. A source of electricity, a battery or house current, is attached to a light bulb. One of the wires is cut, and the insulation is stripped from both ends of the wire. This breaks the circuit.

a.   b.

As long as the two ends do not touch, the bulb will not light. If the separated ends are placed in distilled water or a solution of sugar in water, the bulb remains dark. However, if the bare wires are placed in a solution of salt, the bulb becomes illuminated. Perhaps the light has also gone on in the mind of the experimenter! Pure water and a solution of sucrose in water do not conduct electricity and therefore do not complete the circuit. Sugar and other nonconducting solutes are called **nonelectrolytes.** An aqueous solution of sodium chloride is an electrical conductor, and salt is classified as an **electrolyte.** But what accounts for this difference in properties? That is the next topic we consider.

■ *Ionic Compounds and Their Solutions*

The flow of electric current involves the transport of electric charge. Therefore, the fact that solutions of sodium chloride conduct electricity suggests they contain electrically charged species. These species are called **ions,** from the Greek for "wanderer." When solid sodium chloride dissolves in water, it breaks up into positively charged **cations,** Na^+, and negatively charged **anions,** Cl^-. As these ions wander about in solution, they transport electrical current.

It may be a little surprising to learn that Na^+ and Cl^- ions exist in the salt shaker as well as the soup. Solid sodium chloride is a three-dimensional cubic arrangement of sodium and chloride ions occupying alternating positions (Figure 5.7). These oppositely charged ions attract each other with **ionic bonds** that hold the crystal together. In an **ionic compound** such as NaCl there are no true covalently bonded molecules, only positive cations and negative anions.

Thus far we have described the structure and some of the properties of ionic compounds, but we have not explained why certain atoms lose or gain electrons to form ions. Not surprisingly, the answer involves electronic structure. A sodium atom has only one electron in its outer energy level. A chlorine atom, on the other hand, has seven. For both elements, stability is associated with eight outer electrons. Therefore, an Na atom has a strong tendency to lose its single outer electron and become an Na^+ ion. This is an example of **oxidation,** a process in which a chemical species loses one or more electrons. Similarly, it is energetically favorable for a Cl atom to acquire an extra electron, complete its outer octet, and become a Cl^- ion. Such a gain of one or more electrons by an atom, molecule or ion is called **reduction.**

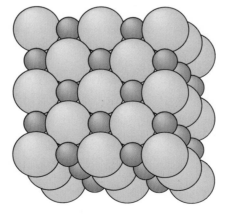

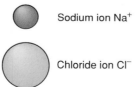

Sodium ion Na^+

Chloride ion Cl^-

Figure 5.7
The arrangement of ions in a crystal of sodium chloride.

When elementary sodium metal and elementary chlorine gas come together, this electron transfer occurs spontaneously with the release of a considerable amount of energy. As a result, a compound, NaCl, is formed from the two elements.

$$2\,Na(s) + Cl_2(g) \rightarrow 2\,NaCl(s) \tag{5.2}$$

Although we write NaCl in the above equation, the assumption is that the formula represents Na^+ and Cl^- ions. The fact that molten salt also conducts electricity is evidence that the ions exist in the liquid state, and presumably in the solid as well. This further indicates that the electrons are actually transferred, not shared as they would be in a covalent compound.

Electron transfer is likely to occur between elements whose electronegativities are significantly different. Note from Table 5.5 that sodium, lithium, magnesium, and other elements on the far left of the periodic table have low electronegativities. These highly reactive metals have a strong tendency to give up electrons and form positive ions. Chlorine, fluorine, oxygen, and other reactive nonmetals from the right side of the periodic table have high electronegativities. This means that they have a strong attraction for electrons and readily form negative ions. Therefore, ionic compounds are likely to be formed when elements at the extreme ends of the periodic table react. Potassium iodide (KI) and calcium chloride ($CaCl_2$) are only two of many. Ordinary table salt is such a good exemplar of ionic compounds that sometimes other similar compounds are also called "salts." Typically, they are water-soluble crystalline solids.

5.7   *Your Turn*

Predict the ions that would be most likely formed by the following atoms.

a. Mg   **c.** Li   **e.** Al
b. Br   **d.** O

Hint: In doing this exercise, it is necessary to know the number of outer electrons in the atoms and to determine the number that must be gained or lost to attain an octet. Figure 2.2 should be of help. For example, an Mg atom, in the second column of the periodic table, has two outer electrons which it readily loses to form an Mg^{2+} ion, which has eight electrons in its outer shell.

5.8   *Your Turn*

Predict the formulas of the ionic compounds that would be formed by the reaction of the following elements.

a. Mg and Cl   **c.** Ca and O
b. K and F   **d.** Sr and Br

Hint: The ratio of ions in an ionic compound is such that the negative charges equal the positive charges. Thus, to solve the problem one first needs to know the charges on the ions that will be formed in the reaction. We have already concluded that Mg forms Mg^{2+} ions and Cl forms Cl^- ions. In order to have an electrically neutral compound, there must be two Cl^- ions for each Mg^{2+} ion. This means that the formula for magnesium chloride is $MgCl_2$.

Many ionic compounds are quite soluble in water. When a solid sample is placed in water, the polar H_2O molecules are attracted to the individual ions. The oxygen atom of the H_2O molecule bears a net negative charge and is attracted to positive cations. Because of their net positive charges, the hydrogen atoms in H_2O are attracted to the anions of the solute. The ions are thus surrounded by water molecules, which screen the attraction of the oppositely charged ions for each other. Anion-cation

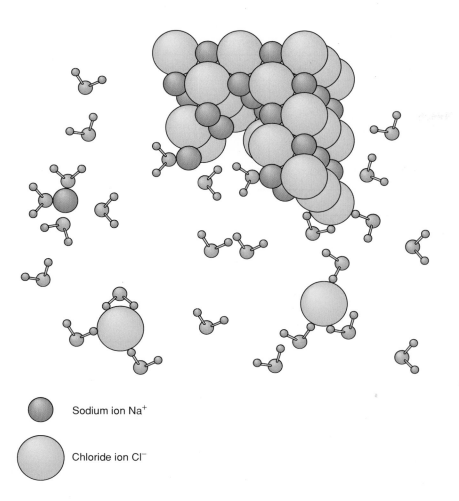

Figure 5.8
The dissolving of sodium chloride in water.

Sodium ion Na⁺

Chloride ion Cl⁻

attraction is decreased, while the attraction between the ions and the H_2O molecules is substantial. The net result is that the ions are pulled out of the solid and into solution. In dissolving, the ionic compound dissociates into its component cations and anions. Equation 5.3 and Figure 5.8 represent this process for sodium chloride and water.

$$NaCl(s) + H_2O(l) \rightarrow Na^+(aq) + Cl^-(aq) \qquad (5.3)$$

The (*aq*) in the equation indicates that the ions are present in an *aqueous* solution.

5.9	***Consider This***
■	Pure water does not conduct electricity. Nevertheless, people are sometimes electrocuted when they step into a puddle into which a live power line has fallen or when an electrical appliance falls into the bathtub. Explain this apparent inconsistency.

Some ionic compounds include **polyatomic ions** that are themselves made up of more than one atom or element. A case in point is sodium sulfate, Na_2SO_4. This compound consists of Na^+ and SO_4^{2-} ions. In the sulfate anion, the four oxygen atoms are covalently bonded to a central sulfur atom. Counting the electrons in the Lewis structure (Figure 5.9) reveals that there are two more electrons than protons. Hence, the ion has a charge of –2. Note that when sodium sulfate dissociates and dissolves in water, the sodium and sulfate ions separate, but the SO_4^{2-} ions remain intact.

$$Na_2SO_4(s) + H_2O(l) \rightarrow 2\ Na^+(aq) + SO_4^{2-}(aq) \qquad (5.4)$$

Table 5.6 is a list of some of the more common polyatomic ions. Most of them are anions, but polyatomic cations are also possible.

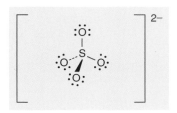

Figure 5.9
Lewis structure of the sulfate (SO_4^{2-}) ion.

Table 5.6	Some Polyatomic Ions		
Name	**Formula**	**Name**	**Formula**
acetate	$C_2H_3O_2^-$	nitrite	NO_2^-
bicarbonate	HCO_3^-	phosphate	PO_4^{3-}
carbonate	CO_3^{2-}	sulfate	SO_4^{2-}
hydroxide	OH^-	sulfite	SO_3^{2-}
hypochlorite	OCl^-	ammonium	NH_4^+
nitrate	NO_3^-		

5.10 Your Turn

Name the compounds having the following formulas:

a. $CaCO_3$ **c.** $NaHCO_3$ **e.** KNO_3
b. NH_4Cl **d.** $(NH_4)_3PO_4$

Ans. **a.** calcium carbonate [One simply uses the names of the ions involved, usually omitting prefixes that would indicate the number of ions.]

5.11 Your Turn

Write the formulas for the following compounds.

a. sodium carbonate (washing soda)
b. ammonium nitrate (an important fertilizer)
c. magnesium sulfate (Epsom salts)
d. calcium phosphate (used as an abrasive in toothpaste)

Ans. **a.** Na_2CO_3. As in 5.8 Your Turn, the formula should be written so the negative and positive charges are equal. Here the ions are Na^+ and CO_3^{2-}, so two Na^+ ions are required for each CO_3^{2-} ion.

■ Covalent Compounds and Their Solutions

Unlike sodium chloride, sucrose (ordinary table sugar) is a **covalent** or **molecular compound.** Like water, carbon dioxide, chlorofluorocarbons, and many of the other compounds you have been reading about, it consists of molecules made up of covalently bonded atoms. The formula for sucrose is $C_{12}H_{22}O_{11}$ and it exists in individual molecules consisting of 45 atoms, connected as shown in Figure 5.10. When sugar dissolves in water, these molecules become uniformly and homogeneously disbursed among the H_2O molecules. As in all true solutions, the mixing is at the most fundamental level of the solute and solvent—the molecular or ionic level. The $C_{12}H_{22}O_{11}$ molecules remain intact; they do not dissociate. Because they, like the H_2O molecules, are not electrically charged, water solutions of sucrose do not conduct electricity. However, the sugar molecules do interact with the water molecules. In fact, solubility is always promoted when there is a net attraction between the solvent molecules and the solute molecules (or ions). This suggests a good solubility rule: "Like likes like." Compounds with similar chemical composition and molecular structure tend to form solutions with each other. The intermolecular attractive forces between similar molecules will be high, thus promoting solubility.

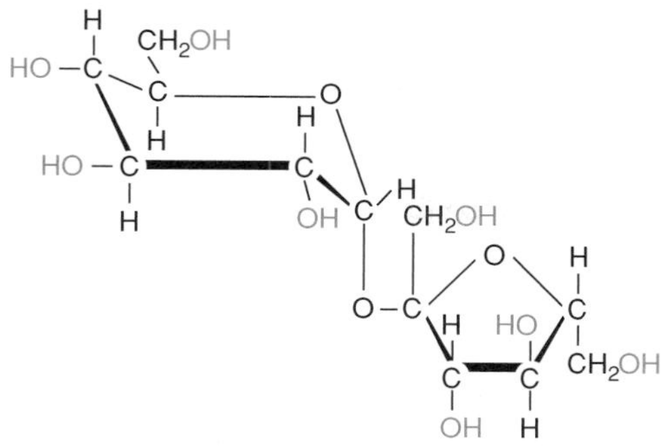

Figure 5.10
Molecular structure of sucrose.

Consider, for example, the following three covalent compounds, all of which have high solubilities in water: sucrose, ethylene glycol (the main ingredient in antifreeze), and ethyl alcohol (the "grain alcohol" found in alcoholic beverages). In order to determine what they have in common with each other and with water, we need to examine their molecular structures. We start with the simplest, ethyl alcohol, C_2H_5OH. Like all alcohols, it contains an –OH group. The oxygen atom is covalently bonded to the hydrogen atom and to a carbon atom. Figure 5.11 illustrates how hydrogen bonds can form between H_2O molecules and the –OH group of a C_2H_5OH molecule. This means that these two polar molecules have a good deal of affinity for each other, a conclusion that is consistent with the fact that ethyl alcohol and water form solutions in all proportions.

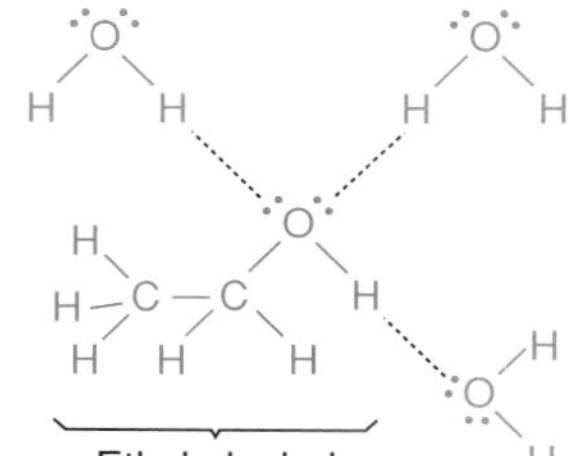

Ethylene glycol is another alcohol, with the formula $HOCH_2CH_2OH$. Here there are two –OH groups available for hydrogen bonding with H_2O. Finally, reexamining Figure 5.10 discloses that the sucrose molecule contains eight –OH groups and three additional oxygen atoms that might also be able to participate in hydrogen bonding. This accounts for the high solubility of sugar in water.

On the other hand, covalent compounds that differ in composition and molecular structure do not attract each other strongly. It has often been observed that "oil and water don't mix." They don't mix because they are very different. Water is a highly polar compound; oil consists of nonpolar hydrocarbons. Placed in contact, they remain aloof and segregated. Energetically speaking, there is nothing to be gained by mixing up the water and the oil. In fact, it would require a good deal of energy to force the two to dissolve. Since nature inevitably takes the easy way out and expends no more energy than is absolutely necessary, the two substances stay separate and do not form a solution. But oily nonpolar compounds generally dissolve readily in hydrocarbons or chlorinated hydrocarbons. Such compounds have often been used as solvents for dry cleaning.

Ethyl alcohol

Figure 5.11
Hydrogen bonding of ethyl alcohol with water.

■ *Water and Energy*

Water's uncommonly high capacity to absorb heat is a significant factor in both nature and industry. On a planetary scale, this property helps determine worldwide climates. By absorbing vast quantities of heat, the oceans and the droplets of water in clouds help mediate global warming. The specific details of these processes are among the uncertainties that complicate efforts to model the greenhouse effect. We do know that heat is absorbed when water evaporates from seas, rivers, and lakes, and is released when it condenses as rain or snow. These phase changes create the great thermal engine that helps drive weather patterns in the short term and regulates climates over longer periods of time. But even in the absence of changes in physical state, liquid water absorbs more energy than the ground because it has a higher capacity to store heat than do rocks and dirt. As the weather turns colder, the ground has less stored heat to lose than the water, and therefore cools more quickly. The water, because of its higher heat capacity, retains more heat and is able to provide more warmth for a longer time to the areas bordering it. Such properties should be familiar to anyone who has ever jumped into a warm lake on a cool fall day.

A quantitative measure of a substance's thermal response to applied heat is called its **specific heat.** Specific heat is defined as the quantity of heat energy that must be absorbed in order to increase the temperature of one gram of the substance by one degree Celsius. The specific heat of water is 1.00 cal/g °C: one calorie of energy will raise the temperature of one gram of $H_2O(l)$ by 1°C. In fact, the calorie was originally defined in this manner. Conversely, when the temperature of one gram of liquid water falls one degree Celsius, one calorie of heat is given off. Because of the relationship between calories and joules, the specific heat of water can also be expressed as 4.184 J/g °C.

This may not sound like a very large value, but liquid water has one of the highest specific heats of any known liquid. Because of this, it is an exceptional coolant, used to carry away excess heat in chemical industry, power plants, and the human body. Most other compounds have significantly lower specific heats. For example, with liquid benzene, C_6H_6, a component of gasoline, only 0.406 calorie need be absorbed to register a 1°C increase in temperature.

The reason for this difference is again associated with structure. The unusually high heat capacity of water is a consequence of strong hydrogen bonding and the resultant degree of order that exists in the liquid. Temperature is a measure of molecular motion—the higher the temperature, the greater the average motion. When molecules are strongly attracted to each other, a good deal of energy is required to overcome these intermolecular forces and enable the molecules to move more freely. Such is the case with water. On the other hand, these forces are much weaker in non-hydrogen bonded liquids, and they are much easier to overcome. Consequently, specific heats are lower.

Specific heat values often enter into calculations of heat flow when heat is absorbed or released. We can demonstrate the process by calculating the quantity of heat that must be absorbed to heat the water used in a 5-minute shower. According to Table 5.1, a typical shower uses 19 L/min. This means that in 5 minutes, 19 L/min × 5 min or 95 L of water will be used. Let's assume that this water enters the water heater at 20°C (68°F) and it is heated to 50°C (122°F). The question is: How much heat is required to bring about this change in temperature?

Logic suggests that the quantity of heat that must be absorbed will depend on the mass of the water, the temperature change, and the specific heat. Increasing the mass of water, the temperature change, or the specific heat will all mean a greater quantity of heat is involved. This implies the following equation, where q designates the quantity of heat absorbed, m represents the mass of the sample, Δt is the temperature change, and *sp ht* designates specific heat.

$$q = \text{heat absorbed} = m \times \Delta t \times sp\ ht \qquad (5.5)$$

Note that this is a perfectly general equation that applies to any substance undergoing a temperature change (in the absence of a phase change such as melting or evaporation). One simply substitutes the appropriate values for the three terms. In our example, $\Delta t = 50°C - 20°C = 30°C$ and *sp ht* = 1.00 cal/g °C. We need to do an additional calculation to find a value for m. Here the key is the fact that 95 L of water are involved. We use this information, plus the density of water, 1.00 g/mL, to determine the mass of water. First we convert the volume from liters to milliliters.

$$\text{volume} = 95\ L \times 1000\ mL/L = 95{,}000\ mL = 9.5 \times 10^4\ mL$$

Then we find the corresponding mass of water.

$$\text{mass} = 9.5 \times 10^4\ mL \times 1.00\ g/mL = 9.5 \times 10^4\ g$$

Using this value for m and the previously identified values for Δt and *sp ht* yields the following expression for the heat absorbed:

$$q = 9.5 \times 10^4\ g \times 30°C \times 1.00\ cal/g\ °C = 2.9 \times 10^6\ cal$$

To put things into perspective, this is identical to 2.9×10^3 kcal or 2900 dietary Calories. This means that heating your daily shower water uses about as much energy as you consume in your daily food intake.

5.12 *Your Turn*

A typical household uses about 15% of its total energy consumption to heat water. Therefore, taking steps to reduce the amount of water heated also lowers the energy used, conserves fuel, and saves money. In 5.2 Your Turn, you calculated that showering daily using a conventional shower head consumes 9200 gallons of water in a year. If, instead, the showers had been taken using a simple low-flow shower head only 5300 gallons would have been used. Assume that in both cases the shower water was heated from 20°C to 50°C and answer the following questions. (Actually, the hot water used in low-flow shower heads is typically somewhat warmer than that used in conventional showers, so the savings are not as great as these calculations will suggest.)

a. How many kilocalories of energy would be saved during one year by a family of four if only low-flow shower heads were used? (1 kcal = 1000 cal, 1 gal = 3.77 L).

Ans. 1.76×10^6 kcal. The general strategy is that employed in the above example, using equation 5.5.

b. If the water is heated electrically and electricity costs 7 cents per kilowatt hour (kWh), how much money would be saved by this family in one year by using low-flow shower heads? (1 kWh = 860 kcal).

Ans. $144

Among other things, the high specific heat of water also helps to keep your overall body temperature regulated to near 98.6°F (37.0°C) regardless of whether you are sleeping, studying, or exercising. A feverish temperature of 103°F is a serious matter because of the enormous thermal energy the body generates in order to raise its temperature by more than 4°F (see 5.13 The Sceptical Chymist).

5.13 *The Sceptical Chymist*

The Sceptical Chymist should check the claim that the body generates "enormous thermal energy" in order to raise its temperature by more than 4°F. How much is considered "enormous"? See for yourself by calculating the quantity of energy that would be required to raise the temperature of a 150-pound person from 98.6°F (37.0°C) to 103.0°F (39.4°C).

Hint: As usual, there are a variety of assumptions that one might make in addressing this problem. We will suggest a few. First of all, we know that the body is about 60% water, so most of the energy will be used to raise the temperature of 150 lb × 0.60 or 90 lb of water. The remaining 40% of the mass (bone, hair, protein, fats, etc.) will have to be heated as well, but far less energy will be required because the specific heat of this stuff is less than that of water. A value of 0.3 cal/g °C is a reasonable guess.

With these assumptions, our final answer, 118,000 cal or 118 kcal, seems quite small—about the energy equivalent of half a slice of bread. But come to think about it, we are forgetting something. A human body, even one that is not feverish, radiates a lot of energy to its surroundings. What effect would this have on the actual amount of energy required to maintain a temperature of 103°F? And what (if anything) does this have to do with the old proverb, "Feed a cold and starve a fever"—or is it the other way around?

The calculations we have been doing have assumed that all of the water remains in the liquid state; it does not freeze or boil. For the sake of completeness (and for use in Your Turn 5.16) let us take a moment to consider the energy changes that occur when water undergoes transformations from one physical phase (solid, liquid, or gas) to another. Perhaps it will be easiest to start with ice. It is self-evident that heat must be absorbed to convert solid water, $H_2O(s)$, to liquid water, $H_2O(l)$. Hydrogen bonds must be broken to free H_2O molecules from the crystal lattice. This is an endothermic process.

$$1440 \text{ cal} + H_2O(s) \, [0°C] \rightarrow H_2O(l) \, [0°C] \qquad (5.6)$$

This change in physical state requires 1440 cal per mole of ice, which corresponds to 80 cal per gram. Therefore, ice is said to have a **heat of fusion** of 1440 cal/mole or 80 cal/g—the heat that must be absorbed to bring about melting or fusion. An equivalent amount of heat is released when water freezes.

During the melting process, the temperature remains constant. As long as solid and liquid H_2O are both present, a thermometer placed in the mixture will register $0°C$. Thus, the temperature of six ice cubes and a little water should be the same as the temperature of two ice cubes and a lot of water. Of course in both cases we are assuming that the solid and liquid water have come to a constant temperature. In an insulated container, such as a vacuum bottle, the two phases are said to be in **equilibrium** when the rate of melting equals the rate of freezing.

Similar considerations govern transformations involving the liquid and gaseous states. At the boiling point of water, $100°C$, the temperature stops rising while heat continues to be absorbed. The energy goes to convert liquid water to gas.

$$9720 \text{ cal} + H_2O(l) \, [100°C] \rightarrow H_2O(g) \, [100°C] \qquad (5.7)$$

The quantity of heat absorbed, 9720 cal/mole or 540 cal/g is the **heat of vaporization,** the heat that must be absorbed to change liquid into vapor. Note that this value is almost seven times the heat of fusion of ice. We have frequently referred to the fact that the forces between H_2O molecules are quite strong in the liquid state. These forces must be overcome in order to physically separate the molecules from each other and convert the liquid into a gas, where intermolecular attractions are very weak. This transformation requires a good deal of energy, which means a high heat of vaporization. In contrast, the difference between the strength of intermolecular forces in the liquid and solid states is not that great. Both ice and liquid water show significant hydrogen bonding. This explains why the heat of fusion is so much less than the heat of vaporization.

5.14   *Consider This*

What difference, if any, do you expect to observe in the temperature of a pot of vigorously boiling water and a pot of slowly boiling water? Make a prediction, offer an explanation, and, if possible, perform an experiment.

■ *Water Sources*

After our review of the properties of water, we now ask a disarmingly simple question: Where does (and where did) this strange stuff come from? Apparently, the Earth was not always as wet as it is now. Scientists believe that much of the water currently on the Earth was originally spewed as vapor from thousands of volcanoes that pocked the young planet. The vapor then condensed and fell as rain. Over the ages, the process was repeated millions of times, as water molecules cycled from sea to sky and back again. Then, more than three billion years ago, the cycle expanded to include primitive plants and, later, animals. This hydrologic cycle continues today, sending surface water, evaporated by the Sun, into the air. At higher altitudes, it cools and concentrates sufficiently to condense into liquid or solid. It falls back to the Earth as rain, snow, sleet, or hail, and the cycle is ready to be repeated.

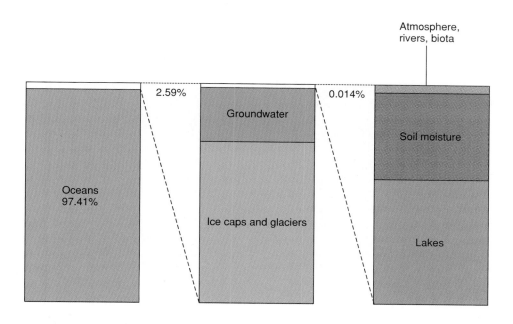

Figure 5.12
Distribution of water on the Earth.
(From J. W. Maurits la Rivière,
"Threats to World Water," in *Scientific American,* September 1989.
Copyright © 1989 Scientific American, Inc. All Rights Reserved.)

During an average year, enough precipitation falls on the continents of the Earth to cover all the land area to an average depth of more than 2.5 ft. Fortunately, it doesn't all come at once, nor is it uniformly distributed. The wettest place in the world is Mount Waialeale on Kauai, Hawaii, where the average annual rainfall is 460 inches. On the other hand, over a 59 year period, Arica, Chile averaged only 0.03 inches of rain per year. In fact, it didn't rain at all for 14 of those years! An average of 1.5×10^{13} liters of water precipitates daily on the continental United States—enough to fill about 400 million swimming pools.

This sounds like a good deal of water, but on a global scale, the amount of *fresh* water is really quite small. The great majority of the world's water, 97.4% of the total, is the undrinkable salt solution that makes up the oceans (Figure 5.12). The remaining 2.6% is all the fresh water we have. Moreover, about 86% of this supply is not directly available because it is frozen in glaciers and the polar ice caps. Underground deposits are the next most plentiful source. Less than 0.01% of the world's total water is conveniently concentrated in lakes, rivers, and streams as fresh water.

This relatively small amount of easily accessible fresh water is repeatedly recycled. We do not really "consume" the compound or destroy it, because matter is conserved during chemical and physical changes. Even when water participates in chemical reactions, its hydrogen and oxygen atoms persist in the new combinations, perhaps some day to recombine as H_2O. In most cases, however, when water is used, it is simply dirtied and returned to reservoirs, lakes, rivers, or **aquifers,** which are large, natural underground reservoirs. Ultimately, it finds its way into your glass or shower.

■ *Potability and Purification*

Common sense and prudence suggest that we must be alert to the supply, sources, and purity of water. Although this indispensable compound is found in all parts of the Earth—from rain forests to the driest desert—there are worldwide concerns about the availability and quality of water. In certain parts of the planet, too much rainfall in a short period of time periodically causes massive flooding; in others, severe and prolonged drought has brought great suffering and death. Even in those areas where water is plentiful, there are still concerns for its potability, that is, its suitability for human consumption. A large supply of water is no guarantee of its drinkability. Coleridge's shipwrecked Ancient Mariner knew this all too well, surrounded as he was by "Water, water, every where, nor any drop to drink." But how pure must water be to be potable?

Figure 5.13
A water purification plant.

As you will soon see (in Chapter 6), atmospheric gases, including oxides of sulfur and nitrogen, can react with water vapor to form acid precipitation. But fresh rainwater, caught in a clean container, is about as pure as natural water gets. Its "flat" taste is evidence of that purity. Once rain reaches the ground, the groundwater percolates through layers of sand, gravel, and rock, which filter out the suspended solids. Thus, aquifers contain water that is generally clear and free of disease-causing microbes. However, it is possible for dissolved compounds from natural and artificial sources to enter the aquifers. For example, the slow trip through the various strata of rock often results in the addition of dissolved minerals to water. Compounds containing calcium, magnesium, iron, and sodium are quite common in many rocky areas of the world. Positively charged cations of these metals and their accompanying anions are released on extended contact with water.

In some locations, water also has high concentrations of dissolved gases such as CO_2 and hydrogen sulfide (H_2S). For centuries, some of these natural waters have been prized for their purported restorative and curative powers. Luxurious resorts sprang up along with the water in Bath (England), Baden-Baden (Germany), White Sulphur Springs (West Virginia), and many other places. Those who cannot afford to "take the waters" at expensive and exclusive spas can drink it from bottles whose labels proclaim sources such as "Saratoga" or "Vichy."

Rain enters public water supplies through reservoirs, lakes, or wells fed by the water table. However, there are some substances often found in community water sources that should be eliminated or at least deactivated before it is used for drinking. Of particular concern are bacteria and organic matter, suspended dirt, and some ions. Fortunately, municipal water treatment facilities remove most of these (Figure 5.13). In the United States, public drinking water supplies are regulated by the Safe Water Drinking Act of 1974, which requires the Environmental Protection Agency to establish and enforce standards of purity and safety. Water entering a typical municipal treatment plant first passes through a screen that excludes objects both natural (fish and sticks) and artificial (tires and beverage cans). Aluminum sulfate or alum, $Al_2(SO_4)_3$, and calcium hydroxide or slaked lime, $Ca(OH)_2$, are then added. These compounds react to form a sticky gel of aluminum hydroxide, $Al(OH)_3$, which collects suspended clay and dirt particles on its surface.

$$Al_2(SO_4)_3(aq) + 3\,Ca(OH)_2(aq) \rightarrow 2\,Al(OH)_3(s) + 3\,CaSO_4(aq) \qquad (5.8)$$

The $Al(OH)_3$ precipitates the suspended particles in a settling tank. Any remaining particles are removed as the water is filtered through gravel and then sand.

The water is then chlorinated to kill disease-causing organisms. Chlorine is usually administered in one of three forms: chlorine gas, Cl_2; sodium hypochlorite,

NaOCl (used in laundry bleach); or calcium hypochlorite, $Ca(OCl)_2$ (used to disinfect swimming pools). The antibacterial agent generated by all three procedures is hypochlorous acid, HClO. The concentration is adjusted so that between 0.075 and 0.60 mg HClO per liter remain in solution to protect the water against contamination as it passes through the pipes to the user.

From a public health perspective, chlorination is undoubtedly the most important step in water purification. Before the practice was adopted, thousands died in epidemics that were spread via polluted water. In a classic study, John Snow was able to trace a cholera epidemic that swept England during the mid-1800s to water contaminated with the excretions of victims of the disease. A more contemporary example occurred in Peru in 1991. This cholera epidemic was traced to bacteria in shellfish growing in estuaries polluted with untreated fecal matter. The bacteria found their way into the water supply where they continued to multiply because of the absence of chlorination.

Many European cities use ozone, the chemical focus of Chapter 2, to disinfect their water supplies. A similar degree of antibacterial action can be achieved with a smaller concentration of ozone, making ozonation more economical than chlorination. Furthermore, ozone is more effective than chlorine against water-borne viruses. A major drawback of ozone is the fact that it decomposes quickly and hence does not protect the water from contamination after it leaves the treatment plant.

Depending on local conditions, one or more additional steps may be carried out after disinfection. Sometimes the water is sprayed into the air to remove objectionable odors and to improve taste. If the water is sufficiently acidic to cause problems such as corrosion of pipes or the leaching of heavy metals from pipes, calcium oxide (lime) is added to partially neutralize the acid. Many municipalities also add about 1 ppm of sodium fluoride, NaF, as a protection against tooth decay.

■ *Hard Water and Soft Soap*

In purifying water for home use, municipal water treatment facilities eliminate or de-activate a number of troublesome or potentially dangerous materials. But many do not attempt to remove the dissolved minerals that make water "hard." The most common contributors to water hardness are the carbonates, sulfates, and chlorides of calcium, magnesium, and iron. Although their presence can create considerable nuisance and expense, these compounds are generally not health threats. Calcium ions, Ca^{2+}, cause the most problems. Therefore, water hardness is usually expressed in parts per million of calcium carbonate by mass. This method of reporting does not mean that the water sample actually contains $CaCO_3$ at the indicated concentration. Rather, it specifies the mass of solid $CaCO_3$ that could be formed from the Ca^{2+} in solution, provided suffi-cient CO_3^{2-} ions were also present. Thus, a hardness of 10 ppm indicates that 10 g of $CaCO_3$ could be formed from the ions present in 1,000,000 g of water. This corre-sponds to 10 mg of calcium carbonate per liter.

Water containing Ca^{2+} can form a hard, insoluble scale of $CaCO_3$ in water heaters, tea kettles, pipes, and industrial equipment. The resulting reduction in heat transfer and water flow can cause serious problems. Iron, in the form of Fe^{3+} (ferric) ions, often gives water a metallic taste and stains fixtures with rust.

Probably the most common manifestation of water hardness is the way in which calcium and magnesium ions interfere with the effectiveness of soaps. Magnesium and calcium are members of the same chemical family, column 2A of the periodic table of elements. They share the tendency to react with soap to form an insoluble compound that separates from solution. This insoluble compound is the stuff of bathtub rings and the scum deposited on clothing washed in hard water. Because much of the soap is tied up in the precipitate, more soap is required to form suds and cleanse things in hard water than is needed in soft water.

Your ancestors made laundry soap by heating lard or other animal fats with lye derived from ashes. Fundamentally, this is still the procedure used for manufacturing

Figure 5.14
Soap and its interaction with grease.

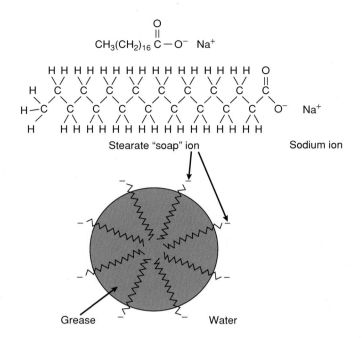

Stearate "soap" ion

Sodium ion

Grease   Water

any true soap; animal or vegetable fats are reacted with sodium hydroxide (NaOH). This caustic compound reacts with the fat to form soap and glycerine. Technically, a soap is the sodium salt of a fatty acid (see Chapter 12). As shown in Figure 5.14, a soap "molecule" contains two parts: a sodium ion (Na^+), and a long hydrocarbon chain with a negatively charged ionic end (the "soap ion"). In aqueous solution, soap releases Na^+ and these negative soap ions. The cleaning ability of soap is associated with the structure of the soap ion. Its long, nonpolar hydrocarbon tail dissolves readily in materials that are predominantly nonpolar, for example grease, chocolate, or gravy. The negatively-charged ionic ends stick out of the surface of a glob of grease because ionic substances are not soluble in nonpolar media. Thus, the grease becomes covered with negative charges. These ionized ends interact favorably with a polar solvent such as water, are solubilized, and get carried away with the rinse water.

The insoluble precipitate arises when soap ions interact with Ca^{2+} or Mg^{2+} ions. Two of the large negatively charged ions react with each of the doubly-charged positive ions, and the result is a solid or scum. One obvious way to avoid the formation of the precipitate is to remove the calcium and magnesium ions, in other words, to "soften" the water. This can be done by adding sodium carbonate (washing soda), Na_2CO_3, along with the soap. The carbonate ions (CO_3^{2-}) react with the Ca^{2+} to form insoluble calcium carbonate that is rinsed away.

$$Ca^{2+}(aq) + CO_3^{2-}(aq) \rightarrow CaCO_3(s) \qquad (5.9)$$

Other water-softening compounds, such as sodium tetraborate or borax ($Na_2B_4O_7$) and trisodium phosphate (Na_3PO_4) work in a similar fashion. Calgon™ water softener contains sodium hexametaphosphate, $Na_6P_6O_{18}$, which ties up calcium and magnesium ions as large, soluble ions.

■ Exchanging Ions

Another way to soften hard water is by removing the interfering ions before they get to the washing machine or shower. This is often accomplished by a process called **ion exchange.** The tank of a water softener is typically filled with a **zeolite**, a claylike mineral made up of aluminum, silicon, and oxygen. These atoms are bonded into a rigid, three-dimensional structure bearing many negative charges. All of these charges

Figure 5.15
Adding salt to an ion exchange water softener.

must be balanced by positive charges. Normally, they are supplied by Na^+ ions associated with the zeolite. However, when a solution containing Ca^{2+}, Mg^{2+}, or Fe^{3+} is passed through the zeolite, these ions replace the Na^+ ions because they are more strongly attracted to the negatively charged zeolite matrix than the Na^+ ions. In other words, "hard water" ions (calcium, magnesium, iron) are exchanged for "soft water" sodium ions. If we represent the zeolite as Z, we can write a representative equation for the process.

$$Na_2Z(s) \quad + \quad Ca^{2+}(aq) \quad \rightarrow \quad CaZ(s) \quad + \quad 2\ Na^+(aq) \qquad (5.10)$$
$$\text{Zeolite (Na form)} \qquad\qquad\qquad \text{Zeolite (Ca form)}$$

The Na^+ ions flow through the tank and into the pipes of the house. Because sodium ions do not interfere with the function of soap or result in the build-up of scale, the problem of hardness is eliminated. To be more exact, the problem of hardness has been left behind on the zeolite ion exchanger. When the exchanger becomes saturated with Mg^{2+}, Ca^{2+}, and other undesirable ions, it is back-flushed with a concentrated solution of sodium chloride (Figure 5.15). The high NaCl concentration makes it possible for the Na^+ ions to displace the Mg^{2+} and Ca^{2+} from the zeolite, reversing equation 5.10. The released ions are flushed down the drain as $MgCl_2$ and $CaCl_2$, and the ion exchanger is left in its fully charged sodium form, ready to soften more hard water.

5.15   *Consider This*

Individuals with high blood pressure (hypertension) or certain heart problems must be concerned about their sodium intake because these conditions are aggravated by high concentrations of sodium ions in the blood. Normally, these people limit their sodium intake by dietary control. List the ways an individual with hypertension, living in a house with a zeolite water softener, could prevent an excessive intake of sodium.

Figure 5.16
Distillation apparatus.

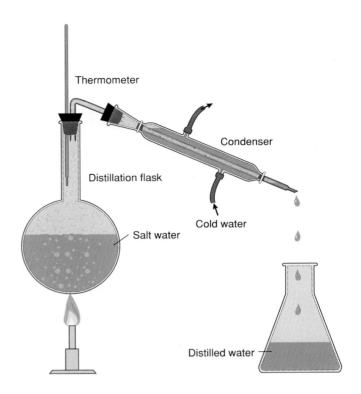

Thermometer

Condenser

Distillation flask

Salt water

Cold water

Distilled water

▣ *Distillation—Another Way to Purify Water*

Distillation is an old and rather common way of purifying water for laboratory and other uses. Distilled water is used in steam irons, car batteries, and other devices whose operation can be impaired by dissolved ions. Distillation is remarkably simple—a liquid is evaporated and then condensed, just as in the natural hydrologic cycle. An apparatus such as that shown in Figure 5.16 is used. Impure water is put into a flask, pot, or other container and heated to its boiling point, 100°C. As the water vaporizes, it leaves behind most of its dissolved impurities. The water vapor passes through a condenser where it cools and reverts back into a liquid, now free of contaminants.

5.16   *Your Turn*

▪

Assume distilled water sells in grocery stores for $0.89 a gallon. Using information given earlier in the chapter about the specific heat and heat of vaporization of water, calculate the cost of the energy required to distill one gallon of water. Assume that the water is originally at 20°C, the source of energy is electricity, the electricity costs $0.07 per kilowatt hour (1 kWh = 860 kcal), and the process is 100% efficient. What percentage of the sales price is represented by the energy cost?

Hint: Here are suggested steps for solving this problem:
1. Find the mass of one gallon of water, remembering that 1 gal = 3.79 L or 3790 mL and that the density of water is 1.00 g/mL.
2. Calculate the number of calories necessary to raise the temperature of this mass of water from 20°C to 100°C, the boiling point. (See previous activities and examples.)
3. The answer to step 2 is just part of the heat that is required, and not even the major part. You must also include the heat necessary to convert the liquid water to water vapor. Do this by using the heat of vaporization: 540 cal must be absorbed to vaporize one gram of water.
4. Find the total quantity of heat required by adding the results of steps 3 and 4. Then convert the sum from calories to kilowatt hours.
5. Finally, compute the cost of this electrical energy. Your answer should be about one-fifth of the purchase price.

In certain water-poor parts of the world, such as desert regions, seawater is distilled on a large scale to furnish potable water. Because of the energy demands of distillation, the costs are extensive, but not when compared with going without drinking water. In order to make the process economically more feasible, heat released during the condensation of the water vapor is recycled to vaporize water in the initial stage of distillation. This technique has been used on ships where "waste" heat from the ship's engines is recycled to vaporize the seawater.

■ *Desalination*

The Ancient Mariner in Coleridge's poem is surrounded by seawater but unable to drink any of it. This is more than just a poetic fantasy, it is a physiological reality. You cannot survive by drinking seawater. Ocean water contains about 3.5% salt compared to only about 0.9% salt in body cells. If a cell were placed in seawater, H_2O would flow from the interior of the cell, through the membrane, and into the seawater. This process is called **osmosis.** It is the natural tendency for a solvent (here water) to move through a membrane from a region of higher solvent concentration to a region of lower solvent concentration. This tendency to equalize concentrations is involved in many biological processes. In this particular instance, the net effect would be that cells would lose water, rather than gain it.

Distillation and ion exchange are two processes used in **desalination,** a broad term describing any process that removes ions from salty water, such as sea- or brackish waters. But an even more widely used procedure for desalination is **reverse osmosis.** As you can gather from the preceding paragraph, if pure water and salt water were placed on either side of a membrane permeable to H_2O molecules but impermeable to Na^+ and Cl^- ions, pure water would flow into the salt water. However, this naturally spontaneous process can be reversed. If sufficient pressure is applied to the saltwater side, water molecules can be forced through the membrane and the ions will be left behind. Figure 5.17 is a schematic representation of this process.

The world's largest desalination plant was very much in the news during the brief Gulf War of 1991. The plant, which is located at Jubail, Saudi Arabia, provides 50% of that country's drinking water by using reverse osmosis to desalinate seawater from the Persian Gulf. It was threatened by a large oil slick released from pipelines passing through Kuwait. Worldwide, desalination plants produce 2 billion gallons of potable water daily. Although most of these installations are in the Middle East, such plants are increasing in the United States. Florida has 109 reverse osmosis desalination facilities, including the one that furnishes the city of Cape Coral with 15 million gallons of fresh water every day from brackish underground supplies. Reverse osmosis desalination

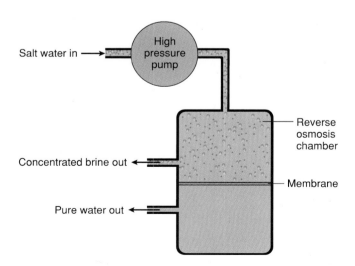

Figure 5.17
Reverse osmosis.

is used to produce pure water for computer chip manufacturers in the "Silicon Valley" and at the Diablo nuclear power station in California. It must be pointed out, however, that reverse osmosis desalination is too expensive for most developing nations.

■ *Water Rights—Who Owns the Water?*

Water is a precious commodity. Because it passes through the hydrologic cycle and falls all over this globe, we seldom think about someone "owning" water or having a "right" to it. Some people in rural or suburban areas do have their own wells or other private sources of clean water. Most of us pay a town or city to furnish water to our houses or apartments. Probably the only time we think about the water supply is when the bills arrive. But earlier in this country's history, bloody battles were fought over the control of streams for irrigation, for livestock, and for personal use. Even today, bitter disputes over water supplies occur in some regions of the world.

Water rights can be intricate when the stuff is relatively abundant and available. They become much more complex and important when the supply of water becomes reduced or uncertain, as in periods of drought. In California, for example, 75% of the rain and snow falls in the northern part of the state, yet three-fourths of the people live in the southern half. In the winter of 1991, as a result of a severe five year drought, the citizens of California were faced with a dilemma: how should the rapidly diminishing water supply be allocated? In particular, who has the strongest claim to the water, farmers or city dwellers? Which group could better cope with a reduction in water allocation and its side effects? Quite obviously, both human and financial considerations are involved. California's thriving irrigation-assisted agricultural industry uses 85% of the state's water and generates over 2.5 billion dollars in cash crops. Major reductions in agricultural water use could have a severe impact on the entire state. On the other hand, Los Angeles is the second largest city in the country, and there are many other highly populous urban areas in southern California.

Under one action in the San Francisco area, the Marin Municipal Water District ordered stringent cutbacks from pre-drought usage levels. The required reductions were 45% for businesses, 50% for institutional accounts, and 85% for irrigation. Residents were limited to 50 gallons per day. In February 1991, irrigation authorities cut off all water to farms, including those in the Central Valley which, through irrigation, produces a major portion of this nation's fruits and vegetables.

The southeastern and southwestern regions of the United States are now beginning to realize that their continued population growth and economic strength are at risk because of potential water shortages. In these areas, available water supplies are almost at their limits.

5.17   *Consider This*

■

For almost 70 years, water from the lower portion of the Colorado River has been diverted through aqueducts to supply farms in the Imperial Valley of California. In addition, the Colorado furnishes 75% of the water needs of the cities of Los Angeles and San Diego. Recently, the Central Arizona Project began diverting water from the Colorado River to supply Phoenix and Tucson, and the heavily-irrigated agricultural area developing between them. This has the possible impact of reducing the amount of water sent to California. Suppose that a decision has been made to reduce the volume of water supplied to California from the Colorado River. You are a member of a municipal water commission in California, and you are charged with responsibility for deciding how to apportion the reduced water supplies between agriculture and municipalities. Identify six major issues associated with these choices that will inform your decision.

In many parts of the United States, rivers have been dammed. The construction of a dam creates a lake behind it such as Lake Mead behind Hoover Dam in Nevada and the string of lakes behind the dams along the Tennessee River. Dams are built as flood control devices, to ensure an adequate water supply, and, in some cases, to generate electricity through hydroelectric power. The lakes have become popular recreational fishing and boating sites with private homes on their shores. Such multiple uses have created a whole set of problems. For example, who has first claim on the water—the power plants, industry, downstream cities and towns, agriculture, or recreational and real estate enterprises? And how are such priorities established? The generation of electricity and the needs of downstream consumers and communities may require the release of water from the lake, but such action would reduce recreational activities and might leave lakeshore property high and dry. Who has the right and under what circumstances, to open the gates and lower the water level? Responsibilities for such decisions are often difficult to assign because they can involve a complicated entanglement of federal, state, and municipal agencies, jurisdictions, and statutes.

■ *Conclusion*

At the core of disputes over water rights are the unique properties of this compound and its irreplaceable role in industry, agriculture, and life itself. It is a superb solvent and an effective heat-transfer agent, and most chemical reactions occur in an aqueous medium. For most purposes, there is simply no substitute for water. But the world's supply of water, though vast, is limited. We have a responsibility to use it prudently and to keep it clean. Fortunately, the appropriate applications of the chemical sciences can do much to retain the magic and the wonder water brings to our planet.

■ *References and Resources*

Brown, L. R. (project director) and Starke, L. (ed.). *State of the World 1986*. New York: Norton, 1986.

Ember, L. R. "Clean Water Act is Sailing a Choppy Course to Renewal." *Chemical & Engineering News,* Feb. 17, 1992: 18–24.

"Ground Water." (Information Pamphlet) Washington: American Chemical Society, 1989.

La Riviere, J. W. M. "Threats to the World's Water." *Scientific American,* Sept. 1989: 80–94.

Thayer, A. M. "Water Treatment Chemicals: Tighter Rules Drive Demand." *Chemical & Engineering News,* March 26, 1990: 17–34.

■ *Experiments and Investigations*

10. Water Hardness

11. Building and Using a Conductivity Tester

12. Solubilities: An Investigation

21. Measurement of Chloride in River Water

■ *Exercises*

1. Express in gallons the 300 L of water required for personal use each day. If the water supply for a family of four were contaminated, how many 55 gallon drums would be needed to hold their H_2O supply for a week? a month? a year?

2. One idea that is often suggested as a way of saving water is to place a brick in a toilet tank. Calculate the volume of such a brick (2 in × 3.75 in × 8 in or 5.5 cm × 9.5 cm × 20 cm) and determine the percentage of water saved this way.

3. The quantity of water wasted in little ways is often ignored. Calculate the volume of water (in liters) that leaks from a dripping faucet in one day and in one year. Assume a rate of one drop per second and a volume of 0.05 mL per drop.

4. Consider the data given in Table 5.4 for Canada, Mexico, and the United States. Account for the different usage patterns (agricultural, industrial, and municipal) in these countries.

5. Methane, CH_4, and water, H_2O, are both compounds of hydrogen plus a non-metallic element. Yet, CH_4 is a gas at room temperature and pressure and H_2O is a liquid. Offer a molecular explanation for this difference in properties.

*6. Calculate the number of moles of water in one liter at 25°C (assume a density of 0.997 g/mL). Determine the number of moles of H_2 gas that could be produced by the complete electrolysis of one liter of water (equation 5.1). If one mole of H_2 occupies about 24.4 liters at 25°C and 1 atm pressure, what volume of H_2 is produced in this electrolysis?

*7. Find the number of H_2O molecules in one cm^3 of water at 0°C (density = 0.99987 g/cm^3). Repeat the calculation for one cm^3 of ice (density = 0.917 g/cm^3).

8. Identify the compounds in the following list that you would expect to dissolve in water. For those that dissolve, indicate whether they do so in an ionic or molecular fashion.
   a. NaOH   b. CCl_4 c. NH_3
   d. C_8H_{18} e. $CaCl_2$

9. Account for the observation that ethyl alcohol, C_2H_5OH, dissolves readily in water, yet dimethyl ether, CH_3OCH_3, which has the same number and kinds of atoms does not.

10. Write equations representing the solution of the following ionic compounds in water.
    a. $KC_2H_3O_2$ b. $Ca(OH)_2$ c. NaOCl
    d. $(NH_4)_3PO_4$ e. $Al_2(SO_4)_3$

11. Calculate the quantity of heat absorbed or released (and specify which) during each of the following transformations.
    a. freezing an ice cube (40 grams)
    b. heating one cup of H_2O (250 g) from 15°C to 100°C
    c. evaporating 5 gallons of H_2O (20,000 g) at 100°C (This is the amount of H_2O that must be evaporated to make one pint of maple syrup.)

12. Discuss the chemical basis for the statement: "Burns from steam are often worse than burns from hot water."

*13. Determine the quantity of heat that must be removed from the air in an average size room (9 ft by 12 ft by 8 ft or 2.7 m × 3.6 m × 2.4 m) to cool it from 90°F (32°C) to 68°F (20°C). Assume that 0.29 calories must be removed to lower the temperature of one liter of air by one degree Celsius. (1 m^3 = 1000 L.)

14. If the 0.014% of Earth's total water supply that is fresh water equals 5.2×10^{15} liters and the world's population is 5 billion people, calculate the following.
    a. the number of liters of fresh water per person
    b. the volume of the Earth's total water supply (in liters)

15. The compounds listed below can all be used to soften water because they react with magnesium or calcium ions to remove them from solution. Write equations for the reactions in which magnesium ions, Mg^{2+} form insoluble precipitates with these compounds.
    a. Na_2CO_3 b. $Na_2B_4O_7$ c. Na_3PO_4

*16. When an ion exchange resin is totally charged with Na^+ ions, it holds 0.35 grams of Na^+ per gram of resin. Calculate the following.
    a. the number of moles of Na^+ ions on 1 gram of resin
    b. the number of moles and number of grams of Ca^{2+} ions that could be held by 1 g of resin
    c. the mass of H_2O with a hardness of 40 ppm of Ca^{2+} that could be deionized by 1 g of this resin

*17. A home water softener contains 100 lb of the ion exchange resin described in Exercise 16. Calculate the minimum mass of solid NaCl (in grams) needed to fully charge it.

18. When confronted with limited water supplies people often develop ingenious ways of recycling and conserving water. Suggest five things that could be done (in addition to those listed in Table 5.2) to recycle or otherwise conserve water.

Neutralizing the Threat of Acid Rain

Acid rain is a litmus test for the nation, a test of how well we choose to confront and solve our environmental problems. To date, the public debate over acid rain has gone as sour as the rain itself. No other controversy in recent times has so divided the nation along regional lines, nor engendered such bitter dispute among powerful competing interests.

Roy Gould, Going Sour: Science & Politics of Acid Rain,
Boston, Birkhauser, 1985

Acid rain has been in the news frequently during the past decade (although the problem is much older than that). Dozens of books have been written on the subject. Dire warnings have been issued about the serious damage already done by acid rain, and there are predictions of more serious effects yet to come. But at the same time, some have stated that there is no problem and that acid rain is a fiction. Still others have counseled caution, claiming that the scientific evidence for damaging effects is inconclusive or that the effects, if any, are mild. The following newspaper article is typical of many that appeared during the 1980s.

April 5, 1988

Acid Rain Ruins State Resources

Boston Globe

Acid rain is devastating the state's lakes and streams, threatening priceless resources such as drinking water reservoirs and historical monuments, and costing Massachusetts residents millions according to a report released yesterday.

State officials say the damage from acid rain and related air pollution has touched all parts of the state: trees are sickening on Mount Greylock; the rainbow trout fishery at Quabbin Reservoir is largely a memory; and Paul Revere's gravestone is a victim of irreparable pollution damage.

Since the federal government has not yet taken action to control acid rain, (Environmental Affairs Secretary James) Hoyte said Massachusetts is preparing to implement a state acid rain law passed in 1985. The Law requires a 30 percent reduction in sulfur dioxide emissions by 1995 . . . But since more than half of the pollution causing acid rain originates outside the state, Hoyte said there is a pressing need for action on the national level. The pollution that causes acid deposition is released when fossil fuels like coal and oil are burned to run factories, automobiles and power plants.

Reprinted courtesy of *The Boston Globe.*

Similar reports of the damaging effects of acid rain have come from many parts of the world. In fact, some of the first evidence of the impact of acid rain was observed in Scandinavia. Many lakes in Norway and Sweden are now effectively "dead," with no fish or other living things. Forests in northern Germany have been severely affected. The beautiful sculptures adorning the exteriors of cathedrals, churches, and other historic buildings throughout Europe are slowly eroding away. In Eastern Europe, the situation may be the most serious of anywhere on the planet, with dying forests, crumbling buildings, and severe air pollution. In all of these cases, acidic rain is one of the major causes of the damage.

Chapter 1 dealt with ambient air quality, which is largely a *localized* problem and most serious in cities. Chapters 2 and 3 focused on two atmospheric phenomena that are *global* in nature: the greenhouse effect and destruction of ozone. Acid rain, on the other hand, tends to be *regional* in character—what goes up comes down in the same general region of the globe. But acid rain does not respect state or national boundaries and this leads to intense political controversies and accusations. Acidic gases that originate in the Midwest are carried to the Northeast by prevailing winds and fall as acidic rain or snow on New York, New England, and eastern Canada. They are also carried to the Southeast, producing acidic precipitation in Tennessee and North Carolina. In Europe, acids generated in Germany, Poland, and the United Kingdom are carried northward into Norway and Sweden.

The problem is compounded by what may well be an environmental paradigm of our times. By trying to use a technological fix for one problem we inadvertently create another. Tall smoke stacks were built to eject pollutants high into the atmosphere where they could spread and thus improve local air quality. But whatever goes up must come down somewhere, sometimes hundreds of miles away. Thus, local pollution problems are converted into regional ones.

■ *Chapter Overview*

The quotation that begins this chapter and the newspaper article from the *Boston Globe* should raise a number of questions in your mind: What is a litmus test? What is an acid? How does rain become acidic? How do acids affect trees, fish, and statues? How does the burning of fossil fuels contribute to acidic rain? These are primarily scientific questions, and in the pages that follow, we will help provide answers. We begin with some phenomenological definitions of important terms such as acid and base, and then consider the ions responsible for acidic and basic properties. Molarity is introduced as a concentration unit used to quantify acid strength, and the concept of pH is defined and applied. Our search for the source of acid rain will lead us to coal-burning power plants and gasoline-burning automobiles. And it will also help us discover why "the public debate over acid rain has gone as sour as the rain itself." Much of the controversy centers around the effects of acid precipitation on materials, visibility, human health, lakes and streams, and trees and forests. We will weigh the evidence for the mechanism and extent of these various effects and the trends they exhibit.

In this chapter you will discover that many of the issues associated with acid rain involve much more than chemistry. The financial impact of acid rain and the money required to curtail it introduce important economic dimensions. But economics soon merges with politics, as states and regions struggle over who should pay for clean air. One nationwide pollution solution is that offered by the 1990 Federal Clean Air Act. A brief discussion of that legislation brings the chapter to its conclusion and a look to the future.

Before we begin, though, one fundamental concept must be clarified. **Acid rain** is the common term used by the news media and politicians. But rain is only part of the story. Snow obviously needs to be included, so **acid precipitation** is a more accurate description. But even this is limited. A more inclusive term is **acid deposition,** used in the newspaper story that began this chapter. For example, acidic deposition includes fog and dew, which are often more acidic than rain or snow. Fog is a suspension of microscopic water droplets, such as found in clouds, and mountain forests are probably most affected by the acids in cloud water. Acidic deposition also takes into account the action of acidic gases and acids attached to solid particles that are deposited on surfaces during dry weather. This so-called "dry deposition" has been shown to be nearly as important as the wet deposition of acids in rain, snow, and fog. In the pages that follow, we consider all of these phenomena.

■ *What Is an Acid?*

The subject of acid rain brings together the atmospheric pollutant gases introduced in Chapter 1 and water, which you have just studied in some depth. Quite obviously, we need to define **acids** in order to understand this linkage. Like many other chemical concepts, acids can be defined either in terms of observable properties or conceptually, using theories of atomic and molecular structure. Chemists usually like to use both, and we shall do so here.

Historically, chemists classified acids by their characteristic sharp or sour taste. Although taste is not a safe way to test for acids, you undoubtedly know the sour taste of vinegar and lemon juice, two common acids. A simple chemical test for acids is called the litmus test. Litmus, a vegetable dye, changes color from blue to pink in the presence of acids. Indeed, the litmus test is so well known that it has become a figure of speech in our culture to describe other kinds of tests, for example, "The litmus test of a true conservative is . . ." It is in this sense that the phrase is used in the quotation that begins this chapter. Another simple chemical test is to add an acid to a carbonate-containing material such as baking soda, marble, or eggshell. A characteristic fizzing occurs due to the release of carbon dioxide gas. (We will return to these reactions later, after our conceptual definition of an acid.)

From the standpoint of chemical structure, acids are substances that release hydrogen ions, H^+, usually in aqueous solution. You will recall from Chapter 5 that an ion is any atom or group of bonded atoms that has a net electric charge. Atoms and molecules normally have equal numbers of protons (positive charges) and electrons (negative charges), so each particle is electrically neutral. But if electrons are gained or lost, the atom or molecule acquires a charge. A hydrogen atom consists of one electron and one proton. If the electron is removed, a proton is all that remains. It is, in effect, a hydrogen atom with a positive charge, hence the designation H^+.

There is a slight complication with the definition of acids as substances that release H^+ ions. Protons are much too reactive to exist by themselves. They always attach to something else. Water is one such substance. As a simple example of this, pure hydrogen chloride is a gas, made up of HCl molecules. But when dissolved in water, each molecule donates a proton to an H_2O molecule, forming an H_3O^+ (**hydronium**) ion and leaving a Cl^- (chloride) ion.

$$HCl(g) + H_2O(l) \rightarrow H_3O^+(aq) + Cl^-(aq) \tag{6.1}$$

The resulting solution is called hydrochloric acid and has the characteristic properties of an acid because of the presence of the H_3O^+ ion. Chemists often write simply H^+ but they understand this to mean H_3O^+. Thus equation 6.1 is typically shortened to

$$HCl(g) \rightarrow H^+(aq) + Cl^-(aq) \tag{6.2}$$

The HCl is said to ionize or dissociate completely in water to produce hydrochloric acid. Essentially no HCl molecules are left undissociated and intact.

6.1 | ***Your Turn***

Write a chemical equation showing the dissociation of the following acids.

a. HNO_3 (nitric acid)
b. HBr (hydrobromic acid)
c. H_2SO_4 (sulfuric acid)

Ans. a. Recheck the polyatomic ions listed in Table 5.6. There you find that the NO_3^- is called the nitrate ion. When HNO_3 dissociates, the hydrogen ion and the nitrate ion separate:

$$HNO_3(aq) \rightarrow H^+(aq) + NO_3^-(aq)$$

▓ *Bases*

No discussion of acids is complete without mentioning **bases (or alkalies),** the chemical opposites of acids. For our purposes, we will define a base as any compound that produces hydroxide ions, OH^-, usually in aqueous solutions. Bases have their own characteristic properties that are attributable to the presence of OH^-. They generally taste bitter and have a slippery feel in water solution. Common examples are solutions of ammonia, NH_3, or sodium hydroxide (lye), NaOH. The cautions on a can of household drain cleaner (mostly lye) give dramatic warning that bases, like acids, can cause severe damage to tissue and textiles.

Sodium hydroxide is an ionic compound, a solid crystalline arrangement of Na^+ and OH^- ions. When it dissolves in water, the sodium and hydroxide ions become separated.

$$NaOH(s) \rightarrow Na^+(aq) + OH^-(aq) \tag{6.3}$$

The source of the OH^- ions produced in an aqueous solution of ammonia, NH_3, is a little less obvious until we note the following reaction.

$$\underset{\text{ammonium ion}}{NH_3(g) + H_2O\ (l) \rightarrow NH_4^+\ (aq) + OH^-(aq)} \tag{6.4}$$

Acids and bases are linked by the reaction of their characteristic ions to form water molecules.

$$\underset{\substack{\text{(from an acid)} \qquad \text{(from a base)}}}{H^+(aq) \quad + \quad OH^-\ (aq) \quad \rightarrow \quad H_2O(l)} \tag{6.5}$$

Note that in equation 6.5 we have only written the ions that participate in the reaction. Even though other ions are necessarily present, we can ignore them because they do not undergo any chemical change. This combination of hydrogen ions and hydroxide ions is called **neutralization.** It is the chemical reaction that occurs when an acid and a base are mixed. As a matter of fact, H^+ and OH^- ions are both present in all water solutions and in pure water as well. Whether a solution is acidic or basic depends on which ion is present in greater concentration.

■ *Molarity*

Because the strength of an acid or a base depends upon the concentration of H^+ or OH^- ions in solution, it is essential to express such concentrations quantitatively. There are several different concentration systems, but all of them relate the quantity of the dissolved substance (the solute) to the quantity of solvent or solution. The system most commonly used for acids, bases, and other chemical solutions is **molarity.** It provides another opportunity to use the mole, a concept introduced in Chapter 3. Molarity, abbreviated M, is defined as the number of moles of solute present in one liter of solution. Written more compactly, molarity is moles solute/L solution, or simply moles/L.

As an example, consider a solution of hydrogen chloride in water, what we call hydrochloric acid. A liter (slightly more than a quart) of a solution containing 1 mole of HCl would be described as having a concentration of one mole per liter. It is said to be "one molar hydrochloric acid," written 1 M HCl. This solution could be made by dissolving one mole of HCl gas (the solute) in enough water (the solvent) to make up 1 liter of solution. Because HCl has a molar mass of 36.5 g (1.0 g for H plus 35.5 g for Cl), 1 L of this 1 M solution contains 36.5 g of HCl . Note that the concentration might be expressed as 1 M or 1.0 M or 1.00 M, depending on how accurately the HCl was weighed and how carefully the volume of solution was measured.

A solution containing 0.500 mole of HCl (18.25 g) in 1.00 liter would be a 0.500 M solution. Similarly, 3.65 g of HCl (0.100 mole) per liter would produce 0.100 M HCl. If, instead, the 0.100 mole of HCl (3.65 g) were present in 2.00 liters of solution, the HCl concentration would be 0.100 mole/2.00 L = 0.050 mole/L = 0.050 M. This illustrates that in using molarity, it is essential to consider the volume of the solution as well as the number of moles of the dissolved substance. Molarity is moles *per liter* and not just the number of moles. Therefore, the number of moles of solute must be divided by the number of liters of solution.

6.2   *Your Turn*

Calculate the molarity of the following solutions.

  a. 0.25 mole of NaOH in 500 mL (0.50 L) of solution
  b. 5.12 g of hydrogen iodide, HI, in 1.00 L of solution
  c. 11.7 g of sodium chloride, NaCl, in 200 mL (0.20 L) of solution

Ans. **a.** Remembering the definition of molarity, we can write an expression for the number of moles of solute divided by the number of liters of solution and actually carry out the division.

   0.25 mole NaOH/0.50 L soln = 0.50 mole NaOH/1.00 L soln

This means that the concentration of the solution is 0.50 M NaOH.

b. Note that it is necessary to first find the number of moles present in 5.12 g HI. The correct molarity is 0.040 M HI.

In expressing the strength of an acid we are really interested in the concentration of the H^+ ions present in solution. Of course it is impossible to have a solution containing only H^+ ions; the solution must be electrically neutral. The sum of the positive charges on all the H^+ ions must be balanced by negative charges supplied by the anions present in solution. Nevertheless, we can focus on the number of moles of H^+ ions present per liter of solution. In chemical shorthand, this is M_{H^+}. The relationship between the molarity of a strong acid such as HCl and the molarity of the H^+ ion in that solution is very simple, the two are equal.

$$\text{molarity HCl} = M_{HCl} = \text{molarity } H^+ = M_{H^+}$$

This identity holds because all of the dissolved HCl molecules dissociate to form H^+ and Cl^- ions (equation 6.2). You get one H^+ ion for each HCl molecule originally dissolved in solution. Thus, in a 1.00 M HCl solution, $M_{H^+} = 1.00$ M and $M_{Cl^-} = 1.00$ M. The same argument applies to OH^- ions. The ionic concentrations in a 0.50 M NaOH solution (6.2 Your Turn) are $M_{OH^-} = 0.50$ and $M_{Na^+} = 0.50$.

6.3   *Your Turn*

Calculate the molarity of the individual ions in the following solutions.

  a. 3 M H_2SO_4 **b.** 0.125 M KOH   **c.** 2.5×10^{-2} M $Ba(OH)_2$

Ans. **a.** When H_2SO_4 dissolves water, it dissociates according to the following equation:

$$H_2SO_4 \rightarrow 2\,H^+(aq) + SO_4^{2-}(aq)$$

This means that two H^+ ions and one SO_4^{2-} ions are formed from each H_2SO_4 molecule. Because the molecules are completely dissociated, the molarity of the H^+ ion will be twice the original molarity of H_2SO_4, and the concentration of SO_4^{2-} ions will be equal to the original molarity.

$$M_{H^+} = 2 \times M(H_2SO_4) = 2 \times 3\,M = 6\,M$$
$$M(SO_4^{2-}) = M(H_2SO_4) = 3\,M$$

b. $M_{OH^-} = 0.125$ M$(K^+) = 0.125$ M

A simple, useful, and very important relationship exists between M_{H^+} and M_{OH^-} in a water solution. The product of the molarity of the hydrogen ion and the molarity of the hydroxide ion has a constant numerical value of 1×10^{-14}. Mathematically this is stated by the following equation.

$$(\text{molarity } H^+)\,(\text{molarity } OH^-) = (M_{H^+})\,(M_{OH^-}) = 1 \times 10^{-14} \qquad (6.6)$$

If we know a value for one of these two molarities, we can use equation 6.6 to calculate the molarity of the other ion. A large value for the H^+ molarity means a low value for the H^- molarity, and vice versa. If the H^+ concentration in a solution is greater than the OH^- concentration, the solution will be acidic; if the OH^- concentration is greater than the H^+ concentration, the solution will be basic. If the two concentrations are equal, the solution is neutral. The acid/base character of any solution in water can thus be simply summarized.

$$M_{H^+} > M_{OH^-} \text{ solution is acidic}$$
$$M_{H^+} < M_{OH^-} \text{ solution is basic}$$
$$M_{H^+} = M_{OH^-} \text{ solution is neutral}$$

M_{H^+} ↑
M_{OH^-} ↓

We now add numbers to this qualitative argument. Consider a 1 M HCl solution. You already know that the H^+ concentration in this solution is also 1 M. We enter $M_{H^+} = 1$ M in equation 6.6 and solve for M_{OH^-}.

$$(1 \text{ M}) (M_{OH^-}) = 1 \times 10^{-14}$$

$$M_{OH^-} = \frac{1 \times 10^{-14}}{1 \text{ M}}$$

$$= 1 \times 10^{-14}$$

The concentration of H^+ in this solution is much higher (10^{14} times higher!) than the OH^- concentration. Therefore, the solution is acidic. Similarly, a 0.1 M HCl solution has an H^+ concentration of 0.1 M (1×10^{-1} M) and an OH^- concentration of 1×10^{-13} M.

$$M_{OH^-} = \frac{1 \times 10^{-14}}{1 \times 10^{-1}}$$

$$= 1 \times 10^{-14+1} = 1 \times 10^{-13}$$

It is still acidic, but not as highly acid as 1 M HCl.

On the other hand, if $M_{OH^-} = 1 \times 10^{-2}$, $M_{H^+} = 1 \times 10^{-12}$.

$$M_{H^+} = \frac{1 \times 10^{-14}}{1 \times 10^{-2}}$$

$$= 1 \times 10^{-14+2} = 1 \times 10^{-12}$$

Because 1×10^{-2} is much larger than 1×10^{-12}, the solution is basic.

6.4	*Your Turn*

■ Calculate the molarity of H^+ and the molarity of OH^- in each of the following solutions, assuming the compounds to be completely dissociated into their ions. Also indicate whether the solution is acidic or basic.

a. 0.005 M H_2SO_4 **b.** 2.0×10^{-3} M HNO_3 **c.** 0.1 M $Ba(OH)_2$

Ans. **a.** $M_{H^+} = 2 \times M(H_2SO_4) = 2 \times 0.005$ M $= 0.010$ M $= 1.0 \times 10^{-2}$
$M_{OH^-} = 1 \times 10^{-14}/1.0 \times 10^{-2} = 1.0 \times 10^{-12}$
$1.0 \times 10^{-2} > 1.0 \times 10^{-12}$, therefore, the solution is acidic.
b. $M_{H^+} = 2.0 \times 10^{-3}$; $M_{OH^-} = 5.0 \times 10^{-12}$; acidic

---

■ pH

The letters **pH** show up almost everyday in articles about acid rain and in advertisements for shampoo, facial care products, and other consumer goods. To understand such articles and advertisements, it is necessary to understand the significance of pH. The notation, always a lower case "p" and an upper case "H" written together, stands

Figure 6.1
The pH scale and common substances.

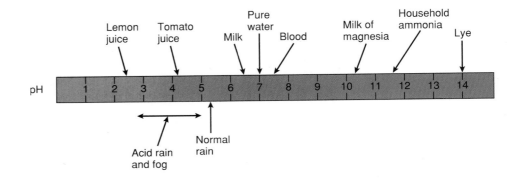

M_{H^+} ↑
pH ↓

for "power of hydrogen." In simplest terms, pH is a number between 0 and 14 (occasionally lower or higher than these limits) that indicates the acidity of a solution. The pH scale has a particular quirk. As the acidity *increases,* the pH number *decreases.* The higher the H^+ concentration, the lower the pH. Therefore, a solution of pH 2 is more acidic than a solution with a pH of 3, but less acidic than one with a pH of 1. The broad scope of pH values includes a significant range of solution types. Solutions with a pH of less than 7.0 are *acidic;* those with a pH of 7.0 are *neutral;* and those with a pH above 7.0 are described as *alkaline* or *basic.* The pH values of various common substances are given in Figure 6.1.

Notice from Figure 6.1 that acid rain has a lower pH value (higher acidity) than does normal rain, but even normal rain is slightly acidic. So is ordinary drinking water, which has a pH of about 6. Pure water has a pH of 7.0, so the obvious inference is that ordinary drinking water and normal rain are not pure H_2O. You will soon see what makes them acidic. You may also be surprised to see how many other acids we eat and drink. To many people, the term "acid" has a bad connotation, implying something dangerous and highly corrosive. To be sure, some acids (such as "battery acid" or sulfuric acid) do have such properties. But they also have very high H^+ concentrations and very low pHs, sometimes negative values less than zero. The naturally occurring acids in foods are much weaker, and often they contribute distinctive tastes. For example, vinegar contains acetic acid and has a pH of about 2.5. Apples (pH about 3.0) contain malic acid, and lemons (pH about 2.3) contain citric acid. Tomatoes are well known for their acidity, but in fact they are usually less acidic (pH about 4.2) than most fruits. Club soda has a pH of 4.8 because of the carbonic acid it contains and the various colas have a pH of about 3.1 because they contain phosphoric acid.

6.5   *Your Turn*

List vinegar, tomatoes, lemons, apples, Coca-Cola®, pure water, and club soda in order of increasing acidity.

6.6   *Consider This*

A legislator from a midwestern state is said to have made an impassioned speech in which he argued that the environmental policy of the state should be to bring the pH of rain all the way down to zero. Assume that you are a legislative aid to this representative and draft a brief memo to your boss on the topic of pH.

By now it should be pretty obvious that pH must be mathematically related to the concentration of hydrogen ions in solution, but the actual relationship may not be obvious. Not surprisingly, it involves molarity. pH is the negative logarithm of the hydrogen ion concentration when the concentration is expressed as moles of H^+ ion per liter of solution, that is, molarity. Written as an equation, the relationship is as follows.

$$pH = -\log(M_{H^+}) \qquad (6.7)$$

Logarithms ("logs" for short) are widely used in situations involving very large numerical ranges. Appendix 3 discusses logarithms in more detail. You will learn there that pH is quite literally a "power"; it is an exponent. Fortunately, logarithms can be obtained rather quickly using a pocket calculator; simply enter the number and press the "log" key.

To illustrate this process, let us calculate the pH of a 0.1 M HCl solution. The solution contains 0.1 mole or 3.65 g of HCl per liter of solution (0.1 mole of H = 0.1 g and 0.1 mole of Cl = 3.55 g). According to equation 6.2, 0.1 mole of HCl dissociates in solution to yield 0.1 mole of H^+ ions and 0.1 mole of Cl^- ions. Thus, there is 0.1 mole of H^+ ions per liter. Using a calculator; we can find the pH of the solution by applying equation 6.7.

$$pH = -\log(M_{H^+}) = -\log(0.1) = -(-1) = 1$$

When we enter the value of the molarity into the calculator and press the "log" key, the display reads "−1." This means that the log of 0.1 is −1; consequently, the pH of a 0.1 M H^+ solution is −(−1) or 1.

The pH of a 0.050 M HCl solution can be calculated in the same way. Entering the H^+ concentration as 0.050 M, the calculator displays −1.3010300 (or something similar) as the "log" value. We will round it off as −1.3. Because pH is $-\log(M_{H^+})$, the pH is −(−1.3) or 1.3. Appendix 3 also describes how to calculate the pH of a solution of known hydrogen ion molarity.

6.7 | ***Your Turn***

Using a pocket calculator, calculate the pH of the following liquids.

   a. orange juice ($M_{H^+} = 3.2 \times 10^{-4}$ M)
   b. wine ($M_{H^+} = 1.6 \times 10^{-3}$ M)
   c. blood ($M_{H^+} = 4.5 \times 10^{-8}$ M)

Ans. a. pH = $-\log(3.2 \times 10^{-4}) = -(-3.5) = 3.5$ **b.** pH = 2.8

6.8 | ***Your Turn***

Hydrogen bromide, HBr, behaves very much like HCl when it is dissolved in water. A solution is prepared by dissolving 8.09 g HBr in enough water to yield 0.50 L of solution (500 mL). Calculate the molarity of the HBr solution, the molarity of H^+ ions in the solution, and the pH of the solution.

Hint: You will need to make use of the molar mass of HBr, which is 80.9 g/mole.

The definition of pH is consistent with the fact that pH decreases as the solution becomes more acidic. Remember that 10 raised to a small negative exponent is larger than 10 raised to a large negative exponent. For example, 10^{-3} or 0.001 is much larger than 10^{-6} or 0.000001. Therefore rain with $M_{H^+} = 10^{-3}$ (pH = 3) is much more acidic than rain with $M_{H^+} = 10^{-6}$ (pH 6). In fact, the pH scale permits us to determine how much more acidic. An important feature of the logarithmic relationship between pH and M_{H^+} is the way in which the pH changes with changes in hydrogen ion concentration.

Figure 6.2

Relationship between pH and hydrogen ion molarity.

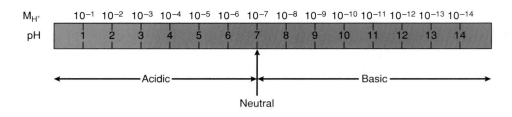

Figure 6.3

Graphical relationship between pH and hydrogen ion molarity.

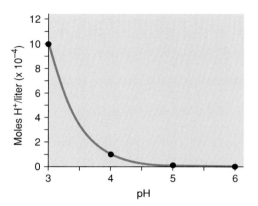

A *ten-fold* change in H^+ concentration is equivalent to a shift of *one* pH unit. Thus, for an H^+ concentration of 0.010 M (1.0×10^{-2} mole/liter), the pH is 2.0; for 1.0×10^{-3} M, it is 3.0; for 1.0×10^{-4} M, it is 4.0 . . . all the way to a pH of 14.0 for an H^+ concentration of 1.0×10^{-14} M (Figure 6.2). Therefore, a solution of pH 4.0 ($M_{H^+} = 1.0 \times 10^{-4}$) is ten times *less* acidic than a solution of pH 3.0 ($M_{H^+} = 1.0 \times 10^{-3}$). On the other hand, a solution of pH 2.0 is ten times *more* acidic than one with a pH of 3.0. Correspondingly, rain with a pH of 3.0 is 1000 times more acidic than rain of pH 6.0.

The way in which pH increases as the H^+ molarity decreases is graphically represented in Figure 6.3. The pH values are given on the horizontal axis and the corresponding H^+ molarities are plotted on the vertical axis. Note the dramatic 100-fold drop in M_{H^+} going from pH 3 to 5.

6.9 *Your Turn*

Using Figure 6.3, find the approximate values for the following solutions.

a. The molarity of H^+ ions in a solution of pH 3.5
b. The pH of a solution containing 6×10^{-4} moles H^+ per liter of solution
c. The pH of a solution with $M_{H^+} = 5 \times 10^{-5}$

Ans. **a.** $M_{H^+} = 3 \times 10^{-4}$

The pH scale is a convenient way to describe the concentrations of acids, but it is equally useful for describing bases. The key is equation 6.6. In pure water, the concentrations of H^+ and OH^- are both equal to 1×10^{-7} M and the pH is exactly 7. In an acidic solution, the H^+ molarity is *greater* than 1×10^{-7} M and the pH is *below* 7. On the other hand, in a basic (or alkaline) solution the *OH^- concentration* is greater than 1×10^{-7}, M_{H^+} must be less than 1×10^{-7} M, and therefore the pH is above 7. This can be summarized as follows:

$M_{H^+} \uparrow$
$pH \downarrow$
$M_{OH^-} \downarrow$

$$M_{H^+} > 10^{-7} \qquad \text{pH} < 7 \text{ solution is acidic}$$
$$M_{H^+} < 10^{-7} \qquad \text{pH} > 7 \text{ solution is basic}$$
$$M_{H^+} = M_{OH^-} = 10^{-7} \qquad \text{pH} = 7 \text{ solution is neutral}$$

Finally, we can put this all together in Figure 6.4 to show the relationship between pH, H^+ molarity, and OH^- molarity.

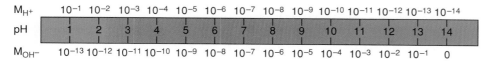

Figure 6.4
Relationship of pH, H⁺ molarity, and OH⁻ molarity.

6.10	**Your Turn**

a. A rain sample has an H⁺ concentration of 0.0001 M (1×10^{-4} M). What is the pH of this sample and what is the molarity of the OH⁻ ion? Is the rain acidic or basic?

b. A sample of water from a lake has a pH of 6.3. What are the H⁺ and OH⁻ molarities? Is the lake very acidic, slightly acidic, neutral, slightly alkaline, or very alkaline?

Ans. a. pH = 4.0; $M_{OH^-} = 1 \times 10^{-10}$; acidic

■ *Measuring the pH of Rain*

The pH of a rain sample or any other solution is usually determined with a pH meter. This device includes a special probe with two electrodes that are immersed in the sample. An electric voltage is produced between the two electrodes and the magnitude of this voltage is proportional to the pH. The meter measures the voltage and converts the result directly to pH, which is indicated on a dial or digital display.

An alternate method of pH measurement is to use indicators, such as litmus, which change color at various pHs. A mixture of indicators can be added to a series of solutions of known pH, thus creating a set of calibrated color standards. The same indicator mixture is also added to a solution of unknown pH, and the resulting color is compared with the standards. If the colors match, the pH of the unknown is equal to the pH of the standard. The indicators can also be used to dye strips of paper that are packaged along with a color chart showing the colors for various pH values. These pH test strips are inexpensive and easy to use for quick measurements of pH. A more elegant version of pH test strips consists of plastic strips with small patches of several indicators permanently bonded to the plastic. These strips can be reused many times.

It is quite easy to measure the pH of rain samples, although certain precautions are necessary in order to obtain reliable results. The use of scrupulously clean containers is crucial and the containers must be placed high enough to prevent "soil splash" which would contaminate the samples. The pH meter must be calibrated carefully to be certain that it reads the correct pH.

Rain pH data have been collected at selected sites scattered around the United States and Canada since about 1970. A more systematic study has been underway since 1978, with over 200 sampling sites at which weekly samples are collected. The pH is measured immediately and then all samples are sent to a central laboratory for further analysis. Figure 6.5 was prepared from such data. It is a map of the average pH of precipitation during 1989. Because this map contains a great deal of useful information, we will return to it several times in this chapter.

From the data of Figures 6.1 and 6.5, it appears that rain is at least slightly acidic. At first thought, this seems unexpected. If rain is pure water (as we tend to assume), then we might expect it to have a pH of 7.0. But pure unpolluted rain always contains dissolved carbon dioxide, CO_2. Recall from Chapter 3 that CO_2 is a natural component of the earth's atmosphere and its presence is essential for holding solar energy close to the earth's surface. Carbon dioxide dissolves slightly in water and reacts to form carbonic acid, H_2CO_3.

$$CO_2(g) + H_2O(l) \rightarrow H_2CO_3(aq) \rightarrow H^+(aq) + HCO_3^-(aq) \qquad (6.8)$$

Figure 6.5

Average annual pH of precipitation in the United States in 1989. These data were collected as part of the United States National Atmospheric Acid Deposition Program.

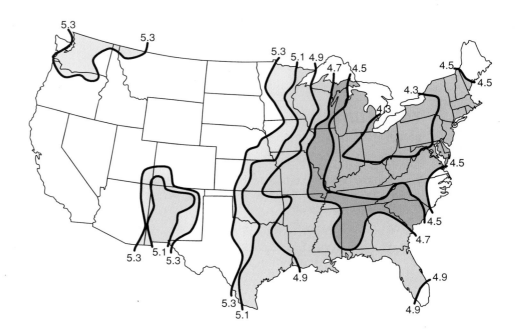

Carbonic acid is a weak acid; only a small fraction of its molecules dissociate into ions. Under normal atmospheric conditions, the H^+ concentration of rainwater is about 0.0000025 (2.5×10^{-6}) M. You can check on your calculator that this is equivalent to a pH of 5.6. But this cannot be the whole story because most rain is more acidic than pH 5.6. This means that the concentration of H^+ is usually greater than 2.5×10^{-6} M. Small amounts of other natural acids such as formic acid, acetic acid, and sulfuric acid are almost always present in rain. When these are added to the carbonic acid, the net result is a pH of about 5.3. This corresponds to the pH listed in Figure 6.1 as characteristic of normal rain. But the figure also suggests that rain frequently has a pH significantly below 5. We are ready to try to find the source of this extra acidity.

6.11   Consider This

Using Figure 6.5 as a reference, answer the following questions.

a. Where in North America is the average pH of precipitation lowest? Where is it highest?

b. What generalizations can you make about the distribution of acid precipitation?

c. What factors can you think of that might account for the variations?

d. If lakes and forests are reported to be affected by acidic precipitation, which parts of the continent should be most affected?

■ In Search of the Extra Acidity

According to Figure 6.5, the most acidic rain falls in the eastern third of the United States, with the lowest pH region being roughly the states along the Ohio River Valley. The extra acidity must be originating somewhere in this heavily-industrialized part of the country. Analysis of rain for specific compounds confirms that the chief culprits are the oxides of sulfur and nitrogen, which are sulfur dioxide (SO_2), sulfur

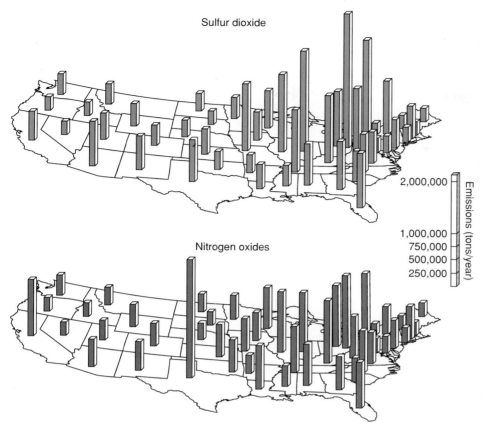

Sulfur dioxide

Nitrogen oxides

Emissions (tons/year)

2,000,000

1,000,000
750,000
500,000
250,000

Figure 6.6

Annual sulfur dioxide and nitrogen oxide emissions by state, 1980. (Source: G. Gschwandtner, *et al.,* "Historic Emissions of Sulfur and Nitrogen Oxides in the United States from 1900 to 1980," draft report to U.S. EPA, 1983.)

trioxide (SO_3), nitric oxide (NO), and nitrogen dioxide (NO_2). These compounds are sometimes collectively designated as SO_x and NO_x and called "sox" and "nox."

If this interpretation of acidic precipitation is correct, then the states with the most acidic rain should also be heavy producers of sulfur and nitrogen oxides. That relationship is generally confirmed by an examination of the maps in Figure 6.6. Emissions of SO_2 run highest in states where there are many coal-fired electric power plants, steel mills, and other heavy industry that uses coal. High NO_x emissions are generally found in states with large urban areas, heavy population density, and much automobile traffic, although power plants are also important contributors to NO_x in the atmosphere.

At this stage, the skeptical reader should be raising an objection to the idea of attributing excess acidity to sulfur oxides and nitrogen oxides. Given the definition of an acid as a substance that contains and releases H^+ ions, how can SO_2, SO_3, NO, and NO_2 qualify? None of these compounds even contains hydrogen. The objection is a reasonable one. The explanation is that they react readily with water to release H^+ ions. Although they are not acids themselves, the oxides of sulfur and nitrogen are **acid anhydrides,** literally "acids without water." When an acid anhydride is added to water, an acid is generated in solution. For example, sulfur trioxide dissolves in water and reacts with the water to form sulfuric acid.

$$SO_3(g) \quad + \quad H_2O(l) \rightarrow H_2SO_4(aq) \qquad (6.9)$$
$$\text{sulfur trioxide} \qquad\qquad \text{sulfuric acid}$$

The sulfuric acid then dissociates to yield two H^+ ions and a sulfate ion, SO_4^{2-}.

$$H_2SO_4(aq) \rightarrow 2\ H^+(aq) + SO_4^{2-}(aq) \qquad (6.10)$$
$$\text{sulfate ion}$$

In a similar, but slightly more complicated way, NO_2 can yield nitric acid, HNO_3, which dissociates into H^+ and NO_3^- ions.

$$4\ NO_2(g) + 2\ H_2O(l) + O_2(g) \rightarrow 4\ HNO_3(aq) \qquad (6.11)$$
$$\text{nitrogen dioxide} \qquad\qquad\qquad \text{nitric acid}$$

$$HNO_3(aq) \rightarrow H^+(aq) + NO_3^-(aq) \qquad (6.12)$$
$$\text{nitrate ion}$$

(You should check to be sure that equation 6.11 is balanced.)

■ Sulfur Dioxide

Having established that sulfur and nitrogen oxides react with water to yield acidic solutions, we intensify our search for the sources of these pollutants. There is an apparent correlation involving acid rain, atmospheric SO_2, and the burning of coal. But why should the combustion of coal yield SO_2, the choking gas formed from burning brimstone (Figure 6.7). To answer this we need to know something about the chemical nature of coal. At first glance, coal appears to be just a black solid, not very different from charcoal or black soot, both of which are essentially pure carbon. When carbon is burned, it forms carbon dioxide and liberates large amounts of heat (which of course is the reason for burning it).

$$C\ (\text{in coal}) + O_2(g) \rightarrow CO_2(g) \qquad (6.13)$$

As you learned in Chapter 4, coal is quite complicated. No two samples have exactly the same composition. But although coal is not a pure chemical compound, we can approximate its composition with the formula $C_{135}H_{96}O_9NS$. In addition to these elements, coal also contains small amounts of silicon and various metals such as sodium, calcium, aluminum, nickel, copper, zinc, arsenic, lead, and mercury. When coal is burned, oxygen reacts with *all* of the elements present to form oxides of those elements. Because carbon and hydrogen are the most plentiful, large quantities of gaseous CO_2 and H_2O are produced. In addition, there is an unburned solid residue consisting of oxides of silicon, sodium, calcium, and the other trace elements mentioned

Figure 6.7

The burning of sulfur in air to form sulfur dioxide.

above. But the sulfur is our primary interest right now. The combustion reaction of sulfur with oxygen yields sulfur dioxide, a poisonous gas with an unmistakable acrid odor.

$$S \text{ (in coal)} + O_2(g) \rightarrow SO_2(g) \qquad (6.14)$$

Sulfur is present in coal because sulfur is present in all living things. Coal was formed 100–400 million years ago from decaying vegetation. When the plants decayed, the sulfur was left behind in the material that eventually became coal. Coals from various parts of the world differ considerably in their sulfur content, but the combustion of almost all coals will produce some sulfur. This fact is central to the acid rain story. In large electric power generating stations and industrial plants, the sulfur dioxide goes up the smoke stack (unless control measures are used) along with the carbon dioxide, water vapor, and various metal oxides. Once in the atmosphere, SO_2 can react with more oxygen to form sulfur trioxide, SO_3.

$$2 SO_2(g) + O_2(g) \rightarrow 2 SO_3(g) \qquad (6.15)$$

This reaction is fairly slow, but it is catalyzed (speeded up) by the presence of finely-divided solid particles. The ash, which goes up the stack along with the SO_2, provides such catalytic sites. Once SO_3 is formed, it reacts rapidly with water vapor or water droplets in the atmosphere to form sulfuric acid (equation 6.9). There are also a variety of other pathways for conversion of sulfur dioxide into sulfuric acid. In reaction 6.15 the oxygen is said to "oxidize" the sulfur dioxide. But other oxidizing agents are also present in the atmosphere. Of particular importance are ozone, O_3, hydrogen peroxide, H_2O_2, and the hydroxyl radical, OH, which is formed from ozone and water in the presence of sunlight. The reaction of SO_2 with OH accounts for approximately 20–25% of the sulfuric acid in the atmosphere. The reaction goes faster in intense sunlight and thus it is more important in summer and at midday.

Although the largest source of sulfur dioxide in the United States is coal combustion for electric power generation, significant quantities of coal are also used in iron and steel production and other industrial processes. The large-scale production of nickel, copper, and certain other metals generates huge quantities of SO_2. The most common ores of these metals are sulfides, which are compounds of the metal plus sulfur. When heated to high temperatures in a smelter, the sulfides are decomposed and sulfur dioxide is released. The world's largest smelter, in Sudbury, Ontario, is used to convert NiS to Ni. The bleak, lifeless lunar landscape in the immediate vicinity of the plant is mute testimony to earlier uncontrolled release of SO_2. Today, even with government controls, the Sudbury smelter emits about 2000 tons of sulfur dioxide per day from its 1250-foot smokestack. The fact that this is the world's tallest smoke stack—equal in height to the Empire State Building—simply means that the emissions are more broadly distributed.

6.12 ***Your Turn***

    a. Assuming the composition of coal is represented by the equation $C_{135}H_{96}O_9NS$, calculate the fraction and percent (by mass) of sulfur in the coal.

    b. A power plant burns one million (10^6) tons of coal per year. Assuming the sulfur content calculated in **a,** calculate the number of tons of sulfur released per year.

    c. Calculate the number of tons of SO_2 formed from this mass of sulfur.

    Ans. **a.** 0.0168 or 1.68% **b.** 1.68×10^4 tons S (16,800 t)

■ *The Acidification of Los Angeles*

The combustion of coal has been indicted as a major environmental offender, contributing sulfur dioxide to the atmosphere and to acid deposition. But we know that SO_2 is not the only cause of acid precipitation and coal is not the only source. Another guilty party has been identified in California. The concentration of SO_2 in the smoggy air above the Los Angeles metropolitan area is relatively low, but so is the pH. For example, in January 1982, fog near the Rose Bowl in Pasadena was found to have a pH of 2.5. Breathing it must have been like breathing a fine mist of vinegar. This level is at least 500 times more acidic than normal, unpolluted precipitation and ten times more acidic than required to kill all fish in lakes. And, in December 1982, fog at Corona del Mar, on the coast south of Los Angeles, was ten times more acidic than that. It registered a pH of 1.5. In both cases, something other than sulfur dioxide was involved.

That acidic "other" is emitted by the millions of cars that jam the Los Angeles freeways day and night. The city literally runs on automobiles, and, as anyone who has ever visited there knows, the quality of the air is not very good. But it is not at all obvious why cars should contribute to acid precipitation. The compounds that make up gasoline blends are essentially all hydrocarbons, and they are mostly converted to CO_2 and H_2O when burned. You recall from Chapter 1 that this conversion is not always complete, and that some CO and unburned hydrocarbon fragments escape in the exhaust. But gasoline contains almost no sulfur, and hence its combustion yields practically no SO_2. Consequently, we must look for another source of acidity.

Nitrogen oxides have already been identified as contributors to acid rain, but gasoline does not contain nitrogen, either. Therefore, logic (and chemistry) assert that nitrogen oxides cannot be formed from burning gasoline. Literally, that is correct—after all, you can't make something out of nothing. Remember, however, that nitrogen is present in air. In fact, 78% of air consists of N_2 molecules. These molecules are remarkably stable and do not readily undergo chemical reactions. That is why nitrogen remains unchanged as we breathe it in and out of our lungs. Nevertheless, it can and does react with a few elements under extreme conditions. One of these elements is oxygen. All that is needed is sufficient energy in the form of high temperatures or an electric spark. Under these conditions, the two elements combine to form nitric oxide, NO.

$$Energy + N_2(g) + O_2(g) \rightarrow 2\ NO(g) \qquad (6.16)$$

Because air is a mixture of nitrogen and oxygen, it is always a potential source for the production of nitric oxide. The energy necessary for the reaction can come from natural lightning or the "lightning" that occurs in an internal combustion engine. Gasoline and air are drawn into the engine cylinders and compressed to a high pressure. Then a spark ignites the rapid burning of the gasoline. The energy released in this process is what provides the motive power of the vehicle. But the unfortunate truth is that the energy also triggers reaction 6.16. Moreover, the high pressure means that the nitrogen and oxygen molecules are closer together and thus even more likely to react.

Reaction 6.16 is not limited to lightning and the automobile engine. Chapter 2 mentioned the concern over NO production in jet aircraft engines. The same reaction occurs when air is heated to a very high temperature in the furnace of a coal-burning electric power plant. Hence, such plants contribute both sulfur and nitrogen oxides to acidify precipitation. On a national basis, stationary sources release more nitrogen oxides than mobile sources, but in urban environments automobiles and trucks account for most of the atmospheric NO.

Unlike nitrogen, nitric oxide is a very reactive substance. In Chapter 2 you read that NO can react with ozone in the upper atmosphere, thus destroying the O_3. Close to the surface of the Earth, it reacts primarily with O_2 to form nitrogen dioxide, NO_2.

$$2\ NO(g) + O_2(g) \rightarrow 2\ NO_2(g) \qquad (6.17)$$

Several other oxides of nitrogen are formed from NO, but the most important is NO_2. It is a highly reactive, poisonous, red-brown gas with a nasty odor. For our purposes, its most significant reaction is the one that converts it to nitric acid, HNO_3. You saw one representation of that conversion in equation 6.11. Actually, a series of steps is involved in the chemistry that takes place in the urban atmosphere above Los Angeles. Unraveling the reactions has proved to be a fascinating scientific detective story. Sunlight is required, and the hydroxyl radical, OH, is involved. This unstable species is formed in a reaction involving ozone, which is especially prevalent in a polluted urban atmosphere. The radicals react rapidly with nitrogen dioxide to form nitric acid.

$$NO_2(g) + OH(g) \rightarrow HNO_3(g) \hspace{2cm} (6.18)$$

As you have already read, when HNO_3 dissolves in water it dissociates into H^+ and NO_3^- ions.

Rain and fog containing nitric acid is just as damaging as rain or fog containing sulfuric acid. Fog is considered to be more serious than rain for direct health effects because people inhale the fog particles into their lungs. The alarmingly low pH values we reported for Los Angeles fog is largely a consequence of high concentrations of nitric acid that had its origins in automobiles.

6.13 *Consider This*

Suppose that smog and very acidic rainfall became the norm rather than the exception in urban settings. Work and recreation activities would have to change in response to this situation. Design a poster for use in elementary schools describing through picture and words some precautions that should be taken when young children play outdoors.

■ SO_x and NO_x

Sulfur oxides and nitrogen oxides are both capable of acidifying the atmosphere and the precipitation that falls from it. It is reasonable to ask which is the greater problem. Figure 6.6 makes it very clear that the emission levels for the two classes of compounds vary across the country and depend on the amount and kind of coal burned and the number of automobiles. Another perspective is provided by Figure 6.8, which shows the relative contributions of the important anthropogenic sources of SO_x and NO_x in the United States. Not surprisingly, almost 80% of the sulfur dioxide emissions can be traced to electric utilities. The largest source of nitrogen oxides is also electric power production, but transportation accounts for about 40%.

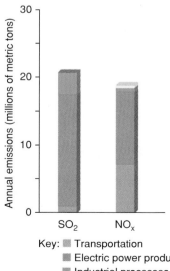

Key: ■ Transportation
■ Electric power production
■ Industrial processes
■ Solid waste and miscellaneous

Figure 6.8
United States emissions of sulfur and nitrogen oxides, 1991 (in millions of metric tons per year). (Source: United States Environmental Protection Agency, *National Air Quality and Emissions Trends Report,* 1991, issued October 1992.)

■ *Table 6.1*	*Estimated Global Emissions of Sulfur and Nitrogen Oxides (in millions of metric tons per year), 1980 Data*	
Source	SO$_2$*	NO$_x$**
Natural:		
oceans	22	1
soil and plants	2	43
volcanoes	19	
lightning	——	15
Subtotals	43	59
Anthropogenic:		
fossil fuels combustion	142	55
industry (mainly ore smelting)	13	
biomass burning	5	30
Subtotals	160	85
Totals	**203**	**144**

*Spiro, et al., "Global Inventory of Sulfur Emissions With 1° × 1° Resolution, "*J. Geophys. Res.*, **97**, No. D5, 6023 (1992).

**U.S. EPA, *Air Quality Criteria for Oxides* of *Nitrogen*, EPA/600/8–91/049aA.

Although human activity introduces vast quantities of sulfur and nitrogen oxides into the atmosphere, it is important to be aware that natural sources can have significant global impact. For example, the eruption of Mount Pinatubo in the Philippines in June 1991 injected between 15 and 30 million tons of SO$_2$ into the stratosphere. The SO$_2$ quickly reacted to form small droplets of sulfuric acid. This H$_2$SO$_4$ aerosol is expected to remain in the atmosphere for one to three years. Guy Brasseur and Claire Granier of the National Center for Atmospheric Research have used a computerized simulation of the atmosphere to predict the climatic effects of this massive eruption. Their calculations suggest that by absorbing and reflecting sunlight, the acidic aerosol cloud will contribute to a cooling of the troposphere and a warming of the lower stratosphere. Moreover, it is possible that a series of chemical reactions associated with this increased SO$_2$ concentration will render the ozone molecules in the stratosphere more vulnerable to destruction by chlorine and chlorofluorocarbons.

The eruption of Mount Pinatubo represents a major perturbation of the average annual emissions of atmospheric pollutants. Table 6.1 includes typical estimates for SO$_2$ and NO$_x$ from natural and artificial sources. Looking at the numbers we see that natural emissions account for about 21% of the sulfur dioxide and 41% of the nitrogen oxides. We also see that sulfur dioxide appears to be a bigger problem than nitrogen oxides. Total global sulfur dioxide emissions from human sources are about twice the mass of nitrogen oxides released. Furthermore, if we look only at fossil fuel combustion, we see that the nitrogen oxide emissions are less than 40% of those of sulfur dioxide. This has important implications for policy planning aimed at reducing acid emissions. Moreover, this balance can change with alterations in fuel sources, industrialization, or standard of living.

■ *The Effects of Acid Precipitation*

The evidence seems persuasive that much of the rain and snow in the United States is more acidic than would be the case for normal, unpolluted precipitation. Fog, dew, and the bottom layers of clouds frequently have a pH of 3.0 or lower. And there is clear indication that, on a regional basis, the acidity of precipitation has increased significantly since the Industrial Revolution. But does it really matter? To answer that fundamental question, we need to know something about the effects of acid deposition and how serious they really are. Clearly, these issues are central to the acid rain debate. Scientific opinion about them is divided, although a consensus is gradually emerging.

In an effort to gain the information necessary to make informed decisions, the United States Congress funded a major national research effort during the decade of the 1980s, called the National Acid Precipitation Assessment Program (NAPAP). Over 2000 scientists were involved, with a total expenditure of $500 million. The project was completed in 1990 and the participating scientists have prepared a 28-volume set of technical reports (NAPAP, *State of the Science & Technology,* 1991). Much of the material in the remainder of this chapter is drawn from the NAPAP reports plus other documents prepared for Congress. We shall first consider possible damaging effects of acid rain on materials, visibility, human health, lakes and streams, and forests. Because the implications are so important, it is crucial to look carefully at the evidence (or lack of it) and not be swayed by emotional statements from special interest groups. Thus armed, we will look at some of the options for cleaning up and reducing the acidic emissions. Finally, we will consider the cost of these measures and the difficult decisions they demand.

■ *Damage to Materials*

Remember Paul Revere's gravestone, mentioned in the newspaper article that began this chapter? It is "a victim of irreparable pollution damage" because it is made of limestone, which is calcium carbonate, $CaCO_3$. Calcium carbonate (marble, eggshells, and seashells are other examples) slowly dissolves in acid.

$$CaCO_3(s) + 2\,H^+(aq) \rightarrow Ca^{2+}(aq) + CO_2(g) + H_2O(l) \qquad (6.19)$$

Many other gravestones in the eastern United States are suffering similar fates. Some are no longer legible. Even more serious is the fact that many priceless and irreplaceable marble and limestone statues and buildings are also being attacked by air-borne acids (Figure 6.9). The Parthenon in Greece, the Taj Mahal in India, and even the United States Capitol show signs of acid erosion. Visitors to the Lincoln Memorial in Washington are told that huge stalactites growing in chambers beneath the Memorial are the result of acid rain eroding the marble.

Another damaging effect of acidic rain is the corrosion of metals. Iron, undoubtedly the most important structural metal, is particularly susceptible. Buildings, bridges, railroads, and vehicles of all kinds depend on iron and steel. Unfortunately, iron will readily corrode or rust by undergoing a reaction with oxygen and water. The reaction requires hydrogen ions, but even in pure water there is sufficient H^+ to promote slow rusting. In the presence of dilute nitric or sulfuric acid, the corrosion is

Figure 6.9

Left: A statue of a Roman emperor at Oxford University, eroded by three centuries of acid precipitation and other atmospheric pollutants. Right: A new copy of the same statue, recently erected in anticipation of reduced acid deposition.

greatly accelerated. The role of H^+ is evident in equation 6.20, which represents the first of a two step process. Iron (Fe) reacts with oxygen and hydrogen ions to yield the Fe^{2+} ion.

$$4 \, Fe(s) + 2 \, O_2(g) + 8 \, H^+(aq) \rightarrow 4 \, Fe^{2+}(aq) + 4 \, H_2O(l) \qquad (6.20)$$

Then Fe^{2+} reacts with more oxygen to produce iron oxide, the familiar reddish brown material we call rust.

$$4 \, Fe^{2+}(aq) + O_2(g) + 4 \, H_2O(l) \rightarrow 2 \, Fe_2O_3(s) + 8 \, H^+(aq) \qquad (6.21)$$

The net result of combining these two reactions is simply the sum:

$$4 \, Fe(s) + 3 \, O_2(g) \rightarrow 2 \, Fe_2O_3(s) \qquad (6.22)$$

6.14 | **Your Turn**

Show that the sum of equations 6.20 and 6.21 is the same as equation 6.22.

Because iron is inherently unstable when exposed to the natural environment, enormous sums of money are spent annually to protect exposed structural iron and steel in bridges, cars, ships, and other applications. Paint is the most common means of protection, but even paint degrades more rapidly when exposed to acid rain and acid gases. Another means of protection is to coat the iron with a thin layer of a second metal such as chromium (Cr) or zinc (Zn). Everyone is familiar with *chrome plated* iron used for automobile bumpers and trim. Iron coated with zinc is called *galvanized iron*. There is widespread evidence that both chrome plated iron and galvanized iron will corrode more rapidly in the presence of acid rain. As a consequence they must be replaced more frequently.

■ Reduced Visibility

Anyone living in the eastern half of the United States is familiar with the summer haze that usually clouds the landscape. (Ironically, you become more aware of it on the occasional really clear day when it does seem that you can see forever.) Travelers crossing the country by jet airplane can easily see the haze covering the East. And visitors to the Great Smoky Mountains National Park can view a prominent display of photographs showing reduced visibility in the mountains. Power plants in the Ohio Valley are identified as the primary cause. This haze in the eastern part of the country has become steadily worse for several decades. It consists primarily of microscopic aerosol particles containing a mixture of sulfuric acid, ammonium sulfate [$(NH_4)_2SO_4$], and ammonium bisulfate (NH_4HSO_4). The haze is most pronounced in summer when there is more sunlight to accelerate the photochemical reactions leading to sulfuric acid, and it is particularly evident when the air is stagnant. As a consequence of this haze, average visibility in the East is now 25 miles or less and occasionally as low as 1 mile. By contrast, visibility in the western mountain states is typically 50–70 miles. It should be noted, however, that even the Grand Canyon is experiencing reduced visibility, probably as a result of SO_2 emissions from the huge Four Corners power plant in the northeast corner of Arizona.

■ Effects on Human Health

Opinion differs on whether acid rain poses a direct human health hazard. It is well known that a fine aerosol mist of sulfuric acid droplets can have very serious respiratory effects, as in the London fog described in Chapter 1. Such conditions do not exist with normal acidic rain. But a fine dust of sulfate particles may be harmful, especially for the elderly, the very ill, and those with serious pre-existing respiratory problems. On the other hand, the pHs of the nitric acid fogs along the southern California coast are so low that they may cause respiratory damage in large numbers of people.

Concern is growing that acids in precipitation are helping dissolve certain toxic heavy metals such as lead, cadmium, and mercury. These elements are naturally present in the environment, but normally they are tightly bound in the soil and rock. Dissolved in acidified water and conveyed into the public supply, these metals can pose serious health threats. Elevated concentrations of heavy metals have already been discovered in major reservoirs in western Europe.

■ *Damage to Lakes and Streams*

When acidic precipitation falls on surface waters, it seems reasonable to predict that the waters will become more acidic. Numerous studies have reported the progressive acidification of lakes and rivers in certain geographic regions, along with reductions in fish populations (Figure 6.10). In southern Norway and Sweden, where the problem was first observed, one-fifth of the lakes no longer contain any fish and half of the rivers have no brown trout. In New England and the Adirondack Mountain region of northern New York, the number of highly acidic lakes is growing. In southeastern Ontario the average pH of lakes is now 5.0, well below the pH of 6.5 required for a healthy lake.

On the other hand, many areas of the Midwest have no problem with acidification of lakes or streams, even though the Midwest is supposed to be the major source of the acids in acid precipitation. This apparent paradox can be explained quite simply. When acidic precipitation falls on a lake, the pH of the lake will drop (become more acidic) unless the lake or its surrounding soils contain bases that can neutralize the acid. The capacity of a lake to resist change in pH when acids are added is called its **acid neutralizing capacity (ANC).**

In order to understand differences in ANC, it is necessary to know something about the surface geology of different areas. In areas where the surface rock is mostly limestone, such as in much of the Midwest, the limestone itself (calcium carbonate) can neutralize acids (equation 6.19). But even more importantly, the lakes and streams have a relatively high concentration of calcium bicarbonate as a result of reaction of the limestone with carbon dioxide and water.

$$CaCO_3(s) + CO_2(g) + H_2O(l) \rightarrow Ca^{2+}(aq) + 2\ HCO_3^-(aq) \qquad (6.23)$$
$$\text{calcium bicarbonate}$$

The bicarbonate ion is a base that will readily neutralize acids:

$$HCO_3^-(aq) + H^+(aq) \rightarrow CO_2(g) + H_2O(l) \qquad (6.24)$$
$$\text{bicarbonate ion}$$

Because the added acid is consumed by this reaction, the pH of the lake will remain more or less constant.

In contrast to this, many lakes in New England and upper New York (as well as in Norway and Sweden) are surrounded by granite, which is a hard, impervious, much less reactive rock. Unless other local processes are at work, there will be very little acid neutralizing capacity in these lakes, and they will most likely show a gradual acidification. Often the decreasing pH is attended by a decline in the population of fish and other aquatic species. Figure 6.11 depicts the effects of pH on various organisms. Healthy lakes have a pH of 6.5 or above. As the pH is lowered below 6.0, various species are affected. Only a few species survive below pH 5.0, and at pH 4.0 a lake is essentially dead.

There is evidence that the adverse effect of increased acidity on fish populations is probably an indirect one involving aluminum (Al). Aluminum is the most abundant metal and third most abundant element in the Earth's crust (after oxygen and silicon).

Figure 6.10
Fish kills can arise from a variety of causes, including acid rain. These dead alewife are on the shore of Lake Michigan.

Figure 6.11
Effect of pH on aquatic life.

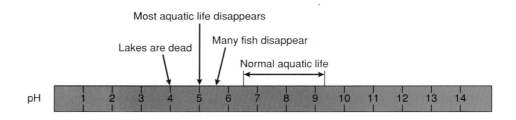

Granite contains aluminum, and soil includes complicated aluminum-silicate structures. Natural aluminum compounds have very low solubility in water, but in the presence of acids their solubilities increase dramatically. Thus, when the pH of a lake drops from say 6.0 to 5.0, the aluminum concentration in the lake may increase 1000-fold. When fish are exposed to a high concentration of aluminum, a thick mucous forms on the gills and the fish literally suffocate.

It is somewhat surprising that the analysis of acidified lakes is a good deal more complicated than simply measuring pH and acid neutralizing capacities. The source of the acid and the change in pH over time is often difficult to establish. Thus, although many reports claim that lakes are becoming more acidic, the evidence is frequently tenuous. We do not know with certainty how many acidic lakes have always been that way and how many have become acidic since the huge increase in SO_2 and NO_x emissions beginning in the 1940s. The difficulties are two-fold. In the first place, no one was aware of a possible problem until about 1970, so very little data were gathered prior to that time. Second, the equipment and procedures now routinely used for measuring pH and ANC were not generally available until the 1970s. This is an example of what is, unfortunately, a frequent problem confronting our technological society. Without the equipment to make measurements and without a reason to suspect that a problem exists, harmful effects of various kinds may go unnoticed for many years.

In spite of these complications, there is general agreement on some points. Approximately 10% of lakes and streams in the eastern United States appear to have been adversely affected by acid precipitation, with a corresponding reduction of aquatic life. Other lakes have low acid neutralizing capacities and are at risk of being damaged if present levels of acidic deposition continue unabated. The already acidified and highly susceptible areas are mostly in the southern Adirondacks, New England, the forested mid-Atlantic highlands, and in the eastern Upper Midwest. A high percentage of lakes in northern Florida are acidic but the NAPAP studies have shown these to be naturally acidic and not a result of atmospheric deposition.

There is also a bit of good news. Evidence suggests that with the proper treatment many acidified lakes can be reclaimed. Recent experiments with adding lime, calcium hydroxide [$Ca(OH)_2$], to acidified lakes are encouraging. Calcium hydroxide is a base, and it reacts to neutralize acid. If large-scale liming were carried out and further input of acids halted, many lakes could be restored to health within a few years.

6.15   *Your Turn*

Write an equation for the reaction of lime with an acid.

■ *The Mystery of the Unhealthy Forests*

Of all the ravages attributed to acid precipitation, the most widely discussed has been damage to trees and forests. Pictures of dead and dying trees have provoked strong emotional responses and the ire of many environmentalists. But the fact is that the effects of acid deposition on trees are less well understood than any other aspect of the acid rain story. Consequently, the topic is fraught with controversy.

At least in certain cases, the reality of forest decline seems indisputable. The phenomenon was first observed in what was then the German Federal Republic (West Germany) in the 1960s, and extensive studies by German scientists have chronicled a steady decline. Fir and spruce trees at high elevation have been especially hard hit. The trees initially show limp branches, then yellowing needles, and then loss of needles. This gradually spreads from the top down until finally the weakened trees are killed by drought, cold, insects, and winds. The decline was especially rapid from 1982 to 1985, when the number of damaged trees (defined as 10% needle loss) throughout West Germany increased from 8% to 52%.

Forest decline appeared to spread rapidly across western Europe during the 1980s, with damaged forests reported in Italy, France, the Netherlands, Sweden, Norway, and Britain. It is uncertain, however, how much of the reported damage actually began during that decade and how much was simply the result of better observing and reporting. For eastern Europe, where air pollution controls have been virtually nonexistent, information has only recently become available. Some reports claim that over 80% of the forests in eastern Germany, Poland, and Czechoslovakia are damaged. In some areas of these countries, forests are said to have been totally destroyed by air pollution.

In North America, forest damage has been most dramatic in portions of the Appalachian Mountains. On Mount Mitchell and in the Great Smoky Mountains in North Carolina and Tennessee, one can now see the stark landscape of dead trees. In the Green Mountains of Vermont half of the red spruce trees on Camel's Hump died between 1965 and 1981. The sugar maple trees in New England and southeastern Canada, famed for autumn color and maple syrup, are claimed to be unhealthy, with some farmers reporting alarming decreases in maple syrup production.

It is tempting to blame acidic precipitation for all these effects. After all, the acids rain down directly on the trees, and the effects are generally most pronounced in regions with highly acidic precipitation. But the story is not that simple. It is difficult to prove cause-and-effect relationships, especially when there are many effects and many possible causes. In some cases (such as the New England and Canadian maple trees) weather-related stresses are probably responsible for most of the damage. In other regions, the cause may be insect infestations (as in the loss of the fir trees in the southern Appalachians) or ozone and other air pollutants (as in the damage to pines in the San Bernardino Mountains). The dead trees on Mount Mitchell and other parts of the southern Appalachians are primarily Fraser firs that were killed by an infestation of the balsam woolly adelgid. Only a few species of trees have shown complete die-back, and in mountainous regions the effect is most serious only at certain elevations. Furthermore, careful field studies have shown that many of the effects are not as serious as had been believed. For example, the famed Black Forest in Germany has experienced a 10% needle loss, but not the severe die-back that had been widely reported in the press. The North American Sugar Maple Decline Project, initiated by the United States and Canada in 1987, has found that only 10% of sugar maples have experienced a 15% crown die-back and that this percentage actually decreased from 1988 to 1989.

Nevertheless, there is strong circumstantial evidence that acidic precipitation is at least a contributing factor to declining health of trees. A careful study of tree growth rings in southern New Jersey pines showed a dramatic reduction in growth since 1965. This change, the greatest growth reduction in the 125-year record, correlates well with the acidity of nearby surface waters. This may be an example of the synergistic effects of acids, ozone, SO_2, and NO_x. According to this theory, which is gaining acceptance, the ozone and nitrogen oxides attack the waxy coating on leaves, permitting hydrogen ions to deplete nutrients. At the same time, acidification of soil beneath the trees mobilizes metals (especially aluminum) that attack the tree roots, thus preventing absorption of nutrients and water. These effects then leave the trees susceptible to destruction by natural factors such as disease, insects, drought, or high winds. A further factor with high elevation forests is that the mountains are often shrouded in fog—so-called "cap clouds"—and, as reported earlier in the chapter, cloud water is often more acidic than even the most acidic rain. There is clear experimental evidence that acid deposition has contributed to the decline of red spruce at high elevation in the northern Appalachians by reducing the species' cold tolerance.

Whatever the complex causes and whatever the extent of involvement of acid rain, the damage to forests is unfortunate and expensive. However, it is important to realize that even in the case of severe decline and total die-back, most of the damaged regions have shown surprisingly good growth of new trees. The new growth appears to be thick and healthy. Nature does indeed regenerate herself.

■ *Costs and Control Strategies*

The acid rain debate has turned increasingly to economic considerations. Policymakers want to know the costs of the damage already incurred, the costs of cleanup, and the projected costs of future abatement. They also want to know who might benefit and, of course, who will pay. Unfortunately, it is extremely difficult to assess accurately the costs of the damage now occurring as a result of acid precipitation. Not only are the cause-and-effect relationships unclear, but many assumptions must be made. One such attempt was published in 1985 by Thomas Crocker, an economist, and James Regens, a natural resources scientist. They acknowledge that the only *conclusive* evidence for damage is to aquatic ecosystems. In spite of this severe caveat, they offer the estimates that appear in Table 6.2. Note that these projections of annual economic losses apply only to the eastern third of the country. The huge numbers involved take on a little more personal perspective when one considers that the damage to buildings in Chicago alone has been estimated to be nearly $300 million a year, or $45 per resident.

While efforts continue to gauge more accurately the current economic burden of acid rain, other studies are underway to evaluate strategies to control acid emissions and to estimate the costs of these measures. Most attention is being focused on ways to limit and eventually reduce the release of sulfur and nitrogen oxides from coal-fired furnaces and gasoline-fueled vehicles. The Clean Air Act of 1990 (already mentioned in Chapter 1) has made the reduction of SO_x and NO_x emissions a major national priority. If all goes well, the bill should have a significant impact by the turn of the century. A variety of techniques have been proposed, bearing a range of price tags.

Since sulfur dioxide from coal accounts for the majority of human-made acidic emissions and because it comes primarily from a limited number of point sources (power plants and factories), this seems like the best place to attack the problem. Three major strategies are under consideration: (1) switch to "clean coal" with low sulfur content, (2) clean up the coal to remove the sulfur before use, and (3) use chemical means to neutralize the acidic sulfur dioxide in the power plant. We will briefly consider the effectiveness and the cost of each of these.

■ *Table 6.2*	*Estimates of 1978 Maximum Annual Economic Losses Caused by Acid Deposition in the Eastern Third of the United States*
Effects Category	**Maximum Annual Losses (in billions)**
Materials	$2.00
Forest ecosystems	1.75
Direct agricultural	1.00
Aquatic ecosystems	0.25
Others	0.10
Total	$5.10

Reprinted with permission from Thomas Crocker and James Regens, in *Environmental Science & Technology* **19,** 112 (1985). Copyright © 1985 American Chemical Society.

Coal switching is an option because coals vary widely in their sulfur content and their heat content. Anthracite or "hard" coal, found mainly in Pennsylvania, yields the greatest amount of energy and the smallest amount of sulfur dioxide per ton of fuel. Unfortunately, the supply is practically exhausted. Bituminous or "soft" coal, which is abundant in the Midwest has nearly the same heat content as anthracite but usually contains 3–5% sulfur. In the western states there are enormous deposits of subbituminous coal and lignite or brown coal, both of which have low sulfur but also a low heat content. This means that more lignite must be burned to obtain the same energy output.

We have already noted that the largest amounts of sulfur dioxide are produced in a small number of states concentrated around the Ohio River Valley. These states have easy access to plentiful supplies of high-sulfur coal. If the coal-burning power plants in this heavily industrialized region were to switch to low-sulfur coal, costs would inevitably increase. Railroad shipment of the western coal to the Midwest and East would roughly double the cost of the fuel. Even so, at first glance, switching appears to be the least expensive of the alternatives considered here, with an estimated average cost (in 1982 dollars) of $250–$500 per ton of SO_2 removed.

But there are hidden costs not included in this estimate. It ignores the social and economic impact on the states that produce high-sulfur coal. It has been estimated that a 50% shift to low-sulfur coal could result in the loss of 20,000 to 50,000 jobs in the high-sulfur coal mining regions of Pennsylvania, Kentucky, West Virginia, and neighboring states. The projected economic impact is $1–$5 billion (in 1981 dollars).

Coal cleaning is relatively easy to do and the technology is available. The coal is crushed to a fine powder and washed with water so that the heavier sulfur-containing minerals sink to the bottom. But the process only removes about half of the sulfur, and it is expensive—$500–$1000 per ton of SO_2 eliminated.

The best alternative to using cleaner coal is to chemically remove the SO_2 during or after combustion in the power plant. The chief method for doing this is called *scrubbing*. The stack gases are passed through a wet slurry of powdered limestone, $CaCO_3$. Calcium carbonate is a base, and it neutralizes the acidic SO_2 to form calcium sulfate, $CaSO_4$.

$$2 \ SO_2(g) + O_2(g) + 2 \ CaCO_3(s) \rightarrow 2 \ CaSO_4(s) + 2 \ CO_2(g) \qquad (6.25)$$

Limestone is cheap and readily available, and the process can be about 90% efficient. Its cost has been estimated at $400–$600 per ton of SO_2 removed. Part of the expense is associated with the disposal of the $CaSO_4$ formed. We simply cannot avoid the law of conservation of matter.

6.16   *Consider This*

Suppose you live in a medium-sized Ohio or Indiana city faced with a problem of excessive sulfur dioxide in the local air, in violation of state and federal regulations. The source of the problem is the local city-owned electric power utility that burns high-sulfur coal obtained within the state. The three choices being considered by the city are to buy low sulfur coal, which must be shipped in from a long distance, install expensive scrubbers, or build a taller smoke stack. The present stack is 150 feet; a 300-foot stack should solve the local air problem. Develop a presentation for local civic groups explaining the problem, outlining the possible solutions, and summarizing the economic and environmental aspects of the choices.

Whatever strategy or combination of strategies we adopt to reduce sulfur dioxide emissions, the costs will be substantial. Experts have estimated that reducing the SO_2 output by 40% would cost the country $1–$2 billion per year. If these costs were passed along to the consumer in the form of electric rate increases, the average increase per customer is projected at about 8%. A 50% reduction would cost $2–$4 billion, and a 70% reduction would require $5–$6 billion each year.

The question remains: Are these costs worth paying? The Japanese have decided that they are. In 1968, the Japanese government issued strict SO_2 controls, and encouraged the use of low sulfur fuels and desulfurization techniques. As a consequence, sulfur dioxide emissions were cut in half by 1975, even though energy use doubled in the same period. Quite obviously, the procedures work. Such dramatic decreases are related to the widespread use of scrubbers on Japanese power plants. In 1982, nearly 1200 scrubbers were in place, compared to about 200 in the United States. To be sure, Japan and the United States are different in many respects, but the Japanese experience clearly indicates that the problems of reducing SO_2 emissions are not primarily technological. They are political.

> **6.17**   *Consider This*
>
> Your city has been warned by the EPA to cut sulfur dioxide and nitrogen oxide emissions or you will lose substantial federal funds. One citizen's group has advocated drastically reducing electricity consumption. A second group advocates reduced use of automobiles within the city limits. Choose one or the other position to support. After two groups are established and the research is done, a debate should be scheduled between the two groups with the goal of coming to a consensus of what action should be taken by the city.

■ *The Politics of Acid Rain*

It is not surprising that legislation to control acid rain has been the subject of intense political maneuvering. Nor is it surprising that the electric power industry, a strong and effective lobby, has resisted legislation and controls. Indeed, some representatives of the electrical utilities have claimed that the effects of acid rain are grossly exaggerated and that there is no evidence that the situation is deteriorating. The causal chain connecting coal-burning power plants, sulfur dioxide emissions, and acid rain has even been questioned. The electric power industry has called for more study rather than more regulations, arguing that it would be a grave mistake to spend billions of dollars before the causes and effects of acid rain were fully understood. The costs and the risks of premature response are held to be too great for society to pay. Environmentalists, on the other hand, have argued that there is already sufficient evidence concerning acid deposition to justify action now. They maintain that to delay may be courting disaster and much higher ultimate costs.

In 1990, the Congress of the United States entered into the controversy by passing major new environmental legislation. The Clean Air Act Amendments were signed into law by President Bush in November of that year. The law seeks to achieve significant improvements in air quality within a decade. This is to be accomplished through a set of revised standards for the emission of toxic substances that contribute to acid precipitation and urban smog. For example, electric power utilities will be required to cut their SO_2 discharge by ten million tons per year and their nitrogen oxide emissions by two million tons per year. By the end of the century, total emissions must be back down to the 1980 levels, which are about half of the 1990 values. The greatest reductions will be required within five years at 111 of the dirtiest power plants in the Midwest and Appalachia.

The new legislation also sets up a unique system of "emissions trading." Under this scheme utilities will receive pollution credits for going beyond their required emission reductions. These credits can be sold to other utilities that do not attain the required levels. There is thus a financial incentive for power producers to achieve significant reductions of acidic oxide emissions. On the other hand, the purchase of credits by those who cannot meet the more stringent standards will permit them to continue operation, possibly with high-sulfur coal. The generally lowered emission levels should benefit the entire nation, and the net national air quality should not be compromised by emissions trading, though local areas of acid rain may persist. The first trade of emissions credits under the provisions of the new law occurred in 1992.

The most difficult aspect of forging a new national policy on acid precipitation has been the issue of who should pay for it. Quite clearly, the costs do not fall evenly across the country. Should the regions that generate the most sulfur dioxide (and the most energy) be forced to bear the expense of shifting to low-sulfur coal and/or installing scrubbers? The public utilities would undoubtedly pass the increased costs on to their customers, an action that might easily result in a 16% electric rate increase in the region. Moreover, the residents would be less able to pay those bills if 50,000 miners were out of work because of the closing of the bituminous coal mines. And to exacerbate the economic impact, the states that would be most affected—Kentucky, West Virginia, and other parts of Appalachia—are already hard hit by unemployment. Nor are urban areas exempt from the high cost of clean air. Newly mandated pollution control devices to remove more nitrogen oxides from automobile exhaust could help the environment, but not the economic environment in Detroit. The new catalytic converters would probably increase the cost of cars, reduce sales, and cost jobs.

6.18 ■ *Consider This*

Without consulting voting records or party affiliations, predict how senators from West Virginia, Wyoming, Texas, and Michigan might have voted on the Clean Air Act of 1990, and explain your reasoning. Then consult the *Congressional Record* to see how they actually did vote.

6.19 ■ *Consider This*

In January 1990, two of the nation's leading newspapers ran editorials about acid rain and the NAPAP study: the *Wall Street Journal* on January 26 and the *New York Times* on January 29. Obtain copies of these editorials (they are included in the *Instructors Resource Guide* that accompanies this text) and read them carefully. Compare and contrast the positions articulated. From this evidence, what can you infer about the general editorial policy of these two influential papers? Then do one of the following.

a. Draft a letter to the editor of one of these newspapers, responding to the editorial, or

b. Write your own editorial on this topic.

■ *Conclusion*

If this chapter has taught anything, we hope it has been skepticism, prudence, and the recognition that complex problems cannot be solved by simple and simplistic strategies. "Acid rain" is not the dire plague described by many environmentalists and some journalists. Careful research has indicated that the damage caused by acid deposition, especially to trees and forests, is less severe than had been supposed. Moreover, the air is getting cleaner, and legislation such as the Clean Air Act of 1990 will help bring about further improvements. On the other hand, to fail to acknowledge the relationships involving the combustion of coal and gasoline, the production of sulfur and nitrogen oxides, and the reduced pH of fog and precipitation, is to deny some fundamental facts of chemistry. As is often the case with sharp and bitter controversy, the truth lies somewhere between the extremes. It is our task to discover it and to act accordingly.

One response that we as individuals and as a society might make to the problems of acid precipitation has hardly been mentioned in this chapter, yet it is potentially one of the most powerful. It is to conserve energy. Sulfur dioxide and nitrogen oxides are by-products of our voracious demand for energy—energy for electricity and energy for transportation. And, of course, carbon dioxide is an even more plentiful product. If our personal, national, and global appetite for fossil fuels continues to grow unchecked, our environment may well become a good deal warmer and a good deal more acidic. Moreover, the problem will be intensified as petroleum and low-sulfur coals are consumed and we become ever more reliant on high-sulfur coal.

There are other sources of energy—nuclear fission, water and wind, renewable biomass, and the Sun itself. All of them are already being used, and their use will no doubt increase. We will explore these sources in subsequent chapters, but we conclude this chapter with the modest suggestion that, for a multitude of reasons, the conservation of energy could have profoundly beneficial effects on our environment.

■ *References and Resources*

"Acid Rain." (Information Pamphlet) Washington: American Chemical Society, 1985.

Ayres, B. D. "Ohio Valley of Tears is Facing More." *New York Times,* April 1, 1990: A20.

Hordijk, L. "Use of the *RAINS* Model in Acid Rain Negotiations in Europe." *Environmental Science & Technology* **25** (1991): 596–603.

MacKenzie, J. J. and El-Ashry, M. T. *Air Pollution's Toll on Forests & Crops.* New Haven, Conn.: Yale University Press, 1990.

Mohnen, V. A. "The Challenge of Acid Rain." *Scientific American* **259,** Aug. 1988: 30–38.

Regens, J. L. and Rycroft, R. W. *The Acid Rain Controversy.* Pittsburgh: University of Pittsburgh Press, 1988.

Reisch, M. S. "SO$_2$ Emission Trading Rights: A Model for Other Pollutants." *Chemical & Engineering News,* July 6, 1992: 21–22.

Smith, W. H. "Air Pollution and Forest Damage." *Chemical & Engineering News,* November 11, 1991: 30–43.

Ulrich, B. "*Waldsterben:* Forest Decline in West Germany." *Environmental Science & Technology* **24** (1990): 436–41.

Worthy, W. "New Flue Gas Desulfurization System to Undergo U.S. Trials." *Chemical & Engineering News,* December 11, 1989: 17–18.

■ *Experiments and Investigations*

2. Preparation and Properties of Atmospheric Gases II

8. Reactions of Acids

9. Analysis of Vinegar

11. Building and Using a Conductivity Tester

22. Determination of the pH of Rain

■ *Exercises*

1. Classify each of the following as acid, base, or neither.

    a. HF   b. NaOH   c. NaCl

    d. CH_4 e. NH_3 f. HNO_3

2. Write equations to represent the ionization of the acids in Exercise 1 in aqueous solution.

3. Write equations to represent the neutralization reactions between the acids and bases in Exercise 1.

4. Calculate the molarity of the solute in each of the following solutions.

   a. 0.30 moles of HCl in 3.0 L of solution

   b. 11.2 grams of potassium hydroxide, KOH, in 1.00 L of solution

   c. 42.6 grams of glucose, $C_6H_{12}O_6$, in 0.40 L of solution

5. Calculate the molarity of each ion in the following solutions.

   a. 0.25 moles of HNO_3 in 0.50 L of solution

   b. 0.45 moles of $Ba(OH)_2$ in 3.0 L of solution

   c. 1.25 moles of $CaCl_2$ in 2.5 L of solution

6. Calculate the molarity of H^+ and OH^- ions in each of the following solutions and classify each solution as acidic, basic, or neutral.

   a. 5.0×10^{-3} moles of H_2SO_4 in 0.25 L of solution

   b. 5.4 grams of NaCl in 0.40 L of solution

   c. 0.12 grams of $Ca(OH)_2$ in 0.20 L of solution

7. Calculate the pH of each solution in Exercise 6.

*8. Convert the following pH values into H^+ concentrations. Calculate the corresponding OH^- concentrations and indicate whether each solution is acidic, basic, or neutral.

   a. 11.0   b. 3.5   c. 7.0

   d. 6.7   e. 7.6   f. 8.9

9. For each of the common substances given in Figure 6.1, determine the H^+ and OH^- concentration from the pH values given. Label each substance as acidic, basic, or neutral.

10. Consult Figure 6.8 to compare the number of tons of SO_2 and NO_x originating from electric utilities and transportation in 1991. Account for the differences in the relative quantities of these two substances generated by the two sources. Describe the ways by which each substance is produced.

*11. Calculate the mass of lime, $Ca(OH)_2$, needed to react with sufficient H^+ in a small lake (3.54×10^8 liters) to raise its pH from 5.0 (where most aquatic life disappears) to a pH of 6.5 (characteristic of normal aquatic life). On the basis of your calculation, is liming a practical and feasible method to neutralize acidic lakes? What problems would you anticipate?

*12. One way to compare the acid neutralizing capacity (ANC) of different substances is to calculate the mass of the substance required to neutralize one mole of H^+. Write balanced equations for the reaction of each of the following substances with H^+ and calculate their ANCs: $NaHCO_3$, NaOH, and $CaCO_3$.

*13. Determine the costs to neutralize one mole of acid with the chemicals listed in Exercise 12. The compounds and their costs per kilogram (from a recent chemical catalog) are $NaHCO_3$, \$8.88; NaOH, \$13.32; and $CaCO_3$, \$18.10.

*14. Calculate the mass of $CaCO_3$ (in tons) necessary to react completely with 1.00 ton SO_2 via equation 6.25. Note that one mole of $CaCO_3$ is required for each mole of SO_2.

15. The estimated costs of reducing SO_2 emissions in the U.S. by various percentages are given below.

Emission reduction (%)	Cost to reduce ($\times 10^9$)
40	1–2
50	2–4
70	5–6

   a. Prepare a graph to represent the relationship between the percentage reduction in emissions and the cost.

   b. Comment on the prospect of achieving 100% reduction, that is, zero emissions.

16. Three strategies for reducing SO_2 emissions are described in the text. They are coal switching, coal cleaning, and stack gas scrubbing. Prepare a list of the advantages and disadvantages associated with each of these three methods. Which option would you advocate and why?

17. The costs associated with three technologies for SO_2 removal are listed below.

   switching: \$250–\$500 per ton of SO_2
   cleaning: \$500–\$1000 per ton of SO_2
   scrubbing: \$400–\$600 per ton of SO_2

   A power plant burning one million tons of coal per year releases 16,800 tons of SO_2 per year (6.12 Your Turn). Estimate the cost of removing the SO_2 by each of the three technologies.

18. Discuss the validity of the statement, "Photochemical smog is a local problem, acid rain is a regional one, and the greenhouse effect is a global one." Describe the chemistry behind each of these air pollution problems and explain why they affect different geographical areas.

7

Onondaga Lake: A Case Study

■

In 1847, the mayor of a middle-sized city in a developing country shared his vision for the lake adjoining his municipality.

> Our beautiful lake will present continuous villas ornamented with shady
> groves and hanging gardens and connected by a wide and splendid avenue
> that shall encircle its entire waters and furnish a delightful drive for the gay
> and prosperous citizens of the town who will, towards the end of each
> summer's day, throng to it for pleasure, relaxation, or improvement of health.
>
> *Harvey Baldwin*

The realization of that vision required capital and the economic factors that create it: a sound tax base, workers and jobs, and industry. All were achieved. Syracuse, New York prospered as new factories were built along Lake Onondaga. A number of exquisite resort hotels soon dotted its shores, and "the gay and prosperous citizens of the town" did indeed throng to it. A thriving fishing industry sprang up, and Onondaga whitefish became a highly esteemed delicacy in some of New York City's more fashionable restaurants. Yet, by the 1920s, the fishing industry and resorts were gone, victims of the lake's rapidly declining water quality. Today, fish taken from the lake are not to be eaten because of mercury contamination, and no one swims in its cloudy waters. Indeed, Stephen Fields, a water scientist, has called Onondaga "one of the world's most polluted lakes."

At first glance, this might seem to be a rather parochial topic, of interest only to residents of upstate New York. But the lessons of Onondaga Lake have significance for the rest of the country—indeed, for the rest of the world. This is an account of the impact of industrialization on a developing society. Initially, technology brought new opportunities and new prosperity to the shores of the lake. But in time, environmental deterioration began to negate those advantages, and the citizens of Syracuse found themselves on the horns of the all too familiar risk/benefit dilemma. It is a dilemma that many other societies have faced in the past or will face in the future. In this chapter we attempt to reconstruct and relate what went wrong. We invite you to wade with us into the murky waters of Onondaga.

■ *Chapter Overview*

This story of salt, soda, and Syracuse is one of industrial development, environmental impact, and technology's response. It begins with a look at the rich natural resources that made the Syracuse region an ideal location for chemical manufacturing. Of particular importance in this industrialization was the growing demand for soda ash or sodium carbonate. The many uses of this compound are described, along with several procedures for its production. Of these, the Solvay process was by far the most significant for Onondaga Lake. The chapter describes the process and the build-up of by-products that ultimately led to serious contamination of the lake. But the Solvay process was not the only contributor of unwanted chemicals to Onondaga. The chlor-alkali process, also described next, released large quantities of toxic mercury to the troubled waters. The chapter and the case study conclude with an overview of the lake's sorry past, a survey of the progress that has been made in halting its pollution, and a look forward to what might be done to restore some of its diminished sparkle.

■ *Salt and Syracuse*

Onondaga Lake is of modest size—about 4.5 miles long and 1 mile wide. It contains 35 billion gallons of water and has an average depth of about 40 feet. As Figure 7.1 indicates, it drains into the Seneca River, which ultimately empties into Lake Ontario. But the relatively small size of Onondaga Lake belies its importance. It provides a textbook case of the impact of industrial processes on the environment and on society.

Figure 7.1
Onondaga Lake.

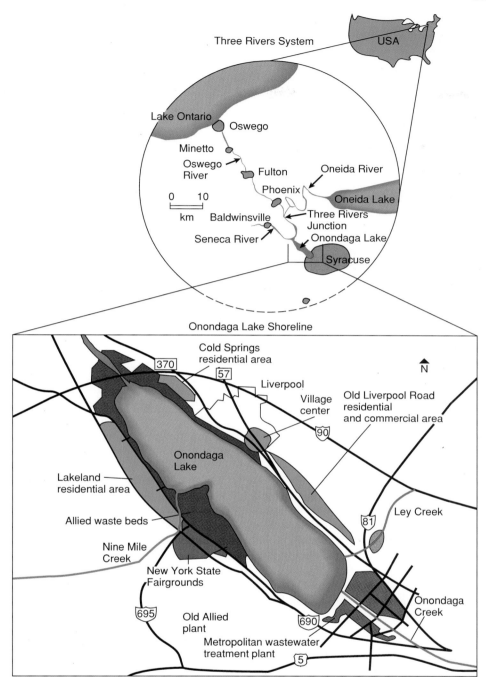

Adapted with permission from *Syracuse Herald American,* August 9, 1987. Map by James Dibeno.

For centuries, the Onondaga Lake region has provided a rich source of important chemicals. The native Americans from whom the lake took its name were the first to discover the springs of brine (concentrated aqueous solutions of sodium chloride) that are common in the area. The first written account of these springs was by a French Jesuit missionary, Father Simon LeMoyne, who visited the Onondaga people in 1654. Soon after, the Indians began a modest salt trade with European settlers. In the years following the American Revolution, Syracuse salt production grew dramatically, and by 1797, almost all of the salt used in the United States came from that area. Significantly, the new republic may also have gained some important intangibles from

Figure 7.2
A restored "boiling block" at the Salt Museum on the shores of Onondaga Lake. The brine was boiled down in the large kettles and the salt crystals were scooped into the baskets.

Figure 7.2
A restored "boiling block" at the Salt Museum on the shores of Onondaga Lake. The brine was boiled down in the large kettles and the salt crystals were scooped into the baskets.

the Onondaga area and its original inhabitants. It was here that the Iroquois League was formed. Some scholars have suggested that this confederation of five Indian nations influenced the federal form of government adopted by the young United States.

Two methods were used to produce salt along the shores of Lake Onondaga. In the original procedure, the brine was boiled down in large iron kettles heated by wood fires (Figure 7.2). As the solution became concentrated, salt crystals formed on the surface of the liquid. These were removed with long-handled scoops and placed in baskets to dry. This process yielded fine grained crystals that could be used in food preservation or further refined for table use. However, as the forests were cleared for farmland and burned for fuel, the cost of firing the "boiling blocks" rose. Coal was sometimes used, but after the Civil War, it, too, became prohibitively expensive. Therefore, the salt manufacturers of Syracuse turned to the Sun for energy and built solar evaporators—50,000 of them at the peak of operation. Most of the coarse Sun-dried salt was used by meat packers and in the churning of ice cream.

Unfortunately for the Syracuse salt industry, rich salt mines were discovered in the west, and other, more sunny climates began to dominate the solar process. In 1926 the last of the Onondaga salt yards was closed. However, a serious negative economic impact was averted because a far more profitable use of sodium chloride had been introduced in 1884. It was the Solvay process for the manufacture of sodium carbonate.

■ Soda and Society

Sodium carbonate, Na_2CO_3, did not originate in Syracuse, New York. The compound was first obtained from naturally occurring deposits in the Middle East or extracted from the ashes of certain seaweeds, hence "soda ash," its common name. It was known in ancient times and has been long recognized as an important chemical. For centuries, sodium carbonate has been used in the manufacture of soap, paper, and glass. These three products may be prosaic, but they have had profound influence on the quality of human life. Soap has significantly improved personal and public health and hygiene;

and cheap paper has made newspapers, magazines, and books widely available. Glass is a particularly interesting case. Its discovery is usually attributed to Phoenician sailors who, around 5000 B.C., may have accidentally produced glass by heating a mixture of sand, limestone, and soda ash. The first samples were, no doubt, very cloudy, and it took years before clear, transparent glass was developed. But you can imagine the social impact when cheap window glass, made with inexpensive sodium carbonate, became widely available for the first time in the nineteenth century.

7.1   *Consider This*

Use appropriate reference sources to find out what materials were used for window "panes" before cheap window glass became available. Also, investigate how sodium carbonate is used in the manufacture of glass.

Sodium carbonate retains its commercial importance today. In 1990 Na_2CO_3 production in the United States ranked eleventh among industrial chemicals, with about 35 kg for every inhabitant of the country. The largest fraction still goes to manufacture glass, but you are probably most familiar with the compound as a water softener. The carbonate ion (CO_3^{2-}) reacts with calcium (Ca^{2+}) and magnesium (Mg^{2+}) ions in hard water to yield insoluble calcium carbonate ($CaCO_3$) and magnesium carbonate ($MgCO_3$) (equation 5.9). Because of this use, sodium carbonate is also known as "washing soda" (Figure 7.3).

The vast amounts of soda ash used today definitely could not be produced by extracting the ashes of seaweeds. As early as two centuries ago, it was obvious that this tedious process and the limited natural deposits were unable to meet the growing demands for the compound. What was needed was an inexpensive source of large quantities of sodium carbonate. Therefore, in the eighteenth century, the government of France offered a prize of 2400 livres (then equal to about eight years' wages for a common laborer) for a method of preparing soda ash from salt. In 1791, Nicholas Leblanc opened a factory that did just that, although not very successfully. In the

Figure 7.3
Sodium carbonate (washing soda) and some related products. Baking soda, or sodium bicarbonate, is also produced in the Solvay process, and both the glass bottle and the paper label are made using sodium carbonate. The carbon dioxide dissolved in the "soda water" could also be generated from sodium carbonate, though it probably is not.

Leblanc process, sodium chloride was reacted with sulfuric acid (H_2SO_4) to produce sodium sulfate (Na_2SO_4) and hydrogen chloride (HCl), which in water solution yields hydrochloric acid.

$$2\,NaCl + H_2SO_4 \rightarrow Na_2SO_4 + 2\,HCl \qquad (7.1)$$

The solid sodium sulfate was then heated with coal and limestone (calcium carbonate). It is impossible to write a simple chemical equation for this reaction because coal is a complex mixture containing mostly carbon, but one of the products is sodium carbonate. Even after it was improved, the Leblanc process remained messy and was plagued by pollution problems. Landowners understandably objected to the corrosive fumes of sulfuric and hydrochloric acid. Clearly, a better process was needed, and others sought it.

■ *Soda from the Solvay Process*

The most significant event in the soda story was probably the development, in 1865, of a new process for the manufacture of sodium carbonate. The inventors were two Belgian brothers, Albert and Ernest Solvay. The **Solvay process** produced soda ash at half the price of the Leblanc process, and it was much cleaner. Hence, it was a great improvement, both economically and environmentally. The overall reaction seems to be simplicity itself. Calcium carbonate and sodium chloride are chemically combined to yield sodium carbonate and calcium chloride.

$$CaCO_3 + 2\,NaCl \rightarrow Na_2CO_3 + CaCl_2 \qquad (7.2)$$

The reactants were available in abundance near Syracuse. Deep brine springs were a ready and reliable source of sodium chloride, nearby rock outcroppings provided a cheap and plentiful supply of limestone, and ice harvested from the lake in winter could be stored and used in the process year-round. Therefore, it is not surprising that in 1884, the Solvay Process Company began making soda ash along the west shore of Onondaga Lake. Production of the compound was prodigious. For example, from 1960 to 1985, 2000 to 2800 tons of soda ash were produced *daily* by Allied Chemical Company, successor to the Solvay Process Company.

The Solvay process contributed much to the Syracuse area—jobs for an ethnically diverse population, a generous tax base, and many of the social benefits that money can buy. Long-time residents of the town of Solvay, where the plant was located, remember with gratitude the excellent public schools in this community of 10,000 residents. The strong academic program was supplemented with extensive athletic, music, art, and other extracurricular programs. Even dental care was free for the students. But with time, it became obvious that the Solvay process was also contributing some less desirable by-products to the town of Solvay, the city of Syracuse, and Onondaga Lake. To understand why, we must take a closer look at the chemistry involved.

One key point is that reaction 7.2 does not occur directly. If one were to mix sodium chloride, calcium carbonate, and water, the NaCl would dissolve, but nothing else would happen. The $CaCO_3$ would remain solid and insoluble. No reaction would take place and no sodium carbonate would be formed. Therefore, a roundabout series of steps is necessary to bring about the overall reaction. It involves a total of at least five individual reactions and six chemical intermediates. These compounds are ammonia (NH_3), ammonium chloride (NH_4Cl), calcium oxide (CaO), calcium hydroxide [$Ca(OH)_2$], carbon dioxide (CO_2), and sodium bicarbonate, ($NaHCO_3$).

A particularly crucial step involves four of these intermediates. It is the reaction that occurs when carbon dioxide (generated by heating limestone) and ammonia are bubbled through a salt solution at $0°C$. Sodium bicarbonate and ammonium chloride are formed according to equation 7.3.

$$NH_3(g) + CO_2(g) + NaCl(aq) + H_2O(l) \rightarrow NaHCO_3(s) + NH_4Cl(aq) \qquad (7.3)$$

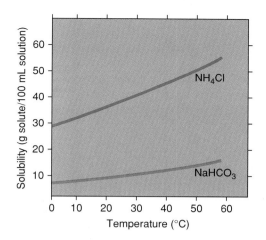

Figure 7.4
Graph of water solubilities of $NaHCO_3$ and NH_4Cl.

Note that in the equation, the sodium bicarbonate is a solid, but the ammonium chloride is identified as aqueous. Both of these compounds form in the water solution, but under the conditions of the reaction, the solubility of the $NaHCO_3$ is less than that of the NH_4Cl. This is evident from Figure 7.4, which is a plot of the water solubilities of these two compounds at various temperatures. As is typical for many solutes, the solubilities of both substances decrease with decreasing temperature. At 0°C, the solubility of sodium bicarbonate is 6.9 g/100 mL solution; that of ammonium chloride is 29.4 g/100 mL solution. This means that most of the sodium bicarbonate crystallizes, separating as a solid, while the ammonium chloride remains behind in solution.

The sodium bicarbonate obtained by this means is heated to about 300°C to obtain sodium carbonate.

$$\overset{\text{heat}}{2\,NaHCO_3(s) \rightarrow Na_2CO_3(s) + H_2O(g) + CO_2(g)} \qquad (7.4)$$

7.2 ◾ ***Your Turn***

Assume that 100 mL of a solution containing 15.0 g $NaHCO_3$ and 15.0 g NH_4Cl at 50°C is cooled to 0°C. Calculate the number of grams of each compound that precipitates. Also compute the percentage of the original compounds left in solution after the crystallization occurs.

Ans. First of all, note from Figure 7.4 that at 50°C the solution is just about saturated with $NaHCO_3$. At this same temperature, the concentration of NH_4Cl is well below its saturation limit of 50.0 g/100 mL. At 0°C, the solubilities of the two compounds are 6.9 g $NaHCO_3$/100 mL and 29.4 g NH_4Cl/100 mL. Because the latter mass is greater than the 15.0 g NH_4Cl originally present in the 100 mL of solution, no ammonium chloride precipitates; 100% of it remains in solution. However, 15.0 – 6.9 or 8.1 g $NaHCO_3$ crystallizes from the solution. The percentage of the original $NaHCO_3$ left in solution is (6.9 g/15.0 g) × 100 or 46.0%.

The Solvay process is very efficient because the intermediates are all reclaimed and reused. Of particular importance is the recycling of ammonia, the most expensive compound in the process. The NH_3 used in equation 7.3 is regenerated from the decomposition of the ammonium chloride formed in that same reaction. Strategies like this made the Solvay process one of the first industrial processes to regenerate intermediates. The recycling of critical materials is important in most industrial processes—at least for companies that want to make a profit! Furthermore, careful recycling can do much to prevent possible environmental harm from the discharge of intermediates.

The waters of Lake Onondaga played a crucial role in the Solvay process. In addition to employing water as the solvent for the various reactions, Allied Chemical used up to a million gallons of lake water daily as a coolant to absorb waste heat generated by the Solvay process. To be environmentally sound, excessively hot coolant water could not be put back into the lake. Rather, the temperature of the outflow water from the plant had to be close to the intake temperature. To meet these restrictions, Allied Chemical spent $2.5 million in 1977 to install a special deep-release thermal diffuser. This device took discharge water and pumped it well out into the lake at a depth of 15 feet. By mixing cold water from deep in the lake with the warm plant water, the discharged water at the surface was only about 3° F warmer than the intake temperature.

7.3 Your Turn

Truly prodigious amounts of heat are transferred to the coolant water from the Solvay process. Suppose that 1×10^6 gallons of water were taken into the plant at 70° F (21.1° C) and discharged through the diffuser deep into the lake at 100° F (37.8° C). One gallon of water weighs about 8 pounds or 3.6 kg. Using information from Chapter 5, calculate the number of kilocalories of heat transferred to the water by the process.

Ans. The quantity of heat involved, 6.0×10^7 kcal, is approximately equal to that released by burning 2400 gallons of gasoline or 14 tons of pine wood!

■ The Solvay Legacy at Onondaga

From what you have read thus far, the Solvay process would seem to be an ideal example of enlightened chemical manufacturing. Two cheap and plentiful naturally occurring substances—salt and limestone—are converted into two useful products—sodium carbonate and calcium chloride. Unfortunately, the demand for calcium chloride is not as great as the demand for sodium carbonate. To be sure, $CaCl_2$ has a variety of uses. It is used in concrete mixtures, in drilling muds for the oil industry, in certain passive solar heating units, as a drying agent, to melt snow and ice on roads and sidewalks, and to control dust. But over the years, the sales of calcium chloride from the Onondaga Lake plant could not keep pace with the sales of soda ash. The excess calcium chloride was permitted, by state and federal regulations, to be released in solution into a tributary of Onondaga Lake. Nor was the Syracuse soda ash plant helped by the 1938 discovery in Wyoming of vast deposits of trona, a mineral with the formula $Na_2CO_3 \cdot NaHCO_3 \cdot 2 H_2O$. The easy availability of this new source of soda ash made serious economic inroads into the competitiveness of the Solvay process in this country. In 1986, Allied Chemical Company closed the plant, resulting in a significant downturn in the Syracuse area economy through the loss of jobs and purchasing power (Figure 7.5).

Soda ash production in Syracuse left another legacy. For many years, nearly 500 tons of unmarketable salts from the Solvay process were dumped into Onondaga Lake *every* day. It was mostly calcium chloride with some unreacted sodium chloride. In 1900 large settling basins were built near the plant and dikes constructed to impound calcium carbonate and calcium sulfate formed in the brine purification and ammonia recovery steps. As production increased, so did the need for waste disposal. In 1901 the company received permission from the state of New York to pump the waste slurries into diked marshes along the shoreline. But materials containing Ca^{2+}, Na^+, and Cl^- ions continued to enter the lake as they were leached from the marshes by rain water. The same year, the state banned ice cutting from the lake because of the high concentration of saline impurities.

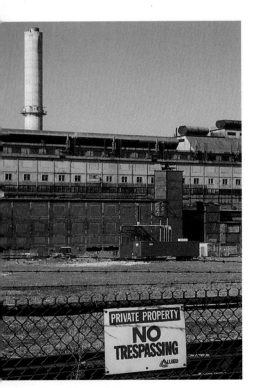

Figure 7.5
The Solvay process plant in Syracuse, New York.

Over the decades, the mounds of minerals in the storage beds rose as high as 70 feet and covered 400 acres (Figure 7.6). On Thanksgiving Day, 1943, the dike holding back the waste beds broke, releasing a salty flood that buried over 20 houses and came up to the gates of the state fairgrounds. Ten years later, Allied Chemical deeded 400 acres of waste beds to the state of New York for one dollar to use for state fairgrounds parking and construction of Interstate 690. This was in exchange for the state agreeing to drop claims against Allied for the 1943 waste spill.

7.4   *The Sceptical Chymist*

In a newspaper article about the Solvay process and disposal of industrial wastes into Onondaga Lake, the author claims that about one ton of waste is generated for every two tons of soda ash produced. Is this 1:2 ratio correct?

Hint: According to equation 7.2, one mole of $CaCl_2$ is produced for every mole of Na_2CO_3. Find the mass ratio of the two compounds that corresponds to this mole ratio. If all the $CaCl_2$ is waste, the ratio of the mass of $CaCl_2$ produced to the mass of Na_2CO_3 formed should be what we are looking for. You will find that this ratio is not 1:2. What conclusions can you draw and how can you explain the discrepancy?

■ *Table 7.1*	*Solubility Rules*
NO_3^-	All **nitrates** are soluble.
Cl^-	All **chlorides** are soluble except AgCl, Hg_2Cl_2, and $PbCl_2$.
SO_4^{2-}	Most **sulfates** are soluble; exceptions include $SrSO_4$, $BaSO_4$, and $PbSO_4$.
CO_3^{2-}	All **carbonates** are insoluble except those of the Group 1A elements and NH_4^+.
OH^-	All **hydroxides** are insoluble except those of the Group 1A elements, $Sr(OH)_2$, and $Ba(OH)_2$. $Ca(OH)_2$ is slightly soluble.
S^{2-}	All **sulfides** are insoluble except those of Group 1A and 2A elements and NH_4^+.

[a]Insoluble compounds are those that precipitate when we mix equal volumes of 0.1 M solutions of the corresponding ions.

Adapted with permission from *Chemical Principles*, 6th edition, W. L. Masterton, E. J. Slowinski, and C. L. Stanitski. © 1985, Saunders College Publishing, Philadelphia.

Most of the substances in the waste beds and the parking lot fill are ionic compounds. Their water solubility is obviously very important to the composition of Onondaga Lake water. Some, including sodium chloride and calcium chloride, are water soluble; others, such as calcium carbonate, are not. In Chapter 5 you learned that ionic compounds are often soluble in water. But many ionic compounds are insoluble or, at best, only slightly soluble in water. The solubility rules given in Table 7.1 are generalizations of the aqueous solubilities of some of the more common ionic compounds.

It is possible to use Table 7.1 and a periodic table of the elements to predict the solubilities of many compounds. For example, calcium nitrate, $Ca(NO_3)_2$, should be soluble, because all nitrates are soluble. On the other hand, copper carbonate, $CuCO_3$, should be insoluble because only the carbonates of the Group 1A elements and NH_4^+ are soluble, and copper is not a Group 1A element. For a similar reason, iron hydroxide, $Fe(OH)_3$, is insoluble in water. Because most sulfates are soluble, sodium sulfate, Na_2SO_4 would be expected to dissolve in water.

7.5 Your Turn

Using the solubility rules of Table 7.1, predict the water solubility of the following compounds and cite reasons for your predictions.

a. magnesium hydroxide, $Mg(OH)_2$ (milk of magnesia)
b. barium sulfate, $BaSO_4$ (used in X-ray diagnosis for ulcers)
c. lithium carbonate, Li_2CO_3 (used to treat manic depression)
d. strontium nitrate, $Sr(NO_3)_2$ (used in fireworks)

Solubility rules can help us understand the chemistry of Onondaga Lake. Over the years, waste beds were added until they covered over 1500 acres (the equivalent of about 1500 football fields). But calcium chloride and sodium chloride continued to wash into the lake. A water engineer from the Department of Environmental Conservation estimates that 2 million pounds of salty water (the equivalent of about 50 truckloads) enter Onondaga daily.

Much of the calcium, carried as Ca^{2+} ions, reacts with the CO_3^{2-} ions that are normally present in the water. The carbonate ions arise when carbon dioxide dissolves in water and the resulting carbonic acid (H_2CO_3) partially dissociates.

$$H_2O(l) + CO_2(g) \rightarrow H_2CO_3(aq) \rightarrow 2\,H^+(aq) + CO_3^{2-}(aq) \qquad (7.5)$$

The Ca^{2+} and CO_3^{2-} ions react to form $CaCO_3$, which has a very low water solubility—only about 0.05 grams per liter at 25°C. Consequently, most of the calcium carbonate precipitates out of solution as a solid.

$$Ca^{2+}(aq) + CO_3^{2-}(aq) \rightarrow CaCO_3(s) \qquad (7.6)$$

The deposition of solid $CaCO_3$ has increased the sedimentation rate of the lake several-fold, and a large $CaCO_3$ delta exists where Ninemile Creek flows into the lake. Estimates based on computerized particle analysis indicate that as many as 17,000 tons of the compounds have been deposited in a single year. Current studies indicate that a layer of calcium carbonate 3–4 feet thick covers the bottom. In short, Onondaga Lake is a saturated solution of calcium carbonate.

But there is at least a little good news in the artificial limestone bottom of the lake. Because of the basic properties of $CaCO_3$, Onondaga is immune from the effects of acid rain. For example, to neutralize fully the 17,000 tons of calcium carbonate that have been put into the lake annually would require enough 0.0005 M sulfuric acid (pH 3.0) to fill 8 million swimming pools. This volume is about 2% of the total average daily precipitation on the continental United States.

7.6 The Sceptical Chymist

Do a calculation to check the claim that to fully neutralize 17,000 tons of $CaCO_3$ would require enough 0.0005 M H_2SO_4 to fill 8 million swimming pools.

Hint: The key to solving this problem is the equation for the reaction of H_2SO_4 and $CaCO_3$.

$$H_2SO_4(aq) + CaCO_3(s) \rightarrow CaSO_4(aq) + CO_2(g) + H_2O(l)$$

Note that the number of moles of H_2SO_4 reacting equals the number of moles of $CaCO_3$ reacting. Using the molar mass of $CaCO_3$ (100.1 g/mole) permits the calculation of the number of moles of $CaCO_3$ in 17,000 tons of the compound. This number is equal to the number of moles of H_2SO_4 required for complete reaction. The concentration of the acid is known: 0.0005 moles H_2SO_4 per liter of solution. Therefore, it should be possible to calculate the total volume of acid necessary to fully neutralize the calcium carbonate. It's up to you to find the volume of a swimming pool.

■ *A Not-So-Great Salt Lake*

The relatively high concentrations of various ions make Onondaga very different from most freshwater lakes. Typically, such lakes have sodium and chloride concentrations below 10 ppm and calcium at less than 50 ppm. In Onondaga, the principal cations are Na^+ (550 ppm) and Ca^{2+} (500 ppm); the chief anion is chloride, Cl^- (425–450 ppm), with sulfate ions, SO_4^{2-} (150 ppm) and nitrate ions, NO_3^- (1 ppm) at much lower concentrations. Sodium, calcium, and chloride ions are clearly the most common ions, accounting for nearly 90% of the salinity in the lake. Recent studies indicate their origins to be largely the manufacturing plants that bordered the lake and not the brine springs as originally thought. Table 7.2 summarizes data for the chief ionic species in Onondaga Lake and includes a number of ions at lower concentration.

■ **Table 7.2**	**Selected Ionic Species in Onondaga Lake**		
Cations	**Concentration (ppm)**	**Anions**	**Concentration (ppm)**
Sodium (Na^+)	550	Chloride (Cl^-)	425–450
Calcium (Ca^{2+})	500	Sulfate (SO_4^{2-})	150
Copper (Cu^{2+})	0.030	Nitrate (NO_3^-)	1.0
Chromium (Cr^{3+})	0.010	Phosphate (PO_4^{3-})	0.35
Cadmium (Cd^{2+})	0.005		

7.7   *Your Turn*

Use the solubility rules of Table 7.1 to explain why the ions in Lake Onondaga water can coexist in solution without precipitating out.

Given the data of Table 7.2, it is not surprising that Syracuse does not take its municipal water supply from Onondaga Lake. Rather, the source is Skaneateles Lake, about 30 miles southwest of the city.

The fact that the phosphate ion concentration is quite low in Onondaga Lake is a consequence of the high concentration of calcium ions. The Ca^{2+} ions react readily with PO_4^{3-} to form calcium phosphate, $Ca_3(PO_4)_2$. This compound is quite insoluble in water, and it precipitates readily if additional phosphates are added to the lake.

$$3\ Ca^{2+}(aq) + 2\ PO_4^{3-}(aq) \rightarrow Ca_3(PO_4)_2(s) \tag{7.7}$$

The phosphate ion is an important contributor to algae growth. The fact that the concentration of PO_4^{3-} is low in Onondaga means that excessive algae growth (eutrophication) is not a problem. Indeed, in 1968, Syracuse and Allied Chemical entered into an agreement according to which excess calcium chloride and calcium hydroxide from the soda works were used to treat the city's wastewater to remove phosphate ions.

In Table 7.2, the concentrations of ions in water are expressed in parts per million (ppm) *by weight or mass*. Thus, to say that Onondaga Lake has a sodium concentration of 550 ppm means that there are 550 g of Na^+ dissolved in 1 million grams (1000 kg or about 1 ton) of water. Note that when we discussed atmospheric pollutants, ppm referred to the number of *molecules* of a particular pollutant out of a total of 1 million gas molecules.

Whether the units are mass or molecules, one part per million is very small, but just how small? Many newspaper and magazine articles (and this textbook) quote values for the concentration of various trace substances in air, water, fish, apple juice—almost anything. Some levels are in the parts per million range, some in parts per billion, some even in parts per *trillion*. The Sceptical Chymist very appropriately asks: "Can I believe those numbers? How were they obtained? How reliable are they? How significant are they? How pure is pure? And do I have to worry about it?"

Obviously, but unfortunately, there is no general answer to these questions. Each set of data must be investigated separately. One important consideration is the technique used to make the measurement. Over the past twenty years amazing advances have been made in the methods used to analyze for small quantities of contaminants. Advances in optics, electronics, and our understanding of chemical and physical phenomena have led to the development of instruments that have increased our ability to detect ever lower concentrations. The limits of detection have gone from parts per thousand to parts per million, parts per billion, and, for some substances, parts per trillion. Routine instrumental analytical methods are able to detect a wide variety of metals at concentrations as low as 2 ppm for mercury and 3 ppb for sodium and magnesium. Special techniques allow analysis at the level of parts per trillion for many metals.

It is difficult to comprehend just how dilute these contamination levels really are, so we again resort to analogies. One part per billion is equivalent to finding a specific blade of grass in 50 football fields. One part per trillion corresponds to finding that particular blade of grass in a field the size of Indiana. It is virtually impossible that any sample of any substance, natural or artificial, can be free of all contamination at this level of sensitivity. Concern for pure water, air, and food is unquestionably justified, but attempting to legislate against an occasional stray atom or molecule is much ado about next-to-nothing.

7.8   *Your Turn*

Because atoms, ions, and molecules are so small and because there are so many of them, the concentration of a contaminant can be very low, but the number of particles can be very large. Assume a sample of water contains 24 ppb of lead. Calculate the number of Pb^{2+} ions in an 8 oz glass of this water. (There are approximately 30 cm^3 per fluid ounce, the density of water is 1.0 g/cm^3, and Avogadro's number is 6.02×10^{23}.)

Ans. 1.7×10^{16} Pb^{2+} ions

■ *Mercury and Onondaga Lake*

Unfortunately, the Solvay story is only part of the sad tale of the pollution of Onondaga Lake. Calcium chloride and calcium carbonate are not the only contaminants in the lake. A greater health threat is the high level of mercury in its water and fish (Figure 7.6).

Mercury (Hg) has fascinated humans since antiquity. The only metallic element that is a liquid at room temperature, it does indeed appear to be "quicksilver." Because of its physical properties (and in some instances, chemical properties), mercury is uniquely suited for use in thermometers, barometers, fluorescent lights, electrical switches, dental fillings, and batteries. Although the inhalation of mercury vapor can be harmful, the element itself and many of its compounds are essentially insoluble in water. Hence, it would seem unlikely that a significant concentration of it could build up in a lake, a fish, or a human being, yet it does. The route by which this occurs was determined about twenty years ago by the sophisticated studies of chemists and biochemists. The pathway involves conversion of metallic mercury by bacterial action to methylmercury ion (CH_3Hg^+). This ion is soluble in water, and thus enters the food chain, ultimately ending up in fish and the people who eat them.

Analyses carried out in 1970 showed that more than 90% of the surface sediments of Onondaga Lake contained metallic mercury at concentrations greater than 0.10 ppm. In that same year, mercury levels in fish from the lake were found to exceed 0.5 ppm, the maximum permissible level at that time. Some fish had mercury concentrations as

Figure 7.6
A warning against fishing in Onondaga Lake.

high as 3.6 ppm—almost entirely in the form of methylmercury ions. As a result of these elevated mercury levels, fishing was banned. Although "catch and release" fishing was reinstituted in 1986, the prohibition against eating the fish continues. And so does mercury in the fish. For example, small-mouth bass sampled in 1989 had mercury concentrations above 2.0 ppm—similar to those in 1970.

The concern over the consumption of mercury-contaminated fish is a consequence of the known toxicity of the element. The physiological effects of mercury poisoning can be severe. In humans, the symptoms include inflammation of the mouth and gums, muscle spasms, severe nausea, diarrhea, kidney damage, blindness, deafness, and damage to the brain and central nervous system. Small amounts of mercury, ingested or inhaled over a long period of time, appear to have cumulative effects. It is very likely that the expression, "mad as a hatter," refers to the fact that people who made hats were routinely exposed to the mercuric (Hg^{2+}) salts used to prepare felt from beaver and other pelts (Figure 7.7). The resulting chronic mercury poisoning was sometimes manifested in personality changes and eccentric behavior. A more recent and more severe case of mercury toxicity occurred in the Minamata Bay area of Japan from 1953–1960. Mercury salts, released into the bay from manufacturing plants, contaminated fish and shellfish. Mercury levels of 5–20 ppm were found in the seafood eaten by the 111 people diagnosed with "Minamata Disease." Of these, 45 died as an apparent result of the poisoning.

Figure 7.7
The Mad Hatter from Sir John Tenniels' illustrations for *Alice's Adventures in Wonderland* by Lewis Carroll.

The mechanism of mercury poisoning is likely similar to that of other **heavy metals,** which are elements with large atomic masses and generally high densities. Included in this somewhat ill-defined set are mercury, lead (Pb), cadmium (Cd), and antimony (Sb). Heavy metals are toxic mainly because their ions interfere with the normal functioning of key **enzymes,** the protein-based biological catalysts that influence the rate and direction of essentially all the chemical reactions of life. In order for many enzymes to function, they must be bonded to Ca^{2+} or Mg^{2+} ions. If these essential ions are replaced by heavy-metal ions, enzymes can cease to operate, perhaps thus shutting down a critical biochemical process. The presence of heavy metals in many bodies of water worldwide and their potentially toxic effects are concerns on a global scale.

Ironically, mercury and mercury compounds have been used for hundreds of years for various medicinal purposes, including the treatment of syphilis and the general prolonging of life. Many of the alleged benefits proved to be illusory at best. But certain mercury compounds, especially mercurous chloride (calomel), Hg_2Cl_2, are still sometimes found in laxatives and antiseptics. In general, the compounds containing Hg^{2+} (mercuric) ions are more toxic that those containing Hg_2^{2+} (mercurous) ions. The transformation of mercury from an important component of many medicines to a severely restricted substance is an example of how the quality of public health depends on our growing knowledge of chemistry.

7.9 ■   ***Consider This***

You have been appointed a member of a regional commission to investigate alleged pollution in a nearby lake. No significant pollution was detected in the lake in the previous water quality study done ten years earlier. Develop a list of five or six important issues, in order of priority, that the commission should address in its investigation of the alleged problem.

■ *The Source of the Mercury: The Chlor-Alkali Process*

Mercury pollution continues to plague Onondaga Lake. A 1987 study by the New York State Department of Environmental Conservation estimated that about 7 million cubic yards of the lake-bottom sediments contain mercury at concentrations greater than 1.0 ppm. The upper layers of this sediment are particularly rich in mercury and other heavy metals, with concentrations exceeding 40 ppm in the top 6 inches. Mercury levels in sediments of two major tributaries are greater than 10 ppm. But where does the mercury come from? Skeptical readers will note that thus far we have offered no evidence to answer that crucial question.

The answer is not obvious. There are no natural deposits of mercury-containing minerals near Onondaga. Nor are mercury or mercury compounds produced in any of the factories that line the lake. However, mercury was used in the chlor-alkali process carried out in a plant formerly owned by Allied Chemical and later by Linden Chemicals and Plastics.

Salt brought the native Iroquois and much later the Solvay process to the shores of Onondaga Lake. That same compound also drew the **chlor-alkali process,** which uses aqueous sodium chloride as the starting material. As the name suggests, the process produces elementary chlorine and an alkali or base—sodium hydroxide, NaOH. Chlorine and sodium hydroxide are two of the world's most important industrial chemicals. They are used in the paper and textile industries and to make hundreds of other chemicals that find their way into countless products. In 1990, the chlor-alkali process produced over 22 billion pounds of sodium hydroxide and an equal mass of chlorine in the United States alone.

The chlor-alkali process involves the electrolysis of a sodium chloride solution (brine). You will recall from previous experience (or Chapter 5) that when a direct electric current is passed through water, the compound breaks down into hydrogen and oxygen gas. However, when a salt solution is subjected to electrolysis, a somewhat different reaction occurs. In addition to H_2O molecules, the solution contains Na^+ and Cl^- ions, and it is these ions that are transformed under the influence of the applied voltage. The ions are converted to atoms. This means that the compound, sodium chloride, is broken down into its constituent elements, sodium and chlorine.

$$2\,NaCl \rightarrow 2\,Na + Cl_2 \tag{7.8}$$

You will recognize equation 7.8 as the reverse of equation 5.2. Sodium and chlorine in their elementary forms readily transfer electrons to yield the ionic compound, sodium chloride. That process gives off energy and happens spontaneously. But salt will not decompose into its elements by itself. To convert NaCl to Na and Cl_2, energy must be supplied. The most convenient way to do so is electrically.

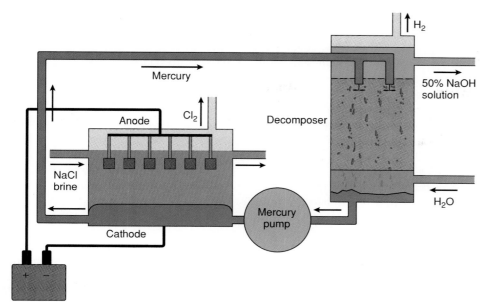

Figure 7.8
Diagram of a mercury chlor-alkali cell.
(From *McGraw-Hill Encyclopedia of Science and Technology,* 5e (1982). Copyright ©1982 McGraw-Hill, Inc. Reprinted by permission.)

In the chlor-alkali process, the chemistry occurs as two linked half-reactions in a large electrolysis chamber (Figure 7.8). A graphite electrode (the **anode**) is the site of the conversion of chloride ions (Cl^-) to elementary chlorine gas (Cl_2), which is collected as it is formed.

$$2\ Cl^-(aq) \rightarrow Cl_2(g) + 2\ e^- \tag{7.9}$$

In this step, each Cl^- ion loses one electron (e^-) to form a neutral chlorine atom. These atoms combine to form Cl_2 molecules. This is another example of oxidation, a chemical change that results in the loss of one or more electrons by an ion, atom, or molecule.

Oxidation must always be balanced by reduction, a process in which the reacting species gains electrons from the accompanying oxidation reaction. In this case, each Na^+ ion is reduced to a neutral Na atom by gaining an electron at the other electrode (the **cathode**). If the newly formed sodium metal were to come in contact with the water of the solution, it would immediately react. But the reaction is prevented by the presence of mercury. The cathode is a large pool of the liquid metal. As sodium metal is produced, it dissolves in mercury to form an **amalgam,** which is a solution in which mercury is the solvent. In the equation for the reaction, the amalgam is symbolized as Na(Hg).

$$2\ Na^+(aq) + 2\ e^- \rightarrow 2\ Na(Hg) \tag{7.10}$$

The amalgam is then sprayed into water where the sodium reacts violently according to equation 7.11.

$$2\ Na(Hg) + 2\ H_2O(l) \rightarrow 2\ NaOH(aq) + H_2(g) \tag{7.11}$$

The products are a 50% aqueous solution of sodium hydroxide (caustic soda) and hydrogen gas, an important by-product. The NaOH can be sold either in solution or evaporated to yield the solid compound.

If equations 7.9, 7.10, and 7.11 are added together, the result is the net over-all reaction for the chlor-alkali process.

$$2\ NaCl(aq) + 2\ H_2O(l) \rightarrow 2\ NaOH(aq) + Cl_2(g) + H_2(g) \tag{7.12}$$

Note that mercury does not appear in this equation. It is neither consumed or produced by the reaction. It simply serves as the cathode and as a solvent for the sodium, and it recirculates to form an amalgam with more sodium. Therefore, it should not be released to the environment. At least that is what is supposed to happen.

In practice, things were a little different at Onondaga Lake. A single electrolysis cell, producing as much as 15 tons of chlorine per day, could contain up to four tons of mercury; and a single plant could consist of dozens of cells. Although the reaction took place in a closed system, some of the mercury escaped through leakage, vaporization, or as the cells were cleaned or replaced.

Thanks to chemical and engineering research, mercury has largely been eliminated from the manufacture of chlorine and caustic. In modern plants, the two half-reactions are separated by barriers made of synthetic resins. Iron cathodes have replaced mercury and new metallic anodes are being used in place of graphite. But these changes came too late to help Onondaga Lake.

It is estimated that from 1946 to 1953, almost 11 pounds of mercury entered the lake each day. From 1953 to 1970, the mercury discharged was approximately twice that. Then, in 1970, the United States Justice Department brought suit against the manufacturers, forcing them to significantly lower mercury emissions. In 1977, Allied Chemicals installed a $1 million facility to reduce mercury output to 0.45 ounces per day on average to meet Environmental Protection Agency guidelines. The plant, sold in 1979 to Linden Chemicals and Plastics, was closed in 1988. In spite of the plant closing, recent studies have indicated that mercury is still flowing into the lake from tributaries, but at levels far reduced from those before 1970.

7.10 The Sceptical Chymist

A newspaper article reported that Onondaga Lake contains as much as 47 tons of mercury dumped as industrial waste. Is this figure reasonable?

Ans. In order to support or refute this report we need to do our own calculation. According to information in the previous paragraph, 11 pounds were released per day from 1946–1953 and 22 pounds were released per day from 1953–1970. The total discharge is determined as follows.

Mercury released 1946–1953:

11 lb/day × 365 days/yr × 8 yrs × 1 ton/2000 lb = 16 tons

Mercury released 1953–1970:

22 lb/day × 365 days/yr × 17 yrs × 1 ton/2000 lb = 68 tons

Total mass mercury released = 16 tons + 68 tons = 84 tons

There is a significant discrepancy between this result and the newspaper estimate of 47 tons. Suggest at least three reasons to explain the difference.

7.11 Your Turn

Assuming that Onondaga does contain 47 tons of mercury, calculate its monetary value if mercury is selling for $120 per pound.

■ Onondaga Lake: Looking Back and Looking Ahead

Given mercury discharges, excess sodium chloride and calcium chloride drainage, and other polluting conditions, it might well appear that the chemical industry is the villain in the tragedy of Onondaga Lake. Some might suggest that all that needs to be done to restore the lake to a much cleaner condition is to banish the factories from its shores.

7.12	*Consider This*

Suppose you and several of your classmates decide that one way to reduce possible tax increases for lake cleanup is to "mine" Onondaga Lake for mercury and sell it. Prepare a feasibility study in which you identify the major economic and environmental factors that must be explored before you and your partners make a decision. You would be well advised to check any calculations that you make. After World War I, the German chemist Fritz Haber proposed a plan to pay German war reparations by extracting gold from seawater. It looked good on paper, but it turned out that Haber's estimate of the gold concentration was ten times too high.

That is a simplistic answer to a very complex problem. Chemical manufacturing is admittedly a major part of the problem, but chemistry is also an essential component in the solution.

In the first place, the 17 major plants that line the lake and its tributaries provide jobs and products desired by the public. They produce essential chemicals, specialty steels and alloys, many metal plated objects, important industrial solvents, and fine china. All of the plants currently operating are subject to discharge permits that regulate the amount and nature of effluents they can release into the lake. Chemistry is certainly required to analyze the effluents and to modify procedures to permit compliance in accordance with limits and other regulations established by federal, state, and county agencies.

7.13	*Consider This*

A lawsuit has been brought against a chemical company charging that the company polluted a body of water in its vicinity. Assume you are the lawyer defending the company in the lawsuit. In order to build the case for the defense, you have scheduled a conference with the company's director of chemical research. Compile a list of five or six pivotal questions that you plan to ask the director.

The dramatic decrease in the amount of mercury released from the chlor-alkali process indicates that the chemical industry has done much to clean up its act in Syracuse. It appears to be willing to do more. A spokesperson for Allied Chemical has said that the company's officers recognize the problems of Onondaga Lake and their responsibility to "do whatever we can to correct them." Moreover, Allied has argued, with considerable justification, that some of the most hazardous pollution of the lake is not the consequence of chemical manufacturing at all. Rather, it stems from the fact that for many years the city of Syracuse and the surrounding county used the lake to dispose of domestic wastes. Inadequate sewage treatment was a way of life for some time.

Raw domestic sewage was dumped directly into Onondaga until 1925, when the first wastewater treatment plant was built. The local chemical industry cooperated with the city in some of its early efforts to purify the wastewater returned to Onondaga Lake. In the 1920s, the Solvay Process Company and the city of Syracuse agreed to mix city sewage sludge with Solvay solid wastes for land storage. The arrangement seems to have worked reasonably well until 1950 when a strike temporarily closed the Solvay plant. With no place to dump its sludge, the city shut down the municipal wastewater treatment plant and dumped its raw sewage directly into the lake for the next four years. Since then, two upgrades to the municipal wastewater treatment plant have been completed, the last in 1979 at a cost of nearly $300 million.

It is obvious that the wastewater treatment plant can only carry out its function if the water passes through it. Unfortunately, this does not always happen. The treatment facility is occasionally bypassed during heavy rains which cause the antiquated sanitary and storm sewers to overflow. This sends raw sewage directly into the lake. These episodic events cause high counts of coliform bacteria that come from human wastes and may cause sickness. Their presence is a major reason why swimming is prohibited. In 1979, at a cost of $10 million, the overflow issue was addressed by removing debris from sewer lines and enlarging and cleaning pipes to enhance their carrying capacities. Although a step in the right direction, it has not completely solved the problem—a problem common to many cities, both large and small.

All such efforts are, of course, attempts to clean up the current input into the lake. None are targeted to remediate the damage done over the last century. That is a far more difficult and far more expensive project. Estimates of the cost of cleaning Onondaga run to $1 billion.

Senator Daniel Moynihan (D-NY) has called Onondaga the most polluted lake in the United States. It has lots of competition for that dubious distinction. At least 10,000 major lakes in the country are said to have serious pollution problems. Responding to the question of whether Onondaga Lake is perhaps at the top of the list, Steve Effler of the Upstate Fresh Water Institute replies, "Do you know of one worse?" Be that as it may, Onondaga is not on the Environmental Protection Agency's list of the most toxic freshwater sites because massive amounts of pollutants are no longer regularly dumped into the lake.

Nevertheless, widespread concern about the fate and future of the lake continues. Thousands of words have been written about Onondaga, and millions of dollars have been poured into it (along with lots of other stuff). The degradation of the lake and efforts to restore it have been the subjects of nearly 200 reports—enough paper to form a four-foot stack. A current study by the U.S. Army Corps of Engineers just to analyze, evaluate, and summarize previous reports and recommend cleanup options will cost over $400,000 by the time it is completed in 1993.

A particular irritant is the fact that no sustained, coordinated study and planning effort was made until recently. More than 30 federal, state, and local agencies have shared responsibility for the lake's management, creating a situation that has led to diffused decision making and inadequate centralization of duties. Recently, however, the Onondaga Lake Management Conference was formed. This group, comprised of local, state, and federal representatives, will seek to coordinate efforts to clean up the lake. Speaking about the newly-formed Conference, Robert Hennigan, environmental studies chairman at the State University of New York College of Environmental Science, says, "This is the first time we've really had top-level officials make a commitment in public to clean up Onondaga Lake, and that's a sign that something's really happening."

Most officials feel that federal monies are needed as an essential part of funding cleanup activities. Senator Moynihan has sponsored legislation proposing $100 million to begin the reclamation. Governor Mario Cuomo expressed his hopes by stating, "There's no reason we can't turn back the clock to a time when the lake was a showcase instead of a terrible sorrow." Lee Flocke, a water quality engineer with the Department of Environmental Conservation sets a more modest goal: "Too much has been done to it for the lake to ever return to its pristine state. I just hope we can have swimming and fishing again in my lifetime."

7.14   *Consider This*

Suppose you are a citizen of some state other than New York. Senator Moynihan's bill proposing a $100 million federal appropriation to help clean up Onondaga Lake is currently being considered by the Senate. Draft a letter to your Senator, stating your position on the legislation.

7.15 **Consider This**

Like Lee Flocke, you too would like to see Onondaga Lake restored to a former level of quality. Assume that you live in Syracuse and as a taxpayer are faced with several options about what level you would support regarding lake cleanup. Each has costs that will require new taxes. Estimated annual tax increases (per taxpayer) to achieve the various quality levels are swimming, $750; fishing, $800; and showcase, $1600. Write a position paper supporting which quality level you choose.

7.16 **Consider This**

Suppose you were an advisor to Mayor Baldwin's successor in 1883, but knew what you have learned in this chapter. A company contacts the mayor wanting to buy lakeshore property for a Solvay process plant. The mayor seeks your advice and asks you to draft a memorandum to him outlining your position in support of or opposition to the company's request. What would your memo say?

■ Conclusion

Well into this century, five smokestacks were part of the Syracuse city crest, proud symbols of the city's manufacturing tradition and legacy. The Solvay Process Company was the first large-scale manufacturing firm on Onondaga Lake. Other chemical manufacturing enterprises also began and drew additional heavy industry to this site. These plants provided jobs, good wages, a moderate tax rate, and, as you have read, many benefits to the community. Moreover, the factories were generally operated in a manner that, at the time, seemed environmentally responsible. Both the Solvay process and the chlor-alkali process, designed to recycle and retain intermediates, marked significant improvements over previous technology.

Paradoxically, the increase in productivity, population, income, and standard of living made possible by industrialization led to the decline of Onondaga Lake. Calcium salts and mercury came from industry, bacteria came from untreated sewage. And, as analytical methods became more sensitive and chemical knowledge gained in sophistication, scientists and the public became more conscious of the hazards associated with these pollutants. Risks that might have been acceptable in a developing country are no longer tolerable in the United States of America in the last decade of the twentieth century.

Today Onondaga Lake is a far cry from the splendid jewel described in 1847 by Mayor Baldwin. Yet, it is deceptive in its appearance. The citizens of the city may not "throng to it for pleasure, relaxation, or improvement of health" but the several parks that line its shores are popular. The authors of this text have picnicked there. Many people visit the restored salt works at the Salt Museum. In spite of the warning signs, the water looks inviting. Every spring, the lake is crowded with the rowing shells of colleges and universities competing in the Intercollegiate Rowing Association Eastern Sprints. And, while the abandoned Solvay plant is a reminder of things past, the new Carousel Center speaks to the future.

It is often futile to look back and ask "What if?" We cannot undo what has already happened. On the other hand, we can influence the future by learning from our past mistakes. Modern chemistry can do much to prevent such disasters from recurring elsewhere. As more and more developing countries weigh the risks and benefits of industrialization and technological innovation, it becomes essential to retell the cautionary tale of Onondaga Lake.

■ *References and Resources*

Effler, S. W. "The Impact of a Chlor-Alkali Plant on Onondaga Lake and Adjoining Systems." *Water, Air, and Soil Pollution* **33** (1987): 85–115.

Greek, B. F. "Squeeze on Soda Ash Capacity Pushes Expansion." *Chemical & Engineering News,* March 12, 1990: 17.

————. "Demand for Soda Ash Holding Up Well During Recession." *Chemical & Engineering News,* Oct. 28, 1991: 9–11.

■ *Experiments and Investigations*

11. Building and Using a Conductivity Tester

12. Solubilities: An Investigation

13. A Study of Onondaga Lake Water

■ *Exercises*

*1. Onondaga Lake contains 35 billion gallons of water with a Ca^{2+} concentration of 500 ppm. Calculate the total mass of Ca^{2+} present in the lake. (To simplify the calculation, assume the density of the water in the lake is 1 g/mL and let 1 gal = 4 L = 4000 mL.)

2. According to the text ". . . from 1960 to 1985, 2000 to 2800 tons of soda ash were produced daily" by the Solvay plant. Compute the total mass of Na_2CO_3 produced during this 25-year period.

*3. In the text, the solubilities of $NaHCO_3$ and NH_4Cl are reported in grams of solute per 100 mL of solution at $0°C$. Calculate the solubility of each of these compounds in moles per liter at this temperature.

4. Write balanced chemical equations for the following reactions that are important in the Solvay process.

   a. the conversion of limestone, $CaCO_3$, to lime, CaO, by heating

   b. the reaction of lime with water to form calcium hydroxide, $Ca(OH)_2$

   c. the reaction of CaO and ammonium chloride, NH_4Cl, to yield $CaCl_2$, NH_3, and H_2O

*5. Write equations to represent the reaction of $CaCO_3$, Na_2CO_3, and $NaHCO_3$ with HCl. What do all of these reactions have in common?

*6. The conversion of $NaHCO_3$ into Na_2CO_3 is represented by equation 7.4. Use that equation to determine the mass of Na_2CO_3 that could be produced from 100 g $NaHCO_3$.

7. Calculate the mass of Na_2CO_3 that can be obtained from 100 g of the mineral trona, $Na_2CO_3 \cdot NaHCO_3 \cdot 2\ H_2O$ if only the Na_2CO_3 initially present in the compound is isolated. How might the yield of sodium carbonate be increased?

*8. The Ca^{2+} concentration in Onondaga Lake is given as 500 ppm.

   a. Convert the Ca^{2+} concentration to moles per liter of solution.

   b. Calculate the number of moles of Ca^{2+} present in the lake, which has a volume of 140×10^9 L.

   c. Assume one gram of an ion exchange resin will remove 5.0×10^{-4} mole Ca^{2+}. Calculate the mass of this resin that would be required to completely remove Ca^{2+} from the lake.

*9. When solutions of certain ionic compounds are mixed, precipitates form. For example, mixing 0.10 M NaCl and 0.10 M $AgNO_3$ results in the precipitation of AgCl, silver chloride, a compound with a very low water solubility. An equation for this reaction can be written using only the ions involved in the precipitation:

$$Ag^+(aq) + Cl^-(aq) \rightarrow AgCl(s)$$

Because $NaNO_3$ is water soluble, its ions remain in solution. Assume 0.10 M solutions of the following compounds are mixed. For each pair, investigate whether the positive and negative ions can react to form an insoluble compound (or compounds). If precipitation does occur, write an ionic equation (as illustrated above) for the reaction. (Make use of the solubility rules of Table 7.1.)

   a. sodium carbonate, Na_2CO_3, and lead nitrate, $Pb(NO_3)_2$

   b. barium chloride, $BaCl_2$, and magnesium sulfate, $MgSO_4$

   c. potassium chloride, KCl, and copper nitrate, $Cu(NO_3)_2$

   d. sodium phosphate, Na_3PO_4, and magnesium chloride, $MgCl_2$

*10. Compare the relative magnitude of concentrations expressed in parts per million by mass and parts per million by molecules by making the following conversions.

    a. 550 ppm Na^+ by mass in H_2O to ppm by molecules (or ions)

    b. 150 ppm SO_4^{2-} by mass in H_2O to ppm by molecules

    c. 9 ppm CO in N_2 by molecules to ppm by mass

11. Briefly describe the process by which "heavy metals" function as poisons.

12. According to one source on mercury poisoning, it takes the human body 70 days to eliminate one-half of any mercury present. How long would be required to reduce body levels to one-fourth and one-eighth of initial levels?

*13. The oxidation-reduction reactions below are important commercially. For each reaction, identify the substance that undergoes oxidation and the substance that undergoes reduction.

    a. $Fe + Cu^{2+} \rightarrow Fe^{2+} + Cu$

    b. $Al_2O_3 + 3 H_2 \rightarrow 2 Al + 3 H_2O$

    c. $Zn + 2 MnO_2 + H_2O \rightarrow Zn(OH)_2 + Mn_2O_3$

    d. $Fe_2O_3 + 3 CO \rightarrow 2 Fe + 3 CO_2$

14. Aluminum metal is produced by the electrolysis of molten Al_2O_3, which is represented by the following equation:

$$2 Al_2O_3(l) \rightarrow 4 Al(l) + 3 O_2(g)$$

    a. Identify the species (atom or ion) being oxidized in the reaction and the species being reduced.

    b. Write equations showing the loss and gain of electrons in the oxidation and reduction reactions.

    c. Explain why significant savings in energy can be realized by recycling aluminum cans.

15. The mercury discharged from the chlor-alkali plant in Syracuse was 11 pounds per day in the 1940s but only 0.45 ounces per day beginning in 1977. Calculate the percentage the later value is of the earlier one. (1 lb = 16 oz)

*16. You have now learned about acid-base, precipitation, and oxidation-reduction reactions. Classify each of the following reactions in one or more of these categories.

    a. $MgCl_2 + 2 NaOH \rightarrow Mg(OH)_2 + 2 NaCl$

    b. $2 H_2 + O_2 \rightarrow 2 H_2O$

    c. $HNO_3 + NH_3 \rightarrow NH_4NO_3$

    d. $H_2SO_4 + Ba(OH)_2 \rightarrow BaSO_4 + 2 H_2O$

17. This chapter is a study of a case where technological innovation and industrialization were initially beneficial, but were later discovered to have some serious negative consequences. Suggest another similar instance, and do the necessary library research to support your choice.

8

The Fires of Nuclear Fission

■

Nuclear phenomena—probably no subject in all of physical science is more likely to provoke an emotional response. The word "nuclear" carries a tremendous baggage of upsetting associations, including the bombing of Hiroshima and Nagasaki, radioactive fallout from bomb tests, radiation-induced cancer and birth defects, the risks of accidents and meltdowns, the difficulties of disposing of radioactive wastes, and the ultimate threat of nuclear annihilation. And yet, many benefits spring from the very heart of matter—the production of electricity by nuclear power plants, the uses of radioactivity and other nuclear phenomena in medicine for the diagnosis and treatment of a wide variety of ailments, and the technological exploitation of nuclear materials in industry. The applications of nuclear phenomena, harmful at one extreme and beneficial at the other, present us with a dilemma of risks and benefits. It is a double-edged sword of Damocles that hangs precariously over our heads.

Certainly the largest, and arguably the most controversial, non-military application of nuclear energy is the generation of electricity by nuclear power plants. When it was first demonstrated that electricity could be obtained from the splitting of atoms, a new age appeared to be dawning. The first commercial nuclear power generating station in this country was completed in 1957 at Shippingport, Pennsylvania, along the Ohio River near Pittsburgh. With great fanfare and a radioactive "magic wand," President Dwight Eisenhower launched this nation on a course of "atomic energy" (Figure 8.1).

This new source held the promise of unlimited, cheap electricity. During the early 1960s proponents of nuclear power suggested that electricity produced by this method would be so inexpensive that it would be inconsequential to even meter the consumers' use of it. There would be plenty of electricity for everyone! Needless to say, the prediction has not come true.

Figure 8.1

Brandishing a radioactive magic wand, President Dwight Eisenhower, in a Denver TV studio on Labor Day, September 6, 1954, activated an automated power shovel 1300 miles away in Shippingport, Pennsylvania, to begin construction on the first commercial nuclear reactor in the United States.

> ### 8.1 ■ Consider This
>
> Look at the photograph of President Eisenhower (Figure 8.1) and read the caption. Analyze the photograph for its public relations value. What are the messages regarding nuclear power conveyed by this picture? How does this conform to public opinion of nuclear power in the 1950s? Would the same type of photograph work today? Explain your answer.

■ Chapter Overview

Over the last forty years, many critical issues have arisen, issues that have provoked serious doubts about nuclear power. Citizens have asked and continue to ask important questions related to it: How does nuclear fission produce electrical energy? What are the safeguards against a "meltdown"? Can a nuclear power plant explode like an atomic bomb? Is there a danger that nuclear fuel can be diverted to make nuclear weapons? What is radioactivity and what are the hazards associated with it? How long will nuclear waste products remain radioactive and how will such wastes be disposed? What is the current status of nuclear energy, nationally and internationally? And finally, how crucial is nuclear power to our future energy requirements? In this chapter, we address all of these questions. In every instance, we combine scientific fundamentals, application technology, and societal implications. Moreover, we try to temper emotionalism with understanding in order to help readers rationally weigh the risks and benefits of nuclear energy and radioactivity. We begin by considering a case study that will reappear throughout the chapter—a nuclear power plant in Seabrook, New Hampshire. But before we start, we ask you to consider your own position regarding nuclear power.

> ### 8.2 ■ Consider This
>
> Without doing any further reading in this chapter or any research on the subject, write down your answers to the following questions.
>
> 1. How does the electricity produced by a nuclear power plant differ from that produced by a coal-burning plant?
> 2. What is the greatest danger associated with nuclear power plants?
> 3. Given a choice between electricity generated by a nuclear power plant and a traditional coal-burning plant, which would you choose and why?
> 4. Would you be willing to live closer to a nuclear plant or a coal-burning plant?
> 5. Would you support the burial of radioactive waste from a nuclear power plant in an appropriate site in your home state?
> 6. If you are opposed to nuclear power, under what circumstances, if any, would you be willing to change your position?

■ The Seabrook Saga

In 1972, the Public Service Company of New Hampshire proposed building a nuclear power plant on the New Hampshire coast at Seabrook. Plans called for twin reactors, the first to become operational in November 1979 and the second to start operating two years later. Total costs for the project were estimated at $973 million. The site was selected for several reasons. Its location on the Atlantic Ocean provides easy barge access for transporting heavy equipment and the vast supply of cooling water that is necessary for the operation of any power plant, nuclear-powered or otherwise, that converts heat into work and electrical power. The underlying rock is a solid and stable foundation for vibration-sensitive machinery. Most important, Seabrook is within 40 miles of

Boston and more than 4 million people addicted to refrigerators, television sets, kitchen ranges, light bulbs, and hundreds of other conveniences that require electricity.

But proximity to a population center also created problems. The proposed facility was met with prompt and vigorous opposition, and groups such as the Clamshell Alliance formed to combat construction of the plant. In January 1974, the Commonwealth of Massachusetts began legal action to block the project, and two years later the citizens of Seabrook voted to oppose the plant. People on both sides of the issue spoke with strongly-held conviction.

> At its core, the nuclear issue is a confrontation between corporate, technocratic domination and decentralized, community independence. The choice is closely linked to a broad spectrum of issues—to unemployment and high electric rates, the exploitation of Third World people and resources, to the plagues of nuclear armaments, environmental chaos, and our soaring cancer rates.
>
> *Harvey Wasserman, Organizer*
> *Clamshell Alliance*

> There's no question that as time goes on even the people who opposed Seabrook will recognize its benefits to their region and their way of life. It will light the homes and run the factories of New England while emitting no greenhouse gases and while displacing 11 million barrels of oil every year.
>
> *Harold B. Finger, President*
> *U.S. Council for Energy Awareness*

Despite objection, construction of the nuclear facility began in July 1976. The first of several protests followed almost immediately. The largest protest occurred in April and May of 1977 when 2000 demonstrators occupied the site and 1400 were arrested (Figure 8.2).

8.3 Consider This

Suppose you are a reporter for the *Boston Globe* assigned to cover the anti-Seabrook demonstration of 1977. You have made arrangements to interview protest leaders and officials of the Public Service Company of New Hampshire. List the questions that you will ask to obtain the information you need to write a balanced article.

Figure 8.2
Protesting the Seabrook, New Hampshire nuclear power plant.

The Seabrook project was plagued with other problems as well, and in 1984 the owners canceled plans to build the second unit. The initial and only reactor at the site was finally completed in July 1986. However, because of changing federal regulations, legal maneuvering, and the bankruptcy of the Public Service Company of New Hampshire, the reactor was not tested until June 13, 1989. This was 17 years from when it was first proposed and almost 10 years past its initial projected operational date. A major factor in the bankruptcy was the $6.45 billion price tag for the power plant—12 times the initial estimate for a single-reactor system.

The political cost of Seabrook was also high, at least for former New Hampshire governor Meldrin Thomson, Jr., whose defeat was probably a consequence of his strong support of the project. One of Thomson's successors, John H. Sununu, fared better. The prominence he gained as a champion of the Seabrook reactor might well have been a reason for his selection as President George Bush's Chief of Staff. In the opposing camp was Governor Michael Dukakis of Massachusetts, whose refusal to submit evacuation plans for Massachusetts towns within 10 miles of the plant gave him a notoriety that may have helped him win the 1988 Democratic presidential primary.

8.4   *Consider This*

Massachusetts Governor Dukakis used a political ploy, the failure to submit federally required evacuation plans, to obstruct the Seabrook plant from coming "on-stream." If you were a Massachusetts voter living within the 10-mile area, how would you feel about such an action? Would you feel differently if you would have to rely on Seabrook for your electricity?

What was apparently the last hurdle for the Seabrook power station was cleared in March 1990, when the Nuclear Regulatory Commission voted to give the plant an operating license. At full capacity the plant will generate 1150 megawatts of power, which equals 1150 million joules every second. A few pounds of uranium daily will produce the same amount of energy that would consume 1,840,000 gallons of oil or 10,000 tons of coal. No carbon dioxide will be added to the atmosphere to contribute to the greenhouse effect, and no sulfur dioxide will be released to create acid rain.

Although Seabrook has sophisticated safeguards to protect the environment and the nearby populace, feelings about the plant still run very high. Whether the people of New Hampshire, Massachusetts, and the other New England states are winners or losers in this drama remains to be seen. In the pages that follow, we will attempt to assemble some of the evidence.

How Does Fission Produce Energy?

The key to answering this question is probably the best known physical relationship of the twentieth century, $E = mc^2$. It dates from the early years of the century and is, of course, one of the contributions of Albert Einstein (1879–1955). The equation summarizes the equivalence of energy E and matter or mass m. The symbol c represents the speed of light, 3.0×10^8 m/s, so c^2 is equal to 9.0×10^{16} m^2/s^2. The fact that this number is very large means that it should be possible to obtain a tremendous amount of energy from a very small amount of matter, whether in a power plant or in a weapon.

For over 30 years, Einstein's equation was a curiosity. Scientists speculated that it probably described the source of the Sun's energy, but as far as anyone knew, no one had ever observed on Earth a conversion of a substantial fraction of matter into energy. Then, in 1938, two German scientists, Otto Hahn (1879–1968) and Fritz Strassmann (b. 1902), discovered what appeared to be the element barium (Ba) among the products formed when uranium (U) was bombarded with neutrons. The observation was unexpected because barium has an atomic number of 56 and an atomic mass of about 137. Comparable values for uranium are 92 and 238, respectively. At first, the scientists were tempted to conclude that the element was radium (Ra, atomic number 88), which is a member of the same periodic family as barium. But Hahn and Strassmann were fine chemists, and the chemical evidence for barium was too compelling.

The German scientists were unsure of the origin of the lighter element, so they sent a copy of their results to their colleague, Lise Meitner (1878–1968), for her opinion. Dr. Meitner had collaborated with Hahn and Strassmann on related research, but she had been forced to flee Germany in March 1938, because of the anti-Semitic policies of the Nazi government. When she received their letter she was living in Sweden. She discussed the strange results with her physicist nephew, Otto Frisch (b. 1904), as the two of them walked along in the snow. Suddenly, the explanation became clear: under the influence of the bombarding neutrons, the uranium atoms were splitting into smaller atoms of lighter elements. The nuclei of the heavy atoms were dividing like biological cells undergoing **fission.**

That word from biology is applied to a physical phenomenon in the letter that Meitner and Frisch published on February 11, 1939, in the British journal *Nature.* In the letter, entitled "Disintegration of Uranium by Neutrons: a New Type of Nuclear Reaction," the authors state the following:

> Hahn and Strassmann were forced to conclude that isotopes of barium are formed as a consequence of the bombardment of uranium with neutrons. At first sight, this result seems very hard to understand . . . On the basis, however, of present ideas about the behavior of heavy nuclei, an entirely different . . . picture of these new disintegration processes suggests itself . . . It seems therefore possible that the uranium nucleus . . . may, after neutron capture, divide itself into two nuclei of roughly equal size . . . The whole "fission" process can thus be described in an essentially classical way.

The letter is just over a page long, but it would be difficult to think of a more important scientific communication. Its significance was recognized immediately, and Niels Bohr (1885–1962), an eminent Danish physicist, brought the news to the United States on an ocean liner. Within a few weeks of Meitner's and Frisch's interpretation, scientists in a dozen laboratories in various countries confirmed that the energy released by the fission of uranium atoms was that predicted by Einstein's equation.

Energy is given off when an atom splits because the total mass of the products is slightly less than the total mass of the reactants. In spite of what you may have been taught, neither matter nor energy are *individually* conserved. Matter disappears and an equivalent quantity of energy appears as the former is converted to the latter. Alternately, one can view matter as a very concentrated form of energy, and nowhere is it more concentrated than in the atomic nucleus. Remember that an atom is mostly empty space. If the electron orbits marking the outer boundary of an atom formed a sphere half a mile in diameter, the nucleus would be the size of a baseball. Because almost the entire mass of an atom is associated with the nucleus, the density of the nucleus is incredibly high. Indeed, a pocket-sized matchbox full of atomic nuclei would weigh over 2.5 billion tons! Given the energy-mass equivalence of Einstein's equation, this means that the energy content of all nuclei is immense.

It is important to realize, however, that only certain elements undergo fission. Furthermore, not every atom of a fissionable element such as uranium is capable of splitting when struck by a neutron. That depends on the relative number of protons and neutrons in the nucleus. Approximately 99.3% of uranium atoms consist of 92 electrons, 92 protons, and 146 neutrons. The relative mass of each of these atoms is the sum of the number of protons and neutrons, 92 + 146, or 238. This mass number is used to identify the isotope as uranium-238 (U-238).

In the notation of nuclear physics, the mass number is written as a superscript preceding the elementary symbol. The atomic number (the number of protons in the nucleus and hence its positive charge) is written as a subscript. Hence, uranium-238 is represented as follows:

$$\text{Mass number = number of protons + number of neutrons = 238}$$
$$\text{Atomic number = number of protons =   92} \quad \text{U}$$

Although U-238 undergoes spontaneous radioactive decay, it is not fissionable. That is a property of uranium-235, an isotope whose atoms consist of 92 electrons, 92 protons, and 143 neutrons. In naturally occurring uranium, only 0.7% or about one atom out of 140 is U-235; the others are U-238 atoms.

A wide variety of possible products or fission fragments can be formed when the nucleus of an atom of U-235 is struck with a neutron. One typical reaction is given by the equation below.

$$_{0}^{1}\text{n} + _{92}^{235}\text{U} \rightarrow _{56}^{141}\text{Ba} + _{36}^{92}\text{Kr} + 3\,_{0}^{1}\text{n} \tag{8.1}$$

This nuclear equation makes use of the notation just introduced. Note that the subscript for a neutron (designated n) is 0, indicating zero charge. The superscript is 1 because the mass number of a neutron is one. In a balanced nuclear equation such as equation 8.1, the sum of the subscripts on the left side equals that of the subscripts on the right side of the equation. Likewise, the sum of superscripts on each side of the equation must be equal. Coefficients in a nuclear equation, such as the 3 preceding the neutron symbol in the products, are treated the same way as in chemical equations: The coefficient multiplies the term following it. In the particular case above, the coefficient indicates three neutrons. Where no coefficient is given explicitly, a 1 is understood. We can check the correctness of equation 8.1 by applying these rules.

	Left		Right				
Superscripts:	235 + 1	=	141 + 92	+	(3×1) =	236
Subscripts:	92 + 0	=	56 + 36	+	(3×0) =	92

8.5 ■

Your Turn

Use the atomic numbers from the periodic table to write nuclear equations for the following fission reactions that occur when an atom of uranium-235 is struck by a neutron.

  a. The conversion of U-235 to Rb-90, Cs-144, and neutrons.
  b. The conversion of U-235 to an element with an atomic number of 30 and a mass number of 72, another element with atomic number 62 and mass number 160, and neutrons.

Ans. **a.** $_{0}^{1}\text{n} + _{92}^{235}\text{U} \rightarrow _{37}^{90}\text{Rb} + _{55}^{144}\text{Cs} + 2\,_{0}^{1}\text{n}$ (Note that two neutrons must be produced in order to balance the superscripts or mass numbers.)
b. Don't forget to look up the symbols of the elements with atomic numbers 30 and 62.

8.6 ■

Your Turn

Strontium-90 (Sr-90) is a radioactive fission product that contaminated milk for some time after atmospheric bomb tests. It can be formed from the neutron-induced fission of U-235 in a reaction that also produces three neutrons and another element. Identify the other product element and write a nuclear equation for the reaction.

Although the sum of the mass numbers of the particles on the reactant side of a balanced fission equation equals the sum of the mass numbers of the particles on the product side, the actual mass does in fact decrease slightly. As a consequence, the total potential energy of the product nuclei is less than the potential energy of the reactants, and the difference is released. When atoms of U-235 split under neutron bombardment, about 0.1% or 1/1000th of the mass disappears and reappears as energy. We can use this information and $E = mc^2$ to calculate just how much energy would be produced by the fissioning of 1.0 kilogram (2.2 pounds) of U-235. Because only 1/1000th of this mass is converted to energy, $m = 1.0$ kg $\times$ 1/1000 = 1.0×10^{-3} kg (or 1.0 g). As you already know, $c = 3.0 \times 10^8$ m/s. Substituting these values:

$$E = mc^2 = 1.0 \times 10^{-3} \text{ kg} \times (3.0 \times 10^8 \text{ m/s})^2$$
$$= 1.0 \times 10^{-3} \text{ kg} \times 9.0 \times 10^{16} \text{ m}^2/\text{s}^2$$
$$= 9.0 \times 10^{13} \text{ kg m}^2/\text{s}^2 = 9.0 \times 10^{13} \text{ J}$$

The unit kg m^2/s^2 may not look familiar, but it is identical to a joule.

To put things into perspective, 9×10^{13} J is the amount of energy released by 33,000 tons (33 kilotons) of exploding TNT or 3300 tons of burning coal. It is enough energy to raise a weight of one million tons six miles into the sky or turn 30,000 tons of water into steam. Yet, all of this comes from one kilogram of U-235, only one gram of which is actually transformed into energy.

One reason why all this energy is accessible is because the fission of a uranium atom releases two or three neutrons, as indicated in equation 8.1. Thus, there is a net production of neutrons. Each of these neutrons can strike another U-235 nucleus and cause it to split. The result is a rapidly branching and spreading chain reaction (Figure 8.3) that can, under certain circumstances, sweep through a mass of fissionable uranium in a fraction of a second. Such a chain reaction will occur spontaneously if a critical mass, about 15 kg (33 pounds), of pure U-235 is brought together in one place. But as you will soon see, the uranium in a nuclear power plant is far from pure U-235.

Figure 8.3

Representation of a nuclear fission chain reaction in uranium-235. (Figure from *Chemistry: Imagination and Implication* by A. Truman Schwartz, copyright © 1973 by Harcourt Brace & Company, reproduced by permission of the publisher.)

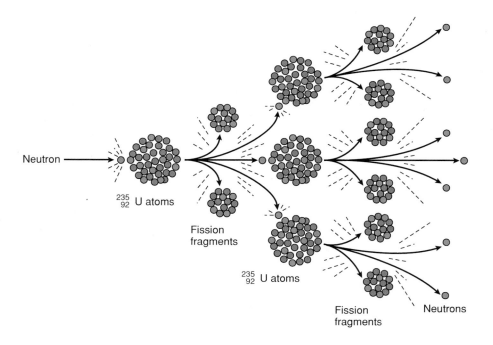

Neutron

$^{235}_{92}$ U atoms

Fission fragments

$^{235}_{92}$ U atoms

Fission fragments

Neutrons

8.7 *Your Turn*

Earlier we reported that at full capacity, the Seabrook plant will generate 1150 million joules of electrical energy per second. Calculate the amount of electrical energy produced per day and the mass of U-235 actually converted to energy per day.

Ans. The first step is to calculate the quantity of energy generated per day. If you make use of the fact that there are 60 seconds in a minute, 60 minutes in an hour, and 24 hours in a day, your answer will be 9.94×10^{13} J/day. The second step is to calculate the mass of matter transformed into energy using the equation $E = mc^2$, and solving for m. Coincidentally, the amount of energy evolved, 9.94×10^{13} J, is approximately equal to the energy calculated above when the equation was first illustrated on page 210. Therefore, you should expect the mass to be near 1.0×10^{-3} kg or 1 g.

How Does a Nuclear Reactor Produce Electricity?

Chapter 4 includes a description of a conventional power generating station in which fuel such as coal or oil is burned to produce heat. The heat is used to boil water into a hot, high-pressure vapor that turns the blades of a turbine. The shaft of the spinning turbine is connected to a large coil of wire that rotates within a magnetic field, thus generating electrical energy. A nuclear power plant operates in much the same way, except that the water is heated by the energy released from the fission of nuclear "fuel" such as U-235. Like any power plant, it is subject to the efficiency constraints imposed by the second law of thermodynamics. The theoretical efficiency for converting heat to work depends on the maximum and minimum temperatures between which the plant operates. This thermodynamic efficiency, typically 55–65%, is significantly reduced by other mechanical, thermal, and electrical inefficiencies.

A nuclear power station consists of two segments: a nuclear reactor and a non-nuclear portion. The latter contains the turbine and the electrical generator. The nuclear reactor is the hot heart of the power plant (Figure 8.4). It is housed in a special steel vessel within a separate reinforced concrete dome-shaped containment building. The uranium fuel is in the form of uranium dioxide (UO_2) pellets, each about the size of a pencil eraser. These pellets are placed end to end in tubes made of a special metal alloy. The tubes, in turn, are grouped into stainless-steel clad bundles.

The rate of fission and the amount of heat generated by it are controlled using a principle employed in the first controlled nuclear fission reaction, which took place at the University of Chicago in 1942. Rods of the element cadmium, an excellent neutron absorber, are interspersed among the fuel elements. As long as the control rods are in place, the reaction cannot become self-sustaining because insufficient neutrons are available. When the rods are withdrawn, the reactor "goes critical," but they can be rapidly reinserted to halt the chain reaction in the event of an emergency.

The fuel bundles and control rods are bathed in what is called the primary coolant. The primary coolant in the Seabrook reactor and in many others is a water solution of boric acid, H_3BO_3. The boron of the boric acid absorbs neutrons and thus serves to control the rate of fission and the temperature. The water solution also serves as a "moderator" for the reactor, slowing the speed of the neutrons and making them more effective in starting fission. Of course, a major function of the primary coolant is to absorb the heat generated by the nuclear reaction. Because the solution is at a pressure

Figure 8.4
Schematic of a boiling-water fission reactor designed by the General Electric Company. (Courtesy GE Nuclear Energy.)

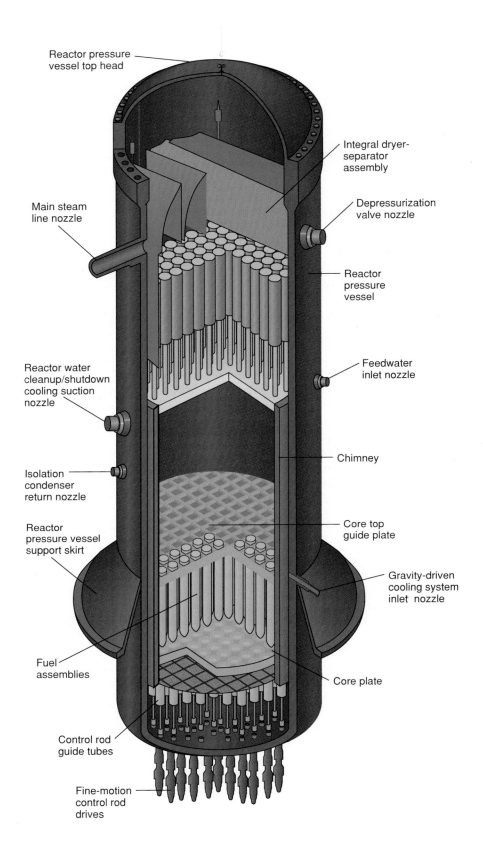

Reactor pressure vessel top head

Integral dryer-separator assembly

Main steam line nozzle

Depressurization valve nozzle

Reactor pressure vessel

Reactor water cleanup/shutdown cooling suction nozzle

Feedwater inlet nozzle

Isolation condenser return nozzle

Chimney

Reactor pressure vessel support skirt

Core top guide plate

Gravity-driven cooling system inlet nozzle

Fuel assemblies

Core plate

Control rod guide tubes

Fine-motion control rod drives

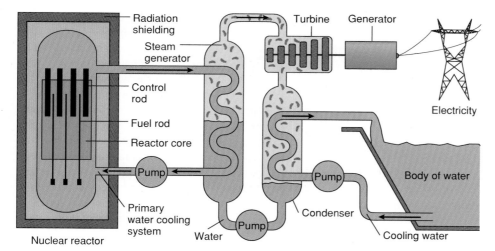

Figure 8.5
Diagram of a nuclear power plant.

Figure 8.6
Cooling tower and containment building at the Union Electric Callaway nuclear power plant.

of more than 150 atmospheres, it does not boil, but it is heated far above its normal boiling point. It circulates, in a closed loop, from the reaction vessel to the steam generators, and back again. This sealed solution thus forms the link between the nuclear reactor and the rest of the power plant. Figure 8.5 provides a general overview of the entire plant.

The heat from the primary coolant is transferred to the water in the steam generators, sometimes called the secondary coolant. At Seabrook, 28,000 gallons of water are converted to vapor each minute. The energy of this hot, pressurized gas is transferred to the blades of a turning turbine and to the attached electrical generator. The water vapor must then be cooled and condensed back into the liquid state before it is returned to the steam generator to continue its heat transfer cycle. In many nuclear facilities this is done using large cooling towers, which are commonly mistaken for the reactors (Figure 8.6). The reactor is actually housed in a relatively small dome-shaped building. Moreover, cooling towers are not an indication of a nuclear power plant. Many fossil-fuel burning plants also use them.

The Seabrook facility does not have cooling towers because ocean water is used to cool the condenser. Each minute, 425,000 gallons of this tertiary coolant flow through a tunnel 19 feet in diameter and 3 miles long, bored through rock 200 feet beneath the floor of the ocean. A similar tunnel from the plant carries the water, now 22°C warmer, back to the ocean. Special nozzles distribute the hot water so that the observed temperature increase in the immediate area of the discharge is about 2°C. Because the primary coolant comes in contact with the steel-clad fuel elements, there is a possibility that the coolant may become radioactive. However, this boric acid solution is kept isolated in a closed circulating system, which makes the transfer of radioactivity to the secondary coolant water in the steam generator highly unlikely. Similarly, the tertiary cooling system does not come in direct contact with the secondary system, so the ocean water is well protected from radioactive contamination. It should be obvious that the electricity generated by a nuclear power plant is identical to the electricity generated by a fossil-fuel plant; it is not radioactive, nor can it be.

8.8   *The Sceptical Chymist*

Again consider the statistics quoted for the Seabrook plant. The energy generated in one day is said to be the equivalent of 10,000 tons of coal. Perform a calculation to determine if this value is consistent with the quoted power rating of 1150 megawatts (1150×10^6 J/s). Assume that burning coal releases 30 kJ or 30,000 J per gram.

Ans. If you did (or read) 8.7 Your Turn, you already know that the electrical energy generated per day is 9.94×10^{13} J. Using that number, the corresponding mass of coal turns out to be well below 10,000 tons. Either the Seabrook people made a mistake or we did. Check our calculation and suggest a reason (or reasons) for the discrepancy. Consider the possibility that someone neglected an important factor. If so, how large was the factor?

■ *What Are the Safeguards against "Meltdown"?*

In 1979, a film called *The China Syndrome* told the story of a near-disaster in a fictitious nuclear power plant. The heat-generating fission reaction almost got out of control. If such a thing were to happen, the intense heat might cause a "meltdown" of the uranium fuel and the reactor housing. Fancifully, the underlying rock might even melt "all the way to China." But in spite of various human and instrumental errors, the safety features of the system worked in the film and fictional disaster was averted. Seven years later, the engineers of the very real Chernobyl power plant in what was then the Soviet Union were less fortunate. The flow of cooling water was interrupted and the temperature of the reactor rose rapidly. Unfortunately, the operators could not regain control of the runaway reaction. Although there was no "atomic" explosion, the effects included the generation of extremely high temperatures, a chemical fire and explosion, the destruction of a major part of the plant, and the release of vast quantities of radioactivity. More than 30 people died as a direct result of the accident, and thousands have been exposed to levels of radiation that could ultimately shorten their lives.

People living within 60 kilometers of the power plant were permanently evacuated soon after the meltdown, but an article in the April 9, 1990 issue of *Time* quotes a local Communist party official as saying that at twice that distance, levels of radioactivity are still nine times the acceptable limit. A new study by a Soviet economist estimates the total cost of the Chernobyl meltdown at $358 billion, a figure that includes the expense of the cleanup and the loss of farm production. The toll on the newborns or the yet-unborn remains unknown. Could it happen here?

America's closest brush with nuclear disaster occurred in March 1979, when the Three Mile Island power plant near Harrisburg, Pennsylvania, lost its coolant and a partial meltdown occurred. There were no fatalities and no serious release of radiation.

In spite of the initial failure, the system held and the damage was contained. Since then, refinements in design and safety have been made to existing reactors and those under construction. Engineers agree that no commercial nuclear reactors in the United States have the design defects that led to the Chernobyl catastrophe.

The Seabrook nuclear power plant has been hailed as an example of state-of-the-art engineering, with multiple safety features. The energetic heart of the station is the 400 ton reaction vessel. Its 44-foot high walls are made of eight-inch carbon steel and it is surrounded by a dome-shaped containment building, a feature the Chernobyl plant did not have, but Three Mile Island did. As the name suggests, this structure is built to withstand accidents of natural or human origin and prevent the release of radioactive material. It is clearly visible in the photograph (Figure 8.7). The inner walls of the building are 4.5 feet thick and made of steel-reinforced concrete; the outer wall is 15 inches thick. Information supplied by New Hampshire Yankee, the company that manages the Seabrook station, states that the containment building is constructed to withstand hurricanes, earthquake, 360-mph winds, and the direct crash of a U.S. Air Force FB-111 bomber.

As discussed in the previous section, the cooling water, the control rods, and the neutron absorbers are designed to prevent temperatures from rising to the point of a meltdown. Indeed, if the primary coolant were lost, the fission reaction would spontaneously slow down. Furthermore, no American nuclear power plants use graphite as a moderator. A large quantity of this common form of carbon was present in the Chernobyl reactor. As the core temperature rose, the graphite caught fire and released enormous amounts of heat. A major chemical explosion followed. The blast blew off the 1000-ton steel plate covering the reactor and spewed radioactive products over a wide area. But there was no real nuclear explosion.

■ Can a Nuclear Power Plant Undergo a Nuclear Explosion?

The question is a very reasonable one. The devastation and destruction that atomic bombs brought to Hiroshima and Nagasaki are painfully etched in the memory of anyone who has even seen the pictures of those cities and their survivors. Therefore, it is reassuring that the answer to the question is "No."

Obviously the purposes of a nuclear power plant and a nuclear weapon are not the same. Correspondingly, the desired rates of their reactions are very different. A nuclear power plant requires a slow, controlled energy release; in a nuclear weapon, the release is rapid and uncontrolled. In both cases, the fuel is U-235 and the reaction is essentially the same, with one important difference. Nuclear power plants typically operate on uranium that is about 3% of the fissionable isotope and 97% U-238. Most of the neutrons given off by fissioning U-235 nuclei are absorbed by atoms of U-238 and elements such as cadmium and boron. As a consequence, the neutron flux cannot build up enough to establish a spontaneously explosive chain reaction, such as that in a nuclear fission bomb.

As we have noted, this will occur only if about 33 pounds of highly purified U-235 are quickly brought together in one place. Fortunately for our troubled world, it is not easy to prepare pure U-235. The separation of this fissionable isotope from the nonfissionable U-238 that makes up 99.3% of naturally occurring uranium requires extensive and expensive processing. Chemical reactions cannot help much in this process because chemical properties are determined by the number and arrangements of the electrons in an atom. Because all the isotopes of any given element, such as uranium, are identical in this respect, they all have essentially identical chemical reactivity.

For more than four decades, uranium isotopes were separated by gaseous diffusion at the Oak Ridge National Laboratory in Tennessee. The method, developed during World War II, takes advantage of the fact that lighter molecules, on average, move faster than heavier molecules. A uranium sample is first reacted with fluorine to form uranium hexafluoride, UF_6, a yellow liquid that boils at 56°C. About 99.3% of the UF_6 molecules contain U-238 atoms and have a molecular mass of 352 [238 + 6(19) = 352]; 0.7% of the molecules contain U-235 and have a molecular mass of 349 [235 + 6(19) = 349]. The process is carried out at a temperature above 56°C, so that all of the UF_6 is in a gaseous state. The average molecule containing a U-235 atom moves only about 0.4% faster than the average molecule containing U-238. But if the diffusion is allowed to occur over and over through a long series of barriers, significant separation of the fissionable and the nonfissionable isotopes can be achieved. Other methods, including centrifugation of UF_6 molecules, have also been developed. Inspectors attempting to determine a nation's nuclear capabilities will often look for the apparatus necessary to concentrate U-235.

■ Could Nuclear Fuel Be Diverted to Make Weapons?

Given the amount of processing that would be required to extract highly purified U-235 from reactor grade fuel, such a diversion from peaceful to military uses would be difficult and costly. A more likely fissionable material for clandestine weapons manufacturing is plutonium-239 (Pu-239). This isotope is formed in a conventional reactor when a nucleus of the plentiful uranium isotope, U-238, absorbs a neutron and subsequently emits two electrons as beta particles. These particles, designed "e" in equation 8.2, will be discussed later when we consider radioactivity.

$$\, _{0}^{1}\text{n} + \, _{92}^{238}\text{U} \rightarrow \, _{94}^{239}\text{Pu} + 2 \, _{-1}^{0}\text{e} \qquad (8.2)$$

This transformation was discovered early in 1940, and the chemical and physical properties of plutonium were determined with an almost invisible sample of the element on the stage of a microscope. The chemical processes devised on such minute samples were scaled up a billion-fold and used to treat the spent fuel slugs from a reactor built on the Columbia River at Hanford, Washington. The reactor was called a **breeder reactor** because it was designed primarily to convert U-238 to fissionable Pu-239 by means of the reaction given in equation 8.2. The plutonium was chemically separated from the uranium and used in the first fission test explosion and the Nagasaki bomb. Because it is fissionable, Pu-239 can also be used to fuel nuclear reactors.

A power plant that creates new fuel as it burns the old seems like a dream come true to an energy hungry planet. France, which generates about 75% of its electrical energy from nuclear reactors, has developed highly sophisticated breeder reactors that permit recovery of plutonium from spent fuel. However, this is another example of the very mixed blessings of modern technology. For one thing, plutonium is one of the most toxic elements known. It is concentrated in the bone, where its long-lived radioactivity does its damage. Moreover, the possibility that the Pu-239 produced in power reactors may wind up in bombs poses an international threat. It has been widely speculated that the 1981 bombing of a nuclear facility in Iraq by war planes from Israel was done to prevent Iraq from being able to produce plutonium-containing nuclear weapons. If Iraq had succeeded in developing a nuclear arsenal, the outcome of Operation Desert Storm in 1991 might have been quite different.

As a consequence of these dangers, the United States for many years banned the reprocessing of commercial fuel elements. That ban was lifted in 1981, but no plutonium is currently being recovered from commercial reactors in this country. Given the potential risks associated with Pu-239 and U-235, it is essential that the supplies and distribution of these isotopes be carefully monitored nationally and around the world.

8.9 ■ *Consider This*

The supply of uranium in the United States (and the rest of the world) is large, but not limitless. By failing to reprocess spent nuclear fuel, we are discarding a potential source of energy that most European countries are tapping. Is the current American practice justified? List arguments on both sides of the "breeder reactor" issue and take a stand.

■ *What is Radioactivity?*

Radioactivity was accidentally discovered in 1896 by Antoine Henri Becquerel (1852–1908). The French physicist found that when a mineral sample was placed on a photographic plate that had been wrapped in black paper, the light-sensitive emulsion became darkened. It was as though the plate had been exposed to light. Becquerel immediately recognized that the mineral itself was emitting a powerful form of radiation that penetrated the light-proof paper. Further investigation revealed that the rays were coming from the element uranium, a constituent of the mineral. Becquerel applied the term **radioactivity** to this spontaneous emission of radiation by certain elements. Subsequent research by Ernest Rutherford (1871–1937) in Canada and England led to the identification of two major forms of radiation. Rutherford named them after the first two letters of the Greek alphabet, **alpha** (α) and **beta** (β). Their properties are summarized in Table 8.1.

Beta radiation consists of negatively charged particles, each with a mass of about 1/2000 on the standard atomic mass scale. These properties are those we attributed to electrons in an earlier chapter. A **beta particle** is thus an electron. Alpha radiation is made up of particles with a mass of 4 units on the same scale and a charge twice as

---

 Table 8.1   **Radioactive Emissions**

Radiation	Mass	Charge	Composition	Symbol
alpha (α)	4	+2	2 protons + 2 neutrons	^4_2He
beta (β)	1/1838	−1	electron	$^0_{-1}\text{e}$
gamma (γ)	0	0	electromagnetic radiation	

large as that of an electron, but with a positive sign. An **alpha particle** is a composite of two protons and two neutrons—the nucleus of a helium atom. It was subsequently discovered that a third form of radiation, **gamma (γ) radiation**, is frequently associated with the emission of an alpha or beta particle. Unlike alpha and beta rays, **gamma rays** do not consist of ordinary particles. Rather, they are made up of high energy, high frequency photons and are part of the electromagnetic spectrum, as are infrared, visible, and ultraviolet light rays.

Whenever an α or β particle is given off during radioactive decay, a remarkable transformation occurs: The elementary identity of the emitting atom is changed. For example, an atom of uranium-238 becomes converted into an atom of thorium-234 (Th-234) when it loses an alpha particle. Such a change might be understood by the ancient alchemists, who sought to transmute lead and other common metals into gold. But according to modern chemistry, elements and atoms are supposed to be unchanging and unchangeable. Yet, there is ample experimental evidence that *whenever an atom emits an alpha particle, it is converted into an atom of the element with an atomic number two less than the original.* Such a transformation can be represented with a nuclear equation, as in the case of alpha emission by uranium-238 to form thorium-234.

$$\,^{238}_{92}U \rightarrow \,^{234}_{90}Th + \,^{4}_{2}He \qquad (8.3)$$

The species undergoing radioactive decay (here U-238) is called the **parent** and the product species is called the **daughter** (here Th-234). Note that the emission of an alpha particle means that the mass number of the daughter isotope is 4 less than the mass number of the parent isotope. The loss of 4 nuclear particles—2 protons and 2 neutrons—accounts for this change.

The Th-234 formed in equation 8.3 is also radioactive. It undergoes beta particle emission to yield protactinium-234 (Pa-234).

$$\,^{234}_{90}Th \rightarrow \,^{234}_{91}Pa + \,^{0}_{-1}e \qquad (8.4)$$

Thus, the emission of a beta particle also results in a change in elementary identity, in this case from Th-234 with atomic number 90, to the element Pa-234 with an atomic number of 91, one greater than that of the parent. *The increase in atomic number means that the number of protons increases by one when a beta particle is ejected from the nucleus.* This suggests that one can regard a neutron as consisting of a proton plus an electron. The loss of an electron (a beta particle) converts a neutron to a proton.

$$\,^{1}_{0}n \rightarrow \,^{1}_{1}p + \,^{0}_{-1}e \qquad (8.5)$$

The total number of neutrons plus protons in the nucleus remains constant (at 234 in this instance) during beta emission.

8.10 *Your Turn*

■

a. One product of nuclear fission in power plants is cesium-137 (Cs-137), a beta emitter. Identify the daughter product (name, symbol, atomic number, and mass number) formed during the decay of Cs-137 and write a nuclear equation for the reaction.

b. Radium-228 (Ra-228) is produced by alpha emission from a parent nucleus. Identify the parent (name, symbol, atomic number, and mass number) from which Ra-228 forms and write a nuclear equation for the reaction.

Ans. **a.** When an atom emits a β particle, the mass number remains unchanged, but the atomic number increases by one, in this case from 55 to 56.

$$\,^{137}_{55}Cs \rightarrow \,^{137}_{56}Ba + \,^{0}_{-1}e$$

Whether an isotope is an alpha emitter, a beta emitter, or nonradioactive depends on the stability of the nucleus. Nuclear stability, in turn, is related to the ratio of neutrons to protons. Radioactive nuclei adjust this ratio by emitting alpha, beta, or other radiation until a stable neutron/proton ratio is achieved. At that point, the nucleus is no longer radioactive. For example, the radioactive decay of U-238 and Th-234 (equations 8.3 and 8.4) are the first two steps in a natural series of 14 steps, ending in the nonradioactive lead isotope, Pb-206. For most elements, the most plentiful isotope is nonradioactive. However, all the isotopes of the elements with an atomic number of 90 or higher are radioactive. Their atoms, with many more numbers of neutrons than protons, are all unstable.

■ *What Hazards Are Associated with Radioactivity?*

It may come as a surprise that the radiation exposure the average American citizen receives from nuclear power plants is about one-tenth the radiation he or she would get during a coast-to-coast jet plane trip. Even under adverse circumstances, radiation exposure from nuclear power plants is low. If you had stood at the gate of the Three Mile Island plant for the first two weeks of March, 1979 (the time of the accident) you would have been exposed to less radiation than from a single chest X-ray. But the evidence of the past makes it clear that it would be a serious mistake to dismiss radioactivity as harmless.

Unfortunately, some of the first scientists to study radioactivity, including Marie Curie (1867–1934), were not fully aware of the potential dangers inherent in the phenomenon. Madame Curie died of a form of leukemia that was very likely induced by her overexposure to radiation. Alpha and beta particles and gamma rays carry a considerable amount of energy. Often the energy is sufficient to produce ionization in atoms and molecules struck by the radiation. As in the case of bombardment by ultraviolet rays, the resulting changes in molecular structure can have profound effects on living things. Rapidly growing cells are particularly susceptible to damage, a fact that has led to the use of radiation as a treatment for some kinds of cancer. But bone marrow and white blood cells are also easily damaged, and anemia and susceptibility to infection are among the early symptoms of radiation sickness. Radiation-induced transformations of DNA can give rise to cancer or genetic mutation.

Today, considerable care is taken to shield medical, laboratory, and other workers from radiation. Protective shielding made of lead and other heavy metals is used to absorb radioactive emissions. However, it is important to recognize that it is impossible to be fully protected from exposure to radioactivity. The Earth itself, the building materials quarried or manufactured from the Earth, and even your best friends are radioactive. Because they contain naturally occurring radioactive atoms, all of these sources emit background radiation. The amount of background radiation you receive depends on where you live, the house you occupy, the number of people you live with, and how close you get to them. The late Isaac Asimov, a prolific science writer, pointed out in one of his many books that a human contains approximately 3×10^{26} carbon atoms, of which 3.5×10^{14} are radioactive carbon-14 atoms. With each breath you inhale 3.5 million C-14 atoms.

8.11	*The Sceptical Chymist*

Assume that Isaac Asimov's figures are correct, and 3.5×10^{14} of the 3×10^{26} carbon atoms in your body are radioactive. Calculate the fraction of carbon atoms that are radioactive carbon-14.

Note: A value for this fraction appears later in this chapter.

■ *Table 8.2*	*Physiological Effects of a Single Dose of Radiation*
Dose (rem)	**Likely effect**
0–25	No observable effect
26–50	White blood cell count decreases slightly
51–100	Significant drop in white blood cell count; lesions
101–200	Nausea, vomiting, loss of hair
201–500	Hemorrhaging, ulcers, possible death
>500	Death

The extent of biological damage that can be caused by radiation depends, in part, on the total amount of energy absorbed. This energy is measured in a unit called the **rad,** short for **radiation absorbed dose.** One rad is defined as the absorption of 0.01 joule of radiant energy per kilogram of tissue. But not all radiation is equally harmful to living organisms; some types are more damaging than others. Therefore, to estimate the potential physiological damage, the number of rads is multiplied by a factor characteristic of the particular radiation involved. This factor is symbolized with an n. Very damaging radiation, such as alpha particles and high energy neutrons, has an n value of 10. Less harmful forms, including beta, gamma, and x-radiation are assigned an n of 1. When n is multiplied by the number of rads, the product is called **rem** for **roentgen equivalent man.**

$$\text{number of rems} = n \times (\text{number of rads})$$

Thus, a dose of 10 rad of alpha radiation ($n = 10$) is equivalent to 100 rad of beta radiation ($n = 1$); each dose equals 100 rem. The number of rems in a dose of radiation exposure is thus a measure of the power of the radiation to cause damage to human tissue. The likely effects of a single dose of radiation at various levels are given in Table 8.2.

Because most doses are less than one rem, a smaller unit, the millirem is commonly employed. One millirem (mrem) is 1/1000th of a rem (1 mrem = 10^{-3} rem). Table 8.3 uses millirems to report the radiation exposure associated with various activities and lifestyle factors. This information is the basis for estimating your personal annual radiation exposure in 8.12 Your Turn. Once you have completed this exercise, you can check Table 8.4 to see how your exposure compares with that of the average American.

8.12   Your Turn

Use Table 8.3 to estimate the approximate dosage of radiation you receive each year.

Note that well over half of the roughly 200 millirems absorbed in a year by a typical resident of the United States comes from natural background sources—mostly cosmic rays, soil, and rock. Of the 67 mrem of artificial radiation absorbed annually, about 60 are attributable to medical procedures such as diagnostic X-rays. As is evident from Table 8.4, the radiation emitted by properly operating nuclear power plants is negligible compared to normal background radiation, including the natural radiation of your own body. About 0.01% of the potassium ions that are essential to your internal biochemistry are a radioactive isotope of mass number 40. These K-40 ions give off about 20 mrem per year, approximately 1000 times more radioactivity than that received by living within 20 miles of a nuclear power plant. In fact, because bananas are rich in potassium, a steady diet of them can contribute significantly to your personal radioactivity.

■ *Table 8.3*	**Personal Radiation Dosage Inventory**

Source	Annual Quantity
1. Location of your town or city.	
a. Cosmic radiation at sea level (U.S. average)	30 mrem[a]
(Cosmic radiation is radiation emitted by stars across the universe. Much of this is deflected by the Earth's atmosphere and ionosphere.)	
b. Add an additional millirem value based on your town or city's elevation above sea level:	

1000 m (3300 ft) above sea level = 10 mrem
      2000 m (6600 ft) above sea level = 30 mrem
      3000 m (9900 ft) above sea level = 90 mrem
      (Estimate the millirem value for any intermediate elevation.)
	___mrem

2. House construction
   Choose the material from which your house is made; enter the correct value.
   (Building materials contain a very small percentage of radioisotopes.)
   Brick, 75 mrem; wood, 40 mrem; concrete, 85 mrem.
	___mrem

3. Ground Radiation from rocks and soil (U.S. average)	15 mrem
4. Food, water, and air (U.S. average)	25 mrem
5. Fallout from nuclear weapons testing (U.S. average)	4 mrem
6. Medical and dental X-rays	
a. Chest X-ray (number of visits times 10 mrem per visit)	___mrem
b. Gastrointestinal tract X-ray (number of visits times 200 mrem per visit)	___mrem
c. Dental X-rays (number of visits times 10 mrem per visit)	___mrem
7. Jet travel (Jet travel increases exposure to cosmic radiation.) Number of flights (five-hour flights at 30,000 ft or 9000 m) times 3 mrem per flight.	___mrem
8. Nuclear power plants If your home is adjacent to a plant site add 1 mrem.	___mrem
Total	___mrem

[a]1 millirem = 10^{-3} rem.

Reprinted from *Chemistry in the Community* (ChemCom), copyright 1988, with permission of the American Chemical Society.

■ *Table 8.4*	**Typical Annual Radiation Exposures in the United States**

Sources	Millirem/Year
Human	
Radioactive fallout	4
Luminous watch dials, TV tubes	2
Nuclear power industry	0.2
Medical Procedures	
Diagnostic X-rays	50
Radiotherapy	10
Internal diagnosis	1
Total from Human sources	67
Natural	
Outside the body	
Cosmic radiation	50
The Earth	47
Building materials	3
Inside the body	
In human tissues	21
Inhalation of air	5
Total from Natural Sources	126
Total exposure	193

Data from W. L. Masterson, E. J. Slowinski, and C. L. Stanitski, *Chemical Principles,* 6th ed. Copyright © W. B. Saunders, Orlando, FL.

Figure 8.8

Dose-response curves for radiation. (The curves level off and then decrease as radiation doses get very high because more cells die than become cancerous.) (Adapted with permission from *The Nuclear Waste Primer.* Copyright © 1985 The League of Women Voters of the United States, Washington, D.C.)

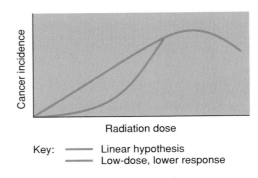

Key: ——— Linear hypothesis
——— Low-dose, lower response

To further put things in perspective, it is useful to note that the immediate physiological effects of radiation exposure are generally not observable below a *single dose* of 25 rems (see Table 8.2). This is over one hundred times the average *annual* exposure. The more a given number of rems is spread out over time, the less harmful it appears to be. However, there is still uncertainty about the long-term effects of low doses of radiation. The assumption is usually made that there is no threshold below which there is no damage. However, the effects are so small and the time span so great that scientists have not been able to make reliable measurements. Moreover, tests with animals are not always reliable because there is considerable species-to-species variation in the effect of radiation.

The issue then is how to extrapolate the known high-dose data to low doses. Two dose-response models are illustrated in Figure 8.8. The assumption of a linear relationship between the incidence of cancer and radiation dose is represented by curve *A*. In this model, doubling the radiation dose doubles the incidence of cancer, tripling causes three times the number of cancers, and so on. This is the relationship currently used by federal agencies in setting exposure standards. Many scientists believe that the biological effect is relatively less at low levels of radiation because of the self-repairing mechanism of cells. This model is represented by curve *B*, which drops below the straight line of curve *A*.

8.13   *Consider This*

You have just encountered two models for the dose-response curve for low-level radiation. In the absence of definitive scientific proof, which model would you advocate and why? Defend your position to someone holding the opposite opinion.

■ *How Long Will Nuclear Waste Products Remain Radioactive?*

Here is one area where popular magazines seldom exaggerate the problem; products formed in nuclear reactors can have dangerously high levels of radioactivity for thousands of years. There is no way to speed up the radioactive decay process. The fact that most of us experience considerably more radioactivity from natural sources than from artificial ones should not lull us into a false sense of security. Waste disposal presents formidable problems because one radioactive isotope often generates others. As noted in the previous section, daughter products from decays act as parents for other radioactive emissions, for example, U-238 to Th-234 to Pa-234.

A property that is particularly acute in the disposal of radioactive waste is the rate at which the level of radiation declines. This can occur very rapidly over a short time,

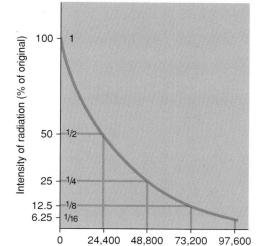

Figure 8.9
Radioactive decay of Pu-239.

or very slowly over a long period of time, depending on the particular isotope. The rate of decay is typically reported in terms of the **half-life,** the time required for the level of radioactivity to fall to one-half of its initial value. For example, plutonium-239, the alpha-emitting fissionable isotope formed in uranium reactors, has a half-life of 24,400 years. This means that it will take 24,400 years for the radiation intensity of a freshly generated sample of Pu-239 waste to drop to one half its original value. At the end of a second half-life of 24,400 years, the radiation will be one-fourth the original level. And in three half-lives (73,200 years), the level will be one-eighth of the original (Figure 8.9).

The half-life of a particular isotope is a constant, and is independent of the physical or chemical form in which the element is found. Moreover, the rate of radioactive decay is essentially unaltered by changes in temperature and pressure. But when various radioisotopes are compared, their half-lives are found to range from millennia to milliseconds. For example, the half-life of uranium-238 (equation 8.3) is 4.5 billion years. (Coincidentally, this is approximately the age of the oldest rocks on Earth, a value obtained by measuring their uranium content.) By contrast, the beta decay of thorium-234 (equation 8.4) occurs very rapidly; its half-life is 24 days. Even faster decay is exhibited by polonium-214 (Po-214), an alpha emitter with a half-life of 0.00016 seconds.

Hydrogen-3 (H-3 or tritium), which is sometimes formed in the primary coolant water of a nuclear reactor, is a beta emitter with a half-life of 12.3 years. Iodine-131 (I-131), with a half-life of 8 days, is used to treat hyperthyroidism in persons with Graves' disease. In this procedure, the orally-administered radioactive iodine concentrates in the overactive thyroid gland, fully or partially destroying it. In most patients, thyroxine, the iodine-containing hormone normally secreted by the thyroid, must then be supplemented with a synthetic substitute. Strontium-90 (Sr-90) proved to be a particularly dangerous isotope in the fallout from nuclear weapons testing because it concentrates in milk and bone. There, its 28.9 year half-life can pose a lifelong threat to an affected individual. Significantly, I-131 and Sr-90 are among the fission products produced in nuclear reactors, and there is concern that many persons living near Chernobyl were exposed to potentially harmful levels of both isotopes. It has been reported that the incidence of thyroid cancer among Chernobyl children is far higher than normal, and iodine-131 has been implicated.

8.14 **Your Turn**

There has recently been considerable publicity about the pollution of indoor air, especially in basements, with radon-222. Rn-222 is a radioactive noble gas element formed by the decay of radium-226 (Ra-226), which is naturally present in hard rocks.

a. Write a nuclear equation for the formation of Rn-222 from Ra-226. (Note that the type of nuclear radiation emitted has not been specified, but you should be able to determine it from the information given.)

b. Rn-222 is an alpha emitter with a half-life of 3.8 days. Give the atomic number, mass number, and chemical symbol of the daughter product formed by the decay of Rn-222.

c. Suppose that the radioactivity from Rn-222 in your basement was measured as 32×10^{-5} rem. If no additional radon entered the basement, how much time would have to pass before the radiation level fell to 2×10^{-5} rem?

Hint: **c.** Note that in dropping from 32×10^{-5} rem to 2×10^{-5} rem, the radiation level is cut in half four times: $32 \rightarrow 16 \rightarrow 8 \rightarrow 4 \rightarrow 2$. This corresponds to four half-lives of 3.8 days each.

Carbon-14 (C-14), with a half-life of 5730 years, is a beta emitter that decays to nitrogen-14 (N-14). This isotope of carbon is well known because it is often used to date the remains of once-living things or objects made from them. The carbon dioxide of the atmosphere contains a constant steady-state ratio of one atom of radioactive carbon-14 for every 10^{12} atoms of nonradioactive carbon-12. Living plants and animals incorporate the isotopes in that same ratio. However, when the organism dies, exchange of CO_2 with the environment ceases. Thus, no new carbon is introduced to replace the C-14 converted to nitrogen-14 by beta decay. As a consequence, the concentration of C-14 decreases with time, dropping by half every 5730 years. In the 1950s, W. F. Libby first recognized that by experimentally measuring the C-14/C-12 ratio in a sample, the time at which the organism died could be estimated. Human remains and many human artifacts contain carbon, and fortunately, the rate of decay of C-14 is a convenient one for measuring human activities. Charcoal from prehistoric caves, ancient papyri, mummified human remains, and suspected art forgeries have all revealed their ages by this technique. The C-14 technique provides ages which agree to within 10% of those obtained from historical records, thus validating the legitimacy of the technique.

8.15 **The Sceptical Chymist**

The Shroud of Turin is a linen wrapping cloth that appears to bear the image of a crucified man. It is held by some to be the cloth in which the body of Jesus Christ was wrapped following his crucifixion. In 1988, several groups of scientists were allowed to take small samples of the cloth to subject it to carbon-14 dating. The results revealed that the C-14/C-12 ratio was approximately 92% of that in living organic matter. Does this result support or refute the possible authenticity of the shroud? Give your reasons.

How Will We Dispose of Waste from Nuclear Plants?

The experience of the past 40 years seems to suggest that the answer to this vitally important question is "Slowly and with difficulty." Whether that is the correct answer is another matter. As long ago as 1985, the League of Women Voters published *The*

Nuclear Waste Primer. The book reported that by 1983, the United States had accumulated 306,000 cubic meters of high-level waste (mostly liquid) from defense programs and 4626 cubic meters of spent fuel rods from nuclear power plants. To give you an idea of the quantity involved, assume the wastes were all placed in barrels, each three feet tall. A single layer of these barrels would cover four football fields. On the one hand, this does not seem like very much for the entire country, but it is a legacy that will be with us for thousands of years. Nearly all of the military wastes were stored at government facilities and most of the fuel rods were still at the power plants. Since 1983, more radioactive waste has been added, but little progress has been made toward finding a safe and permanent storage site.

After about three or four years of use, the U-235 concentration in the fuel elements of a power reactor, initially at 3%, drops to the point where it is no longer effective in sustaining the fission process. Nevertheless, the fuel rods are still "hot"—both in temperature and in radioactivity. They contain various isotopes of uranium, plutonium-239 formed by the capture of neutrons, and a wide variety of fission products such as iodine-131, cesium-137, and strontium-90. Remotely controlled machinery, operated by workers protected by heavy shielding, removes the rods from the reactor and replaces them with new fuel elements. Thus, approximately one-fourth of the rods are replaced annually, on a rotating schedule. The spent rods are transferred to deep pools where they are cooled by water containing a neutron absorber. For example, the storage facility at Seabrook is a 34-foot deep steel-lined concrete pool in a secure building. It has the capacity to hold up to 25 years' worth of waste.

The on-site storage of high-level radioactive waste for 25 years is hardly ideal. Yet, no long-term storage facility currently exists in the United States. The absence of such a repository is becoming a significant impediment to the use of nuclear power. In fact, in the 1970s, some states passed laws prohibiting the construction of any new nuclear power plants until the federal government demonstrated that radioactive wastes could be disposed of safely and permanently. The Department of Energy has contracted with electric utility companies to begin accepting spent fuel elements in 1998, but progress in preparing a national underground disposal site has been painfully slow. Geologists have been studying the problem for over 30 years. The site selected for radioactive waste disposal must safely contain the radioactive material for an extended period of time and prevent it from entering the underground water supply. It is estimated that the beta-emitting fission products need to be isolated for 300 to 500 years, which is about 10 half-lives for species such as Sr-90 and Cs-137. Uranium and heavier elements such as plutonium typically have much longer half-lives, but their radiation intensity is much lower. Most plans call for encasing the spent fuel elements in ceramic or glass, packing the product in metal canisters, and burying them deep in the earth.

The idea is to carve out a chamber at least 1000 feet below ground in an appropriate rock formation. Salt, basalt, tuff, granite, and shale have all been considered. Salt domes, which are geological formations entirely of salt, are particularly attractive because they are very stable, extremely dry, and self-sealing if cracks should appear. Granite and basalt always contain cracks, but they have a great capacity to chemically absorb most wastes. A 1983 report by the National Academy of Sciences concluded that it is possible to identify rock formations from which a drop of water would require millions of years to travel to the point where it could become incorporated into living things. Supporting evidence is provided by a self-sustaining natural fission reactor in Gabon, Africa. Measurements taken there indicate that fission products have moved less than six feet in two million years.

Geological formations suitable for radioactive waste disposal have been identified in Nevada, Texas, Washington, Utah, and Mississippi, but the political problems are at least as thorny as the technical ones. The federal government must deal with state legislatures and Indian tribes whose land rights are affected. The "not in my backyard" (NIMBY) syndrome has even led a number of states to adopt legislation prohibiting the disposal of high-level nuclear waste within their boundaries. Nevertheless, underground disposal still seems to be the method of choice.

A five-year waste isolation pilot project is currently underway in southeastern New Mexico, in salt beds 2150 feet below ground. If the tests prove successful, the site may be used to store transuranium wastes. Meanwhile, tunnels are being dug 1400 feet beneath Yucca Mountain, Nevada. When completed (the target date is 2003), the site will be the largest radioactive storage facility in the country, with a capacity of 63,500 tons of high level waste. It has been estimated that at least 20 years will be required just to transport the waste to the Nevada site.

Disposal in deep-sea clay sediments, under 3000–5000 m of water has been and still is being investigated. Proposals to bury the radioactive waste under the Antarctic icesheet or to rocket it into space have largely been discredited. But one thing is sure: whatever disposal methods are ultimately adopted, they must be effective over the long term. Radioactive wastes will be with us for a long time, and a solution to the problem must occur more rapidly than the decay of plutonium-239.

8.16   *Consider This*

The Department of Energy recently asked 13 experts to design a system of markers to warn future generations of the existence of the underground nuclear waste repository being dug in New Mexico. The markers should last for 10,000 years, over four times the age of the pyramids of Egypt, and their message must be intelligible to Earthlings of the year 12,000. Try your hand at this formidable challenge, keeping in mind the changes that have occurred in *Homo sapiens* during the past 10,000 years.

8.17   *Consider This*

Some industrialized nations have proposed a novel method of foreign aid—with some strings attached. The producers of nuclear waste would ship their radioactive residues to developing countries and pay them to dispose of it. Suppose you are the president of a developing nation that has just received such an offer. The foreigners are willing to pay handsomely, and your country desperately needs money for schools and medical care. You have called in your science advisor to help you decide what response would be in the best interest of your nation. What questions will you ask the science advisor?

■ *Nuclear Power in the United States and Worldwide*

Cost overruns, changing federal regulations, years of litigation, and public opposition have halted the construction of new fission power plants in the United States. At present, only four plants are either being built or in various stages of testing and start-up. All plants ordered since 1974 have been canceled. The Shoreham reactor on Long Island has been completed, but by agreement it sits forever unfueled and unused. Currently, 110 nuclear power plants are operating in the United States at sites indicated in Figure 8.10. These plants collectively produce only 19.1% of our electrical energy. By contrast, the burning of coal is used to generate 57% of our electrical energy, along with acid-rain producing sulfur oxides and the greenhouse gas, carbon dioxide.

Globally, about 17% of the electricity produced and consumed is generated in nuclear power plants. If this amount of energy were to be replaced, it would require the entire annual coal production of the United States or the former Soviet Union. Thus, there is already a relatively small but significant international reliance on nuclear energy. Table 8.5 gives nuclear power statistics, as of December 1989, for some of the countries that are among the largest users, and a few others. Several important points are evident from the table. Note, for example, that France, Belgium, South Korea,

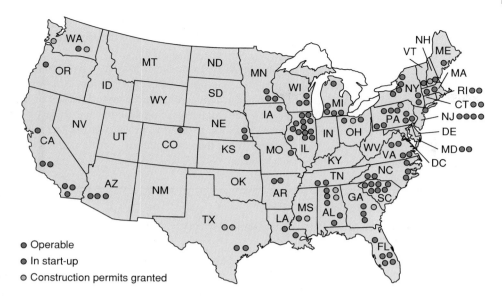

Figure 8.10
Map showing locations of nuclear power plants as of December 1988. (Reprinted with permission from Rosenbaum, Walter A., *Environmental Politics & Policy,* 2/e, Washington, D.C.: CQ Press, 1991.)

- Operable
- In start-up
- Construction permits granted

Table 8.5	**Nuclear Power Statistics for Selected Countries as of December 31, 1989**			
Country	**Electricity Generated**		**Reactors**	
	percent	MW[a]	operating	construct[b]
France	74.6	52,588	55	9
Belgium	60.8	5,500	7	0
South Korea	50.2	7,220	9	2
Hungary	49.8	1,645	4	0
Sweden	45.1	9,817	12	0
Switzerland	41.6	2,952	5	0
Spain	38.4	7,544	10	0
West Germany	34.3	22,716	24	1
Japan	27.8	29,300	39	12
United Kingdom	21.7	11,242	39	1
United States	19.1	98,331	110	4
Canada	15.6	12,185	18	4
Soviet Union	12.3	34,230	46	26
Argentina	11.4	935	2	1
Netherlands	5.4	508	2	0
Brazil	0.7	626	1	1
Romania	0	0	0	5

[a]Megawatts
[b]Number of reactors under construction.
Data from Wolfe Häfele, in *Scientific American,* September 1990, p. 138.

Hungary, Sweden, and Switzerland obtain more than 40% of their electricity from nuclear power plants. The percentages of nuclear-generated electricity in Spain, West Germany, Japan, and the United Kingdom are also above the international average; but the United States barely beats it.

The absolute number of operating reactors alone does not tell the complete story. France, for example, has 55 nuclear power plants, compared to twice that number in the United States. These French reactors generate about half the electricity produced by the American nuclear plants. Nevertheless, nuclear fission supplies almost three-fourths of the electricity used in France, but only 19.1% of the needs of this country.

The Swiss also have a high nuclear dependency, producing nearly 42% of their electricity with only five reactors, less than 5% of the total number of fission-fueled power plants in the United States.

Also noteworthy are the countries not listed in Table 8.5. Thus far, only industrialized nations have been able to afford major exploitation of nuclear fission. Although China and Iran both have reactors under construction, they have none that are currently operative. Mexico and Pakistan each have one operating reactor, and India obtains 1.6% of its electricity from its seven reactors. Ironically, in spite of the fact that much of the world's uranium comes from Africa, the great majority of the nations on that continent use no nuclear energy whatsoever.

8.18 Consider This

Using data from Table 8.5 data and other sources, answer the following questions.

a. What is the per capita amount of electrical energy (Megawatts/person) from nuclear power plants for the United States, France, Brazil, and Japan?
b. For the countries listed in Table 8.5, what general relationship, if any, appears to exist between standard of living and the percentage of electricity generated by nuclear power?

8.19 Consider This

Again use the data of Table 8.5 to answer the following questions.

a. Select three countries in which the percentage of electricity generated by nuclear power plants is less than the global average of 17% and suggest reasons why this might be the case.
b. Select three countries in which the percentage of electricity generated from nuclear power plants is greater than 40% and suggest reasons why this condition might exist.

■ Living with Nuclear Power: What Are the Risks and Benefits?

As of 1989, there were 426 nuclear reactors operating in over 30 nations. It is obvious from Table 8.5 and the accompanying text that countries differ markedly in their use of nuclear power to generate electricity. What is not so obvious are the reasons for this variability. Some nations have fiscal problems so severe that it is difficult or impossible for them to fund the construction or expansion of nuclear power facilities. For others, an adequate supply of relatively cheap electricity is available from water power, fossil fuels, or other sources. Therefore, there is little need to use atomic energy. Nuclear power provides a means from some countries to gain independence from the need to import fuels. In still other countries, such as France, a conscious choice has been made to use nuclear energy to produce the bulk of electrical energy. And, just the opposite conclusion has been reached by other nations. In Sweden, which currently obtains 45.1% of its electricity from fission, a referendum has called for the halt of nuclear power generation by 2010.

Regardless of whether a country contains many nuclear powered electrical generators or only a few, associated risks and benefits must be weighed. Such risk-benefit analyses are never easy, though in a sense we do it every day. We commonly regard risk as the probability of being injured or losing something, but there are many types of risk. They can be voluntary, such as those associated with bungee jumping, or involuntary, such as inhaling someone else's cigarette smoke. When we drive a car we control the risks (at least to some extent), but we have no control over the increased risk of radiation exposure at high altitudes or of a commercial plane crash. Counterbalancing risks are benefits such as the improvement of health, increased personal comfort or satisfaction, saving money, or reducing fatalities. Everyday living inevitably involves risks and their related benefits: crossing a street, riding a motorcycle or in a car, cooking a meal or eating one, and even the simple act of getting up in the morning. Because there is some element of risk in everything we do, we almost automatically make judgments about what level of risk we consider acceptable. The instinct for survival is such that most people do not intentionally put themselves at high risk, even when the potential benefit is also high. On the other hand, there is an alarming increase in the number of people who expect "zero risk" in whatever they do or in whatever surrounds them, although it is a virtual impossibility.

In the case of nuclear energy, we are dealing with social benefits in relation to technological risks, but we must not make the mistake of only considering the risks and benefits that relate directly to fission. We must also weigh the risks associated with the alternatives—the coal-fired power plants that nuclear reactors are designed to replace. Recall, for example, the estimate in Chapter 4 that over 100,000 workers have been killed in American coal mines since 1900. Table 8.6 summarizes the risk of fatalities from the annual operation of a 100-megawatt power plant using either coal or nuclear power. The conclusion is that, at least for the hazards identified here, the risk associated with nuclear power is considerably less than that of coal-burning plants. However, as we have noted, nuclear energy carries tremendous emotional overtones, made of mystery, misunderstanding, and mushroom clouds. The risks of radiation, involuntary and uncontrolled, and the possibility of a major disaster, however remote, loom large in human consciousness. The accidents at Three Mile Island, Chernobyl,

■ Table 8.6	Risks from Coal and Nuclear-Powered Electricity Generation	
Hazard Type	**Coal**	**Nuclear**
Routine occupational hazard.	Coal-mining accidents and black-lung disease constitute a uniquely high risk.	Risks from sources not involving radioactivity dominate.
Deaths	2.7	0.3 to 0.6
Routine population hazards.	Air pollution produces relatively high, though uncertain, risk of respiratory injury. Significant transportation risks.	Low-level radioactive emissions are more benign than corresponding risks from coal. Significant transportation risks incompletely evaluated.
Deaths	1.2 to 50	0.03
Catastrophic hazards (excluding occupational).	Acute air pollution episodes with hundreds of deaths are not uncommon. Long-term climatic change, induced by CO_2 is conceivable.	Risks of reactor accidents are small compared to other quantified catastrophic risks. The problem lies in as yet unquantified risks for reactors and the remainder of the fuel cycle.
Deaths	0.5	0.04
General environmental degradation.	Strip mining and acid run-off; acid rainfall with possible effect on nitrogen cycle, atmospheric ozone; eventual need for strip mining on a large scale.	Long-term contamination with radioactivity; eventual need for strip mining on a large scale.

(Deaths are the number expected per year for a 100-megawatt power plant. In all cases, 6000 man-days lost are assumed to equal one death.)

Source: Modified from *Perilous Progress: Managing the Hazards of Technology,* by Robert W. Kates, Ed., 1985. Boulder, Colorado: Westview Press.

and other nuclear reactor sites make us wary. We have limited trust in technology, and perhaps even less in people. We are apprehensive about human error in the design, construction, and management of nuclear power plants. After all, human errors and technicians' responses to them were the weak points in the prescribed safety procedures at Three Mile Island and Chernobyl.

> ### 8.20   *Consider This*
>
> It is very likely that coal-burning power plants release radioactive isotopes to the environment. Explain how this could be.

■ *What Is the Future for Nuclear Fission?*

This final question is, in many ways, the most difficult one posed in this chapter. The answer is very uncertain. But, as the cover of *Time* magazine on April 29, 1991 asked, "Do we have a choice?" Then President George Bush's proposed energy plan relied heavily on licensing a new generation of nuclear power plants. The National Academy of Science, in an April 1991 report, called for the rapid development of such new plants as a means of reducing the emission of carbon dioxide and other greenhouse gases. And yet, the people of this country are not completely at ease with such a forecast. Their mixed feelings on the subject are typified by a poll taken in 1991. Among the respondents, 40% agreed that the United States should rely mostly on nuclear energy as the source to meet our increased energy needs in this decade, 25% favored oil, 22% coal, and 5% "other." However, when asked whether they favored building more nuclear power plants in this country, 32% were strongly opposed, 20% somewhat opposed, 22% somewhat supportive, and 18% strongly supportive. The chief executive officer of Carolina Power & Light Company suggests that once citizens understand that new plants of some kind must be constructed in order to meet growing electrical power demands and avoid blackouts, they will see the environmental advantages of nuclear power over conventional fuels.

Nuclear power has been touted as a way to reduce global warming and acid rain. Because heat from the fission of uranium produces the steam used to generate electricity, a nuclear power plant releases no carbon dioxide, a major greenhouse gas. Nor does it emit the acidic oxides of sulfur and nitrogen. Burning sufficient coal to generate 1000 megawatts of electrical power releases 250 tons of SO_2 and 80 tons of NO_x. Given the publicity dedicated to acid precipitation and the greenhouse effect during the 1980s, the public seems ready to reexamine increased power production from fission. But whether the risks associated with nuclear power outweigh those of global warming and acid rain is a difficult question for which there are no clear-cut answers, in spite of extended study and debate. Proponents line up on each side of the argument.

In August 1988, during a summer of record heat, 15 U.S. senators co-sponsored a bill to fund research to combat the greenhouse effect by developing carbon dioxide-free energy sources, including safer and more cost-effective nuclear power plants of standardized design. Alan Crane, an energy-policy specialist, appeared before a House subcommittee hearing and spoke of global warming and nuclear energy risks in these terms:

> There is significant, though not yet quantifiable, risk that the resulting climate changes will wreak devastating changes in agricultural production throughout the world, among other problems. Such changes could lead to the death of far more people and cause far greater environmental damage than any nuclear reactor accident, and appear to be considerably more likely.

Others conclude that nuclear power can reduce global warming only slightly and suggest that it would be much less expensive to invest in enhanced energy efficiency. Bill Keepin and Gregory Kats of the Rocky Mountain Institute have estimated that to reduce carbon dioxide emissions significantly through the use of nuclear power would require the completion of a new nuclear plant every 2 days for the next 38 years. Oak Ridge National Laboratory staff members reported that before any massive replacement of fossil fuel with nuclear power could occur, several new techniques would have to be developed. These include commercial-scale recycling of nuclear fuels, breeder reactors to extend existing fuels, and possibly uranium recovery from seawater.

Hans Blix, the director general of the International Atomic Energy Agency, in describing the international dimensions of the problem, noted that developing nations will not likely build nuclear plants in the near future: "If nuclear power is to be relied on to alleviate our burdening of the atmosphere with carbon dioxide, it is therefore to the industrialized countries that we must first look. They are in the position to use these advanced technologies—and they are also the greatest emitters of carbon dioxide."

The original nuclear era, synonymous with the growth of nuclear power in the United States, began in the early 1960s and lasted until 1979. It fell victim to stabilized demand for electricity, which was brought on by enormous oil price hikes in 1975, and the Three Mile Island crisis in 1979. As a consequence of that accident, the required number of nuclear plant personnel and their training requirements grew significantly. In addition, the mandatory retrofitting of existing nuclear facilities to enhance their safety added significantly to the cost of an already capital-intensive industry. As a result, the electric-power industry became understandably reluctant to invest further in proposed and planned nuclear facilities.

If a second nuclear era occurs, it likely will be characterized by a period of cheap nuclear energy delivered by reactors that hold public confidence in terms of their safety and economy. Some experts believe that such a rebirth is possible through the development of smaller, more efficiently designed reactors in the 500–600 megawatt range, rather than the current 1000–1200 MW facilities. These new reactors would be of a standardized, easily replicated design, have a longer operational lifetime (60 years versus the current 30), and be demonstrably safer and more economical in operation. One proposal advocates a series of developmental stages over a 13–15 year span. The initial stage would involve experimental construction, a prototype operational plant seven years later, and commercial use after another six years. If such work began in 1992, it might be possible to begin a second nuclear era by the second decade of the twenty-first century. But in all of this, the unsolved problem of the safe disposal of radioactive waste remains perhaps the greatest impediment.

8.21   *Consider This*

Make two lists, one of the benefits and the other of the risks associated with nuclear fission reactors. On the basis of this analysis, write an editorial in which you propose a 20-year national policy on fission-fueled power plants.

■ *Conclusion*

Almost 40 years have passed since President Eisenhower waved his radioactive magic wand to begin the construction of the first commercial nuclear reactor in the United States. The glittering promise of a boundless supply of unmetered electricity, drawn from the nuclei of uranium atoms, has proved illusory. But the needs of our nation and our world for safe, abundant, and inexpensive energy are far greater today than they were in 1954. So scientists and engineers continue their atomic quest. Where it will lead is uncertain, but it is clear that society will have a major voice in ultimately making the decision. Reason and not emotionalism must govern our actions.

■ References and Resources

Ahearne, J. F. "The Future of Nuclear Power." *American Scientist* **81,** Jan./Feb. 1993: 24–35.

"Clean Power's Strange Enemies." (editorial) *New York Times,* March 12, 1990: A16.

Golay, M. W. "Longer Life for Nuclear Plants." *Technology Review,* May/June 1990: 25–30.

Greenwald, J. "Time to Choose." *Time,* April 29, 1991: 54–61.

Häfele, W. "Energy from Nuclear Power." *Scientific American* **263,** Sept. 1990: 136–44.

League of Women Voters Education Fund. *The Nuclear Waste Primer.* New York: N. Lyons Books, 1985.

Meitner, L. and Frisch, O. "Disintegration of Uranium by Neutrons: A New Type of Nuclear Reaction." *Nature,* Feb. 11, 1939: 239–40.

Murray, R. L. *Understanding Radioactive Waste* (3d ed.). Columbus, Ohio: Battelle Press, 1989.

Taylor, J. S. "Improved and Safer Nuclear Power." *Science* **244,** 1989: 318–25.

Toufexis, A. "Legacy of a Disaster" (Chernobyl). *Time,* April 9, 1990: 68–69.

Wald, M. L. "License is Granted to Nuclear Plant in New Hampshire." *New York Times,* March 2, 1990: A1.

Wasserman, H. "The Clamshell Reaction." *The Nation,* June 18, 1977: 744–49.

———. "The Clamshell Alliance: Getting it Together." *The Progressive,* Sept. 1977: 14–18.

■ Exercises

1. Give the number of protons and neutrons in each of the following stable nuclei.

   a. $^{15}_{7}N$ b. $^{79}_{35}Br$ c. $^{208}_{82}Pb$

2. What generalization can you propose about the relative number of protons and neutrons in stable nuclei as the atomic number increases?

*3. A certain element consists of two stable isotopes with the masses and percent abundances given below. Determine the average atomic mass and identify this element.

mass	% abundance
10.0129	19.78
11.00931	80.22

4. Einstein's equation, $E = mc^2$, applies to ordinary chemical changes as well as to nuclear reactions. Use the equation to calculate the mass that, if converted to energy, would be equivalent to the energy released when 1 g of CH_4 is burned as described in Chapter 4. ($1 J = 1 kg m^2/s^2$.)

5. Write equations to represent the following nuclear reactions.

   a. reaction of U-235 with a neutron to form Br-87, La-146, and neutrons

   b. reaction of Pu-239 with a neutron to form Ce-146, another nucleus, and three neutrons

   c. neutron-induced fission of U-235 to form one nucleus with 56 protons, a second with a total of 93 neutrons and protons, plus two extra neutrons

6. Identify the equations below that represent chain reactions. State the characteristic that they have in common.

   a. $^{235}_{92}U + ^{1}_{0}n \rightarrow ^{141}_{56}Ba + ^{92}_{36}Kr + 3\,^{1}_{0}n$
   b. $^{27}_{13}Al + ^{4}_{2}He \rightarrow ^{30}_{15}P + ^{1}_{0}n$
   c. $^{40}_{20}Ca + ^{1}_{0}n \rightarrow ^{40}_{19}K + ^{1}_{1}H$
   d. $Cl + O_3 + O \rightarrow 2\,O_2 + Cl$

*7. The text states that in the Seabrook plant, every minute 28,000 gallons of water are converted to vapor and 425,000 gallons of cooling [ocean] water are heated by 22°. (1 gal = 3.79 L = 3790 mL.)

   a. Calculate the number of calories required to bring about each of these transformations. (Hint: See Chapter 5.)

   b. Which number is larger? By how much?

   c. What happened to the energy difference between the two?

8. It is generally believed that terrorists would be more likely to construct a nuclear bomb with Pu-239 reclaimed from breeder reactors than with U-235. Use your knowledge of chemistry to offer reasons for this.

9. Identify the types of radiation that typically accompany the changes described below.

   a. the mass number of an emitting nucleus changes

   b. the atomic number of the emitting nucleus changes

   c. both the atomic number and the mass number change

   d. neither the atomic number nor the mass number change

10. Write nuclear equations to represent the following transformations.

   a. I-131 (used in nuclear medicine and tracer studies) releases a beta particle

   b. U-238 decays with the loss of an alpha particle

   c. F-19 (used in nuclear medicine) emits a positive electron or positron (^0_1e)

   d. Mo-98 is bombarded with neutrons to produce Tc-99 and another particle

11. Equation 8.2, which represents the conversion of U-238 to Pu-239 actually occurs in the steps represented below. Determine the atomic and mass numbers of species X and use a periodic table to identify the element.

$$^1_0n + {}^{238}_{92}U \rightarrow {}^{239}_{92}U \rightarrow X \rightarrow {}^{239}_{94}Pu + {}^{0}_{-1}e$$
$$+$$
$$^{0}_{-1}e$$

12. Use the masses of the nuclear species given to find the masses of the reactants and products in each equation below.

e^-	.00055	Th-234	233.99409
n	1.00867	Pa-234	233.99215
H	1.00728	U-238	238.00018
He	4.00150	Pu-239	239.00061

   a. U-238 → Th-234 + He

   b. Th-234 → Pa-234 + e^-

   c. n + U-238 → Pu-239 + 2 e^-

   For each of these reactions, calculate the mass change. Also specify whether energy is released or gained and calculate how much.

13. Offer reasons for the following observations.

   a. All nuclei with atomic numbers ≥ 90 are radioactive.

   b. Neutrons are more effective than protons at penetrating nuclei, but protons are easier to accelerate.

   c. Alpha particles are less penetrating but more damaging to human tissue than beta particles.

14. Prepare a graph of the quantity of radiation experienced versus height above sea level (Table 8.3). Suggest why the graph appears as it does.

15. Use the data of Table 8.4 to calculate the percentage of radiation exposure the average United States citizen receives from

   a. natural sources inside the body

   b. human-made sources

   c. the nuclear power industry

16. Determine the fraction of a radioactive isotope that would remain after 3 half-lives, 5 half-lives, and 10 half-lives.

17. A hospital orders iodine-131 for thyroid gland therapy from a radioactive isotope supply house. How many atoms should the suppliers ship if the hospital needs 3×10^{18} atoms of I-131 and the shipping time is one half-life (8 days)? Two half-lives? What is the mass (in grams) of the 3×10^{18} iodine atoms?

18. It is suggested that the fission products from nuclear plants should be isolated for 10 half-lives. Calculate the fraction of a radioactive species that remains after that length of time.

19. Suppose the half-life of carbon-14 were very short (one second) or very long (one million years). How would these values affect the potential usefulness of the isotope for dating human artifacts? Explain your answer.

20. Consider the following statement: Resolved—the United States of America should increase its reliance on nuclear power to generate electricity. Decide which side of the issue you support and develop a compelling one-page magazine advertisement to convert others to your position.

9

Solar Energy: Fuel for the Future

■

Here comes the sun, doo da doo doo,
Here comes the sun, and I say "It's all right."
Little darling, it's been a long, cold, lonely winter;
Little darling, it feels like years since it's been here.

"Here Comes the Sun" by George Harrison.

It is unlikely that the Beatles were attempting to deliver a commentary on the future of energy production in the 1970s when they performed this song. Yet, as we approach the year 2000, these words take on added meaning. "Here comes the Sun" may yet become a rallying cry for expanding our use of humanity's oldest and most important energy source.

In two earlier chapters we considered the two sources of energy that are extensively used in the United States and in most other countries. In Chapter 4 we discussed the energy obtained from the changes that occur in chemical bonds when fossil fuels are burned. Chapter 8 focused on the energy released when certain atomic nuclei are induced to undergo fission. The convenience of these forms of energy is undeniable; they are concentrated and fairly easy to use. But the supplies of both fossil and fission fuels are limited, and when they are gone, they are gone forever. There is no practical way to replace them!

There is, however, another source of energy in almost limitless supply—the Sun. Our local star has been providing energy to the Earth for billions of years and is expected to continue to do so for billions more. Solar energy is, in fact, simply another type of nuclear energy, one that depends on the fusion of small atomic nuclei to yield larger ones rather than on the fission of large nuclei to form smaller ones. We will return to a discussion of nuclear fusion in the Sun (and on Earth) later in this chapter, but first, we offer an overview of what follows.

■ *Chapter Overview*

The chapter begins with a brief survey of the essential role played by the Sun in the Earth's energy balance, a consideration of some of the ways in which this energy has been exploited, and an accounting of the amount of solar energy potentially available. The simplest and most direct way to use some of this energy is by the passive heating of houses and buildings. A more versatile approach is to use sunlight to produce fuels—biomass or hydrogen. We consider both the energetics and the economics of this technology. Of particular importance are the innovations being made by chemists in their efforts to use solar energy to decompose water into hydrogen and oxygen. Once the hydrogen is obtained, it can either be burned as a fuel or used to generate electricity in a fuel cell. The discussion of fuel cells leads to a related aside on the electrochemical cells and batteries that power many modern devices. The most promising technology for using solar energy is probably the photovoltaic cell—a device for the direct conversion of the Sun's radiation to electricity. The text describes the role of semiconductors in solar cells, the principles governing their operation, and the current and potential applications of photovoltaics. We then turn to the process responsible for the energy of the Sun, nuclear fusion, and describe various efforts to duplicate the process on Earth. The chapter ends with an account of the most recent (and most controversial) of these attempts: cold fusion.

■ *Here Comes the Sun*

Like all living things, we humans owe our very existence to the Sun. But beyond that, solar energy in its various forms has played a significant role in the progress of humanity. It was solar energy that warmed our early ancestors before they learned to use fire. It was also solar energy, in the form of wind, that drove sailing ships on their journeys across the seas, whether carrying ancient Phoenicians, European explorers, or American merchants. And it was solar energy, as both wind and water power, that provided western Europe with a transition between the Middle Ages, which relied chiefly on animal and human power, and the fossil-fuel economy that began with the Industrial Revolution.

It may seem strange that we refer to water and wind as forms of solar energy, but in fact they both derive their energy from the Sun and would soon disappear without sunlight. Wind is a consequence of the movement of air between high and low pressure areas of the atmosphere—differences produced in part by uneven heating by the Sun. Water power is the result of water moving toward the sea from higher elevations, where it has fallen as precipitation. The supply of moisture in the atmosphere is replenished on a continuous basis as the Sun evaporates liquid water on the Earth. Fully 20% of the solar energy reaching the surface of the planet is spent to drive this hydrologic cycle.

But the Sun shines only during the day, the wind blows only sporadically in most locations, and water flowing with sufficient force to power hydroelectric plants is available only in certain places. Therefore, these forms of energy are limited in their applicability. On the other hand, fossil fuels can be burned to produce heat, light, and motion at any time; and if they are not found where they are needed, they can be transported there. In addition, as you have already seen, they represent very concentrated energy sources. Consequently, once fossil fuels were discovered, their tremendous potential energy was soon realized and exploited. It must be recognized, though, that the widespread use of fossil fuels is a very recent phenomenon in human existence, covering only the last 200 years. This represents only 0.2% of the 100,000 years that our species has existed. Relatively speaking, if these 100,000 years are equated to one day, we have been using fossil fuels for less than 3 minutes and have been producing power by means of nuclear fission for approximately 25 seconds.

Recent years have seen a rebirth of interest in solar energy. The reasons for this new sunrise are implicit in several of the preceding chapters. Chapter 4 made it clear that the supplies of fossil fuels will someday be exhausted or become so expensive to obtain that they will no longer be burned. We have also seen that the combustion of fossil fuels is a major cause of environmental pollution, producing toxic substances such as nitrogen oxides and sulfur oxides (Chapters 1 and 6), as well as carbon dioxide, which is the major contributor to global warming (Chapter 3). Furthermore, Chapter 8 suggested that the early promise of large quantities of cheap energy from nuclear fission has not been realized and that its social and environmental costs are so high that its widespread use seems unlikely. In contrast to the gloomy prospect presented by these existing sources of energy, solar energy is available almost everywhere on Earth, is non-polluting, and is virtually inexhaustible.

The quantity of energy that reaches the Earth from the Sun is almost unbelievable. Every year, 5.6×10^{21} kJ of solar energy fall on the planet's surface—15,000 times the world's present energy supply. But even if we should decide that the time has come to "go solar," the task is not simple. Solar energy is not in a concentrated form and cannot be stockpiled readily. Therefore, its widespread use depends on devising effective schemes for capturing, storing, and transforming it into more useful forms. The remainder of this chapter is a review of various ways that have been developed or are being developed to accomplish these tasks.

■ *Heat from Sunlight*

The most obvious and familiar form of solar energy is heat. If we could save the heat from the summer to warm us during the winter, we could conserve the portion of the fossil fuels we now burn for that purpose. Unfortunately, the technology for the long-term storage of heat does not yet exist. However, storage for short periods of time is not difficult. Many of the innovations for the solar homes we will discuss are designed to store solar energy during the day and use it for heating during the night. This is an example of the passive use of solar energy—strategies for directly trapping the heat of the Sun by some convenient means and releasing it later.

It is important to keep in mind that not all types of solar storage are the products of human invention, and those that are must conform to natural laws. We have already discussed some of the mechanisms by which energy is stored in natural systems. For example, you know from Chapter 3 that the energy stored in the vibrating molecules of certain gases, such as CO_2, H_2O, and the CFCs, can contribute to global warming. In Chapter 5 we considered some of the effects of the relatively high specific heat of water (1 cal/g °C), a property that has even influenced agricultural production. For example, the natural passive solar storage in the Great Lakes and the Finger Lakes is partially responsible for the wine industry of upstate New York.

You may recall that the energy absorbed by a body of water (a cupful or Lake Ontario) is equal to the mass of the water m times the temperature change Δt times the specific heat of water $sp\ ht$.

$$q = \text{heat absorbed} = m \times \Delta t \times sp\ ht \qquad (9.1)$$

In the fall, when the temperature of the land and air around the lakes begins to drop, the water slowly releases its stored heat. As a result, the temperature in the area remains warm longer than it would be otherwise, leading to an extension of the growing season. This permits the cultivation of certain crops such as grapes and other fruits that could not be grown at these northern latitudes in the absence of passive solar storage.

A major application of passive solar storage by humans has been in the construction of solar-heated homes. Small-scale changes in house design, made at relatively little expense, often have dramatic paybacks. The changes involve capturing the Sun's heat in an enclosed air space or in other materials such as stone, water, or certain chemicals that can be used as heat sources later. You are already aware of the effectiveness with which an enclosed air space can capture heat. You have, no doubt, entered a car that has been in direct sunlight with all of its windows closed. The metal, the plastic, the fabric, and the air itself can be painfully hot—sometimes fatally hot to children or pets. Similarly, the energy storage capacity of other materials is apparent if you sit on a rock or stone bench that has been exposed to the Sun's rays for a few hours.

Homes that feature passive solar heating capitalize on this property. Some recently designed models promise that heating costs will be one-half or less than those for conventional homes, even in northern regions. One such design is "Advanced House," located in Toronto and pictured in Figure 9.1. This prototype features a "sun space" in which air is warmed by sunlight before it is circulated through the house. The wall adjoining the sun space is made of brick, which also has a high heat capacity. At night the house is warmed, in part, by the heat stored in this wall during the day. The effectiveness of the sun space is increased by orienting it toward the south to collect the maximum amount of sunlight. In addition, the windows are designed and constructed to maximize the amount of heat trapped and minimize its loss. They consist of three panes separated by spaces filled with argon, which has a low heat conductivity. In addition, the panes are coated with a material that does not affect the transmission of visible light but does inhibit the transfer of heat. Tin oxide is most commonly employed for this purpose, though some other substances have also been tested. The performance of such windows is referred to in terms of "R-numbers," where R-2 represents a window that has twice the resistance to heat flow as a single pane window. Some windows have been developed that have R-values of 15 or greater.

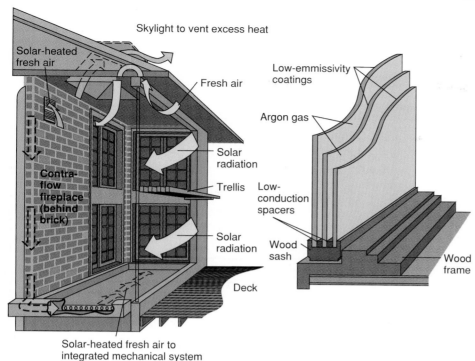

Figure 9.1
Diagram of "Advanced House" and its "sun space." The drawing on the right illustrates the triple-glazed windows used in the house. (Drawings by Mario Ferro for *Popular Science*, December 1990. Reprinted by permission.)

9.1 ■ *Consider This*

Suppose you are a telephone salesperson for a company called "Windows of Nature" that manufactures insulating, energy-efficient solar windows. The three panes of glass are coated with tin oxide and the spaces between them are filled with argon gas. These windows permit solar radiation to enter the house, but trap heat within it. They can thus effect a considerable savings in energy bills, but they are expensive. Your job is to present a one-minute sales pitch that will convince a prospective customer to replace the windows in his or her home with "Windows of Nature." What will you say?

"Advanced House," like most solar homes, also incorporates other energy-saving devices that result in an energy bill that is only about one-fourth that of a conventional home in the same location. The walls and ceilings are heavily insulated to inhibit the loss of heat. Moreover, the air within the house is circulated to prevent heat from collecting near the ceilings (which would leave cold the living space nearer the floor). Finally, the humidity is maintained at an appropriate level, because moist air feels warmer than dry air of the same temperature. Although the cost of constructing such homes is greater than that for building conventional ones, the money saved in fuel costs pays for the investment within ten years or less, depending on the cost of fuel.

Chapter 4 concluded with some mention of the energy and fuel savings that are possible through the design of homes and commercial buildings. Home heating and other domestic energy uses cost approximately $120 billion per year, or about 20% of the total United States energy bill. Commercial buildings, including hospitals, offices, and stores, contribute another $80 billion. In addition to its economic impact, the energy used for these purposes is expensive environmentally, representing the release of

500 million tons of carbon dioxide per year. Some of the energy-saving strategies employed in model solar homes have also been applied to new and existing commercial buildings. Indeed, the heating requirements for commercial buildings have decreased by 50% or more since the 1960s and 1970s. New office buildings designed and constructed with energy conservation in mind have even lower requirements—as little as 20% of the energy consumed by a building of comparable floor space in 1970.

9.2 ■ *Consider This*

All buildings leak energy to some extent. Examine the residence hall, apartment building, or house in which you live. List the places where heat energy escapes. For example, perhaps the front door of your dormitory doesn't close completely or the caulking around the windows in your house is cracked, permitting heat to escape. Once you have completed your survey, choose three items and describe what could be done to decrease the energy loss.

9.3 ■ *Consider This*

Again, considering the building in which you live, describe at least three ways solar energy could be used to meet (or partially meet) energy needs currently supplied by traditional energy sources.

■ *Water and Wind*

Although passive solar heating holds promise for fuel conservation, there are many energy demands that cannot be met by such methods. Modern technological societies are, by definition, active. For use in transportation, energy must be portable, concentrated, and convenient, like gasoline. For many other purposes, we need energy that can be transmitted over long distances and is available at the throw of a switch in a widely versatile form. Clearly, this latter case describes electrical energy. At first thought, using sunlight to produce fossil-fuel substitutes or to generate electricity seems a remote and impractical dream. Yet, in fact, both have already been accomplished.

Waterfalls and windmills, both indirectly powered by solar energy, have long been used to drive electric generators. Huge dams and their associated hydroelectric plants produce just over 8% of the electrical energy used in the United States. In addition, there are occasional proposals to use the tides as a source of energy for the generation of electricity.

Currently, wind power supplies only a tiny fraction of our electricity—often to individual homes in rural areas. However, there has been a recent resurgence of interest in large-scale wind-powered electrical generation. Figure 9.2 is a photograph of an array of windmills near Palm Springs, California. Hundreds of these windmills cover 12 square miles of the California desert and produce sufficient electricity to meet the needs of the residents and businesses in the surrounding valley.

Although the increased use of wind and water power can help reduce our dependence on fossil fuels, both are very indirect ways of capturing the vast supply of solar energy that reaches us daily. Therefore, much research activity is being devoted to developing more direct means of using sunlight as an energy source. We first consider the possibility of synthesizing fuel from sunbeams, something that nature has been doing successfully for a very long time.

Figure 9.2
Wind turbines for generating electricity, Palm Springs, California.

■ *Fuel from Sunlight*

The Sun is the ultimate energy source for the food we eat, the wood in our houses and furniture, the paper in this book, and the flowers, grass, and trees that beautify our planet. All of this is a consequence of photosynthesis, the elegant chemical process that captures solar energy and uses it to create grains of wheat, oak trees, and violets out of carbon dioxide and water. Surprisingly, only 0.06% of the solar radiation reaching the surface of the Earth is used to power photosynthesis. Nevertheless, sunlight stored in chemical bonds remains our major energy source. The Suns of a million yesterdays are rekindled whenever we burn gas, oil, or coal. But as we have seen, our limited supply of fossil fuels is being rapidly depleted. And as these fuels are burned, they contribute vast quantities of carbon dioxide, sulfur dioxide, and nitrogen dioxide to the atmosphere.

One partial solution to the problems of fossil fuel is to make use of photosynthesis to generate renewable fuels. Because of the biological sources of these substitutes, they are generally termed biomass. To be sure, biomass releases carbon dioxide when burned, but CO_2 is absorbed from the atmosphere when plants grow, so a rough parity is maintained. Wood is an obvious example of a biologically derived fuel of considerable historical importance. But wood cannot be a large-scale modern energy source because it simply takes too long to grow trees. Therefore, attention has been focused on fuels that might be derived from rapidly growing plants that could be harvested and replanted each year. Ethanol (ethyl alcohol), made by the fermentation of corn, sugar cane, or other renewable plant products, is a particularly promising alternative to fossil fuels. But as you learned in Chapter 4, there are obvious limitations to the amount of cropland that can be diverted from producing food to producing fuel.

9.4 ■	**Consider This**
	In assessing the promise of any alternative fuel, one must compute the energy cost of producing that fuel. If more fossil-fuel derived energy is required to produce a gallon of a fossil-fuel substitute than that substitute can provide, the process is energetically unfeasible. Although the Sun provides the energy to grow corn, identify some of the other energy costs of producing ethanol and suggest the likely immediate sources of this energy.

Although biomass cannot meet all of the planet's energy needs, photosynthesis provides a tantalizing example of what might be done. If chemists could copy nature, they might be able to design other processes that use the energy of sunlight to create fuels. The prospect of partial energy independence from green plants has prompted exciting research projects aimed at generating the simplest of elements, hydrogen, from the most common of compounds, water.

■ *Water: "An Inexhaustible Source of Heat and Light?"*

In Jules Verne's 1874 novel *Mysterious Island,* a shipwrecked engineer speculates about the energy resource that will be used when the world's coal supply has been used up. "Water," the engineer declares, "I believe that water will one day be employed as fuel, that hydrogen and oxygen which constitute it, used singly or together, will furnish an inexhaustible source of heat and light."

Is this simply science fiction, or is it energetically and economically feasible to break water into its component elements? Can hydrogen really serve as a useful fuel? And what does this have to do with solar energy? In order to answer these questions and to assess the credibility of the claim by Verne's engineer, a Sceptical Chymist needs to examine the energetics of the reaction represented by equation 9.2.

$$2H_2(g) + O_2(g) \rightarrow 2\,H_2O(l) \tag{9.2}$$

Experiment shows that the reaction as written above gives off 572 kJ. This quantity of energy is released when two moles of liquid water are formed from the combination of two moles of hydrogen and one mole of oxygen. It follows that the burning of one mole of H_2 will yield one mole of H_2O and $\frac{572}{2}$ or 286 kJ of energy. This indicates a highly exothermic reaction.

The data of the previous paragraph suggest that hydrogen might indeed be a good fuel. In order to find out just how good, it would be useful to compare hydrogen with other fuels such as coal, oil, or gas. Because heats of combustion are often reported in kJ per gram of fuel, we need to obtain a similar value for hydrogen. Fortunately, the calculation is easy, and we have all the necessary information readily at hand. We have just determined that burning one mole of hydrogen gas yields 286 kJ. We also need to find the number of grams of hydrogen corresponding to one mole of the gas. The mass of one mole of H_2 molecules (its molar mass or MM) is twice the mass of one mole of hydrogen atoms. Because the periodic table indicates that hydrogen has an atomic mass of 1.00, the mass of a mole of hydrogen atoms is 1.00 g. It follows that:

$$MM\ H_2 = 2 \times MM\ H = 2 \times 1.00\ \text{g/mole}\ H = 2.00\ \text{g}$$

To obtain the energy released per gram of hydrogen, we perform a simple operation that corresponds to dividing by 2.

$$\text{Energy} = 286\ \text{kJ/mole}\ H_2 \times \frac{1\ \text{mole}\ H_2}{2.00\ \text{g}} = 143\ \text{kJ/g}$$

In comparison, coal has a heat of combustion of 30 kJ/g, octane (a major component in gasoline) is 46 kJ/g, and methane (natural gas) is 54 kJ/g. Clearly, hydrogen has the potential of being a powerful energy source. In fact, it is used (along with oxygen) to launch the space shuttle and other rockets.

9.5   *Your Turn*

■

Calculate the number of moles and grams of H_2 that would have to be burned to yield an American's daily energy share of 250,000 kcal. (1 kcal = 4.18 kJ)

Ans. 3650 moles or 7300 g

Having established that hydrogen can be used as a fuel, we turn to a very important question: How can we obtain a sufficient supply? On the one hand, things look promising because hydrogen is the most plentiful element in the universe. Over 93% of all atoms are hydrogen atoms. Although hydrogen is not nearly that abundant on Earth, there is still an immense supply of the element. But essentially all of it is tied up in chemical compounds. Hydrogen is too reactive to exist for long in its pure, elementary form in the presence of all the other elements and compounds that make up the atmosphere and the Earth's crust. Therefore, to obtain hydrogen for use as a fuel, it is necessary to break down hydrogen-containing compounds, and this requires energy.

■ Splitting Water

The traditional laboratory method of generating hydrogen involves the action of certain acids on certain metals. Perhaps the most common combination is sulfuric acid and zinc.

$$H_2SO_4(aq) + Zn(s) \rightarrow ZnSO_4(aq) + H_2(g) \tag{9.3}$$

Although convenient on a small scale, this reaction is far too expensive to scale up for industrial applications. Therefore, we turn to the most abundant earthly source of hydrogen—water.

It should come as no surprise that if the formation of one mole of water from hydrogen and oxygen releases 286 kJ, an identical quantity of energy must be absorbed to reverse the reaction. (Figure 9.3 summarizes the processes.) All we need to do is to find a source of the 286 kJ. The most convenient method of decomposing water is by **electrolysis,** which is the passage of a direct current of electricity of sufficient voltage to break the O–H bonds (see Figure 5.2).

$$286 \text{ kJ} + H_2O(l) \rightarrow H_2(g) + 1/2\ O_2(g) \tag{9.4}$$

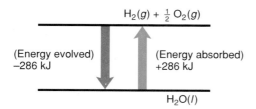

Figure 9.3

Energy differences in the hydrogen–oxygen–water system.

One answer often leads to another question, in this case, "How will the electricity be generated?" Currently, most of it is produced by the burning of fossil fuels in conventional power plants. If we only had to deal with the first law of thermodynamics, the best we could possibly do would be to burn an amount of fossil fuel equal in energy content to the hydrogen produced in electrolysis. But as you recall from Chapter 4, we must also deal with the consequences of the second law of thermodynamics. Because of the inherent and inescapable inefficiency associated with transforming heat into work, the maximum possible efficiency of an electrical power plant is about 60%. When we add the additional energy losses due to friction, incomplete heat transfer, and transmission over power lines, it would require at least twice as much energy to produce the hydrogen than we could obtain from its combustion. This is comparable to buying eggs for seven cents each and selling them for five; it's no way to do business. Furthermore, most methods of generating electricity have a variety of negative environmental effects. At one time it was thought that the "cheap" extra electricity from nuclear fission could be used to produce hydrogen to fuel the economy, but that energy utopia has hardly been realized. It should be apparent, therefore, that electricity generated from fossil fuels or nuclear fission does not offer a practical way to split water to produce hydrogen for use as a fuel.

A second possibility is to use heat energy to decompose water. The process by which most of the commercial hydrogen supply is currently produced does, in fact, use heat. You have already encountered it, in Chapter 4, in the discussion of substitutes for fossil fuels. Hot steam is passed over coke (essentially pure carbon) at 800°C.

$$H_2O(g) + C(s) \rightarrow H_2(g) + CO(g) \tag{9.5}$$

The mixture of hydrogen and carbon monoxide that is produced can be burned directly, it can serve as the starting material for the synthesis of hydrocarbon fuels and other important compounds, or the hydrogen can be separated and used as needed. The reaction is being studied in an effort to find catalysts that will make it possible to carry it out at lower temperatures.

The direct thermal decomposition of H_2O is not commercially promising. In order to obtain reasonable yields of hydrogen and oxygen, temperatures of over 5000°C would be required. The attainment of such temperatures is not only extremely difficult, it would also use enormous amounts of energy—at least as much as would be released when the hydrogen was burned. Thus, we have again reached a point where we would be investing a great deal of time, effort, money, and energy to generate a quantity of hydrogen that would, at best, return only as much energy as we had invested, and in practice a good deal less.

■ *Hydrogen from Sunlight*

Happily, all is not lost in our efforts to attain a hydrogen economy. We may yet be saved by the sunrise. Indeed, sunlight appears to be the only energetically and economically feasible way of obtaining hydrogen from water. However, this active use of solar energy is not without formidable technical problems. A major difficulty is the strength of the O–H bonds in the H_2O molecule (459 kJ/mole). In order for absorbed radiation to break these bonds, the light must be of very high energy and therefore very short wavelength (less than 2.61×10^{-7} m, or 261 nm). Although ultraviolet radiation in this wavelength range regularly strikes the top of the Earth's atmosphere, it is absorbed by oxygen and ozone (Chapter 2). The solar energy that does reach the Earth's surface is mainly in the visible and infrared regions of the spectrum. The energy per photon is only about one-half of that needed to break the O–H bonds. Because the interaction of matter and radiation is normally an either/or process involving one molecule and one photon, water molecules cannot be directly decomposed by the sunlight penetrating the atmosphere.

Fortunately for life on Earth, nature is a very sophisticated chemist with a remarkable catalytic system that uses visible light to split H_2O. The key substance in photosynthesis is chlorophyll, a compound of carbon, hydrogen, oxygen, nitrogen, and magnesium that is found in all green plants. Through a complex, but now well-understood series of reactions, the energy of two photons of red light is used to break the O–H bond in water. Although elementary oxygen is a product of photosynthesis, hydrogen gas is not formed. Instead, the hydrogen atoms participate in a chain of reactions and are eventually incorporated into glucose, $C_6H_{12}O_6$. Nevertheless, there is

9.7 Your Turn

Prove that radiation must have a wavelength of less than 2.61×10^{-7} m to have sufficient energy to break an O–H bond, which has a bond strength of 459 kJ/mole.

Soln. To solve this problem, it is necessary to use concepts introduced in Chapter 2. Recall that a chemical bond can break if it absorbs a photon having energy greater than the energy of the bond. The energy per photon is related to wavelength by the following equation.

$$E = \frac{hc}{\lambda} = 6.63 \times 10^{-34} \text{ J} \times \frac{3.00 \times 10^8 \text{ m/s}}{\lambda}$$

The correct values for h (Planck's constant) and c (the speed of light) have been included. Substituting $\lambda = 2.61 \times 10^{-7}$ m yields $E = 7.62 \times 10^{-19}$ J for the energy of a *single photon.* But the O–H bond strength is given as 459 kJ/mole, and one mole of O–H bonds consists of 6.02×10^{23} bonds. One photon would be required to break each of these bonds, and one mole of photons would be needed to break one mole of bonds. To calculate the energy of a mole of these photons, we multiply the energy per mole by Avogadro's number.

$$\text{Energy/mole} = 7.62 \times 10^{-19} \text{ J/photon} \times 6.02 \times 10^{23} \text{ photons/mole}$$
$$= 4.59 \times 10^6 \text{ J/mole} = 459 \text{ kJ/mole}$$

Because radiation with wavelengths shorter than 2.61×10^{-7} m has more energy per photon (and per mole of photons) than this limit, it is capable of breaking O–H bonds and dissociating water.

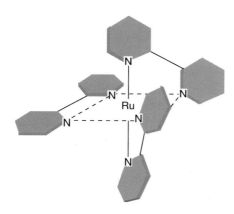

Figure 9.4

The structure of the ruthenium tris(bipyridine) ion. A carbon atom is located at each undesignated corner of each of the six hexagonal rings, and hydrogen atoms are also omitted. (From William W. Porterfield, *Inorganic Chemistry: A Unified Approach,* 2d ed. Copyright © 1993 Academic Press, Inc., Orlando, FL. Reprinted by permission.)

hope that by mimicking some aspects of photosynthesis, scientists can succeed in using sunlight to generate hydrogen and oxygen from water. What is needed is a substitute for chlorophyll.

One compound that has been studied extensively in this regard is ruthenium tris(bipyridine)chloride, $Ru(bpy)_3Cl_2$ [bpy = bipyridine]. The compound, which is called "rubippy" for short, has the structure shown in Figure 9.4. A ruthenium ion, Ru^{2+}, is at the center of the large molecular ion, bonded to six nitrogen atoms. Each of the nitrogen atoms is in a ring with five carbon atoms that are not indicated in the figure. The structure of rubippy is a little like that of chlorophyll, which includes a magnesium ion, Mg^{2+}, bonded to four nitrogen atoms.

Figure 9.5

Diagram of normal and excited energy states in Ru(bpy)$_3$Cl$_2$. The six dots on the left represent six electrons occupying the ground state of ruthenium. When a 450 nm photon is absorbed, one electron is transferred from the Ru^{2+} to a bipyridine ring, leaving a positive "hole" behind. The excited electron can then be transferred to another chemical species.

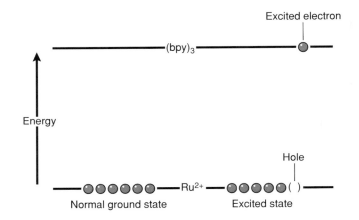

Ruthenium tris(bipyridine) chloride strongly absorbs radiation at a wavelength of 450 nm (1 nm = 10^{-9} m), which is in the middle of the visible portion of the spectrum. This wavelength corresponds to a blue-green color. It may seem a bit surprising to learn that rubippy is orange, but remember that the color of an object is due to the light that is "left over" or reflected after absorption has occurred. What happens within the Ru(bpy)$_3^{2-}$ ion when this light is absorbed is the critical step. The ion, of course, includes many electrons that occupy certain energy levels, similar to the energy levels in an atom. The normal distribution of electrons within these levels is called the "ground state." There are six electrons in the highest occupied energy level in a Ru^{2+} ion. When a photon corresponding to a wavelength of 450 nm is absorbed, one of these electrons is "excited." It acquires sufficient energy to move from the ruthenium ion to one of the bipyridine rings—a process diagrammed in Figure 9.5. This transfer places the electron in a relatively high energy condition that persists for about a microsecond (10^{-6} second). Although a microsecond is a short time by human standards, it is long on the time scale of molecular processes. It is plenty of time for the Ru(bpy)$_3$Cl$_2$ to come in contact with many other atoms, molecules, or ions while the electron is still in its excited state. The high energy state of this electron can lead to its transfer to another chemical species. Rubippy thus acts as an electron pump powered by solar energy.

Suppose, for example, that the excited rubippy encounters a water molecule. The electron could be transferred to the H$_2$O, according to the following equation.

$$H_2O + e^- \rightarrow OH^- + 1/2\ H_2 \tag{9.6}$$

Note that this converts the molecule into a hydroxide ion and a hydrogen atom (half a hydrogen molecule). We do not normally write equations involving half molecules, and for reasons that will soon become apparent, we multiply equation 9.6 by four. The result is equation 9.7.

$$4\ H_2O + 4\ e^- \rightarrow 4\ OH^- + 2\ H_2 \tag{9.7}$$

The reaction represented by equations 9.6 and 9.7 involves the gain of an electron by a water molecule. Remember that such a process is called reduction. In terms of the transfer of electrons, it corresponds exactly to what happens during electrolysis.

It is significant that Ru(bpy)$_3$Cl$_2$ is also capable of producing O$_2$. In this process H$_2$O is oxidized; it loses electrons that are transferred to the positive "holes" created on the ruthenium ions when the initial photo-induced excitation occurred. The reaction is represented in equation 9.8.

$$2\ H_2O \rightarrow O_2 + 4\ H^+ + 4\ e^- \tag{9.8}$$

The reduction and oxidation reactions occur simultaneously, so we add equations 9.7 and 9.8 to obtain the overall reaction.

$$6\ H_2O + \cancel{4\ e^-} \rightarrow 2\ H_2 + O_2 + \underbrace{4\ H^+ + 4\ OH^-}_{} + \cancel{4\ e^-} \tag{9.9}$$

$$-4\ H_2O \qquad\qquad\qquad -4\ H_2O$$

$$2\ H_2O \rightarrow 2\ H_2 + O_2 \tag{9.10}$$

Note that 4 H$^+$ + 4 OH$^-$ equal 4 H$_2$O, which is subtracted from both sides of equation 9.9. Moreover, 4 electrons appear on both sides of the arrow, so we cancel them. The net result, equation 9.10, represents the reaction scientists have been trying to achieve.

From this brief overview, it would seem that rubippy offers everything we could want to launch the hydrogen age—a means of absorbing sunlight and using its energy to split water into H$_2$ and O$_2$. Regrettably, the compound also possesses several limitations that have, to date, prevented its commercialization for the large-scale production of hydrogen. Perhaps the major drawback is the fact that the maximum efficiency of hydrogen production thus far attained is less than 1%. Moreover, Ru(bpy)$_3$Cl$_2$ is expensive and has been made mainly on a small scale for research studies. Third, the compound is somewhat unstable and undergoes various reactions that break it down and lower its ability to promote the desired reactions.

Even in the most successful trials, rubippy and sunlight alone have not been sufficient to produce H$_2$ from H$_2$O. It has been necessary to introduce other chemicals to act as "relays" in the electron transfer process. In addition, the reaction now requires a catalyst, such as finely divided platinum metal, which adds one more complication and considerable additional cost. However, the fact that this process can be carried out at all, coupled with its tremendous potential for revolutionizing the energy economy, means that "water splitting" is a very hot research topic. Success may still be a long way off, but the fascinating chemistry involved, to say nothing of the social consequences, justifies a significant investment of time, talent, and money. The nineteenth-century fiction of Jules Verne has repeatedly been turned into twentieth-century fact by modern science. Perhaps you will someday see water become "an inexhaustible source of heat and light."

The Hydrogen Economy

If and when we succeed in developing improved methods for producing hydrogen cheaply and in large quantities, we will still face significant problems in storing and transporting it. Although elemental H$_2$ has a high energy content per gram, it occupies a very large volume—about 12 liters per gram at normal atmospheric pressure and room temperature. If it is to be stored and transported in its gaseous state, large, heavy-walled metal cylinders will be required, thus eliminating much of the advantage of the favorable energy/mass ratio. To save space, gases are typically converted to the liquid state under high pressure, as in the case of "bottled gas" or "liquid propane." But hydrogen must be cooled to –253°C before it liquifies. This means that keeping it in liquid form requires very low temperatures and very high costs.

A number of attempts have been made to circumvent these problems by storing the H$_2$ in other forms. One of these, proposed only recently, involves absorbing gaseous H$_2$ on a solid, such as activated carbon, which has a high surface area and can therefore hold a great deal of H$_2$ at only moderate pressures. The carbon can be heated as needed to release the H$_2$ for combustion. A slightly different approach involves reacting the H$_2$ with certain metals to produce compounds called hydrides, which are relatively stable and have a reasonably high storage capacity. For example, if 10 liters (about 10 quarts) of H$_2$ gas at 25°C and 1 atmosphere were reacted with lithium metal, the resulting LiH would occupy a mere 4.3 mL, or slightly less than a teaspoon. When such hydrides are reacted with H$_2$O, they produce H$_2$, which can than be burned in the usual fashion. Prototype vehicles that operate on this principle have been built and used in a number of locations. Clearly, either approach would improve greatly the safety and convenience of handling H$_2$, and perhaps be the decisive factor in determining the extent of its acceptance as a fuel.

Even if we manage to solve all of the production, storage, and transport problems identified above, we must consider how best to extract energy from our hydrogen. The most obvious way would be to burn it in power plants, vehicles, and homes. A pure stream of hydrogen burns smoothly, quietly, and safely in air, delivering 143 kJ per gram and forming only non-polluting water as an end product. But when hydrogen is mixed with oxygen, a small spark is often sufficient to detonate a devastating explosion.

The vivid photographs of the exploding space shuttle Challenger and the burning hydrogen-filled dirigible Hindenburg are unforgettable reminders of the risks associated with the many benefits of a hydrogen economy.

 9.8   *Consider This*

Specify the technical problems that must be solved in order to achieve a "hydrogen economy." Then specify the economic, political, or social problems that you envision.

■ *Fuel Cells: A Slow Burn*

Suppose someone were to suggest that there is a way to combine H_2 and O_2 to form H_2O without the hazards of combustion. Suppose further that this individual claimed that the reaction could be carried out without allowing the hydrogen or oxygen to come in contact with each other. The Sceptical Chymist might well dismiss such assertions as sheer nonsense—an outright impossibility. And yet, sometimes what appears to be completely contrary to common sense can in fact happen in the natural world. The operation of a **fuel cell** is a case in point. In a fuel cell, a chemical reaction corresponding to combustion takes place, but no burning actually occurs. Instead, the energy is released as electricity. Such devices are routinely used as sources of electrical energy in the United States space program. The space shuttle carries 3 sets of 32 cells each that use hydrogen as a fuel, but never burn it.

The principles governing the operation of a fuel cell are fundamentally the same as those that apply to a conventional flashlight battery. The idea is to physically separate a spontaneous chemical reaction into two parts, each of which occurs in a segregated region of a cell. One of the reactions is always oxidation, in which some chemical species loses electrons. The other reaction is a reduction, involving the gain of electrons.

In the hydrogen fuel cell, hydrogen gas is oxidized in the following reaction.

$$2\,H_2(g) + 4\,OH^-(aq) \rightarrow 4\,H_2O(l) + 4\,e^- \tag{9.11}$$

Equation 9.11 implies that two H_2 molecules dissociate and the individual H atoms give up one electron each for a total of four electrons.

$$2\,H_2 \rightarrow 4\,H \rightarrow 4\,H^+ + 4\,e^-$$

The four H^+ ions then combine with the four OH^- ions to form four molecules of H_2O.

$$4\,H^+ + 4\,OH^- \rightarrow 4\,H_2O$$

Note that equation 9.11 is balanced with respect to elements, atoms, *and charge*. There is a charge of −4 on both sides of the arrow. However, the equation represents a half-reaction. Oxidation cannot occur alone; it is like one hand clapping. Oxidation must always be balanced with a reduction reaction. That process (another half-reaction) is represented by equation 9.12.

$$O_2(g) + 2\,H_2O(l) + 4\,e^- \rightarrow 4\,OH^-(aq) \tag{9.12}$$

Here oxygen is reduced. Each of the O atoms initially present in O_2 molecules acquires two electrons. This means that in the four OH^- ions, each oxygen is effectively in a O^{2-} state.

The overall reaction is the sum of the oxidation and reduction half-reactions, 9.11 and 9.12.

$$2\,H_2 + 4\,\cancel{OH^-} + O_2 + 2\,H_2O + \cancel{4\,e^-} \rightarrow 4\,H_2O + 4\,\cancel{OH^-} + \cancel{4\,e^-} \tag{9.13}$$

The four electrons lost in the oxidation half-reaction equal the four electrons gained in the reduction process, as they must in a balanced equation. The 4 e^- and the 4 OH^- that appear on both sides of the arrow in equation 9.13 can be canceled, and 2 H$_2$O can be subtracted from both sides. This simplifies the equation to 9.14.

$$2\,H_2(g) + O_2(g) \rightarrow 2\,H_2O(l) \qquad\qquad (9.14)$$

This is clearly the equation for the burning of hydrogen in oxygen, but in a fuel cell, it is "burning" without flame and with relatively little heat.

Although the oxidation and reduction half-reactions are physically segregated in a fuel cell, they must be connected in such a way that the electrons released during oxidation are conducted to the reduction reaction. This is accomplished by using **electrodes,** electrical conductors placed in the cell as sites for chemical reaction. The electrode at which oxidation takes place is called the **anode.** The electrons given up in the process flow from the anode through a wire to the **cathode.** At the cathode, the electrons are used in the reduction half-reaction. This flow of electrons is what we call electricity, and the apparatus is an example of an **electrochemical cell.** An electrochemical cell is a device that converts the energy released in a spontaneous chemical reaction into electrical energy. It is the opposite of an **electrolytic cell** in which electrical energy is converted to chemical energy. The chlor-alkali process, described in Chapter 7, uses electrolytic cells to bring about the nonspontaneous conversion of sodium chloride to elementary sodium and chlorine.

The electrons flowing from the anode to the cathode of a fuel cell can be intercepted and used to do work, which is the whole point of an electrochemical cell. On the space shuttle these electrons are used to illuminate bulbs, power small motors, and operate computers. The tendency of the electrons to flow through the external circuit depends on the difference in electrical **potential** between the anode and the cathode. That in turn depends upon the chemistry that is taking place. The difference in electrode potential is the **voltage** of the cell, and it is measured and reported in **volts (V).** The greater the difference in potential, the higher the voltage. The rate at which the electrons flow is called the **current,** and it is measured in **amperes (amps).** You are probably already familiar with these terms because they are used to describe household electricity—110 volts and 15 amps for many purposes.

The fuel cell diagrammed in Figure 9.6 operates at 1.23 volts. The overall reaction (equation 9.14) releases 286 kJ of energy per mole of water formed, but instead of liberating most of this energy in the form of heat, 70–75% of it appears as electrical energy. The direct production of electricity eliminates the inefficiencies associated with using heat to do work to produce electricity. The electrodes in the hydrogen-oxygen fuel cell are often made of porous carbon. In addition, the anode contains nickel and the cathode contains both nickel and nickel oxide (NiO). Potassium hydroxide, KOH, dissolved in water serves as the electrolyte—the electrically conducting solution that surrounds the anode and cathode. KOH is an ionic compound; it consists of K$^+$ and OH$^-$ ions. These ions are responsible for the electrical conductivity of the solution.

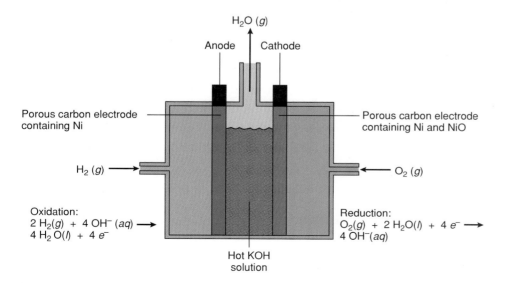

Figure 9.6

A hydrogen-oxygen fuel cell.

Further inspection of equations 9.11 and 9.12 reveals that OH⁻ ions are produced at the cathode and consumed at the anode. Therefore, these ions must move from the cathode compartment to the anode compartment to complete the circuit and make possible the operation of the cell.

A new generation of hydrogen fuel cells uses a solid polymer electrolyte that currently costs about $6 per gram or $6000 per kilogram. On the other hand, the technology has permitted a 40-fold savings in the platinum used to catalyze the electrochemical reaction. Today only 4 g of platinum are required for the fuel cell necessary to power a car. This adds only $50 to the cost of the vehicle. Mazda has several prototypes under development, and hopes to have commercial models on the market by the year 2010.

■ *Cells and Batteries*

Although you may not have had personal experience with fuel cells, the chances are very good that you have a close relative of a fuel cell on you at the present time. If you are wearing a battery-powered watch or carrying a pocket calculator or portable cassette player, you have a system for the direct conversion of chemical energy to electrical energy. Most people refer to these devices as batteries, and we shall discuss them in this section, not because they involve solar energy, but because they are similar to fuel cells in some significant respects. It goes without saying that batteries are very important in today's society because they afford a convenient, transportable source of stored energy.

The first thing to note is that most batteries are not batteries at all. The word "battery" refers to a collection of similar objects—as in a battery of cannons. A standard flashlight "battery" is more correctly called a cell, or an electrochemical cell. A collection of several cells wired together constitutes a true battery. The means by which electrochemical cells produce electricity is fundamentally the same as the way in which fuel cells operate. The energy released by a spontaneous chemical reaction is liberated as electricity. Again, two half-reactions are involved, one oxidation and the other reduction. Electrons are generated by the oxidation reaction, collected at the anode, and flow through an external wire to the cathode, where reduction occurs. The difference in voltage or potential between the two electrodes is proportional to the energy evolved.

The two half-reactions for the familiar alkaline cell, pictured in Figure 9.7, are given below.

$$\text{Anode (Oxidation): } Zn(s) + 2\,OH^-(aq) \rightarrow Zn(OH)_2(s) + 2\,e^- \tag{9.15}$$

$$\text{Cathode (Reduction): } 2\,MnO_2(s) + H_2O(l) + 2\,e^- \rightarrow Mn_2O_3(s) + 2\,OH^-(aq) \tag{9.16}$$

As in the fuel cell, the overall cell reaction is the sum of the two half-reactions.

$$Zn(s) + 2\,MnO_2(s) + H_2O(l) \rightarrow Zn(OH)_2(s) + Mn_2O_3(s) \tag{9.17}$$

9.9	*Your Turn*

Prove to yourself that equations 9.15 and 9.16 are balanced with respect to number and identity of atoms and *electrical charge*.

The voltage produced by this cell is 1.54 volts. The voltage of a cell depends primarily on which elements and compounds are participating in the reaction. It does not depend on factors such as the overall size of the cell, the amount of material which it contains, or the size of the electrodes. This is apparent from the fact that all alkaline cells, from the tiny AAA size to the large D cells, give this same 1.54 volts. On the other hand, the current or electron flow does depend on the size of the cell. Larger cells generate larger currents. And, because power is obtained by multiplying the voltage by the current, larger cells are more powerful than smaller ones.

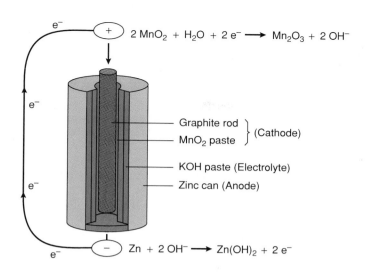

$$2\ MnO_2 + H_2O + 2\ e^- \longrightarrow Mn_2O_3 + 2\ OH^-$$

Graphite rod
MnO$_2$ paste } (Cathode)

KOH paste (Electrolyte)
Zinc can (Anode)

$$Zn + 2\ OH^- \longrightarrow Zn(OH)_2 + 2\ e^-$$

Figure 9.7

Diagram of an alkaline cell.

9.10   *Consider This*

An indirect cost of using "batteries" (electrochemical cells) to power small appliances is the cost of disposal, both in filling our landfills and in discarding non-renewable metal resources. In an attempt to ascertain the magnitude of this problem, do the following.

a. Make a list of things that you own that run on disposable batteries.
b. Estimate the number of times you replace the batteries in these appliances in one year.
c. Expand your estimate to include other members of your family. In other words, how many batteries does your family discard in one year?
d. Estimate the number of batteries disposed of per year in this country.

Table 9.1	Some Common Electrochemical Cells and Their Associated Voltages

Type	Voltage	Rechargeable
Dry cell	1.5	No
Alkaline	1.54	No
Mercury	1.3	No
Lithium-iodine	2.8	No
Lead storage	2.0	Yes
Nickel-cadmium	1.46	Yes

Many different electrochemical cells have been developed for different specific purposes. Some of these are listed in Table 9.1, along with their voltages and an indication of whether they are rechargeable. Because mercury batteries can be made very small, they are used widely in watches, camera equipment, hearing aids, calculators, and other devices that use transistors and integrated circuits that do not draw large currents. Unfortunately, the toxicity of mercury (Chapter 7) makes the disposal of these cells a potential hazard. The Environmental Protection Agency estimated that in 1989, 88% of the 1.4 million pounds of mercury in urban trash came from single-use batteries. Burning trash may exacerbate and extend the problem by releasing mercury vapor into the atmosphere.

Lithium-iodine cells are so reliable and long-lived that they are used to power cardiac pacemakers. A battery implanted in the chest can last as long as 10 years before it needs to be replaced. In fact, the widespread use of such pacemakers today has been due, in large part, to the improvements that have been made in the batteries used to power them, rather than in the pacemakers themselves.

Most electrochemical cells convert chemical energy into electrical energy with an efficiency of about 90%. This may be compared with the much lower efficiencies that typically characterize the conversion of heat to work (30–40%). However, it is important to remember that much energy is required to manufacture electrochemical cells. Metals and minerals must be mined and processed, and the various components must be assembled. Moreover, a battery has a finite life. Sooner or later, the chemical reaction will reach completion, the voltage will drop to zero, and electrons will no longer flow. The battery will be "dead" and ready for disposal, and disposal is a formidable problem. In February 1993, *National Geographic* reported that some 2.5 billion household batteries are purchased each year in the United States at a total price of $3.3 billion. Of these, over 90% are single-use batteries that find their way into landfills or incinerators.

On the other hand, some batteries are rechargeable, a characteristic that greatly extends their lifetimes and increases their cost-effectiveness. The best known example of a rechargeable battery is the lead storage battery used in almost all American cars. It is a true battery because it consists of six cells, each generating 2.0 V for a total of 12.0 V. The overall cell reaction is given by equation 9.18. It is a *storage* battery because it is a device for storing energy.

$$\underset{\text{lead}}{Pb(s)} + \underset{\text{lead oxide}}{PbO_2(s)} + \underset{\text{sulfuric acid}}{2\,H_2SO_4(aq)} \rightarrow \underset{\text{lead sulfate}}{2\,PbSO_4(s)} + \underset{\text{water}}{2\,H_2O(l)} \qquad (9.18)$$

The anode is made of lead and the cathode of lead oxide. The liquid electrolyte is a concentrated solution of sulfuric acid. Despite the weight of the lead and the corrosive properties of the acid, the lead storage battery is dependable and long lasting. The key to its success is the fact that reaction 9.18 is readily reversible. As it spontaneously proceeds in the direction indicated by the arrow, the reaction produces the energy necessary to power a car's starter, headlights, and various devices. But as the reaction proceeds, the battery "discharges." The electrical demands of a modern car are so great that in a short time the reaction would be pulled far to the right, significantly reducing the voltage and the current. To counter this, the battery is attached to a generator, or alternator, turned by the engine. The alternator generates direct current electricity, which is run back through the battery. This input of energy reverses the reaction represented by equation 9.18 and recharges the battery. In a high-quality lead storage battery, this process of discharging and recharging can go on for five years or more. Because of this dependability, lead storage batteries are sometimes used in conjunction with wind-turbine electrical generators. The generator charges the batteries when the wind is blowing, and the batteries provide electricity when the wind stops.

Rechargeable nickel-cadmium or "nicad" cells are even longer lasting, although they are more expensive. Because they are lighter and do not contain sulfuric acid, they are used in various portable tools and appliances. Portability is, in fact, one of the great benefits of electrochemical cells and batteries. It is extremely unlikely that they could ever make a significant impact on large-scale energy production, but they are already being used in ground transportation. Battery-powered vehicles are frequently found where fumes from internal combustion engines cannot be tolerated. Forklifts in warehouses, passenger carts in air terminals, wheelchairs, and golf carts are typically powered by linked lead storage batteries. The relatively low power-to-weight ratio, though, makes this means of locomotion uneconomical over long distances.

Such applications require batteries having higher power-to-weight ratios. One promising development makes use of the reaction of aluminum with oxygen in the air to produce aluminum hydroxide.

$$4\ \text{Al}(s) + 3\ \text{O}_2(g) + 6\ \text{H}_2\text{O}(l) \rightarrow 4\ \text{Al(OH)}_3(s) \tag{9.19}$$

Although the use of these and other new batteries is currently restricted to experimental vehicles, the day may come when motorists will stop at recharging stations rather than at filling stations. If a group of regional officials have their way, that day will come in 2007 in southern California. In an effort to remedy what has been called the worst air pollution in the United States, a tentative plan has been adopted that would require that all cars in the Los Angeles basin be converted to electric power or other clean fuel by that year. Even the proponents of the plan admit that it will be impossible to achieve without major technological breakthroughs. Therefore, the state of California has taken a somewhat more gradual approach. New legislation requires that by 1998, 2% of the cars sold in the state must meet "zero emission" standards, with the number of non-polluting cars increasing to 10% by 2003. Figure 9.8 is a photograph of a battery-powered California car charging up outside the offices of the Sacramento Municipal Utilities District. If you look carefully, you will note that it is plugged into the Sun via a bank of photovoltaic cells.

Figure 9.8

A battery-powered electric car being charged from photovoltaic cells at the Sacramento Municipal Utilities District.

> **9.12   *Consider This***
>
> The costs of a complete ban on gasoline or diesel-powered vehicles in southern California by the year 2007 are unknown, but everyone agrees that they would be staggering. Assume you are an economic consultant to the mayor of Los Angeles. Prepare a list of expenses and hidden costs that might be anticipated in conjunction with the conversion to electric-powered cars, trucks, and buses. You are not expected to include numbers in your list. (If you can do so, you are either an economic genius or in possession of a very clear crystal ball.)

> **9.13   *The Sceptical Chymist***
>
> Just how environmentally friendly is the conversion to battery-operated cars proposed for southern California? The electricity required to recharge the batteries must come from somewhere. Given current technology, what is the most likely source and what are the environmental implications? What are the risks and benefits of a conversion to battery-powered cars to Los Angeles, to the state of California, and to the world?

■ *Photovoltaics: Plugging in the Sun*

Thus far we have considered two ways of using the Sun's energy: passively absorbing it as a source of heat, and using it to decompose water to yield hydrogen, which can then be burned as a fuel or used to generate electricity in a fuel cell. For many purposes it would be even more advantageous if the solar energy could be directly converted into electricity, without the intermediary of hydrogen. This is the function of **photovoltaic cells.** Such devices have already demonstrated their practical utility for both large- and small-scale electrical generation, and they may represent the best hope for capturing and using solar energy.

Electricity involves a stream of electrons, flowing from a region of higher potential (voltage) to one of lower voltage. In order for a photovoltaic cell to generate electricity, light must induce such a flow. This flow depends on the interaction of matter and photons of radiant energy—a topic already treated in considerable detail in Chapters 2 and 3. You will recall that the portion of the Sun's radiation reaching the Earth's surface is mainly in the visible and infrared regions of the spectrum, having a maximum intensity near 500 nm. Light of this wavelength is green in color and has an energy of about 4×10^{-19} J per photon, corresponding to 240 kJ per mole of photons. A photovoltaic cell must be made of atoms or molecules that will release electrons when struck by radiation of approximately this same wavelength.

Among the substances that will do this is a class of materials known as **semiconductors**—materials that do not normally conduct electricity well, but will do so under certain conditions. One of the first semiconductors identified was the element silicon, which you may recognize as being used in transistor radios, pocket calculators, and digital watches. A crystal of silicon consists of an array of silicon atoms, each bonded to four others by means of shared pairs of electrons, as represented in Figure 9.9a. These shared or bonding electrons are normally fixed in place and are unable to move about through the crystal. Consequently, silicon is not a very good electrical conductor under ordinary circumstances. However, if an electron absorbs sufficient energy, it can be excited and released from its bonding position, as indicated in Figure 9.9b. Once freed, the electron can move throughout the crystal lattice, thus making the silicon an electrical conductor. The energy required for silicon to release an electron from a bond is 1.8×10^{-19} J per photon, which is equivalent to radiation with a wavelength of 1000 nm. Visible light has a wavelength range of 350 to 700 nm. Recall that the shorter the wavelength of radiation, the greater the energy per photon. Therefore,

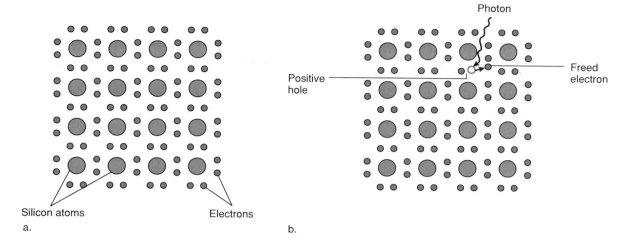

Silicon atoms   Electrons

a.   b.

Figure 9.9

(a) Schematic of bonding silicon.
(b) Photon-induced release of a
bonding electron in a silicon
semiconductor.

photons of visible sunlight are sufficiently energetic to excite electrons in silicon semiconductors. Indeed, many pocket calculators are now powered with solar cells, thus eliminating the added expense of buying batteries.

The fabrication of photovoltaic cells is not without some major problems. In the first place, purifying silicon to the appropriate degree is fairly expensive, although the starting material (common sand) is cheap and abundant. A second complication is that the direct conversion of sunlight into electricity is not very efficient. A solar cell could, in principle, transform up to 44% of the radiant energy to which it is sensitive into electricity. But more than a third of this (16% of the total) is lost to internal cell processes. This leaves a theoretical efficiency limit of 28%. In practice, efficiencies between 10 and 20% have been achieved.

In Chapter 4 we lamented the 30–40% efficiency of converting heat to work in a conventional power plant. It might seem that we should be even more distressed at the lower limits that can be achieved by photovoltaics. However, remember that the Sun is an essentially unlimited energy source, and there is a good deal of empty space on the planet that is well suited for large arrays of solar cells and for little else. Moreover, the fact that solar energy is free of many of the environmental problems associated with burning fossil fuels or with nuclear fission adds impetus to research and development.

The first use of solar cells was to provide electricity in NASA spacecraft, where cost seems to be of little concern and the intensity of radiation is so high that the low efficiency is not a serious limitation. But because most commercial applications must be more cost-conscious, a great deal of effort is being directed to increasing the efficiency of solar cells and lowering manufacturing costs. One promising innovation is replacing crystalline silicon with the amorphous form of the element. Photons are more efficiently absorbed by the less highly ordered atoms, a phenomenon that permits reducing the thickness of the silicon semiconductor to 1/60th of its former value. The cost of materials is thus significantly reduced. More common is the "doping" of silicon with other elements. This process consists of intentionally introducing about 1 ppm of elements such as gallium (Ga) or arsenic (As) into the silicon. These two elements and others from the same periodic families are used because they differ from silicon by a single outer electron. Silicon (atomic number 14) has 4 electrons in its outer energy level, gallium (atomic number 31) has 3, and arsenic (atomic number 33) has 5. Thus, when an atom of As is introduced in place of Si in the silicon lattice, an extra electron is added. The replacement of an Si atom with a Ga atom means that the crystal is now one electron "short."

The extra electrons in arsenic-doped silicon are not confined to bonds between atoms. Rather, they can move easily through the lattice, thereby increasing the electrical conductivity of the material over that of pure silicon. Silicon doped in this manner is called an **n-type** semiconductor because its electrical conductivity is due to **negative** carriers—electrons. On the other hand, for each silicon atom replaced with a gallium ion, an electronic vacancy or "hole" is introduced into what is normally a

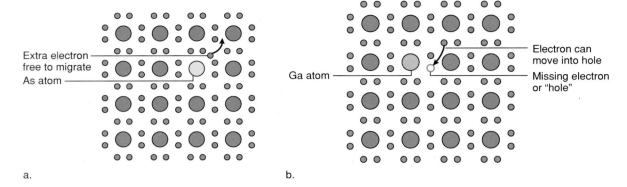

a.   b.

Figure 9.10

(a) An arsenic-doped n-type silicon semiconductor. (b) A gallium-doped p-type silicon semiconductor.

two-electron bond. When an electron moves into this hole, a new hole appears where the mobile electron formerly was located. Thus, holes and electrons move in opposite directions. Because holes can be regarded as **positive** carriers of electricity, gallium-doped silicon is called a **p-type** semiconductor. Figure 9.10 illustrates both n- and p-type semiconductors.

Both types of doping increase the conductivity of the silicon because less energy is needed to get extra electrons or holes moving. This means that photons of low energy (in other words, light of longer wavelength) can induce electron release and transport in doped crystals.

Sandwiches of n- and p-type semiconductors are used in transistors and many of the other miniature electronic devices that have revolutionized communication and computing. Similar structures are central to the direct conversion of sunlight to electricity. A photovoltaic cell typically includes sheets of n- and p-type silicon in close contact (Figure 9.11). An electric field is built up between the electron-rich n-semiconductor and the electron-deficient p-semiconductor. This voltage difference accelerates the electrons released when sunlight strikes the doped silicon. The electrons move into the electron-rich region and the holes into the electron-deficient layer. The result is a direct current of electricity that can be intercepted to carry on essentially all the things that electricity does. As long as the cell is exposed to light, the current will continue to flow powered only by radiation.

In addition to making an effort to improve the performance of silicon semiconductors by doping, scientists have also been searching for other substances that would exhibit the same or better semiconductor properties. Among promising substitutes are germanium, an element found in the same family of the periodic table as silicon, and compounds that have the same number of outer electrons as these elements. Included in this latter list are gallium arsenide (GaAs), indium arsenide (InAs), cadmium selenide (CdSe), and cadmium telluride (CdTe). Some of these new semiconductors have enhanced the efficiency of radiation-to-electricity conversion and made possible the tuning of solar cells to certain regions of the spectrum.

Figure 9.11

Schematic diagram of a solar cell.

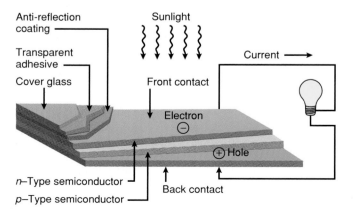

9.14   *Your Turn*

Using the periodic table as a guide, demonstrate why the electron arrangement and bonding in GaAs would be the same as in Ge.

As a result of these and other advances in photovoltaic technology, the cost of producing electricity in this manner has dropped dramatically from $3 per kilowatt-hour in 1974 to 30 cents per kilowatt-hour in 1990. Although electricity produced by this technology is still not competitive with that produced from fossil fuels, the long-range prospects for solar energy are encouraging. Its cost is decreasing while the cost of electricity generated from fossil fuels is increasing. In 1974, the cost of electricity from conventional power plants was 5–10 cents per kilowatt-hour; today it is 20 cents per kilowatt-hour. Moreover, the increasing cost of clean fuel and the added expense of pollution controls will drive the cost still higher. Given this situation and the expected improvements in the performance and decreases in the cost of solar cells, electricity from photovoltaic systems could become competitive with that from fossil fuels before the end of the century. Writing in the April 1987 *Scientific American,* Yoshihiro Hamakawa estimates that at that time the cost of electricity from a coal-fired plant will be close to 35 cents per kilowatt-hour, while the cost of electricity generated by solar cells will be about 8 cents per kilowatt-hour.

Photovoltaic technology is already in use in a number of countries. The largest solar installation, located in Carrisa Plains, California, was built by ARCO Solar, Inc. and Pacific Gas and Electric Company. It generates about seven megawatts at peak power. While this generating capacity is small compared to that of fossil fuel, nuclear, and hydroelectric plants, much larger solar installations are expected in the future. A 200-megawatt plant, which could provide household power for 300,000 people, could be erected on a square mile of land at relatively modest capital expense and minimal maintenance.

Alternatively, photovoltaic cells can be used to generate electricity for private homes. In Gardner, Massachusetts, New England Electric has installed arrays of solar cells on 30 individual houses. Each solar panel covers approximately 25% of the roof area and generates 2.2 kilowatts, enough to provide the needed power and heat for the house. The initial cost of the solar cells is $10,000 per house, but their lifetime is estimated at 30 years, with little upkeep. The fact that the annual gas and electric bill for a single family home in northern United States is about $1500 suggests that photovoltaics may soon be in wider distribution.

Photovoltaic electricity has already found a variety of other uses, including applications in the aerospace industry and electrical generation in remote regions of the Earth. For example, the highway traffic lights in certain parts of Alaska, far from power lines, operate on solar energy. There are even several automobile races for solar-powered cars. Sunrayce USA runs from Florida to Michigan, and another crosses Australia from Darwin to Adelaide. Such races were originally considered as unusual as the strange-looking vehicles that took part in them. But the fact that solar-powered cars were able to average more than 40 miles per hour, despite cloudy and rainy weather, in the last Cross-Australia race, suggests that they may no longer be just a curiosity. In addition, an airplane powered only by battery-charging amorphous silicon solar cells has flown more than 2400 miles in less than 120 hours, and even a solar-powered boat has made its appearance.

The direct conversion of sunlight to electricity obviously has many advantages. In addition to freeing us from our dependence on fossil fuels, an economy based on solar electricity would decrease the environmental damage that frequently occurs when these fuels are extracted or transported. Further, it would help to lower the levels of atmospheric pollutants such as sulfur and nitrogen oxides, and it would also help avert the dangers of global warming by decreasing the amount of carbon dioxide released into the atmosphere. It is, of course, likely that fossil fuels will continue to be the preferred form of energy for certain applications. Nevertheless, the future looks sunny for solar energy. Indeed, the only better source of energy might be if some modern Prometheus could steal a part of the Sun and bring it down to Earth.

■ *Stealing the Sun*

Essentially all of the energy used by the inhabitants of the Earth is nuclear in origin. The energy stored in the fossil fuels we burn and the foods we eat originated in the furnace of the Sun, where it was born in a nuclear reaction. Every second, 5 million tons of the Sun's matter are converted into 3×10^{23} kilojoules of energy. Yet, despite the profligate rate at which our star is consuming its substance, astrophysicists estimate that it is only about halfway through its expected 9-billion-year life cycle.

The primary reaction that fuels the fires of the Sun is believed to be the combination or **fusion** of four hydrogen atoms to yield one helium atom and two **positrons**. A positron has a mass equal to that of an electron, but carries a positive charge instead of a negative one.

$$4\,^1_1H \rightarrow\,^4_2He + 2\,^0_1e \tag{9.20}$$

As in the case of nuclear fission, the mass of the products is less than that of the reactants, and the difference is manifested as energy. The fusion of one gram of hydrogen will release as much energy as the combustion of 20 tons of coal. Gram for gram, this is eight times as energetic as the fission of uranium-235, making it the most concentrated energy source known. If we could only capture and control this same reaction on Earth, our energy problems would be solved forever!

For one thing, there would be no need to worry about running out of fuel. Hydrogen is the most abundant element in the universe; and even on Earth, the waters of the oceans contain a hydrogen supply that is, for all practical purposes, limitless. Furthermore, because the fusion reaction produces only a stable, nonradioactive, and nonreactive element (helium), concerns about air pollution or storing long-lived radioactive waste products would disappear. There are, however, some formidable technical problems that must be overcome before we can create and control a mini-Sun here on Earth. Temperatures higher than 100,000,000°C are required to overcome the repulsion of positively charged hydrogen nuclei and cause them to fuse. The problem is thus twofold: how to create such extreme temperatures and how to contain the incredibly hot matter produced.

9.15   *The Sceptical Chymist*

Check the claim that the fusion of one gram of hydrogen will release as much energy as the combustion of 20 tons of coal. To do so, you will need to make use of the fact that when 1.00 g of hydrogen undergoes fusion (as in equation 9.20), 6.72×10^{-3} g of matter is converted into energy. Burning coal releases 30 kJ per gram.

Hint: Remember that $E = mc^2$.

To be sure, humans have already carried out fusion reactions on Earth, but hydrogen bomb tests hardly could be considered controlled. In such thermonuclear devices the heat necessary to trigger fusion is generated by the fission of uranium-235 or plutonium-239. The fusing species are two isotopes of hydrogen: deuterium, which has one proton, one electron, and one neutron per atom; and tritium, which has one proton, one electron, and two neutrons per atom. Deuterium is symbolized as $_{1}^{2}H$ or $_{1}^{2}D$ while tritium is represented as $_{1}^{3}H$ or $_{1}^{3}T$. When atoms of these two isotopes fuse, a helium atom and a neutron are formed.

$$_{1}^{2}H \quad + \quad _{1}^{3}H \quad \rightarrow \quad _{2}^{4}He \quad + \quad _{0}^{1}n \qquad (9.21)$$

$$\text{deuterium} \quad \text{tritium} \quad \text{helium} \quad \text{neutron}$$

This reaction is being investigated because it is thought to be somewhat easier to initiate than that involving four hydrogen atoms. However, it does suffer from the disadvantage that tritium, a radioactive isotope with a half-life of 12.3 years, does not occur in nature in any appreciable amount. But it can be produced by bombarding lithium-6 ($_{3}^{6}Li$) with neutrons.

$$_{3}^{6}Li + _{0}^{1}n \rightarrow _{1}^{3}H + _{2}^{4}He \qquad (9.22)$$

Deuterium alone is being investigated as an alternative fusion fuel. It forms either tritium and normal hydrogen or helium-3 ($_{2}^{3}He$) and a neutron.

$$_{1}^{2}H + _{1}^{2}H \rightarrow _{1}^{3}H + _{1}^{1}H \qquad (9.23)$$

$$_{1}^{2}H + _{1}^{2}H \rightarrow _{2}^{3}He + _{0}^{1}n \qquad (9.24)$$

The other major problem associated with using nuclear fusion as an energy source is controlling the reaction once fusion has begun. For almost 40 years scientists have been trying to accomplish this task but success has evaded them. The principal difficulty is one of containing the hot gases, because most materials that might be used as containers vaporize at temperatures above 6000°C. A special bottle must be created to enclose this genie. Two schemes are currently under investigation: magnetic containment and inertial confinement.

The magnetic containment method involves injecting atoms of the appropriate hydrogen isotopes into a donut-shaped chamber, called a **tokamak,** where they are subjected to strong magnetic and electrical fields (Figure 9.12). The electrons are stripped from the atoms to produce a new state of matter, a gas-like **plasma** which consists of electrons and positively charged ions. The electric field heats the plasma until the nuclei fuse. This incredibly hot matter is contained by lines of magnetic force that prevent it from touching and vaporizing the metal walls of the container. Unfortunately, vast amounts of energy are required to generate this plasma. Although the tokamak at Princeton University has achieved temperatures high enough to initiate fusion, the energy output has not yet equaled the energy input, to say nothing of producing excess energy.

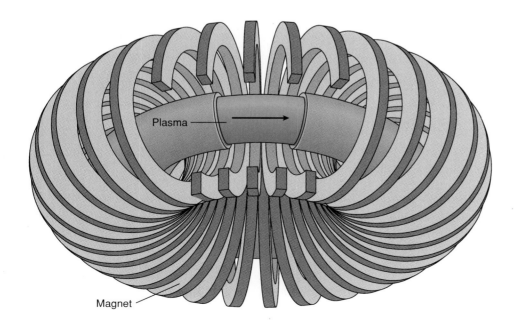

Figure 9.12

Magnetic containment of a fusing plasma in a tokamak.

Figure 9.13

Diagram of an inertial confinement laser fusion device. (From R. Wilson and W. J. Jones, *Energy, Ecology, and the Environment*. Copyright © 1973 Academic Press, Inc., Orlando, Florida. Reprinted by permission.)

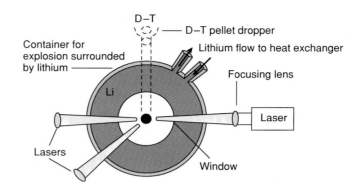

In **inertial confinement,** diagrammed in Figure 9.13, tiny glass spheres are filled with deuterium or deuterium-tritium mixtures at a pressure of several hundred atmospheres. These fuel pellets are then subjected to ultraviolet radiation generated by high-energy lasers. Under this photon bombardment, the atoms are squeezed together tightly enough, it is hoped, to cause them to fuse. Up to the present time, this goal has not been achieved and inertial confinement remains unfulfilled as a means of controlling nuclear fusion.

Despite the failure to achieve a controlled, sustained, energy-producing fusion reaction, research continues because the potential benefits, like the fuel supply, are almost boundless. Although the natural abundance of deuterium is only about 1.6 parts per million, just one cubic kilometer of seawater would furnish enough deuterium to produce the energy equivalent of 2 trillion barrels of oil, roughly the planet's total oil reserves. Considering the vastness of the oceans, there is enough readily accessible deuterium to supply the Earth's energy needs until the dying Sun engulfs our blue-green marble.

9.16 ■ *Consider This*

Let's think big! Suppose you are Secretary of Energy and you have $5 billion in research and development funds to distribute among the following three alternative energy technologies: solar-powered hydrogen production, photovoltaic cells, and nuclear fusion. Make your distribution and defend it.

■ *Drama in Dallas*

The 140,000-member American Chemical Society holds two national meetings each year. Typically, about 10,000 chemists gather to transact the business of the society, present formal papers, exchange information about the latest research results, examine new equipment and new books, and visit with friends and colleagues. Reporters usually cover a few of the sessions that seem particularly relevant, and they interview some of the scientists involved. But the banks of television cameras and the rows of reporters in the arena of the Dallas Convention Center at noon on Wednesday, April 12, 1989, signaled that the symposium about to take place was of much greater public interest than the typical concatenation of chemists. Over 6000 people, including several authors of this text, jammed the circular auditorium. There was a sense of high historical drama in the air. We were there to learn about what might be one of the most important scientific discoveries of the century. We were waiting to hear B. Stanley Pons, professor of chemistry at the University of Utah, speak on "Electrochemically Induced Nuclear Fusion."

Only a few weeks earlier, in a hastily called press conference, Pons and Martin Fleishmann of the University of Southampton had announced to the world that they had succeeded in achieving the fusion of deuterium at roughly room temperature. The two chemists claimed to have accomplished, with conventional glassware and simple instruments, what physicists had unsuccessfully tried to do for almost 40 years with the expenditure of billions of dollars and trillions of joules. If Pons and Fleishmann were indeed correct, the energy problems of the Earth might be effectively over. No wonder the news of cold fusion spread like wildfire, communicated by fax machines, electronic mail, and the *Wall Street Journal* rather than by peer-reviewed journals such as *Nature* or *Science*.

The atmosphere in the Dallas Convention Center was, appropriately enough, electric and energized. The first speaker was Harold P. Furth, director of the Princeton University Plasma Physics Laboratory, who described some of the plasma research being done there. He showed a slide of the huge Princeton tokamak. And he emphasized the potential importance of fusion by noting projections that fossil fuel supplies would be essentially exhausted by the year 2350. Two other scientists spoke about electrochemistry, and then it was time for the featured speaker.

In her introduction of Dr. Pons, the moderator dramatically announced that a message had just been received from the Soviet Union that the cold fusion experiment had been successfully repeated at the University of Moscow. The usually sedate chemists went wild! But Pons sounded appropriately professorial as he described the Utah experiments, showed slides of the simple equipment, and reported on the excess energy that the research team had observed when they passed an electric current through deuterium oxide. Members of the audience greeted Pons's remarks with either hope or skepticism—sometimes both! Nevertheless, we all rushed from the auditorium to snatch up papers describing this incredible claim.

9.17 *The Sceptical Chymist*

Suppose you were one of the reporters in the Dallas Convention Center on April 12, 1989. What questions would you have asked Stanley Pons?

■ *Fusion in a Flask?*

In retrospect, the story of cold fusion was probably too good to be true. The euphoria of the spring of 1989 has faded. The cartoonists and the late-night television comedians have forgotten low-temperature fusion. But the chemists and physicists still remember, and some research continues. Most of the efforts to duplicate the work of the Utah group have proved unsuccessful. Consensus, always a strong factor in the scientific community, seems to be that whatever produced the excess heat in the Salt Lake City experiments was not nuclear fusion. Nevertheless, the case of low-temperature fusion is such a fascinating study of the sociology of science that it deserves some attention.

Perhaps what made many chemists optimistic and most physicists skeptical was the fact that the procedure employed by Pons and Fleishmann was one that has been part of chemistry for two centuries. Their approach was to electrolyze deuterium oxide, also known as heavy water. Deuterium oxide, D_2O, is just like ordinary H_2O, except that it contains atoms of deuterium instead of the more common isotope of hydrogen. Pons and Fleishmann dissolved a little lithium deuteroxide, LiOD, in the D_2O to make it conduct electricity, and then applied a direct current across the two electrodes. Platinum (Pt) was used for the anode or positive electrode and palladium (Pd)

"Don't you remember? We were at Herb and Sally's, and Herb said he knew how to achieve fusion at room temperature, using only gin and vermouth."

(Drawing by Handelsman; © 1989 *The New Yorker Magazine,* Inc.)

for the cathode or negative electrode. As one would expect from years of similar experiments, oxygen gas was evolved at the anode and deuterium (hydrogen) gas at the cathode.

Anode (Oxidation): $4\ OD^- \rightarrow 2\ D_2O + O_2 + 4\ e^-$ (9.25)

Cathode (Reduction): $4\ D_2O + 4\ e^- \rightarrow 2\ D_2 + 4\ OD^-$ (9.26)

Net: $2\ D_2O \rightarrow 2\ D_2 + O_2$ (9.27)

This is all conventional chemistry, and, as we have seen, conventional chemistry predicts that energy must be absorbed to break down water (or heavy water) into its constituent elements. That is the purpose of the applied electric field. However, Pons and Fleishmann reported that after long periods of operation (200–250 hours), the palladium cathode began to emit heat energy. The quantity of heat evolved in some experiments was reported to be four times the electrical energy put into the apparatus. And in one trial, a substantial portion of the palladium, which has a melting point of 1554°C, melted and vaporized.

No one has ever seen the law of conservation of energy broken, and this can be no exception. The heat must come from somewhere, and there is no obvious chemical phenomenon that can account for the quantity observed by the Utah–Southampton team. Recombination of D_2 and O_2 to form D_2O would evolve energy, but only in an amount equal to that absorbed to break down the D_2O. Fleishmann and Pons suggest that the energy release is a consequence of one or both of the two reactions that occur when deuterium nuclei fuse (equations 9.23 and 9.24).

In their initial report, Pons and Fleishmann argued for this mechanism by citing their identification of neutrons, tritium, and helium-3 as reaction byproducts. It was subsequently determined that the initial neutron readings were erroneous, but the presence of tritium has been confirmed by a number of investigators. Some also report detecting low neutron fluxes, but in no case has the number of neutrons produced been large enough to account for the quantity of energy released.

If fusion reactions are indeed occurring, it is necessary to explain how the experimental design makes it possible to overcome the strong repulsion between deuterium nuclei. It has generally been assumed that this can happen only at extremely high temperatures, hence the techniques of magnetic containment and inertial confinement. Pons and Fleishmann have argued that their procedure works because of the great tendency of palladium to absorb hydrogen isotopes. As the D atoms are formed at the cathode, many of them are absorbed by the electrode itself, possibly as D^+ ions. Pons estimates that the cathode might contain as many as two deuterium atoms for each palladium atom and that the compression of deuterium in the palladium matrix corresponds to that attained by a pressure of 10^{27} atmospheres. The nuclei are thus brought into such intimate contact that they fuse with the attendant conversion of matter into energy.

The unprecedented idea of simple, low-energy chemical processes bringing about nuclear phenomena is not the only problem associated with the cold fusion saga. Much of the controversy relates to the mores of scientific research—how it is conducted and how it is communicated. The fact that the initial announcement of the Pons–Fleishmann results was made at a press conference, and not by publication in a refereed journal, is contrary to accepted practices in the scientific community. A draft of a preliminary paper was widely circulated via electronic mail and fax machines, but critics have argued that the information conveyed in that paper was not sufficiently detailed to enable anyone to duplicate the experiment. The contrast with the discovery of nuclear fission 50 years earlier is instructive (see Chapter 8). Although some of our modern means of communication were not available in 1939, the news of the Meitner–Frisch interpretation of the Hahn–Strassmann results also spread rapidly. The initial paper was composed over the telephone, and information about the research was brought to the United States by Niels Bohr before the paper was published. Publication in the prestigious international journal, *Nature,* came within two months of the announcement. By then, scientists in a number of countries had already succeeded in carrying out experiments that confirmed the release of the energy predicted by Meitner and Frisch.

Pons and Fleishmann have not been that lucky. Although they corrected some of their first statements and provided additional information about their procedures, it has not been possible, even for those who claim to observe the evolution of heat, to obtain consistent results. Nature should be more reliable than that.

There have also been charges that the Utah team was careless in measuring temperature and radiation. Many physicists seem to be congenitally scornful of the experimental work of chemists. *Nature* even mounted an editorial campaign to discredit the research of Pons and Fleishmann. Countercharges have been leveled against the experimental techniques of detractors, and demands have been made for the retraction of published papers.

It should be obvious that the motivation of the scientists involved is somewhat more than a disinterested search for the truth. A significant factor is the drive for priority, recognition, and honor that motivates many scientists. But there is more than a Nobel Prize at stake here. If low-temperature nuclear fusion is real and if it can be commercialized, it is a fabulously valuable discovery. At least one patent application was filed by the University of Utah, and a National Cold Fusion Institute was established in Salt Lake City. At this writing, the jury is still out, but whatever the verdict, this cautionary tale has tremendous implications for science and for society.

9.18   *Consider This*

Do some background reading and compare the stories of the announcement of cold fusion in 1989 and the announcement of nuclear fission in 1939. Pay particular attention to the audience to whom the information was initially transmitted and the means employed for transmittal. What effects did existing communications technology have on the two events? How do these two cases relate to the accepted standards of scientific research and communication and to the public's right to know potentially important information?

■ *Conclusion*

We began this chapter by noting the energy debt our planet and its inhabitants owe to the Sun. Its ancient investments—fossil fuels—are fast disappearing and we must seek alternatives for tomorrow. Once again, we are looking to our star for energy or for an example. It is fortunate that what seems to be the most promising method of capturing and transforming solar energy is also the most direct. Photovoltaic cells convert sunlight into electricity. There is no need for inefficient intermediate steps in which fuels are synthesized and burned, and the resulting heat energy is transformed first into mechanical and then into electrical energy. Of course, for some purposes, fuels are more convenient than electricity, and perhaps advances in research will make it economically and energetically feasible to use solar radiation to decompose water into hydrogen and oxygen, its reactive components. And maybe someday, chemists, physicists, and engineers will succeed in simulating the Sun by stimulating nuclear fusion—if not in a flask, in a magnetic donut or a glass bead. Experts predict that the most environmentally benign approach to energy generation will involve a mixture of sources and strategies, including passive solar heating, hydroelectric, wind, biomass, hydrogen, and photovoltaics, with limited reliance on natural gas as a back-up to the intermittent energy sources.

But the laws of thermodynamics and human nature are such that these transformations will not occur spontaneously. Solar-based energy alternatives cannot be developed without hard work and the investment of intellect, time, and money. Yet, in the United States, the amount of effort devoted to research on new energy sources appears to be directly proportional to the cost of oil. When international crises drive up the price of petroleum, there is a sudden flurry of official interest in energy conservation and the development of alternate technologies. When oil supplies are plentiful and prices are low at the gasoline pump, few seem to care about preparing for the time when fossil fuels will be depleted or too dirty or too expensive to burn. We need to establish priorities and act on them. We have been the beneficiaries of a bountiful nature, but we have an obligation to assure sources of energy for unborn generations.

■ References and Resources

Dagani, R. "Nuclear Fusion: Utah Findings Raise Hopes, Doubts." *Chemical & Engineering News,* April 3, 1989: 4–6, and the following subsequent articles by the same author in the same journal: "Cold Fusion: Race to Clarify Utah Claims Heats Up." April 10, 1989: 6–7; "Cold Fusion: ACS Session Helps Shed Some Light." April 17, 1989: 4–6; "Fusion Confusion: New Data, But Skepticism Persists." April 24, 1989: 4–5; "Fusion Controversy: Congress Excited, But Doubts Grow." May 1, 1989: 6–7; "Hopes for Cold Fusion Diminish as Ranks of Disbelievers Swell." May 22, 1989: 8–20; "Advocates, Skeptics Alike Still Puzzled by Cold Fusion." April 16, 1990: 28–30.

Gilland, B. "Energy for the 21st Century: An Engineer's View." *Endeavour* (New Series) **14,** (1990): 80–86.

Hubbard, H. M. "Photovoltaics Today and Tomorrow." *Science* **244,** April 21, 1989: 297–304.

Johansson, T. B.; Kelly, H.; Reddy, A. K. N.; and Williams, R. H. *Renewable Energy: Sources for the Future.* Washington: Island Press, 1993.

Krieger, J. "Development Efforts Target Advance Electric Auto Batteries." *Chemical & Engineering News,* Nov. 16, 1992: 17–18.

Nadis, S. "Hydrogen Dreams." *Technology Review,* Aug./Sept. 1990: 20–21.

Weinberg, C. J. and Williams, R. H. "Energy from the Sun." *Scientific American* **263,** Sept. 1990: 146–55.

White, D. C.; Andrews, C. J.; and Stauffer, N. W. "The New Team: Electricity Sources Without Carbon Dioxide." *Technology Review,* Jan. 1992: 42–50.

■ Exercises

1. The text states that "every year, 5.6×10^{21} kJ of solar energy fall on the planet's surface—15,000 times the world's present energy supply." It further reports that 20% of this energy goes to drive the hydrologic cycle and 0.06% is used in photosynthesis. Use this information to calculate the following.

   a. the world's present energy supply in kJ and in tons of coal, assuming all the energy is in that form (One ton of coal is equivalent to 2.7×10^7 kJ.)

   b. the quantity of energy that drives the hydrologic cycle

   c. the amount of energy used in photosynthesis

2. A swimming pool with a surface area of 5 meters by 10 meters contains 88,000 liters of water. The temperature of the water in this pool increases by 2°C when heated by the Sun for 6 hours. Calculate the energy absorbed.

3. The specific heats and densities of some materials are given below.

material	sp. ht. (cal/g°C)	density (g/cm³)
brick	0.220	2.0
concrete	0.270	2.7
steel	0.118	7.9
water	1.00	1.0

   a. Calculate the temperature change produced in 100.0 g of each substance by the absorption of 1 kcal of energy.

   b. Which substance would store energy most efficiently

      i. on the basis of mass?

      ii. on the basis of volume?

   c. Explain any differences between b.i. and b.ii.

4. Explain the similarities that exist between the function of the windows in Advanced House and an important property of CO_2 in the atmosphere.

*5. Photosynthesis is represented by the following equation:

$$6\ CO_2 + 6\ H_2O \rightarrow C_6H_{12}O_6 + 6\ O_2$$

   a. Based on the information provided in Exercise 1, calculate the total mass of glucose, $C_6H_{12}O_6$, (in grams) produced through photosynthesis in one year. The formation of 1 mole of $C_6H_{12}O_6$ requires the absorption of 2800 kJ.

   b. Note that 6 moles of CO_2 (264 g) are required for each mole of $C_6H_{12}O_6$ (180 g) formed. Use this information to calculate the total mass of CO_2 removed from the atmosphere by green plants in one year.

6. The text suggests a "rough parity" exists between CO_2 absorption and CO_2 evolution as biomass fuels are produced and consumed. Critically analyze this statement. On balance, would you expect more CO_2 to be absorbed from the atmosphere or released to it? Defend and explain your answer.

7. Use the bond energies of Table 4.1 to calculate the energy change associated with the following reaction:

$$2\ H_2O(g) + O_2(g) \rightarrow 2\ H_2O(g)$$

   Also compute the energy per gram of water vapor formed. (Note that the values listed in Table 4.1 apply to gases, not liquids or solids.)

8. In the lithium–iodine cell, Li is oxidized to Li^+ and I_2 is reduced to $2\ I^-$.

   a. Write equations for the two half-reactions and the overall reaction in this cell.

   b. Identify the half-reaction that occurs at the anode and the half-reaction that occurs at the cathode.

9. The unbalanced equations for the half-reactions in the lead storage battery are given below.

$$Pb(s) + SO_4^{2-}(aq) \rightarrow PbSO_4(s)$$
$$PbO_2(s) + 4\ H^+(aq) + SO_4^{2-}(aq) \rightarrow PbSO_4(s) + 2\ H_2O(l)$$

   a. Balance both equations with respect to charge by adding electrons as needed.

   b. Indicate which half-reaction represents oxidation and which represents reduction.

   c. One of the electrodes is made of lead, the other of lead oxide. Specify which is the anode and which is the cathode.

*10. Electric power (in watts or joules/second) is equal to the product of the current (in amps) and the difference in potential (in volts).

$$watts = joules/second = amps \times volts$$

   Use this definition in the following calculations.

   a. Calculate the electrical energy (in joules) generated by a 12-volt lead storage battery that produces a current of 1 amp for 100 hours.

   b. A gallon of gasoline yields 1.4×10^8 J. Compute the volume of gasoline that must be burned to generate the same quantity of energy calculated in part a.

   c. Assume gasoline costs $1.199 per gallon and a storage battery costs $50. Calculate and compare the cost of the gasoline that equals the energy output of the battery. Of course, the battery can be recharged, the gasoline cannot be reburned.

11. Determine the number of 12-volt batteries needed to produce the quantity of energy used by the average U.S. citizen each day (250,000 kcal) if each battery produces a current of 4.17 amperes per hour (see Exercise 10). [1 cal = 4.18 J.]

12. Show that radiation with energy of 1.8×10^{-19} joules per photon corresponds to a wavelength of 1000 nm. What would be the energy of a mole of these photons? In what region of the spectrum is this radiation found?

*13. The text describes houses in Gardner, Massachusetts that have all their energy needs supplied by photovoltaic cells generating 2.2 kilowatts (kW) per home. Electrical energy is usually measured and sold in kilowatt hours (kWhr). Solar panels operating at a power of 2.2 kW for a 24-hour day will generate 2.2 kW × 24 hr or 52.8 kWhr. This energy comes from the Sun, and both the intensity of the solar radiation and the hours of sunlight per day vary with location. The total amount of energy received per day per square meter will be the product of these two terms. Listed below are values for solar intensity and hours of sunlight for three different parts of the country.

Gardner, MA:	50 watts/m² for 8 hours
Washington, DC:	65 watts/m² for 9 hours
Conway, AR:	75 watts/m² for 10 hours

Use these data to calculate the area of solar cells (in m²) that would be required to generate 52.8 kWhr per day in these three cities. Assume the cells operate at an efficiency of 15%.

14. Again consider a house using 52.8 kWhr of electrical energy per day. What is the minimum number of gallons of fuel oil that would be saved per day if all this energy were provided by sunlight and photovoltaic cells rather than by burning fuel oil? Why is the answer a *minimum*? (1 gal of fuel oil is equivalent to 1.53×10^8 J.)

15. The series of reactions represented below is required to convert sand, SiO_2, into silicon that is pure enough for semiconductor use. Balance the equations.

    a. $SiO_2 + C \rightarrow Si(impure) + CO_2$
    b. $Si(impure) + Cl_2 \rightarrow SiCl_4$
    c. $SiCl_4 + Mg \rightarrow Si(pure) + MgCl_2$

16. A typical silicon chip such as those in digital watches and pocket calculators weighs approximately 2.3×10^{-4} gram. How many chips can be made from 1.0 kilogram of silicon?

17. As discussed in the text, other elements (for example, Ga or As) are often added to silicon at the 1 ppm (by mass) level to improve its electrical conductivity. For a chip weighing 2.3×10^{-4} g, calculate the number of grams of either gallium or arsenic that would have to be introduced to achieve this concentration. How many atoms does this represent?

18. Consult a periodic table to identify other elements that could be used as substitutes for arsenic in an n-type semiconductor or for gallium in a p-type semiconductor.

19. Listed below are nuclear masses in grams per mole.

^1_1H 1.00728	^1_0n 1.00867
^2_1H 2.01355	^3_2He 3.01493
^3_1H 3.01550	

    Use these values to calculate the mass differences and the energies released in the fusion reactions represented by these equations.

    a. equation 9.23   b. equation 9.24

20. Offer explanations for the following observations.

    a. Nuclear fusion appears to be possible only at high temperatures, whereas nuclear fission occurs spontaneously at room temperature.

    b. Nuclear fusion releases more energy per gram than nuclear fission.

    c. Electricity produced by solar energy is expected to decrease in cost by the end of the century; electricity produced from fossil fuels is expected to increase in price.

21. The "splitting" of water with $Ru(bpy)_3Cl_2$, the operation of a fuel cell, and the operation of a photovoltaic cell all involve the transfer of electrons. Discuss this common feature and explain how it is central to each of these processes.

10

The World of Plastics and Polymers

Dustin Hoffman as "The Graduate," probably pondering his future in plastics.

Ben, "The Graduate," played by a young Dustin Hoffman, has just returned home from college, picking up his luggage to "The Sounds of Silence," sung by Simon and Garfunkel. Now a party in his honor is underway. His parents' upper middle class home is filled with their upper middle class friends, all fawning over Ben's academic and extracurricular accomplishments. Suddenly Mr. McGuire appears, looking appropriately earnest in his dark blue suit, and calls Ben out for a serious conversation:

McG: Ben, come with me for a minute. I want to talk to you. I just want to say one word to you. Just one word.

Ben: Yes, sir.

McG: Are you listening?

Ben: Yes, I am.

McG: Plastics!

Ben: Exactly how do you mean?

McG: There's a great future in plastics. Think about it. Will you think about it?

Ben: Yes, I will.

With the prophetic word, "plastics," ringing in our ears, we leave Ben to the seductive wiles of Mrs. Robinson and cut to another, even more imaginary scene. You and Dustin Hoffman are sitting in a kitchen. Suddenly, all of the plastic in the room disappears. As a result, you and Hoffman find yourselves in your underwear. (We have assumed it to be made of cotton. Nylon underwear would earn the film an X rating.) Around you, various liquids and solids pour out of the cabinets as their containers disappear. Foodstuffs flow from the refrigerator because, being mostly plastic, it has largely vanished. Countertops, floor coverings, and the paint on the walls and woodwork have all disappeared. At this point, a fire breaks out. The plastic insulation on the electrical wiring has vanished and the wires have shorted out.

10.1	***Consider This***
■	Using your imagination, continue the above scenario and describe other events that could occur if all plastics were to suddenly disappear.

■ *Chapter Overview*

We begin the chapter by asking you to observe some of the properties of common plastics. This leads to a section of definition and description, and an introduction to the ways in which chemists can vary the properties of polymers by modifying their molecular structure. Because polyethylene is the simplest and most widely used plastic, it is subjected to a fairly detailed analysis. Its composition, structure, production, properties, modifications, and uses are all considered. This study leads to briefer treatments of the five other plastics that complete the "Big Six." In this context, both addition and condensation polymerization are described. A brief, but significant, aside addresses proteins, an important class of natural polymers, and nylon, a related synthetic polymer. As an example of the impact of plastics on modern life, we consider the many ways in which sports and recreation have been transformed by the introduction of synthetic materials.

The phenomenal success of plastics and their widespread distribution has not been without cost. Therefore, the final third of the chapter is devoted to the raw materials that go into the manufacture of plastics and the problems associated with the disposal of used plastics. Using the issue of plastics versus paper as a recurring case study, we examine disposal options including incineration, biodegradation, recyling, and source reduction. Not surprisingly, we find that there are no easy solutions to the problems posed by these useful and ubiquitous forms of matter.

■ *Plastics and Properties*

The best way to begin your study is by collecting various types of plastic and making some observations.

10.2	*Consider This*

Gather up a variety of plastic items from home or your dorm room—plastic bags, soda bottles, whatever happens to be at hand. Make a list of the objects and note the properties of the plastics. Include color, transparency, flexibility, elasticity, hardness, tensile strength, and other properties that could be used to classify and identify the plastics. Try to draw conclusions about which objects are made from the same material.

You have no doubt discovered from this activity (or from prior experience) that plastics exhibit a wide range of properties. We can illustrate this with a few objects that might well be found in your room. The ubiquitous Styrofoam cup is white, opaque, light, soft, and easily deformed and torn, but it is an excellent heat insulator. An audiocassette box is transparent and almost glass-like in its clarity, hard, and brittle. The tape in it is flexible and very strong; it is difficult to stretch or to tear it. The plastic that makes up most soft-drink bottles is transparent and has moderate hardness and flexibility. A typical plastic bag is light, transparent, and flexible. When a strip cut from a bag is pulled, the plastic stretches and often "necks down." This refers to the dramatic increase of the length of the plastic strip as the width and thickness decrease. The necking effect is not reversible, as is the stretching of a rubber band. A fairly strong pull is required to start the necking process, but once it begins, less force is needed to keep it going. A little shoulder forms on the wider part of the strip and the narrow neck almost seems to flow from it. Eventually, however, the plastic tears. Finally, in our brief survey, a plastic milk bottle is somewhat opaque or at least translucent and although it can be deformed, it is not as soft and flexible as many plastics.

Investigations of this sort yield useful information about the properties of plastics. Additional data can be obtained in the laboratory by quantitative determination of density, hardness, tensile strength, melting point, and so on. But it is sometimes difficult to relate these properties unambiguously to the chemical composition of the plastics. If you were trying to sort the items described in the previous paragraph for recycling, you might find it hard to do so. In fact, it may be surprising that objects as different as the coffee cup and the cassette box are made of the same plastic—polystyrene. The audiotape and the soda bottle are compounded primarily of polyethylene terephthalate, and the plastic bag and the milk bottle are both polyethylene.

It follows that the properties of a plastic must be a consequence of more than just its chemical composition—the ratio of the elements that make up the material. How the atoms of those elements are linked together is an important factor. Indeed, the great variety in the properties of plastics are all consequences of variations in molecular structure. Therefore, a major goal of this chapter is to correlate the molecular structures, properties, and uses of plastics. But first we ask an even more fundamental question: What is a plastic?

According to a standard dictionary definition, plastic is an adjective meaning "capable of being molded" or a noun referring to something that is capable of being molded. More specifically, the Merriam-Webster *Seventh New Collegiate Dictionary* mentions "any of numerous organic synthetic or processed materials that are molded, cast, extruded, drawn, or laminated into objects, films, or filaments." Plastics come in a large variety of types, with different characteristics and applications. You are familiar with the common or brand names of dozens of examples—Teflon®, polyurethane, Saran®, Styrofoam®, Formica®, rayon, and nylon to list only a few. In this text we will reserve the term *plastics* for such synthetic substances—all creations of the chemist and all polymers. What they have in common is evident at the molecular level.

 10.3   *Consider This*

Keep a journal to determine how much plastic you throw away in one week. Record every plastic item you discard during this time, including plastic-coated materials. In your journal, classify the plastic according to its various uses, such as food packaging, bottles, etc.

■ Polymers: The Long, Long Chain

All plastics are made up of long chains of atoms covalently bonded together. Like a linked strand of paper clips, the molecular chain in a plastic consists of subunits that are repeated many times. This subunit is called a **monomer** (from *mono* meaning "one" and *meros* meaning "unit"). Many monomers join together to form a long molecular strand called a **polymer** (*poly* means "many"). These polymer molecules can be very long indeed. Sometimes they involve thousands of atoms, and molecular masses can reach over a million. No wonder that polymers are sometimes referred to as **macromolecules.**

Although this chapter will focus primarily on synthetic polymers, it is important to note that many polymers occur in nature. Natural polymers are found in wood, wool, cotton, skin, hair, starch, and even some minerals, such as asbestos. Polymeric molecules give strength to an oak tree, delicacy to a spider's web, softness to goose down, and flexibility to a blade of grass. Much of the motivation for the synthesis of plastics has been a desire to reproduce such properties in artificial materials. Indeed, many synthetic polymers were originally created as substitutes for expensive or rare naturally occurring materials or to improve on nature.

As a matter of fact, the first commercial plastic, celluloid, was developed in response to a $10,000 prize offered for a synthetic substitute for ivory in billiard balls. In 1870, a printer named John Hyatt obtained a patent for a mixture of cellulose nitrate, alcohol, and camphor that was heated, molded, and allowed to harden. Cellulose nitrate, made by treating cotton with nitric and sulfuric acids, is better known as "gun cotton." It is highly flammable and, under some conditions, sufficently explosive to be used in smokeless gunpowder. Such properties are somewhat less than desirable in billiard balls, though the story of exploding pool balls may be apocryphal. Back in 1870, the primary motivation for seeking substitutes for ivory was probably economic. Today, the motivation has changed. Elephant herds have been drastically depleted by poachers collecting tusks for the ivory trade. In response, the United States government has banned the import of ivory. New plastics provide the starting materials, not only for billiard balls, but for many art objects as well.

To some people, the word "plastic" may carry the connotation "cheap" or "tacky." But the fact remains that synthetic polymers have revolutionized modern life. Few advocates of natural materials would be willing to give up nylon stockings, synthetic rubber tires, "fake" furs that spare endangered species, and the dozens of other plastic objects that have become an accepted part of today's lifestyle. As chemists have developed new polymers, the variety of properties and uses have expanded dramatically. For example, plastics have become increasingly important in automobile manufacturing. Some new plastics are stronger than steel and much more resistant to corrosion. Hence, they can be substituted for steel and other metals and materials in various parts of a car. Because the plastic is considerably less dense than structural metals, such substitution has led to significant reductions in vehicle weight. As a result, new cars have generally become more fuel efficient because more plastics have been used in their manufacture.

This is, of course, only one of many examples. Plastic packaging reduces weight, eliminates breakage, and helps to save fuel during shipping. Plastic construction materials have replaced wood in some applications, and plastic pipes substitute effectively for lead, iron, copper, and tile. And, as we will see later, recreation has been revolutionized by the introduction of synthetic polymers. All of this adds up to a formidable economic opportunity. Ben's friend gave him good advice. As Figure 10.1 indicates, from 1935 to 1985, production of plastics in the United States increased 500-fold. In 1988 alone, 57 billion pounds of plastics were produced in this country. Indeed, since 1976, the United States has manufactured a larger volume of synthetic polymers than the volume of steel, copper, and aluminum combined.

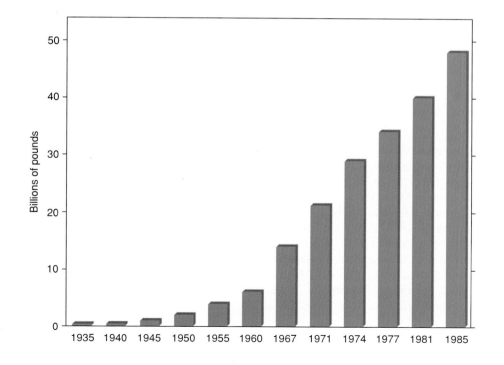

Figure 10.1

Annual United States production of plastics from 1935 to 1985, in pounds. (Data from Joseph Alper and Gordon L. Nelson. *Polymeric Materials: Chemistry for the Future.* Washington: American Chemical Society, 1989, p. 3.)

The widespread applicability of plastics is a consequence of the ability of chemists to modify the properties of plastics in particular ways by altering their molecular structures. Such activities provide meaningful and gainful employment to many of our colleagues. As a matter of fact, more chemists are employed in the polymer/plastics sector than in any other branch of the chemical industry. In a sense, these scientists "design" the desired properties of plastics into their constituent molecules. In doing so, they follow one or more of only six general strategies for modifying polymer chains—a remarkably small number when you consider the thousands of plastics known. Yet these six ways produce stunning results.

The strategies involve one or more alterations of the following molecular features of the polymer chain:

1. the length of the chain (the number of monomer units),
2. the three-dimensional arrangement of the chains in the solid,
3. the branching of the chain,
4. the chemical composition of the monomer units,
5. the bonding between chains, and
6. the orientation of monomer units within the chain.

All of these options will be illustrated in the pages that follow. We begin with the most common plastic of all.

■ Polyethylene: The Most Common Plastic

You probably encounter **polyethylene** (or polythene, if you are in the British Isles) every day of your life. Nearly 10 million tons of it are produced in the United States each year. It is found in grocery bags for fruits and vegetables, dry-cleaner garment bags, squeeze bottles, TV cabinets, toys, and hundreds of other objects. This wide variety of uses suggests a similarly wide range of properties for this single polymer. Yet all polyethylene is made from the same starting material—ethylene, C_2H_4.

Ethylene is a compound obtained from petroleum. At ordinary temperatures and pressures it is a gas. However, it was discovered in the 1930s that, with the initiation of a catalyst, individual ethylene molecules will bond to each other to form a polymer. The key to this behavior is the structure of the polyethylene molecule. The two carbon atoms are linked with a double bond that is capable of reacting with another ethylene molecule. This reaction is represented in equation 10.1.

$$n \quad \begin{array}{c} H \\ \diagdown \\ \diagup \\ H \end{array} C = C \begin{array}{c} H \\ \diagup \\ \diagdown \\ H \end{array} \quad \longrightarrow \quad \left[\begin{array}{cc} H & H \\ | & | \\ -C - C - \\ | & | \\ H & H \end{array} \right]_n \qquad (10.1)$$

The notation signifies that *n* ethylene molecules combine to form a long polymeric chain of *n* monomeric units. *The numerical value of* n *and hence the length of the chain varies with the reaction conditions (strategy one)*, but it is often in the hundreds or thousands. Moreover, it does not have a fixed value; a typical synthetic polymer is a mixture of individual molecules of varying length and mass. Molecular masses are generally between 10,000 and 100,000. In every case, however, the carbon atoms are attached to each other by single bonds, and the hydrogen atoms are bonded to the carbon atoms. A polyethylene molecule is thus a macromolecular version of a hydrocarbon molecule such as those in petroleum.

The polymerization process can be described in terms of what happens to the electrons in ethylene. The reaction is initiated by a catalyst that is a free radical—a molecule with at least one unpaired electron. In Figure 10.2, a representation of the

10.4	***Your Turn***

The average molar mass of a sample of polyethylene is 84,500. What is the average value of *n* in the polymer? In other words, how many monomer units are present in an average polyethylene molecule? How many atoms?

Soln: The monomeric unit in polyethylene is CH_2CH_2, with a molar mass of 28.0 g. Therefore, *n,* the number of monomer units, is obtained by an operation that is equivalent to dividing the molar mass of the polymer by the molar mass of the monomer.

$$n = 84{,}500 \text{ g/mole polymer} \times \frac{1 \text{ mole monomer}}{28.0 \text{ g}}$$

$$= 3000 \text{ mole monomer/mole polymer}$$

This means that there are 3000 monomer units in the average polymer molecule. To answer the second part of the question, you need to make use of the fact that there are six atoms in each CH_2CH_2 monomer.

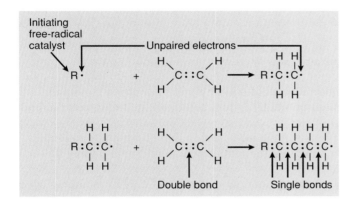

Figure 10.2
The polymerization of ethylene.

polymerization of polyethylene, the free radical is represented by "R·" (the dot indicates an unpaired electron). The radical reacts readily with a CH_2CH_2 molecule. One of the two bonds between the carbon atoms breaks, and one of the electrons from that bond pairs with the unpaired electron of the radical to form a covalent bond. The new molecule that is formed, RCH_2CH_2·, is another free radical because it carries an unpaired electron left over from the broken carbon-carbon bond. It can therefore react with another ethylene molecule that bonds to the carbon atom at the reactive, growing end of the polymer. This process is repeated many times over in many chains at the same time. Occasionally, the active ends of two free radical polymers will interact to form a bond and stop the chain growth. The result of all this is that gaseous ethylene is converted to solid polyethylene.

Many of the properties of polyethylene are related to the presence of these long molecular chains. Relatively speaking, they are very long indeed. If a polyethylene molecule were as wide as a piece of spaghetti, the molecular chain could be as much as half a mile long. To continue the analogy, in the polyethylene used to make plastic bags these chains are arranged somewhat like spaghetti on a plate. The strands are jumbled up and not very well aligned, though there are quasi-crystalline regions where the molecular chains are parallel. Moreover, the polyethylene molecules, like spaghetti strands, are not bonded to each other. Now recall what happens when a polyethylene strip is stretched. As the strip narrows and necks down, the previously mixed-up

Figure 10.3

Molecular rearrangement as polyethylene is stretched.

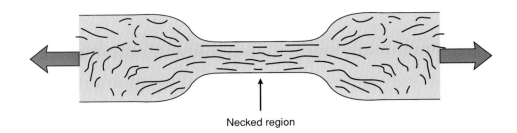

Necked region

molecules move. They shift, slide, and became aligned parallel to each other and the direction of the pulling force (Figure 10.3). In some plastics, such stretching or "cold drawing" is carried out as part of the manufacturing process in order to obtain ordered polymer chains. *This is an example of the general strategy of altering the three-dimensional arrangement of the chains (the second in our list of six).* Of course, as the force and stretching continue, the polymer eventually reaches a point where the strands can no longer realign, and the plastic breaks. Paper, another polymeric material, tears when pulled because the strands (fibers) in paper are rigidly held in place and are not free to slip like the long molecules in polyethylene.

■ *Low- and High- Density Polyethylene: Chain Branching*

The third of the strategies to control the molecular structure and physical properties of polymers is to regulate the branching of the polymer chain. This approach is used to produce two general types of polyethylene. The version found in plastic bags is low-density polyethylene or LDPE. It is soft, stretchy, transparent, and not very strong. This low density form was the first type of polyethylene to be manufactured. Study of its structure reveals that the molecules consist of about 500 monomeric units and that they are highly branched. The central molecular chain has many side branches, not un-like a tree trunk.

About twenty years after the discovery of LDPE, chemists were able to adjust reaction conditions to prevent branching and make another form of polyethylene called high-density polyethylene (HDPE). In their Nobel Prize-winning research, Karl Ziegler and Giulio Natta developed new catalysts that enabled them to make linear polyethylene chains consisting of about 10,000 monomer units. Because these long chains are not impeded by side branches, they can be arranged parallel to one another. The structure of HDPE is thus more like a regular crystal than the amorphous tangle of the polymer chains in LDPE. The highly ordered structure of HDPE gives it greater density, rigidity, strength, and a higher melting point than LDPE. Furthermore, the high-density form is opaque and the low-density form tends to be transparent. Figure 10.4 provides a de-tailed view of molecular structure in linear and branched polyethylene. Figure 10.5 is a representation on a somewhat larger scale.

The differences in properties of high- and low-density polyethylene give rise to different applications. HDPE is used to make toys, gasoline tanks, radio and television cabinets, heavy-duty pipes, and the opaque grocery bags often used as a substitute for paper. One new use has been spurred by the AIDS epidemic. Surgeons who break their skin during an operation on an HIV-positive patient run the risk of acquiring the HIV virus through contact with the patient's blood. Allied-Signal has produced a lin-ear polyethylene fiber called Spectra that can be fabricated into liners for surgical gloves. Spectra gloves are said to have 15 times the cut resistance of medium-weight leather work gloves, but they are so thin that a surgeon can retain a keen sense of touch. A sharp scalpel can be drawn across the glove with no damage to the fabric or the hand inside. Such strength is in marked contrast to the properties of the common plastic bag, which is made of low-density polyethylene. LDPE is also used for the covers of disposable diapers, squeeze bottles, cling wrap for foods, and plastic flowers.

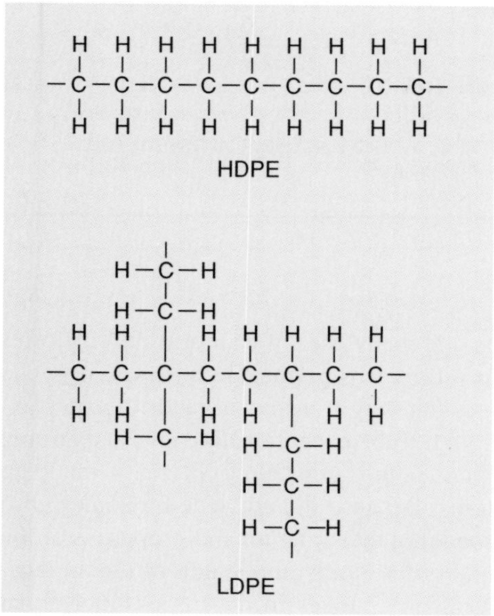

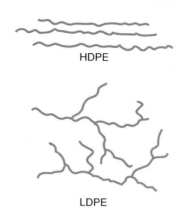

Figure 10.4

Detail of bonding in molecules of high-density (linear) polyethylene and low-density (branched) polyethylene.

Figure 10.5

Representations of high-density (linear) polyethylene and low-density (branched) polyethylene.

It would be a mistake, though, to conclude that polyethylene is restricted to the extremes represented by highly branched and strictly linear forms. By modifying the extent and location of branching in LDPE, its properties can be varied from soft and wax-like (coatings on paper milk cartons), to stretchy (plastic food wrap), to fairly rigid (plastic milk bottles). Unfortunately, the consumer is sometimes unaware of the consequences of such molecular tinkering. For example, the higher melting point of HDPE (130° C) permits plasticware made from it to be washed in automatic dishwashers. But objects made of LDPE, with a melting point of 120°C, melt in dishwashers.

Finally, we should not forget that one of the first and most important uses of polyethylene was a consequence of the fact that it is a good electrical insulator. During World War II, it was used by the Allies as insulation to coat electrical cables in aircraft radar installations. Sir Robert Watt, who discovered radar, described polyethylene's critical importance in these words.

> The availability of polythene [polyethylene] transformed the design, production, installation, and maintenance problems of airborne radar from the almost insoluble to the comfortably manageable. . . . A whole range of aerial and feeder designs otherwise unattainable was made possible, a whole crop of intolerable air maintenance problems was removed. And so polythene played an indispensable part in the long series of victories in the air, on the sea, and on land, which were made possible by radar.[1]

10.5   *Consider This*

List some of the properties that make polyethylene the most widely used plastic.

[1.] Quoted by J. C. Swallow in "The History of Polythene," from *Polythene—The Technology and Use of Ethylene Polymers (2d ed.)* A. Renfrew (ed.). London: Iliffe and Sons, 1960.

■ The Big Six

In spite of polyethylene's range of properties and many uses, it cannot fill all the roles we assign to plastics. It melts at a low temperature, it is permeable to gases, it swells in the presence of oil or organic solvents, it is not very transparent, and it is very expensive to make polyethylene crystalline enough to be exceptionally rigid and strong. A serious limitation is the fact that polyethylene is the simplest of polymers, made up of carbon chains with attached hydrogen atoms. Because the ethylene monomer is so simple, the only options chemists have for changing the structure and properties of the polymer are to alter molecular branching and, within limits, the length of the molecular chain. These alterations can, in turn, result in changes in the three-dimensional arrangement of the chains in the solid. As we have just seen, this has been done brilliantly. *But to obtain greater variety in properties and greater control over them, chemists have made frequent use of one of the most important strategies of molecular manipulation—the use of different monomers to form different polymers (item four on our list).*

Today, more than 60,000 plastics are known. Most have been developed for special purposes ranging from fry pan coatings to resins for restoring antiques. Yet, the two types of polyethylene (LDPE and HDPE) and four other polymers make up the bulk of the plastics you regularly encounter. Approximately 20 million tons of these six polymers are made annually in the United States, and they account for about 63% of all plastics used in this country. In addition to polyethylene, the other four plastics are **polypropylene, polystyrene, polyvinyl chloride (PVC),** and **polyethylene terephthalate (PET).** All are ultimately based on petroleum.

Table 10.1 is a summary of information about "The Big Six." Six different monomers are involved. Ethylene, vinyl chloride, styrene, and propylene molecules are similar in that they each contain two carbon atoms connected by a double bond. In ethylene, two hydrogen atoms are attached to each of the doubly-bonded carbon atoms. But in vinyl chloride, styrene, and propylene molecules, one of the hydrogen atoms has been replaced with something else. In the case of vinyl chloride, the substituent is a chlorine atom. In styrene it is a phenyl group, $-C_6H_5$, consisting of six carbon atoms bonded in a ring with a hydrogen atom attached to five of them. The replacement in propylene is a methyl group, $-CH_3$. These "side chains" introduce variety into the monomers and the polymers formed from them. Moreover, the substituents give the chemist greater latitude in designing plastics for particular uses. Polyethylene terephthalate appears to be a special case, and we will return to it, but only after spending more time with the other members of this sextet.

Table 10.1 also lists some of the more important properties of these six polymers. They are all thermoplastic (in other words, they can be melted and shaped), and all tend to be flexible. Three of them, the two polyethylenes and polypropylene, have both crystalline and amorphous regions. The regions of structural regularity convey toughness and resistance to mechanical abrasion and make the polymers opaque. The amorphous regions promote flexibility. The other three polymers—polyethylene terephthalate, polystyrene, and polyvinyl chloride—are not crystalline. Their molecular chains are bonded together tightly but more or less randomly. *This process, called*

■ Table 10.1 ▮ The Big Six (This table also includes the code identifying the polymer.)

Polymer	Monomer	Properties of Polymer
Polyethylene (LDPE) ⟲4⟳ LDPE	Ethylene $\displaystyle \mathop{H}_{H}\!\!>\!C\!=\!C\!<\!\mathop{H}_{H}$	Opaque, white, soft, flexible, impermeable to water vapor, unreactive toward acids and bases, absorbs oils and softens, melts at 100° – 125°C, does not become brittle until −100°C, oxidizes on exposure to sunlight, subject to cracking if stressed in presence of many polar compounds.
Polyethylene (HDPE) ⟲2⟳ HDPE	Ethylene $\displaystyle \mathop{H}_{H}\!\!>\!C\!=\!C\!<\!\mathop{H}_{H}$	Similar to LDPE, more opaque, denser, mechanically tougher, more crystalline and rigid.
Polyvinyl chloride ⟲3⟳ V	Vinyl chloride $\displaystyle \mathop{H}_{H}\!\!>\!C\!=\!C\!<\!\mathop{H}_{Cl}$	Rigid, thermoplastic, impervious to oils and most organic materials, transparent, high impact strength.
Polystyrene ⟲6⟳ PS	Styrene $\displaystyle \mathop{H}_{H}\!\!>\!C\!=\!C\!<\!\mathop{H}_{⬡}$	Glassy, sparkling clarity, rigid, brittle, easily fabricated, upper temperature use 90°C, soluble in many organic materials.
Polypropylene ⟲5⟳ PP	Propylene $\displaystyle \mathop{H}_{H}\!\!>\!C\!=\!C\!<\!\mathop{H}_{CH_3}$	Opaque, high melting point (160°–170°C), high tensile strength and rigidity, lowest density commercial plastic, impermeable to liquids and gases, smooth surface with high luster.
Polyethylene terephthalate ⟲1⟳ PETE	Ethylene glycol $HOCH_2CH_2OH$ Terephthalic acid $HOOC\!-\!⬡\!-\!COOH$	Transparent, high impact strength, impervious to acid and atmospheric gases, not subject to stretching, most costly of the six.

10.7 ■ Consider This

For each of the following uses, specify the desirable properties of a plastic and, using the information in Table 10.1, suggest the most suitable polymer or polymers.

   a. a bottle for salad oil
   b. a bottle for a carbonated beverage
   c. a gallon milk bottle
   d. a disposable coffee cup
   e. a dishwasher-safe coffee cup
   f. a tool box

cross-linking, was number five in the list of strategies identified for modifying polymer structure and properties. As a consequence of the cross-linking, the chains cannot move or slip. This linkage is somewhat like the arrangement of strands in a net, but there is a wide variety of randomly sized holes. The bonding between the chains means that the bulk plastic is rigid and hard to stretch. Another property of these amorphous polymers is their transparency and clarity. This range of properties means that different polymers are differently suited for specific applications. Consider This 10.7 gives you an opportunity to match uses, properties, and polymers.

Whatever use is made of them, the six plastics also generally have small amounts of other materials added to them. Because all six are colorless, coloring agents are often added. Plasticizers, substances that improve the flexibility of the polymer, are commonly added, as are a variety of other substances that enhance the performance and durability of the plastic. Indeed, the smell associated with certain plastics (and new cars) is sometimes due to escaping plasticizers.

■ Addition Polymerization: Adding up the Monomers

We have already noted that ethylene, vinyl chloride, and propylene molecules each contain a carbon-carbon double bond. All of these monomers polymerize by a process call **addition polymerization.** In every case, the reaction involves the un-pairing and re-pairing of electrons described above for polyethylene. The monomers simply add to the growing polymer chain in such a way that the product contains all the atoms of the starting material. No other products are formed, and no atoms are eliminated. Thus, vinyl chloride molecules become bonded together to form **polyvinyl chloride (PVC).** In the process, the double bonds disappear, and the polymer contains only single bonds.

$$\text{(10.2)}$$

Ethylene and polyethylene are made up only of carbon and hydrogen atoms. But the fact that a vinyl chloride molecule contains a chlorine atom introduces an opportunity for variability in the structure of polyvinyl chloride. Let us arbitrarily think of the carbon atom bearing two hydrogens (CH_2) as the "head" of a vinyl chloride molecule and the chlorinated carbon atom (CHCl) as its "tail." (We could just as easily have made the reverse assignments.) The presence of the chlorine atom creates an asymmetry in the molecule. Because of this, when vinyl chloride molecules add to each other to form polyvinyl chloride, the molecules can be oriented in three possible arrangements: alternating head-to-head and tail-to-tail; repeating head-to-tail; and a random distribution of heads and tails. Figure 10.6 should help make this more obvious.

In a head-to-head/tail-to-tail arrangement of PVC, chlorine atoms are on alternate carbons. In the head-to-tail structure, chlorine atoms are on adjacent carbons. And in the random polymer, an irregular mixture of the previous two types occurs. In each case, the properties are somewhat different *You will recall that controlling molecular orientation within the chain was strategy number six in our list of methods to influence polymer properties.* The head-to-tail arrangement is the usual product for polyvinyl chloride. Depending on its formulation, PVC can be stiff or flexible. The former finds use in phonograph records, pipes, house siding, toys, furniture, and various automobile parts. The flexible version is familiar in wall coverings, upholstery, shower curtains, garden hoses, insulation for electrical wiring, and packaging films.

The familiar plastic foam hot beverage cup is the most common example of **polystyrene.** The styrene monomer, like vinyl chloride, has a substituent (here the

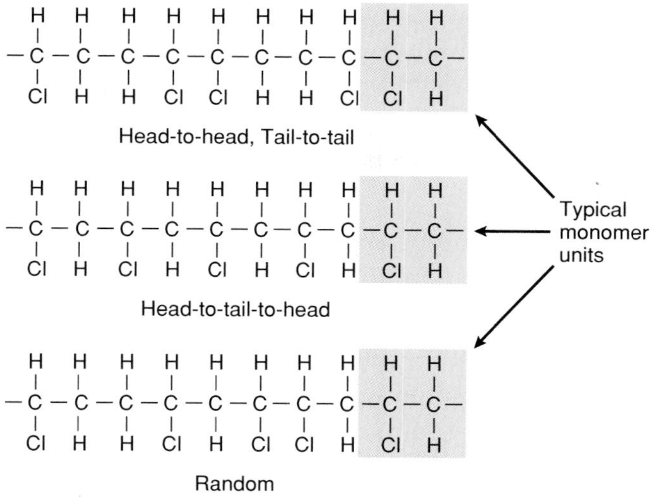

Figure 10.6
Three possible arrangements of monomers in PVC.

10.8 ***Your Turn***

A molecule of polyvinyl chloride consists of 15,000 monomer units. Calculate the molar mass of this polymer.

Ans. 937,500 g

C_6H_5 ring) in place of a hydrogen atom on one of the doubly bonded carbons. Under appropriate catalytic conditions, styrene polymerizes to polystyrene, usually with the head-to-tail arrangement. The, by now, familiar type of addition equation applies; here n = about 5000.

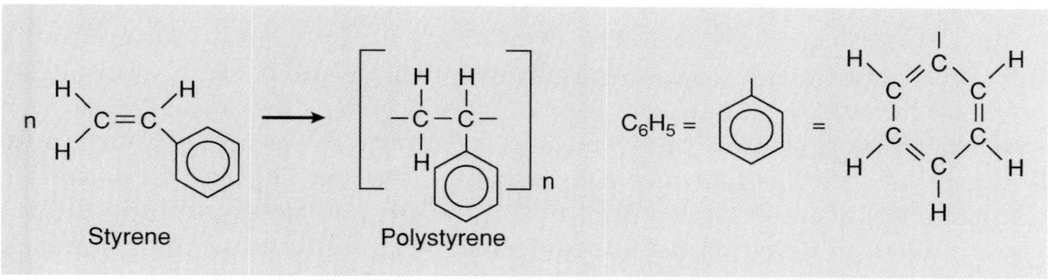

(10.3)

We noted earlier that the hard, brittle, transparent audiocassette boxes are chemically almost identical to light, white, opaque foam coffee cups. Both are polystyrene. Styrofoam is made by expansion molding. Polystyrene beads containing 4–7% of a low-boiling liquid are placed in a mold and heated using steam or hot air. The heat causes the liquid to vaporize and the expansion of the gas also expands the polymer. The expanded particles are fused together into the shape determined by the mold. Because it contains so many bubbles, this plastic foam is not only light, but it is also an excellent thermal insulator. Until relatively recently, chlorofluorocarbons were used as foaming agents, but concern over the involvement of CFCs in the destruction of stratospheric ozone (Chapter 2) led to their replacement in 1990. Hydrocarbons are now frequently used for this purpose. The hard, transparent version of polystyrene is made by molding the melted polymer without the foaming agent. It is used to fabricate wall tile, window moldings, and radio and television cabinets.

Polypropylene, like PVC and polystyrene, is also formed by an addition reaction, in this case using propylene monomers. A particularly useful form of polypropylene has the monomeric units bonded in a head-to-tail fashion. This regularity imparts a high degree of crystallinity and makes the polymer strong, tough, and able to withstand high temperatures. Its uses reflect these properties. Polypropylene is found in indoor-outdoor carpeting, videocassette boxes, and cold weather underwear. Strength and chemical resistance make polypropylene a good choice for automobile parts, battery cases, and other applications where toughness is required.

10.9 Your Turn

Using structural formulas, write the equation for the formation of polypropylene from propylene.

10.10 Your Turn

Teflon® is a polymer used in nonstick cookware and baking utensils. Discovered in a serendipitous accident, this addition polymer is made from the monomer tetrafluoroethylene. Write the formula for the monomer.

Hint: Think about the structure of ethylene and the meaning of the prefix *tetra*. Then write a structural formula for a portion of a Teflon molecule.

■ Condensation Polymers: Bonding by Elimination

Unlike the other polymers described above, **polyethylene terephthalate** is not formed by an addition reaction. Rather, it is produced via a **condensation reaction**, also called a step reaction. Many polymers are formed by condensation reactions: natural ones such as cellulose, glycogen, wool, silk, and proteins; and synthetics like nylon, Dacron®, Formica®, Kevlar®, and Lexan®. This class of polymers and this type of reaction are so important that we will take some pains to introduce and illustrate it before we consider PET in detail.

In condensation polymerization, monomer units join by eliminating a small molecule, often water. Thus, a condensation polymerization has two products—the polymer itself plus the small molecules split out during the polymer's formation. Polyethylene terephthalate is compounded of two monomers, ethylene glycol and terephthalic acid, and hence it is called a **copolymer.** Ethylene glycol, the chief ingredient in automobile antifreeze, is a dialcohol. Its formula, $HOCH_2CH_2OH$, reveals that it has an alcohol group (–OH) on either end of the molecule. A molecule of terephthalic acid, $HOOCC_6H_4COOH$, has an acidic group (–COOH) at either end.

Key to the polymerization process is the fact that the acid and alcohol groups can interact to eliminate a water molecule and form a class of compounds known as **esters.** The reaction of ethylene glycol and terephthalic acid is represented by equation 10.4, and the ester linkage is enclosed in a box.

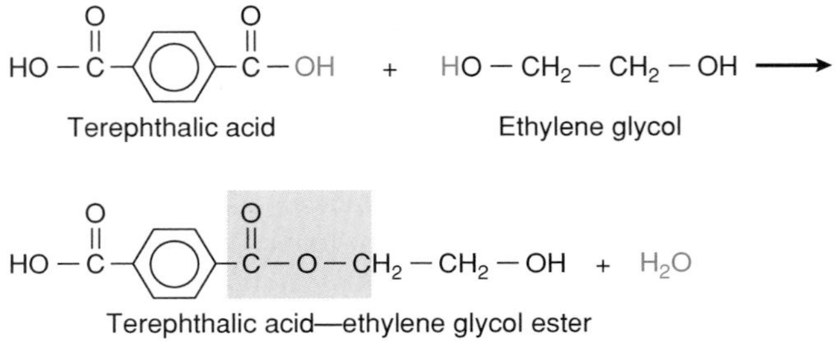

Terephthalic acid   Ethylene glycol

(10.4)

Terephthalic acid—ethylene glycol ester

Note that the product molecule in equation 10.4 has a –COOH group on one end and an –OH on the other. The acid group can react with an alcohol group of another ethylene glycol molecule and the alcohol group of the growing polymer can react

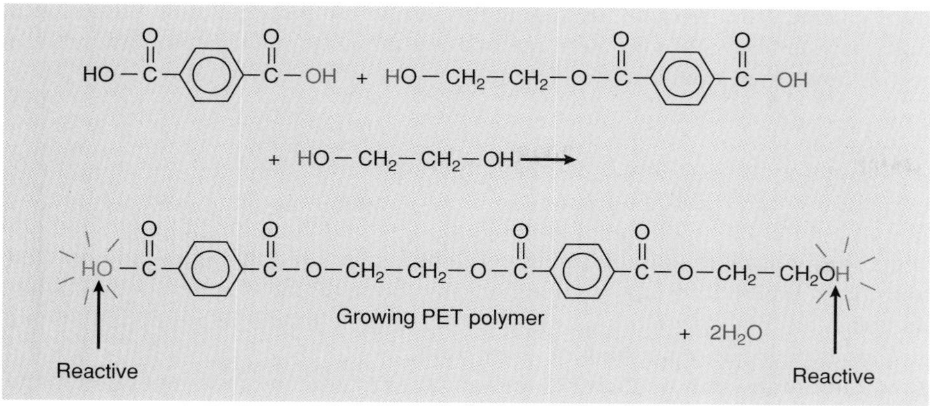

Figure 10.7
A growing PET molecule.

with an acid group of another molecule of terephthalic acid. This process, represented in Figure 10.7, occurs many times over to yield a long polymeric chain of polyethylene terephthalate.

PET is classified as a **polyester** because it contains many ester linkages. Since their introduction, polyester fibers have found many uses in fabrics and clothing. The polymer is perhaps most familiar under the trade name Dacron. This polyester is frequently mixed with cotton, wool, or other natural polymers, but it has many other uses. Indeed, over 5 million pounds of PET are produced annually in the United States. Narrow, thin-film ribbons of it (under the trade name Mylar) are coated with metal oxides and magnetized to make audio- and videotapes. Dacron tubing is used surgically to replace damaged blood vessels, and artificial hearts contain parts made of PET. The most common use for this plastic is in two-liter soft drink bottles (Figure 10.14), developed by the late Nathaniel Wyeth, a chemical engineer and the brother of the famous painter Andrew Wyeth. Both scientific and artistic creativity run in this remarkable American family.

■ *Polyamides: Natural and Nylon*

No discussion of condensation polymerization can be complete without including one of the most important classes of natural polymers and the synthetic substitute that brilliantly duplicates some of the properties of the natural material. The naturally occurring polymer is **protein.** Actually there are a wide variety of these biological macromolecules that make up our skin, hair, muscle, and enzymes. All proteins are polyamides, which are polymers of **amino acids.** As the name suggests, molecules of amino acids contain both amine groups ($-NH_2$) and acidic groups ($-COOH$). A general formula for an amino acid is given below. The amine and acid groups are both attached to the same carbon atom. In addition, a hydrogen atom and another group (represented by an R) are bonded to the same carbon.

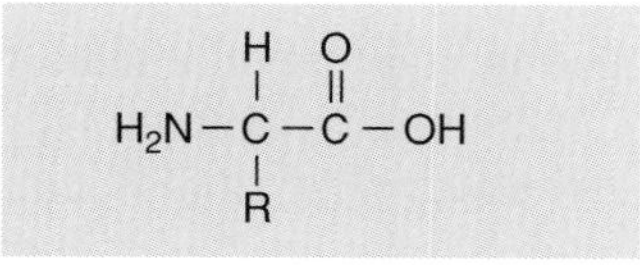

The 20 amino acids found in most proteins differ in the identity of the R group. In some amino acids, R consists of carbon and hydrogen atoms, as in alanine, where R is a methyl group, $-CH_3$. In others, R also includes oxygen, nitrogen, or sulfur atoms. Some R groups are acidic and some are basic.

Chapter 12 includes a good deal of additional information about amino acids and proteins. At present we will focus on some fundamentals. The crucial point in the formation of the protein polymer is the fact that the $-COOH$ group of one amino acid can react with the –NH_2 group of another. In this reaction, an H_2O molecule is eliminated and a peptide bond is formed. The reaction is represented by equation 10.5, and the peptide bond is enclosed in a box.

(10.5)

$$H_2N-\underset{\underset{R_1}{|}}{\overset{\overset{H}{|}}{C}}-\overset{\overset{O}{\|}}{C}-OH \ + \ H_2N-\underset{\underset{R_2}{|}}{\overset{\overset{H}{|}}{C}}-\overset{\overset{O}{\|}}{C}-OH \longrightarrow H_2N-\underset{\underset{R_1}{|}}{\overset{\overset{H}{|}}{C}}-\boxed{\overset{\overset{O}{\|}}{C}-\underset{\underset{H}{|}}{N}}-\underset{\underset{R_2}{|}}{\overset{\overset{H}{|}}{C}}-\overset{\overset{O}{\|}}{C}-OH \ + \ H_2O$$

In the sophisticated chemical factories called biological cells, this reaction is repeated many times over to form long polymeric chains. Given the fact that there are 20 different amino acid building blocks, a great variety of proteins can be synthesized. Some proteins are made up of hundreds of amino acids.

Chemists are often well advised to attempt to replicate the chemistry of nature. In the 1930s, a brilliant chemist working for the DuPont Company set out to do just that. Wallace Carothers (1896–1937) was studying a variety of polymerization reactions, including the formation of peptide bonds. Instead of using amino acids, Carothers tried combining adipic acid, $HOOC(CH_2)_4COOH$, and hexamethylene diamine, $H_2N(CH_2)_6NH_2$. Note that a molecule of adipic acid has an acidic group on both ends and the hexamethylene diamine molecule has a basic amine group on each end. As in the case of protein synthesis, the acid and amine groups reacted to eliminate water and form peptide bonds. But in this instance, the polymer consisted of alternating adipic acid and hexamethylene diamine monomers.

(10.6)

$$HO-\overset{\overset{O}{\|}}{C}-(CH_2)_4-\overset{\overset{O}{\|}}{C}-OH \ + \ H_2N-(CH_2)_6-NH_2 \longrightarrow HO-\overset{\overset{O}{\|}}{C}-(CH_2)_4-\overset{\overset{O}{\|}}{C}-\underset{\underset{H}{|}}{N}-(CH_2)_6-NH_2 \ + \ H_2O$$

Adipic acid   Hexamethylene diamine   Part of a nylon molecule

Reactive   Reactive

DuPont executives decided the new polymer had promise, especially after company scientists learned to draw it into thin filaments. These filaments were strong and smooth, and very much like the protein spun by silkworms. Therefore, it was as a substitute for silk that "Nylon" was first introduced to the world. The world greeted it with open arms and open pocketbooks. Four million pairs of nylon stockings were sold in New York City on May 15, 1940, the first day that they became available. But in spite of consumer passion for "nylons," the supply soon dried up, as the polymer was diverted from hosiery to parachutes, ropes, clothing, and hundreds of other wartime uses. By the time World War II ended in 1945, nylon had repeatedly demonstrated that it was superior to silk in strength, stability, and resistance to rot. Today this polymer, in its many modifications, continues to find wide applications in clothing, sportswear, camping equipment, the work room, the kitchen, and the laboratory.

■ *Plastics and Recreation*

The first plastic was developed in response to a recreational need—a substitute for ivory in billiard balls. Today, recreation has been revolutionized by polymers. There is hardly any sport or recreational activity that does not use plastics. Football is played on artificial turf by players wearing plastic helmets and padding and nylon pants. Tennis balls, tennis racket frames, and racket strings are all made from synthetic polymers. Ice skaters can skate without ice—on rinks of Teflon or high density polyethylene. Most modern canoes are made of polymers, not birchbark, wood, or aluminum. And an athletic shoe may contain as many as five different types of polymer: in the sole, the padding, the upper portion, the trim, the laces, and even the lace tips. Professional baseball, that bastion of conservatism, still clings to natural polymers in the form of wooden bats, leather gloves, and a ball made of horsehide covering woolen yarn and a cork center. But even here, double-knit polyester uniforms have replaced the hot scratchy wool worn by Babe Ruth and Joe DiMaggio.

Figure 10.8
An example of polymers at play.

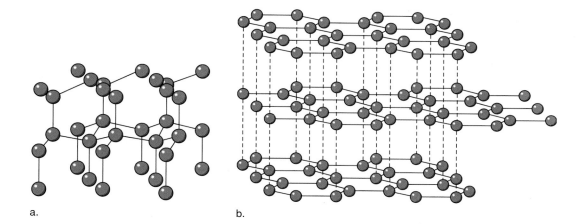

a.   b.

Figure 10.9

Structures of diamond (a) and graphite (b). (Adapted, with permission, from *Chemistry* by J. W. Moore, W. G. Davies, and R. W. Collins, McGraw-Hill, 1978.)

As a case in point, consider downhill (alpine) skiing, a recreational sport enjoyed by thousands worldwide. Not too many years ago, a skier would take to the slopes in woolen clothing, leather ski boots, and wooden skis. The wool got wet, the leather cracked or mildewed, and the skis broke. Now, skiers keep warm and dry with synthetic materials such as polypropylene, nylon, Thinsulate®, and Gore-Tex®. Their boots are preformed hardened plastic that easily withstand the repeated twisting and turning of downhill skiing while providing excellent ankle support and protection. And the skis themselves are made of plastic resins reinforced with carbon fibers.

Composite materials of this sort have found wide applications in other recreational gear—in tennis rackets, fishing rods, and golf-club shafts—where strength and flexibility are essential. Both the carbon fibers and the matrix in which they are embedded are polymers. All of the polymers we have studied exist because carbon atoms have a remarkable tendency to bond to each other. They combine in chains, in rings, in mesh-like networks, and in three-dimensional structures.

Even pure carbon possesses these characteristics, which are evident in its two allotropes—graphite and diamond. In diamond, the upscale allotrope, each of the carbon atoms is covalently bonded to four others as illustrated in Figure 10.9. The shared electron pairs are tightly held between adjacent atoms. In the low-priced form, graphite, the atoms are arranged at the corners of six-membered rings. These rings are

Figure 10.10
Structure of C_{60}, buckminsterfullerene. Again, the spheres indicate the carbon atoms and the lines the bonds.

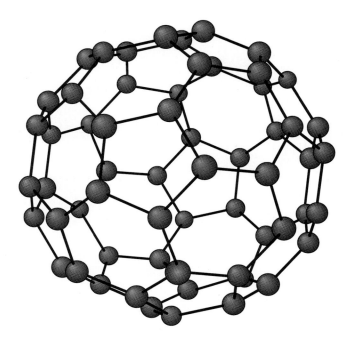

interconnected so that a sheet of graphite looks like chicken wire. Although the bonding between adjacent atoms is strong, the forces between the sheets are weak. Hence, it is easy to rub off layers of carbon. You do so every time you use a "lead" pencil. The "lead" isn't lead at all, but graphite with a binder. Not too surprisingly, the carbon fibers in skis are more like graphite than like diamond. They are long chains of hexagonally arrayed carbon atoms, well aligned in a polymeric resin. The carbon provides the strength and flexibility, and the resin holds the composite together.

Until a few years ago, this brief discussion of diamond and graphite would have told almost the entire story of carbon and its structures. But recently, chemists have become very excited about a new form of carbon called "buckyballs." In spite of the name, these balls have nothing to do with sport and recreation, at least not yet. But they have a good deal to do with the arrangement of atoms. Chemists have succeeded in forming and isolating carbon molecules composed of 60 atoms. The shape of one of these C_{60} molecules is beautiful to behold, (Figure 10.10). It looks like a soccer ball, with the carbon atoms bonded in 20 six-membered rings and 12 five-membered rings. This new form of carbon bears the fanciful name of **buckminsterfullerene,** after the late imaginative and visionary designer and thinker, Buckminster Fuller. Fuller was a pioneer in designing geodesic domes, rigid structures with the same three-dimensional geometry as his namesake molecules. Right now, fullerenes are mostly chemical curiosities, but chemists are already thinking of dozens of potential uses (see Chapter 13).

It may even be that buckyballs will someday find their way into athletics and recreation. In any case, there is no doubt that polymers will continue to play an ever important part in professional sports and leisure time activities. But as these and other uses of plastic proliferate, so do the plastics themselves. Therefore, it is essential that we reconsider the sources and the disposal of polymeric materials.

10.11 ■ *Consider This*

Regardless of which leisure or recreational activity you enjoy, from photography to football, synthetic polymers have had an influence on your hobby or sport. Study the equipment used in your favorite activity and list the places where plastics appear. Also identify the properties of the polymers that make them especially suitable for the intended uses.

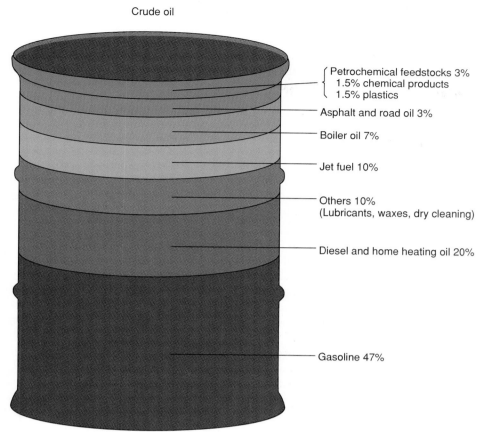

Crude oil

Petrochemical feedstocks 3%
  1.5% chemical products
  1.5% plastics

Asphalt and road oil 3%

Boiler oil 7%

Jet fuel 10%

Others 10%
(Lubricants, waxes, dry cleaning)

Diesel and home heating oil 20%

Gasoline 47%

Figure 10.11
End-uses for products made from the refinement of one barrel of crude oil. (Source: Data from United States Energy Information Administration.)

◼ *Plastics: Where from and Where to?*

Given the constraints of the law of conservation of matter, it is obviously important that we pay close attention to the raw materials that are incorporated into plastics and the disposal or recycling of this matter after its use. You have already read that the source of most synthetic polymers is petroleum. Crude oil is, of course, a mixture of many compounds that is refined into various fractions on the basis of boiling point and molecular mass. Figure 10.11 graphically depicts those fractions and their primary uses. Not surprisingly, given our discussion in Chapter 4, the great majority of petroleum is burned as fuel. Only 3% is feedstock for manufacturing polymers and other chemicals.

This 3% is essential to current methods of manufacturing the polymers that have reshaped modern life. But the planet's supply of petroleum is limited and non-renewable, a fact that creates a serious dilemma. You already know from previous chapters that petroleum is not an ideal energy source. Its combustion releases carbon dioxide that can contribute to global warming, and unburned fragments and other compounds that give rise to smog and air pollution. But compared to coal, petroleum is quite clean and convenient. Therefore, we return to a question posed earlier: "To burn or not to burn?" What are the risks and benefits—the economic and social trade-offs—in using oil as a source of energy or a source of matter? Should at least some fraction of petroleum be reserved for use in the synthesis of materials that cannot now be made from any other source? If not, the age of plastics may prove very short.

It is important to note that chemistry may rescue society from this dilemma. In principle, at least, polymers can be made from any carbon-containing starting material. Crude oil is simply the most convenient and the most economical. But it might also prove possible to convert renewable biological materials such as wood, cotton fibers, straw, starch, and sugar into polymers. After all, chemists at the Arthur D. Little

Figure 10.12

What's in our garbage?
(Source: Data from "Characterization of Municipal Solid Waste in the United States: 1990 Update, Executive Summary." June 1990, United States Environmental Protection Agency.)

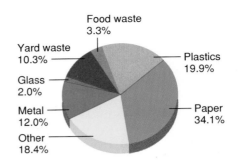

Company once actually made a silk purse out of a sow's ear. But new methods and new technologies would have to be developed, and the cost of the research and the manufacturing might be substantial. Moreover, it would be essential to estimate the supply of the starting materials and the demand for the finished products. There are implications for land use, crop productivity, the environment, and no doubt much more.

There seems to be a good deal more concern about where plastics go than where they come from. Much of the plastic we use eventually ends up in a landfill, along with lots of other types of municipal and domestic solid wastes, in the usual "out of sight, out of mind" approach. As a nation, we daily discard enough trash to fill two Superdomes. Of course that's just a ballpark figure, but the Environmental Protection Agency (EPA) estimates that about 75% of all municipal solid waste is put into landfills. Of the remainder, 10% is recycled and 15% is incinerated. Figure 10.12 provides information about the contents of a typical landfill given in terms of percent by volume. It is the volume of the buried materials, not the weight, that causes landfills to reach their capacity.

You will note from Figure 10.12 that about 20% of municipal solid waste is plastic, about 43% of which is used in packaging. But the largest percentage of municipal solid waste (34.1%) is paper and paper products. This raises a question that has been much in the news: Which constitutes the lesser environmental burden, paper or plastic? The section that follows is a real-life glimpse into this controversy.

10.12 *Consider This*

Suppose that the federal government has decided that nonessential uses of plastic is more than a waste—it is a crime. Anyone found using or possessing nonessential plastics will be subject to a heavy fine. Your room is due to be inspected by the authorities in one hour. What will you dispose or what will you keep? Be prepared to defend your choices.

■ *Paper or Plastic? The Battle Rages*

The rather melodramatic title of this section appeared recently as a headline in the *Syracuse Herald-Journal.* It introduced a heated exchange of opinions, forcefully expressed by readers. At issue was whether grocery bags should be made of paper or plastic. The actual letters that follow are selected from many more, and reprinted here to illustrate this controversy in the context of a supermarket.

In these environmentally conscious times I am often appalled at the number of people who choose plastic bags over paper ones at the supermarket. They must be aware that plastic bags are neither biodegradable nor as easily recycled as the paper variety.

It's common knowledge that plastic bags are not biodegradable, and although some forms of plastic are recyclable, no recycling center in the local area takes plastic bags. What most consumers fail to realize is that the production and processing of plastic involves a great amount of highly toxic chemicals.

Improper land disposal of hazardous wastes, emissions of toxic chemicals into the air, and discharges of toxic industrial effluents into waterways as a result of plastic production seriously threaten the public health and the environment.*

The local supermarket took the opposite position, and argued that by using plastic bags it was acting in an environmentally responsible manner. As evidence, it printed the following message on its grocery bags:

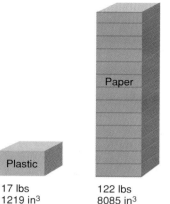

Plastic
17 lbs
1219 in³

Paper
122 lbs
8085 in³

Figure 10.13

A scale representation of the volumes occupied by 1000 plastic bags and 1000 paper bags.

Thank you for using plastic bags. If all of the Wegmans shoppers using plastic bags last year had insisted on paper, they would have increased the amount of solid waste by over eight million pounds and taken up nearly seven times more space in landfills.

1000 plastic bags equal	1000 paper bags equal
17 lbs and 1219 cubic inches	122 lbs and 8085 cubic inches*

A second letter to the editor, based on information such as that given above, provides the perspective of a consumer and an employee.

As an employee of Wegmans Food markets you may determine that my opinion is biased and it certainly is. . . . Clearly the plastic bags take up less space than paper. . . . In our backroom of the store, an entire pallet of paper bags takes up as much space as the plastic, however, there are only approximately 10,000 paper bags on a pallet compared to approximately 30,000 plastic sacks. . . . (T)he cost certainly is a benefit. Plastic bags cost 1.5 cents whereas paper ones cost 3 cents. If all customers would insist on plastic bags, the savings would certainly be passed on to the consumer.

Finally, I would like to address the landfill issue. As mentioned on our plastic bags, the use of paper sacks fills landfills much faster than the use of plastic bags. Plastic bags take up considerably less space than paper. . . . While environmental groups claim that paper bags degrade at a rapid rate, they are simply misleading the public.*

There is much in these letters to engage the Sceptical Chymist, but we may not yet have enough information to pass critical judgment on the many complex issues involved. Indeed, we may not achieve such knowledge and wisdom within the limitations of this chapter. Nevertheless, in the next section we press on in our efforts to become better informed.

10.13 | ***The Sceptical Chymist***

Analyze these two letters with their opposing views. Pay particular attention to the initial assumptions, the evidence cited, the logic used, and the conclusions drawn. Which makes the more compelling case and why? On the basis of these two letters only, which position would you support?

---

*Courtesy *Syracuse Herald-Journal,* Syracuse, N.Y.

■ *Disposing of Plastics*

Every year, about 60 billion pounds of plastic are produced in the United States—nearly 250 pounds for every man, woman, and child. Most of this ultimately finds its way into landfills. Given this huge quantity, there is little consolation in the fact that the landfills contain considerably more paper than plastic. The reduction of the amount of plastic going into landfills remains a high priority. Four strategies suggest themselves: **incineration, biodegradation, recycling,** and **source reduction.** In the paragraphs that follow, we will examine all of these approaches and attempt to weigh their relative merits.

Since "The Big Six" and most other polymers are primarily made of carbon and hydrogen, it would seem that **incineration** would be an excellent way to dispose of waste plastics. The chief products of their combustion are carbon dioxide, water, and energy. Although some plastics tend to burn incompletely in their usual form, they can be made to burn efficiently by shredding them and appropriately adjusting the combustion conditions such as the temperature and the oxygen supply. In some cases, fuel oil or other combustibles are mixed with the polymer waste. Because of their composition, plastics have a high fuel value. A sizable fraction of the energy obtained from the garbage-burning power plant described in Chapter 4 comes from polymers.

But incineration of plastics is not without some drawbacks. The repeated message of Chapters 1, 2, and 3, that burning does not destroy matter, applies here as well. Effluent gases may be "out of sight," but they had best not be "out of mind." Of special concern are chlorine-containing polymers such as polyvinyl chloride, which release hydrogen chloride during combustion. Because HCl dissolves in water to form hydrochloric acid, a strong and corrosive compound, the smokestack exhaust could make a serious contribution to acid rain. Moreover, some plastics are printed with inks containing heavy metals such as lead and cadmium. These toxic elements concentrate in the ash left after incineration and thus contribute to a secondary disposal problem. On balance, however, if carefully monitored and controlled, incineration can lead to a large reduction in plastic waste, generate much-needed energy, and have little negative impact on the environment.

Another potential strategy for disposing of plastic wastes is to enlist bacteria to do the job—in other words, to employ **biodegradation.** The problem is that bacteria and fungi do not find most plastics very appetizing. Because these microorganisms evolved in our natural environment, they possess enzymes to break down naturally occurring polymers into simpler molecules. Indeed, many strains of bacteria use cellulose from plants or proteins from plants and animals as their primary energy sources. You have already encountered several instances of such processes in this text. In Chapter 3 you read about the release of methane by belching cattle. Actually, the methane is produced when bacteria decompose cellulose in the cow's rumen. In the same chapter, we also mentioned that methane is generated by natural decomposition of organic material in landfills.

Ironically, the very properties of inertness and durability that make plastics so desirable as replacements for natural materials also create serious problems with their disposal. Thus, as in the case of the highly stable chlorofluorocarbons, a virtue becomes a liability. To be sure, synthetic polymers, like their natural counterparts, are based on carbon, hydrogen, oxygen, and nitrogen. Yet, subtle differences in the way in which these atoms are bonded to one another make most synthetic polymers nonbiodegradable.

Of late, scientists have been attempting to engineer biodegradability into artificial polymers. Certain bonds or groups are introduced into the molecules to make them susceptible to fungal or bacterial attack, or to decomposition by ultraviolet light. One strategy has been to incorporate starch, a naturally degradable biopolymer, into plastic formulations. Another example is the research project launched by the Procter & Gamble Company to evaluate the biodegradation of the absorbent filler used in disposable diapers. Although some progress is being made in achieving biodegradability in polymers, a recent EPA report raises cautions:

> Before the application of these technologies can be promoted, the uncertainties surrounding degradable plastics must be addressed. First, the effect of different environmental settings on the performance (e.g., degradation rate) of degradables is not well understood. Second, the environmental products or residues of degrading plastics and the environmental impact of degradables on plastic recycling is unclear.[2]

Part of the difficulty is that even natural polymers do not decompose as completely in landfills as was suggested earlier in this chapter. Modern waste disposal facilities are covered and lined to prevent leaching into the surrounding ground. This creates anaerobic (oxygen-free) conditions that impede bacterial and fungal action. As a result, many supposedly biodegradable substances decompose slowly or not at all. Recent excavations of old landfills have found 37-year-old newspapers that are still readable and five-year-old hot dogs that, while hardly edible, are at least recognizable.

Given the problems associated with land disposal of natural and synthetic polymers, attention has logically turned to **recycling** both. Although recycling does not literally dispose of plastics as does incineration or biodegradation, it helps to reduce the amount of new plastic entering the waste stream. Soft drink bottles are the most recycled plastic material, with about 20% of them being melted and reused. But less than 3% of plastic jugs made from high density polyethylene are recycled. Altogether, only about 1% of all plastics are recycled, compared to nearly 30% of aluminum.

10.15   *Consider This*

Find out what chemical and economic factors might account for the significant difference in the percentage of aluminum recycled compared to the percentage of plastic.

In spite of this low overall rate, increasing amounts of "post-consumer" plastics are being recycled. For example, national supermarket chains are now recycling their polyethylene grocery bags. In fact, if you read the labels on plastic materials, you will increasingly find them made of a mixture of virgin and recycled plastics. Recycling centers pay about 8 cents per pound of PET, and the cleaned, recycled PET sells for 38 cents per pound. This is roughly half the price of about 67 cents per pound for virgin PET. Of course, the laws of supply and demand work here as they do throughout the economy. During the Persian Gulf war, when petroleum prices rose significantly, virgin PET prices rose accordingly to about 80 cents per pound. At the same time, recycled PET suppliers raised their prices to 48 cents per pound in response to the market.

Another major recycling initiative involves Styrofoam (polystyrene). More than 200 million pounds of polystyrene foam food and beverage containers were recycled in 1989. The 1995 goal of the National Polystyrene Recycling Company is to recycle 25% of the plastic foam used in the United States for food service and beverage packaging. This ambitious project, if successful, would recycle a quarter billion (250,000,000) pounds of polystyrene a year.

---

[2] From EPA Report to Congress "Methods to Manage and Control Plastic Wastes," February, 1990.

The Sceptical Chymist

One of the major uses of polystyrene foam has been for "clamshell" fast food containers (though many chains have now discontinued their use). If each container weighs 10 grams and if they are assumed to be the only polystyrene source, how many of these containers would have to be recycled in a single year to achieve the goal of 250,000,000 pounds? Is this a reasonable and attainable goal?

Ans. 11×10^9 or 11 billion

For recycling to be economically self-sustaining, a number of factors must be coordinated. The first is supply. A relatively large volume of material must be consistently available at designated locations for reprocessing. This creates the rather formidable task of collection, along with the associated problem of separation of the plastics either before collection or before reuse. The codes that have started appearing on plastic objects (Table 10.1) are provided to help facilitate this sorting. Once the entropy and disorder have been overcome, the reprocessing is relatively simple. Almost any polymer that is not extensively cross-linked can be melted. If the supply of waste is homogenous, that is, if it contains only one type of plastic, it can be melted and used directly to fabricate new products. Alternatively, it can be solidified, pelletized, and stored for future use.

Many plastic objects, however, are intentionally constructed of several different polymers. A good example is the familiar two-liter soft drink bottle (Figure 10.14). The bottle itself is made of polyethylene terephthalate, with a high-density polyethylene bottom reinforcement and a cap of cross-linked HDPE. In addition, the cap liner and the adhesives for the bottom and the label may be still other polymers. When heterogeneous plastic objects such as these bottles are melted, the product tends to be

Figure 10.14

The anatomy of a two-liter soft drink bottle.

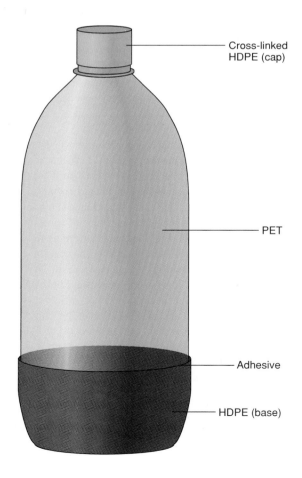

Cross-linked HDPE (cap)

PET

Adhesive

HDPE (base)

dark with varying properties, depending on the nature of the mixture. Although this re-processed material does not have outstanding working properties, it is good enough for general lower grade uses such as parking lot bumpers, disposable plastic flower pots, and cheap plastic lumber.

The problems are more than simply technical. Uses for recycled plastics are obviously important because without a product and a market, recycling programs are doomed to fail. In fact, recycling laws in a number of cities have not been implemented and enforced because one of the links in this polymeric chain of supply, collection, sorting, processing, manufacturing, and marketing is missing. Without all of these, the system will not work, unless it is heavily subsidized, and municipalities have been unwilling to provide the necessary funds.

10.17   *Consider This*

A number of national supermarket chains ask customers to return their plastic grocery bags for recycling. If there is such a program in your area, find out what happens to the returned plastic after it leaves the supermarket site. How is the plastic used after reprocessing and recycling?

10.18   *Consider This*

Laws have been passed making recycling mandatory in several states and the District of Columbia. Because of these laws, the volume of material recycled has increased substantially. Yet in spite of legislative mandates, recycling has not yet proved financially self-sustaining. As a legislator attempting to rectify this situation, you decide to draft or redraft recycling laws so that they contain financial incentives, not legislative mandates. Describe the incentives you would create.

The remaining option, **source reduction,** appears to be simplest and most direct—simply decrease the quantity of plastics produced and consumed. However, even this seemingly innocuous option is far more complicated than it appears. The problem is that something else is generally used to replace the plastic, and this substitution can be fraught with hidden pitfalls. In making choices between alternative materials, the decisions must be informed by the source and nature of feedstocks, the method of manufacturing, waste products produced during manufacturing and their disposal, and many other factors. Energy costs as well as economic costs must be taken into account. How much energy must be expended in the entire life cycle of a product from raw material to final disposal?

Obviously, the identification and proper weighing of all of the possible variables is a complex and difficult task. But when the job is done properly, one sometimes discovers that attemps to reduce the amount of plastic waste by substitution may actually increase the overall amount of waste and the associated negative environmental impact. As an example, take the replacement of a plastic cup with one made of paper. Each occupies about the same volume in a landfill, where both will probably remain undecomposed for a long time. But it is likely that a larger quantity of potentially harmful emissions enters the air and water from the production of the paper than the plastic. Moreover, the net energy input required for the paper is higher than for plastic. This kind of cradle-to-grave evaluation is essential if we are to obtain a reliable test of which material is more environmentally sound. Popular opinion must be supported by fact.

> **10.19   Consider This**
>
> If you did 10.2 Consider This, examine your list of discarded plastic objects and suggest alternatives to plastics for each of these items.

Of course, the best method of source reduction is not to replace plastics, but to do without them whenever possible. One correspondent in the great plastic versus paper battle said it well:

> There is a danger in this grocery bag controversy of losing sight of issues of greater importance. One of these is the matter of legitimate, responsible use of resources. Plastics are made from one of the most precious resources, one which cannot be renewed or replaced. In many respects we should regard it as more precious than gold or diamond. There are products essential to human health and well-being which can be made only from petroleum. There are also non-essential, wasteful uses of this priceless commodity. Where did we get the idea that it is our right to waste millions of barrels of oil each year exceeding the speed limit? Who said we're justified in manufacturing and using plastic items like shopping bags, burger boxes, and disposable diapers which are instant garbage? How did we get hooked on the consumer habits that are destroying not only a level of comfort we take for granted, but the very air and water we need to survive?*

■ Conclusion

Quite clearly, the letter that concludes the last section goes well beyond a choice of plastic or paper shopping bags. Once more we come to an issue of lifestyle. Over the past 50 years, chemists have created an amazing array of polymers and plastics—new materials that have made our lives more comfortable and more convenient. Many of these plastics represent a significant improvement over natural polymers. Furthermore, many of the products we take for granted today would be impossible without synthetic polymers and plastics. There would be no audio- and videotape, no compact discs, no kidney dialysis apparatus and no heart/lung machines. We have become dependent on plastics, and it would be difficult if not impossible to abandon their use.

Chemical industry has given consumers what they want. But there now appears to be rather more of it than we would like—mountains of soft drink bottles and miles of plastic bags. At times the world seems to be filling up with polystyrene peanuts. We must learn to cope with this glut of stuff while saving matter and energy for tomorrow. Mr. McGuire was right, there still is a great future in plastics. To create a new world of plastics and polymers will require the intelligence and efforts of policy planners, legislators, economists, manufacturers, consumers, and, above all, chemists. Perhaps Ben should go back to college, and this time major in chemistry.

■ References and Resources

Alper, J. and Nelson, G. L. *Polymeric Materials: Chemistry for the Future.* Washington: American Chemical Society, 1989.

Baum, R. M. "Systematic Chemistry of C_{60} Beginning to Emerge." *Chemical & Engineering News,* Dec. 16, 1991: 17–20.

Borman, S. "New High-Strength Fiber Finding Innovative Uses in Protective Clothing." *Chemical & Engineering News,* Oct. 9, 1989: 23–24.

Curl, R. F. and Smalley, R. E. "Fullerenes." *Scientific American,* Oct. 1991: 54–63.

*Courtesy *Syracuse Herald-Journal,* Syracuse, N.Y.

Gibbons, A. "Making Plastics that Biodegrade." *Technology Review*, February/March 1989: 69–73.

Methods to Manage and Control Plastic Wastes. Report to Congress: Executive Summary. Washington: U.S. Environmental Protection Agency, 1990.

Reisch, M. S. "Electronic Uses Spur Growth of High-Performance Plastics." *Chemical & Engineering News*, Sept. 4, 1989: 21–48.

Sicilia, D. B. "A Most Invented Invention." (polypropylene) *Invention & Technology*, Spring/Summer 1990: 45–50.

Stone, R. F.; Sagar, A. D.; and Ashford, N. A. "Recycling the Plastic Package." *Technology Review*, July 1992: 48–56.

Thayer, A. M. "Solid Waste Concerns Spur Plastic Recycling Efforts." *Chemical & Engineering News*, Jan. 30, 1989: 7–15.

———. "Advanced Polymer Composites Tailored for Aerospace Use." *Chemical & Engineering News*, July 23, 1990: 37–58.

■ Experiments and Investigations

16. The Ubiquitous Styrofoam® Cup

17. Classification of Polymers

■ Exercises

1. Consider the polymerization of ten ethylene monomers to form a segment of polyethylene:

$$10 \; H_2C = CH_2 \rightarrow (CH_2-CH_2)_{10}$$

Use the bond energies of Table 4.1 to calculate the energy change during this reaction. Is the reaction endothermic or exothermic? Should heat be supplied or removed to promote this reaction?

2. Determine the number of CH_2CH_2 monomeric units in one molecule of polyethylene with a molar mass of 40,000 g.

*3. The current U.S. production of polyethylene is 10 million tons per year. Calculate the number of moles of ethylene necessary for this production. What volume would this C_2H_4 occupy at a pressure of one atmosphere and a temperature of 25°C if one mole occupies 24.5 liters?

4. Suggest why a free radical with an unpaired electron should be an effective catalyst for the polymerization of ethylene.

5. The text suggests that a polyethylene molecule as wide as a piece of spaghetti would be about half a mile in length. If the spaghetti is 1 millimeter in diameter, calculate the ratio of its length to its width.

6. Calculate the percentage by mass of each of the following.

   a. chlorine in polyvinyl chloride
   b. hydrogen in polystyrene
   c. oxygen in polyethylene terephthalate
   d. nitrogen in nylon

7. Describe briefly how each of the six strategies for modifying the molecular structure of polymers would be expected to change the properties of material produced.

8. Of the three orientations of polyvinyl chloride in Figure 10.6, which arrangement would you expect to be most likely to give rise to the flexible form of PVC and which would be most likely to give rise to the more rigid variety? Explain your answer.

9. Calculate the molar mass of a polystyrene molecule consisting of 5000 monomer units.

*10. Of the "Big Six" polymers, which contains the highest percentage by mass of carbon? What is that percentage?

11. One limitation of the "Big Six" is the relatively low temperatures at which they melt, 90–170°C. Suggest ways to raise these temperature limits while maintaining the other desirable properties of these substances.

12. Butadiene, $CH_2=CH-CH=CH_2$, is polymerized to make buna rubber. Write an equation representing this process. Is this an example of addition or condensation polymerization?

13. Nylon 6 is made by the polymerization of

$$H_2N-(CH_2)_5-\overset{\displaystyle O}{\overset{\|}{C}}-OH.$$

Write an equation for the polymerization of two such monomeric units. Is this an addition or a condensation polymerization?

14. What structural features must a monomer possess in order to undergo addition polymerization? What characteristics are necessary for condensation polymerization?

*15. A tripeptide is made up of three amino acids. The identity and properties of a tripeptide depend on the amino acids present and their sequence. Assume three amino acids are available: alanine (Ala), glutamic acid (Glu), and lysine (Lys).

    a. Write down all the possible tripeptides that could be formed from these amino acids if each amino acid could be used only once per tripeptide, e.g., Ala-Lys-Glu.

    b. Write down all the possible tripeptides that could be formed from these amino acids if each amino acid could be used more than once per tripeptide.

16. Insulin is a protein made up of 51 amino acids. Assuming an average amino acid to have a molar mass of 120 g, calculate the molar mass of insulin if the mass of H_2O eliminated is ignored. Repeat the calculation taking into consideration the mass of H_2O that is elminated on polymerization.

17. Assuming the average amino acid to have a molar mass of 120 grams, and ignoring the mass of the water eliminated on polymerization, estimate the number of amino acids in the following proteins. Note the molar masses (MM) of the proteins.

    a.   lipase   MM =   6700 g
       (an enzyme found in milk)

    b.   crotoxin   MM = 29,900 g
       (rattlesnake venom)

    c.   fibrinogen   MM = 340,000 g
       (involved in blood clotting)

18. Natural polymers include cotton, rubber, silk, and wool. Consult other sources to identify the monomer unit in each of these polymers and specify which are addition and which are condensation polymers.

*19. Assume that the average household discards 1.0 kg (2.2 lb) of plastics each week. Suppose a town of 5000 households and 100 linear miles of streets sets up a once-weekly plastics collection system. Estimate the weekly energy and monetary costs of this system. Consider the number of trucks required, the number of workers, the fuel consumed, and other relevant variables. Do not include processing costs. Specify the other assumptions and approximations you used to obtain your answers.

20. The text offers four strategies for addressing the problem of disposing of plastics: incineration, biodegradation, recycling, and source reduction. Which strategy do you favor and why?

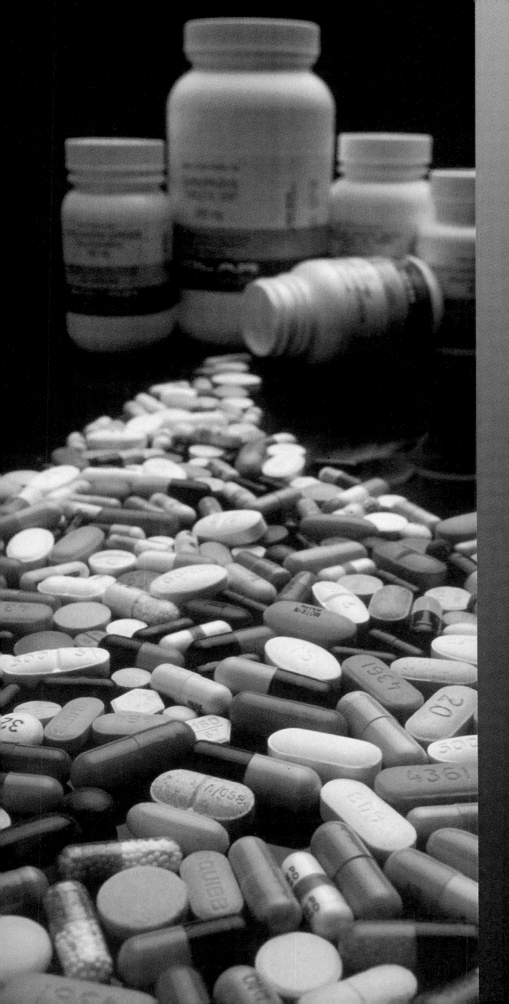

11

Designing
Drugs and
Manipulating
Molecules

■

In the fourth century B.C., Hippocrates, perhaps the most famous physician of all time, described a "tea" made by boiling willow bark in water. The concoction was said to be effective against fevers. Over the centuries, that improbable folk remedy ultimately led to the synthesis of a true "wonder drug"—one that has aided millions of people. Indeed, the name of the drug is so familiar that we have intentionally tried to hide its identity in the brief history that follows.

■ *The Origin of a Miracle Drug*

One of the first modern investigators of willow bark tea was Edmund Stone, an English clergyman. His report to the Royal Society set the stage for a series of further chemical and medical investigations. Chemists were subsequently able to isolate small amounts of yellow needle-shaped crystals of a pure compound from the willow bark extract. Because the tree species was *Salix alba,* this new compound was named *salic*in. Experiments showed that salicin could be chemically separated into two compounds. Clinical tests provided evidence that only one of these components reduced fevers and inflammation. It was also demonstrated that the active component was converted to an acid in the body. Unfortunately, the clinical testing revealed some troubling side effects. The active component not only had a very unpleasant taste, its acidity also led to acute stomach irritation.

Although the active acid was used as a treatment for pain, fever, and inflammation, chemists set out to modify its structure in order to form a related compound that still had the desired medicinal properties, but without the undesirable taste or stomach distress. Because the active ingredient was known to be an acid, the first modification attempt took a very simple approach. The acid was reacted with a base to produce a **salt,** an ionic compound formed by such a neutralization reaction. A number of simple salts were made with bases such as sodium hydroxide or calcium hydroxide.

$$2 \text{ RCOOH} + \text{Ca(OH)}_2 \rightarrow \text{Ca(OOCR)}_2 + 2 \text{ HOH} \qquad (11.1)$$

Active acid   Calcium salt of acid

The –COOH is the acidic portion of the molecule; the R represents a group of carbon and hydrogen atoms that are not directly involved in the acid-base reaction. The derivatives (salts) had less pronounced side effects than the parent compound. Thus, chemists correctly concluded that the acidic part of the molecule was responsible for these effects. The next step was to seek a structural modification that would lessen the acid strength of the compound without destroying its medicinal effectiveness.

One of the chemists working on the problem was Felix Hofmann, an employee of a major German chemical firm. Hofmann's motivation was more than just scientific curiosity or assigned work. His father suffered nausea from the acidic compound he took for his arthritis. The molecular modification achieved by the younger Hofmann greatly reduced the nausea and other adverse reactions. The resulting compound was a stable solid that reverted to the active acid form once in the body.

Extensive hospital testing of Hofmann's compound began along with a simultaneous development for its large-scale manufacture by a well-known pharmaceutical company. The new drug itself could not be patented because it was already in the chemical literature. However, the company hoped to recoup its investment by patenting the manufacturing process. Clinical trials showed the drug to be nonaddicting and relatively nontoxic. Its toxicity is classed as low by ingestion, but 20–30 grams ingested at one time may be lethal. At the suggested dose of 325–650 mg (0.325–.650 g) every 4 hours, it is a remarkably effective antipyretic (fever reducing), analgesic (antipain), and anti-inflammatory agent. Data from clinical tests uncovered the side effects noted in Table 11.1. The drug was also found to increase blood clotting time and to cause at least some small, almost always medically insignificant, amounts of stomach bleeding in about 70% of users.

■ Table 11.1	Side Effects of the "Wonder Drug"

The severity scale ranges from 1—life threatening, seek emergency treatment immediately to 5—continue the medication and tell physician at next visit.

Symptoms	Frequency	Severity
Drowsiness	rare	4
Rash, hives, itch	rare	3
Diminished vision	rare	3
Ringing in the ears	common	5
Nausea, vomiting, abdominal pain	common	2
Heartburn	common	4
Black or bloody vomit	rare	1
Black stool	rare	2
Blood in the urine	rare	1
Jaundice	rare	3
Anaphylaxis (severe allergic reaction)	rare	1
Unexplained fever	rare	2
Shortness of breath	rare	3

Adapted from H. W. Griffith. *The Complete Guide to Prescription and Non-Prescription Drugs.* Tucson, Arizona: HP Books, 1983.

11.1   *Consider This*

■

In the United States, the final step for approval of a drug is the submission of all clinical test results to the Food and Drug Administration (FDA) for a license to market the product. Suppose you were part of an FDA panel that had to rule on the safety of the drug described in Table 11.1. Should it be allowed on the market as an over-the-counter drug or a prescription drug, or should approval be withheld? What further information might you require before making a ruling?

■ *The World's Most Used Drug*

Perhaps you have already guessed the identity of the miracle drug related to willow bark tea. Its chemical names, 2-(acetyloxy)-benzoic acid or (more commonly) acetyl salicylic acid, may not help much. But the power of advertising is such that, had we revealed that the firm that originally marketed the drug was the Bayer division of I. G. Farben, we would have let the tablet out of the bottle. The compound in question is aspirin, the world's most widely used drug.

Admittedly, we have compressed the time somewhat. Most of the development, testing, and design of aspirin occurred in the eighteenth and nineteenth centuries. Stone's letter to the Royal Society was written in 1763, and Felix Hofmann's modification of salicylic acid to yield aspirin was done in 1898. Furthermore, the clinical testing of aspirin was somewhat less systematic than our account implies. But the basic facts and the steps that led to aspirin's full development are essentially correct. We must also add one more very important fact: aspirin did not have to receive drug approval before being put on the market; no such certifying process was in place at that time. Had such approval based on clinical test results been necessary, it is quite likely that aspirin might only be available on a prescription basis.

11.2 ■ *Consider This*

The price of aspirin tablets varies considerably, though the compound itself is a pure substance. The heavily-marketed brands can cost five to ten times as much as less well-known or no-name brands. Suppose the FDA proposed to establish a standard price for all aspirin tablets. Speculate on who might favor such a ruling and who might be opposed. Cite arguments on both sides and state your own opinion.

■ *Chapter Overview*

Drugs or pharmaceuticals are substances that prevent, moderate, or cure illnesses. We began the chapter with a drug that has probably been used by all readers of this book. We chose aspirin not only because of its familiarity, but because its discovery and development demonstrate how a new drug often comes into existence. The section that follows the overview generalizes this process and cites several other examples. All of the drugs mentioned (indeed, most of the drugs used today) contain the element carbon. Therefore, we embark on an excursion into the realm of carbon compounds—organic chemistry. Although the principles involved should be familiar, you will encounter new features such as isomers and functional groups. Both are of great importance in linking molecular structure to drug function. An example of characteristic drug activity is again provided by aspirin, but a more general treatment of the topic introduces the concept of the fit between the drug and the biochemical site at which it acts. Sometimes, activity depends on the subtle property of optical isomerism.

We next turn to the steroids, one of the most interesting and important families of biologically active compounds. Members of the family that are treated in some depth are cholesterol, sex hormones, contraceptives, aborting agents, and anabolic steroids. These compounds, or at least their uses, are familiar to almost everyone because they have often been steeped in controversy. By setting the steroids in their social context we explore a number of these issues. Finally, we turn to the topic of drug testing, and survey the lengthy and demanding process that is necessary in order to obtain approval to sell, distribute, and use a new drug. Here again, we will find that risks and benefits contend.

■ *Drug Discovery and Development*

The curative powers of certain chemicals have been discovered by various means, from lucky accidents to systematic investigation and from folk remedies to targeted research. However, the development of drugs, particularly in modern societies, generally follows a common set of steps. Both discovery and development are summarized in Table 11.2, but more detailed commentary follows.

Folk Remedies

In aspirin you have already encountered an example of a drug derived from a folk remedy. Throughout recorded history and in all societies, the study of substances effective against illness and disease has been an important activity. Traditional healers, from Greek philosophers to tribal shamans, have been keen observers of the effects of plant extracts on their patients. Some of these extracts have proven useful in modern medicine. In addition to aspirin, the best known are probably quinine, morphine, and mind-altering drugs. One argument for the preservation of tropical rain forests is that they may contain plant species with still undiscovered medicinal properties.

Table 11.2   **Drug Discovery and Development**

1. Discovery
   a. From folk remedies or traditional remedies. Examples: aspirin, quinine, digitalis.
   b. Accidental discovery. Examples: antibiotics, LSD, some tranquilizers.
   c. Planned investigations. Examples: sulfa drugs, anti-cancer drugs.
   d. Specified, targeted drug design and modification, planned screening. Examples: analgesics, steroids, modern narcotics, antihistamines, and most derivative drugs of all kinds.
2. Development
   a. *In vitro* testing (literally "in glass," testing outside the body), often coupled with screening, drug modification, and research on the mode of drug action.
   b. Chemical modification of existing drug to improve efficacy.
   c. *In vivo* testing (testing in living organisms, primarily in animals).
   d. After suitable approvals, carefully controlled human trials.
   e. Various levels of drug approvals and reviews from clinical trials to acceptance for general use.

On the other hand, curative powers have been attributed to many more folk remedies than the number from which actual healing results. The spectacular medical and rejuvenative properties attributed to powdered rhinoceros horn or dried bear spleen are without medicinal validity and pose serious threats to endangered species. Moreover, natural folk remedies can do considerably more harm than good. For example, tea made from sassafras contains safrole, a known carcinogen.

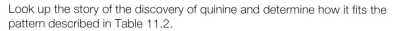

11.3   *Consider This*

Look up the story of the discovery of quinine and determine how it fits the pattern described in Table 11.2.

Accidental Discovery

Some drugs have been discovered accidentally, through the correct interpretation of a lucky observation or curiosity about an unusual occurrence. The most famous, and perhaps luckiest, discovery of this type was that of penicillin by Alexander Fleming. Fleming's curiosity was aroused by the chance observation that in a container of bacterial colonies (a petri dish), the area contaminated by the mold *Penicillium notatum* was free of bacteria. He correctly concluded that the mold gave off a substance that inhibited the growth of the bacteria. That substance turned out to be penicillin. A careful reconstruction has indicated that at least six critical, but lucky, events had to occur sequentially in order for the discovery to be made. They are summarized below.

1. In a laboratory one floor below Fleming's, a colleague worked on molds used to make vaccines in the treatment of allergies. Mold spores from that lab were carried through the air, entered Fleming's laboratory, and settled on a dish of *staphylococci* bacteria, an organism sensitive to penicillin.
2. The mold was a rare strain of *Penicillium notatum* that produced relatively large amounts of penicillin.
3. The dish was left at room temperature rather than put into an incubator.
4. The dish was left in the unheated laboratory during Fleming's vacation, a period of unusually cool weather that retarded bacterial growth, but not mold growth.
5. The weather subsequently changed and the temperature increased, allowing the bacteria to grow. By this time, however, the mold had produced enough penicillin to kill any bacteria near it, thus generating a bacteria-free region or zone of inhibition around it.

6. The sixth lucky accident had to do with poor housekeeping by an overworked laboratory staff. On September 3, 1928, Fleming was being visited by D. M. Pryce, a former colleague. A pile of unwashed petri dishes became the focus of their attention. This is an account of what happened:

> The dishes on top of the pile had not yet been soaked in the disinfectant, and it was these that Fleming showed Pryce to illustrate the work he had done with *staphylococci* before departing on holiday. Most of them were contaminated by yeasts and molds, which was hardly surprising since they had been lying in the pile for several weeks. Fleming, however, spotted something unusual on one of them, namely a zone of inhibition of *staphylococcal* growth around a mold.[1]

Fleming's insightful response to the unexpected illustrates the often misquoted maxim of the great French scientist, Louis Pasteur: "In the fields of observation, chance favors only the prepared mind." Most versions of this famous aphorism neglect the "only." But it was *only* because Fleming's mind was prepared that he was able to capitalize on this chain of unlikely events.

Drugs through Planning and Design

More recently, new drugs have come about through careful planning and designed syntheses. Typically, a researcher considers a group of compounds that, because of some shared properties, hold some promise as a drug against a particular condition or infectious agent. The researcher then plans a series of systematic investigations to further study the chemical nature of the compound and its potential as a drug. For example, after it became clear that molds produce antibiotic compounds, several pharmaceutical companies began programs to collect mold samples from all over the world and screen them for antibacterial activity. Though costly and time consuming, the approach proved fruitful by producing a number of potent antibiotics and other pharmaceuticals. Cyclosporin, a major anti-tissue rejection drug that has revolutionized organ transplant surgery, was discovered in this way.

Medicinal chemists, specifically trained in the nature and actions of drugs, have developed methods to synthesize molecules for specific applications. These scientists have become adept at creating new compounds having novel structures that may be active drugs, or making small, but intentional changes in rather complex molecules in order to convey or enhance biological activity. In a sense, medicinal chemists are molecular architects who "customize" molecules for use against a particular medical problem. To better understand this design process, we must first make a brief excursion into organic chemistry.

■ *A Brief Excursion into Organic Chemistry*

Organic chemistry is the study of carbon and its combination with a relatively small number of other elements, principally hydrogen, but also oxygen, nitrogen, sulfur, chlorine, phosphorus, and bromine. About 10 million organic compounds have been identified, but we will concentrate on particular portions of only a few molecules and stress their important role in interactions with living systems. Molecular shape will prove to be of special significance.

Chemists name organic compounds using a formal, stylized set of nomenclature rules set down by an international committee. However, many organic compounds have been known for a long time by common names such as alcohol, sugar, morphine, and aspirin. When a headache strikes, even chemists do not call out for 2-(acetyloxy)-benzoic acid; they simply say "give me some aspirin!" Likewise, prescriptions specify

---

[1.] From W. Sneader, *Drug Discovery: The Evolution of Modern Medicines.* New York: Wiley, 1985: p. 300.

penicillin-N rather than 6[(5-amino-5-carboxy-1-oxopentyl)amino]-3,3-dimethyl-7-oxopentyl-4-thia-1-azabicyclo[3,2,0,]-heptane-2-carboxylic acid. Mouthfuls like this are the cause of great merriment to those who like to satirize chemists, but they are important and unambiguous to those who know the system. You can rest easy because in this chapter we will use common names in almost all cases.

The incredible variety of organic compounds exists because of the remarkable ability of carbon atoms to bond in multiple ways. They can bond with other carbon atoms or with atoms of other elements. To better understand such possibilities, we need a few basic "ground rules" for bonding in organic molecules. The most fundamental generalization is one you used as early as Chapter 2—the **octet rule.** Each carbon atom shares in eight electrons. These electrons are paired to form covalent bonds and can be grouped in four different bonding patterns: (a) four single bonds, (b) two single bonds and one double bond, (c) one single bond and one triple bond, or (d) two double bonds. These arrangements are illustrated in Figure 11.1. Other elements exhibit different bonding behavior. A hydrogen atom is always attached to a molecule with a single covalent bond. An oxygen atom in a molecule typically has two pairs of bonding electrons, either in the form of two single bonds or one double bond. A nitrogen atom shares in three pairs of bonding electrons and hence can form three single bonds, one triple bond, or one single and one double bond.

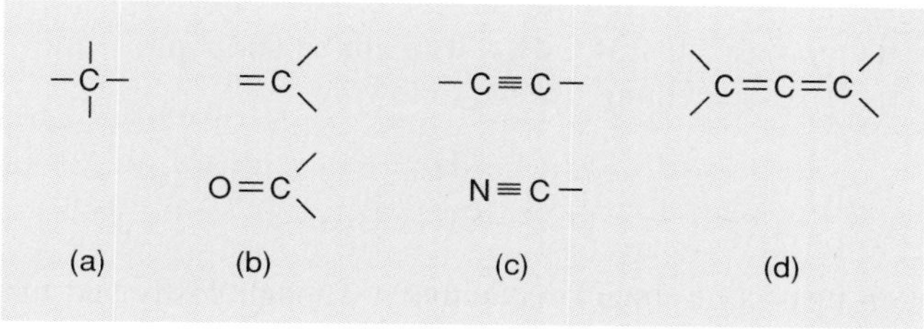

Figure 11.1

Representation of carbon with some of its bonding possibilities.

Molecular formulas, such as C_4H_{10}, indicate the kinds and numbers of atoms present in a molecule, but do not show how the atoms are arranged. In order to get that higher level of detail, structural formulas are used. These representations show the atoms and their arrangement with respect to each other in a molecule. In the case of C_4H_{10} (butane, a hydrocarbon used in cigarette lighters and camp stoves) a structural formula can be written as follows.

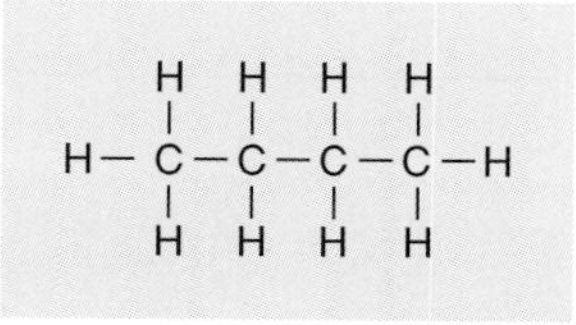

Notice that in this representation, the bonding and position of each atom relative to all others is specified. But a drawback to writing structural formulas, at least in a textbook, is that they take up considerable space. Instead, modified or condensed structural formulas can be used to convey the same information. In these, carbon-to-hydrogen bonds are not drawn out explicitly, but simply understood to be single bonds. Condensed structural formulas for C_4H_{10} are given below.

$$CH_3-CH_2-CH_2-CH_3 \qquad\qquad CH_3CH_2CH_2CH_3$$

This representation implies that carbon atoms are bonded directly to other carbon atoms in a straight chain. The hydrogen atoms do not intervene in the chain. Rather, two or three are attached to each carbon atom, depending upon its position in the molecule.

One reason why there are so many different organic molecules is because the same number and kinds of atoms can be arranged in unique ways called **isomers.** Isomers are compounds with the same chemical formula but different molecular structures and properties. You have already encountered isomers in Chapter 4 in the discussion of various components of petroleum. We will illustrate the idea with C_4H_{10}. One way the atoms can be arranged is given above, the linear isomer called normal-butane or n-butane. However, another arrangement is possible in which the four carbon atoms are not in a "straight" line. This other isomer is represented by the following structural formula.

$$
\begin{array}{ccc}
& H & H & H \\
& | & | & | \\
H- & C- & C- & C-H \\
& | & | & | \\
& H & | & H \\
& & H-C-H \\
& & | \\
& & H
\end{array}
$$

This isomer is known as iso-butane, and its formula can also be written in condensed form.

$$CH_3-\underset{\underset{CH_3}{|}}{CH}-CH_3 \qquad\qquad CH_3CH(CH_3)CH_3$$

The parentheses around the CH_3 indicate that its carbon is attached to the carbon to its left. It introduces a "branch" into the molecule.

These are the only isomers of C_4H_{10}. It might be tempting to draw another structural formula that looks something like the following.

$$\underset{CH_3-\underset{|}{CH}-CH_3}{\overset{CH_3}{|}}$$

However, a bit of inspection should reveal that this is simply the previous structure for iso-butane written upside down, and not a different compound. Molecular models of both isomers of C_4H_{10} are pictured in Figure 11.2. As the number of atoms in a hydrocarbon increases, so do the number of possible isomers. Thus, there are 18 isomers of C_8H_{18} and 75 isomers of $C_{10}H_{22}$.

11.4	*Your Turn*
	Pentane, C_5H_{12}, exists in three isomers. Draw structural formulas for each of them.

Figure 11.2

Photographs of molecular models of the two isomers of butane, C_4H_{10}: normal-butane (left) and iso-butane (right).

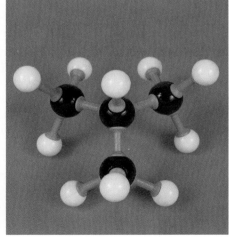

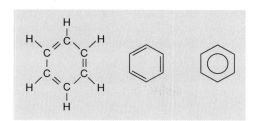

Figure 11.3
Three representations of benzene, C_6H_6.

Carbon atoms are not limited to being bonded in a linear or branched fashion. In many molecules, including aspirin, carbon atoms are arranged in a ring. Such rings most commonly contain five or six carbon atoms. In aspirin, the carbon atoms are joined in a six-member hexagonal ring called a benzene ring after the compound with the formula C_6H_6. The Lewis octet rule applied to C_6H_6 predicts alternating single and double bonds between adjacent carbon atoms. The complete representation of this structure appears on the left of Figure 11.3. But the benzene ring is so widely distributed in organic molecules that symbols for individual carbon and hydrogen atoms are generally not written. The form in the center indicates the individual bonds, but is a bit misleading. Experiment shows that the bonding electrons in benzene are uniformly distributed around the ring and shared equally by all six carbon atoms. This is conveyed by the circle within the hexagon in the representation on the right.

■ *Functional Groups*

Another very important concept in the properties of drugs and in the subdiscipline of organic chemistry is the idea of **functional groups.** These are certain arrangements of atoms that appear over and over again in carbon-containing molecules. Functional groups convey characteristic properties to the molecules that contain them. Indeed, these groups are so important that we often focus our formulas on them and symbolize the remainder of the molecule with an R. The R is generally assumed to include at least one carbon atom that is connected to the functional group, but it can be practically anything. You already encountered some functional groups in Chapter 10. An alcohol is ROH, as in methyl (wood) alcohol, CH_3OH, and ethyl (grain) alcohol, CH_3CH_2OH. The presence of the –OH group makes the compound an alcohol.

Similarly, acidic properties are conveyed by a carboxylic acid group,

$$\begin{matrix} & O \\ & \| \\ -C&-OH, \end{matrix}$$

commonly written as $-COOH$. The hydrogen is released in solution as an H^+ ion. Thus, we represent an organic acid with the general formula, RCOOH. In acetic acid, the acid in vinegar, R is $-CH_3$, a methyl group. There are a dozen or so common functional groups, and you will meet a few of them on a "need to know" basis.

The presence and properties of functional groups are responsible for the action of all drugs. Aspirin has three such subunits, boxed and numbered in Figure 11.4.

You will recognize that box 1 encloses a benzene ring. Its presence makes aspirin soluble in lipids, which are fatty compounds that are important cell membrane components. The other two portions, boxes 2 and 3, are responsible for the drug activity. You have just been reminded that the $-COOH$ group indicates an organic acid. The other functional group (box 3) is an ester. You will recall from our discussion of polyethylene terephthalate in Chapter 10 that an ester is formed by the reaction of an alcohol and an acid. Water is eliminated in the process.

Felix Hofmann prepared aspirin by modifying the structure of salicylic acid. But note that he did not modify the carboxylic acid group on the molecule. Salicylic acid

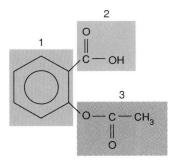

Figure 11.4
Structural formula of aspirin.

also contains an alcoholic OH group, and it was this part of the molecule that Hofmann reacted with acetic acid via equation 11.2. The product was an ester of acetic acid and salicylic acid, which accounts for one of aspirin's names—acetyl salicylic acid.

(11.2)

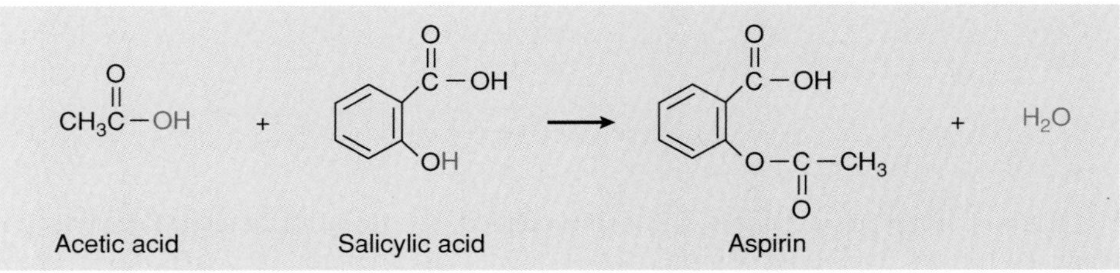

Acetic acid   Salicylic acid   Aspirin

Because aspirin retains the $-COOH$ group of the original salicylic acid, it still has some of the undesirable acidic properties of the parent compound. However, the presence of the ester group reduces the strength of the acid group and makes the compound more palatable and less irritating to the stomach lining. Once aspirin is ingested and reaches the site of its action, reaction 11.2 is reversed. The ester splits into acetic acid and salicylic acid, and the latter compound exerts its antipyretic and analgesic properties.

Functional groups sometimes play a role in the solubility of a compound, an important consideration in the uptake, rate of reaction, and residence time of drugs in the body. The general solubility rule, "like likes like" applies in the body as well as in the test tube. Functional groups containing oxygen and nitrogen atoms usually increase the polarity of the molecule and enhance its solubility in a polar substance such as water. Those compounds whose molecules do not contain such groups, but consist primarily or exclusively of carbon and hydrogen atoms, are more likely to dissolve in nonpolar solvents. Therefore, drugs with significant nonpolar character tend to accumulate in cell membranes and fatty tissues, which are themselves largely hydrocarbon.

Research has produced about 40 other aspirin-like compounds. Two of these, ibuprofen and acetaminophen (Tylenol®), are more specific in their mode of action than is aspirin. Their structural formulas are given in Figure 11.5.

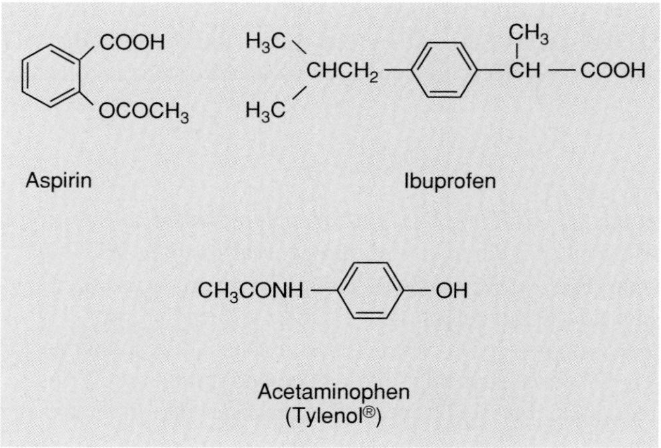

Figure 11.5
Structural formulas of analgesics.

Aspirin   Ibuprofen

Acetaminophen
(Tylenol®)

11.5	***Your Turn***
	Look at the structural formulas given in Figure 11.5. Identify the structural features and functional groups that aspirin, ibuprofen, and acetaminophen have in common.

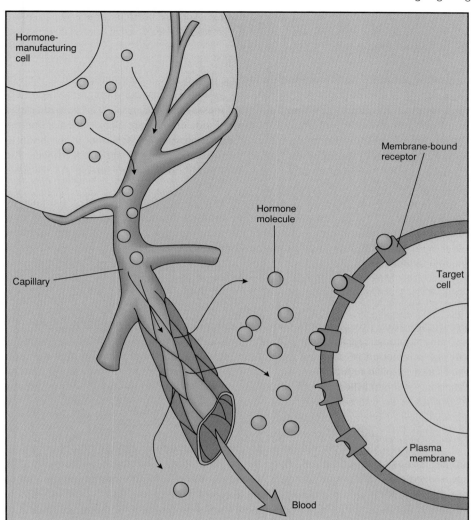

Figure 11.6
Chemical communication in the body. Hormone molecules travel from the cell where they are made, through the bloodstream, to the target cell.

■ *How Aspirin Works*

A complex internal means of communication is essential for higher-order living systems such as human beings. You may think of your body's communication system as made up almost entirely of electrical impulses traveling along nerves. This is certainly true for activities such as movement, reflex actions, breathing, and heartbeats. However, most of the body's internal "messages" occur not by electrical impulses, but through chemical processes. In fact, your very first communication with your mother was a chemical signal saying "I'm here; better get your body ready for me." It is much more efficient to release chemical messengers into the bloodstream, which then circulates them to appropriate body cells, than to "hardwire" each individual cell with nerve endings. Figure 11.6 is a representation of such chemical communication.

These chemical messengers are called **hormones** and they are produced by the body's endocrine glands. Hormones encompass a wide range of functions and a similarly wide range of chemical composition and structure. Thyroxine, an iodine-containing amino acid, is one of the simpler ones, but is essential for regulating metabolism. The chemical breakdown of "blood sugar" or glucose requires insulin. This hormone, a small protein of only 51 amino acids, is secreted by the pancreas. Persons who suffer from diabetes are often required to take daily injections of insulin. Yet another well-known hormone is adrenaline or epinephrine, a small molecule that prepares the body to "fight or flee" in the face of danger. And the hormonal messages that are so compelling in adolescents are carried by steroids, a sexy set of molecules that we will revisit in a few pages.

Aspirin and other drugs that are physiologically active but not anti-infectious agents are almost always involved in altering the chemical communication system of the body. A significant problem is that this system is very complex, allowing many compounds to be used to send more than one message simultaneously. The wide range of aspirin's therapeutic properties, as well as its side effects, are clear evidence that the drug is involved in several chemical communication systems. It works in the brain to reduce fever, it relieves inflammation in muscles and joints, it apparently lowers the chances of stroke, and it seems to reduce the likelihood of heart attack. As we have seen, it also can cause stomach irritation.

In large measure, the versatility of aspirin and similar nonsteroidal anti-inflammatory drugs is related to their remarkable ability to block the actions of other molecules. Research on the activity of aspirin indicates that one of its modes of action involves blocking a particular **enzyme,** prostaglandin synthase. This enzyme, like all others, is a biochemical catalyst. It is a protein that influences the rate and direction of a chemical reaction. Most enzymes speed up reactions and channel them so that only one product (or a set of related products) is formed. In the case of prostaglandin synthase, the reaction is the synthesis of a series of hormone-like compounds called **prostaglandins.** Prostaglandins cause a variety of effects. They produce fever and swelling, increase sensitivity of pain receptors, inhibit blood vessel dilation, and regulate the production of acid and mucous in the stomach. By preventing prostaglandin production, aspirin reduces fever and swelling. It also suppresses pain receptors and so functions as a painkiller. Because the benzene ring conveys high fat solubility, aspirin is also taken up into cell membranes. In certain specialized cells, the drug blocks the transmission of chemical signals that trigger inflammation. This process also appears to be related to aspirin's effectiveness as a pain reliever.

The aspirin substitutes exhibit these same properties in varying degrees. For example, because acetaminophen blocks prostaglandin synthase, but does not affect the specialized cells, it reduces fever but has little anti-inflammatory action. On the other hand, ibuprofen is a better enzyme blocker and specialized cell inhibitor. Consequently, it is both a better pain reliever and fever reducer than aspirin. Ibuprofen has fewer functional groups than aspirin, which may be the reason why ibuprofen has fewer side effects. Fewer functional groups also makes it less polar and more lipid soluble than aspirin. Its anti-inflammatory activity is five to fifty times that of aspirin.

All of these related anti-inflammatory drugs appear to affect the way cell membranes respond to stimuli. Research has shown that this is yet another possible mode of action for aspirin and its chemical relatives. On the other hand, aspirin is unique among these three compounds in its ability to inhibit blood clotting. This property has led to the suggestion that low regular doses of aspirin can help prevent strokes or heart attacks. Of course, these anticoagulation characteristics also mean that aspirin is not the painkiller of choice for surgical patients or those suffering from ulcers. That is why "more hospitals use Tylenol®."

■ *Drug Function and Drug Design*

The modern approach to chemotherapy and drug design probably began early in this century with Paul Ehrlich's search for an arsenic compound that would cure syphilis without doing serious damage to the patient. His quest was for a "magic bullet" that would affect only the diseased site and nothing else. He systematically varied the structure of many arsenic compounds, simultaneously testing each new compound for activity and toxicity using experimental animals. He finally achieved success with Salvarsan 606, so named because it was the 606th compound investigated. Since then, medicinal chemists have adopted Ehrlich's strategy of carefully relating chemical structure and drug activity. The goal remains to produce a drug that meets the desired therapeutic need while exhibiting minimum side effects.

Drugs can be broadly classified into two groups: those that produce a physiological response in the body and those that kill or inhibit the growth of substances that cause infections. You have already learned that aspirin falls in the first group. So do synthetic hormones and psychologically active drugs. These drugs typically initiate or block a chemical action that generates a cellular response, such as a nerve impulse or the synthesis of a protein. Antibiotics exemplify drugs that kill foreign invaders. They do so by inhibiting an essential chemical process in the infecting organism. Thus, they are particularly effective against bacteria.

Although drugs vary in their versatility, many of them act only against particular diseases or infections. This specificity is consistent with the relationship that exists between the chemical structure of a drug and its therapeutic properties. Both the general shape of the molecule and the identity and location of its functional groups are important factors in determining its physiological efficacy. This correlation between form and function can be explained in terms of the interaction between biologically important molecules. Although many of these molecules are very large, they often contain a relatively small **active site** or **receptor site.** For example, the active site of an enzyme is the region where the catalytic activity occurs. The process can be illustrated with one of the enzymes that breaks down proteins as part of the digestive process. The enzyme forms a temporary chemical bond with the protein molecule. The protein is called the **substrate,** a general term for the species upon which the enzyme acts. In this particular geometrical arrangement, the functional groups in the active site of the enzyme facilitate the breaking of the protein chain. Similarly, many hormones trigger chemical reactions by specific interactions with receptor sites on cell membranes.

These interactions have been likened to the relationship of a key to a lock. Just as specific keys fit only specific locks, a molecular match between substrate and receptor site is required for physiological function. The process is illustrated in Figure 11.7.

If a perfect lock-and-key match were required in the body, it would mean that each of the millions of physiological functions would have a unique receptor site and a specific molecular segment to fit it. Simple logic suggests that such rigid demands would not promote cellular efficiency. Consequently, the lock-and-key model, although a good starting point that works in a limited number of cases, must be modified.

Using another analogy, a receptor site is like a size 9 right footprint in the sand. Only one foot would fit it exactly, and many feet (all left feet and all right feet larger than size 9) would not fit it at all. But many other right feet could fit into the print reasonably well. So it is with receptor sites and the molecules or functional groups that bind to them. Some active sites can accommodate a variety of substrates—including drugs. Indeed, the way most drugs function is by replacing a normal protein, hormone, or other substrate in the invading organism. The presence of the drug molecule thus prevents the enzyme, cell membrane, or other biological unit from carrying out its chemistry. As a result, the growth of an invading bacterium is inhibited, or the synthesis of a particular molecule is turned off.

Generally speaking, the drug that best fits the receptor site has the highest therapeutic activity. In some cases, however, a drug molecule does not need to fit the receptor site particularly well. The bonding of functional groups of the drug to the receptor site may even alter the shape of the drug, the site, or both. Often what counts is for the drug to have functional groups of the proper polarity in the right places. Thus, one important strategy used by medicinal chemists in designing drugs is to determine the specific part of the molecule that gives the compound its activity. Chemists then synthesize a molecule having that specific active portion, but with a much simpler non-active remainder. In effect, chemists custom design the molecule to meet the requirements of the receptor site.

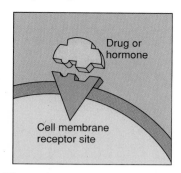

Figure 11.7
Lock-and-key model of biological interaction.

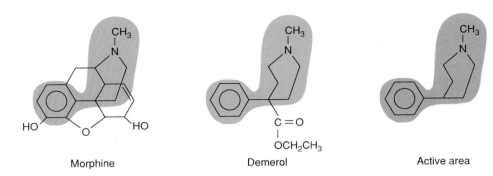

Morphine   Demerol   Active area

An outstanding example of this approach is provided by opiate drugs such as morphine. Morphine, a very complex molecule, is difficult to synthesize. However, the particular portion of the molecule responsible for opiate activity has been identified and is highlighted in Figure 11.8. The flat benzene ring fits into a corresponding flat area of the receptor, and the nitrogen atom binds the drug molecule to the site. Incorporating this particular portion into other less complex molecules, such as demerol, conveys opiate activity.

The discovery that only certain functional groups are responsible for their therapeutic properties of pharmaceutical molecules has been an important breakthrough. Sophisticated computer graphics are now used to model potential drugs and receptor sites. Thanks to these drawings, with their three-dimensional character, medicinal chemists can "see" how drugs interact with a receptor site. Computers can then be used to search for compounds that have structures similar to that of an active drug. Chemists can also modify structure in the computer models and visualize how the new compounds will function. These revolutionary techniques hold great promise for speeding up drug design and development.

11.6 ■ **Consider This**

Aspirin and other drugs have a large and profitable market around the world. But development and marketing of critical drugs needed by a small number of people suffering from rare diseases can be an enormous drain on a pharmaceutical company. Should the government step in and require successful drug companies to contribute a percentage of their profits to a fund for research on these "orphan drugs"? As president of a pharmaceutical company, draft a letter to your senator, stating your position on this issue.

■ *Left- and Right-handed Molecules*

Drug design is complicated when drug-receptor interaction involves **optical isomerism.** Optical isomers have the same chemical formula, but they differ in their molecular structure and their interaction with light, hence the name. Although the nature of that interaction is rather complex, it is related to a familiar consumer product. You may even own a pair of Polaroid® sunglasses. The lenses of these glasses permit only the passage of light waves vibrating in a single plane. All other waves are filtered out (Figure 11.9). When a beam of this plane-polarized light is passed through a solution of an optically active compound, the plane of vibration is rotated. If the rotation is to the right, the isomer is said to be the **dextrorotary** or "right-handed" form. Conversely, the **levorotary** isomer rotates the plane of vibration to the left.

Research has shown that optical isomerism most frequently arises when four *different* atoms or groups of atoms are attached to a central carbon atom. A compound having such a carbon atom can exist in two different molecular forms that are *mirror*

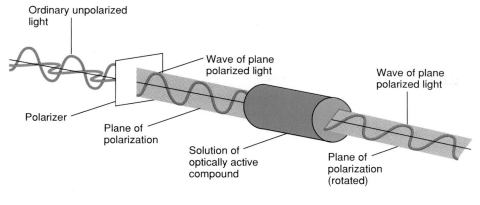

Figure 11.9

An optically active compound (an optical isomer) rotating the plane of polarized light. (Figure from *Chemistry: Imagination and Implication* by A. Truman Schwartz, copyright © 1973 by Harcourt Brace & Company, reproduced by permission of the publisher.)

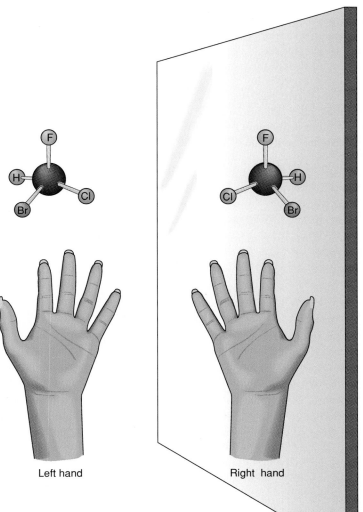

Figure 11.10

Mirror images of molecules and hands. In the molecule CHClFBr, all four of the attachments to the central carbon atom are different.

images of each other. These are the optical isomers, and they rotate light in opposite directions—one isomer to the right and the other to the left. Mirror images should not be completely unfamiliar to you because you carry two around with you all the time— your hands. If you hold them so that the palms face each other, you can recognize them as being mirror images; what is on the left side of one is on the right side of the other. Your left hand looks like the reflection of your right hand in a mirror; they are not identical. Figure 11.10 illustrates this for both a hand and a molecule. Note that the four atoms or groups of atoms bonded to the central carbon atom are in a tetrahedral arrangement. Their positions correspond to the corners of a three-dimensional solid figure with equal triangular faces.

It turns out that many biologically important molecules, including sugars and amino acids, exhibit optical isomerism. This is significant because, although most chemical and physical properties of a pair of optical isomers are very nearly identical, their biological behavior can be profoundly different. You can illustrate this relationship between optical isomerism and biological activity by taking things into your own hands. Your *left* hand will represent an optically active drug molecule. Start by folding the middle two fingers down onto the palm. Then, put your thumb, your index finger, and your little finger down on a sheet of paper in a triangular arrangement. Your palm represents the central carbon atom and the three fingers and the wrist represent the four different substituents. Mark the location of your three fingers and your wrist on the paper, specifically noting where the index finger, little finger, and thumb touch. The marked paper represents a receptor site for a left-handed molecule. Now, fold your *right* hand in a similar position and try to fit the index and little finger, thumb and wrist onto this receptor site. You will soon find that, no matter how hard you try, your right hand cannot fit into the pattern that is appropriate for your left hand. For example, if your index finger is in the correct position, your thumb and little finger will be interchanged. Similarly, right-handed optical isomers will not fit into receptor sites for left-handed molecules (Figure 11.11). It follows that any drug containing a central carbon atom with four different atoms or groups attached to it will be optically active and have optical isomers, only one of which will usually fit into a particular receptor site.

Figure 11.11

Model representation of an optical isomer of a substrate molecule binding to an asymmetric site.

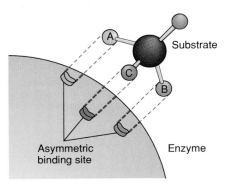

The extreme molecular specificity created by optical isomerism makes the medicinal chemist's job more complex. A drug molecule must include the appropriate functional groups, and these groups must be arranged in the biologically active configuration. Often the "right" and "left" isomers are made together, but only one isomer is pharmaceutically active. For example, many opiate drugs exist in optical isomers, only one of which may have opiate activity. Levomethorphan, the sinister, left-handed isomer of methorphan, is an addictive opiate. On the other hand, its dexterous mirror image is a non-addictive cough suppressant. This permits the use of dextromethorphan in many over-the-counter cough remedies.

11.7	*Consider This*
■	The discussion of the complications of optical isomerism probably sounds familiar to left-handed people. Southpaws often encounter difficulties living in a right-handed world. List five ways in which our society favors the right handed.

■ *Steroids: Cholesterol, Sex Hormones, and More*

Sometimes a single structural feature appears, with modification, in a wide variety of biomolecules and related drugs. This structural ubiquity is beautifully illustrated by one of the most important biochemical families—the **steroids.** These compounds cover an amazingly broad range of form and function—from cholesterol to sex hormones. Moreover, as we will soon see, some steroids have been associated with a good deal of controversy.

The common molecular basis of these compounds is a marvelous example of the economy with which living systems use certain structures for many different purposes. Cells operate rather efficiently by combining small molecular fragments in controlled ways to synthesize large molecules. Once such a molecular process is established, a cell continues to use it, incorporating molecular fragments into a variety of compounds. The process is rather like having a standardized house plan that can be reproduced readily—a unit that gains individuality by changes in the types of windows and doors or by the interior decorations. The common characteristic of the steroids is a molecular framework consisting of 17 carbon atoms arranged in four rings—"three rooms and a garage" if you like. This steroid nucleus is illustrated below. Recall that in such a representation, carbon atoms are assumed to occupy the corners of the rings, but they are not explicitly drawn.

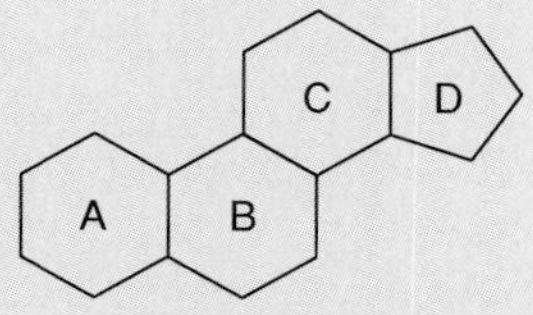

The steroid nucleus

It is evident from the structural formula that the steroid nucleus has three 6-carbon rings and one 5-carbon ring, designated as A, B, C, and D. This framework serves as the basic unit for many physiologically active molecules that differ in the functional groups attached to certain locations on the four rings. Adding extra carbon atoms and/or functional groups at a critical position on the ring can cause enormous differences in physiological effects. The range of these physiological effects is indicated in Table 11.3.

Figure 11.12 shows some of the molecular structures responsible for this variety of function. The system used to represent these structures concentrates on the carbon backbone of the molecule. A carbon atom is assumed to occupy each bend in a sequence of line segments, and a line protruding from a molecule also represents a carbon atom unless the symbol for another atom is attached to it. Hydrogen atoms, which

■ Table 11.3   *Steroid Functions*

Function	Example
Regulation of secondary sexual characteristics	Estradiol and testosterone (an estrogen and an androgen)
Reproduction and the control of the reproductive cycle	Progesterone and other gestagens
Regulation of metabolism	Cortisol and cortisone derivatives
Digestion of fat	Cholic acid and bile salts
Cell membrane component	Cholesterol

Figure 11.12
Molecular structures of some
important steroids.

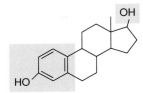

Female sex hormone
Estradiol

Male sex hormone
Testosterone

Metabolic regulator
Cortisone

Pregnancy hormone
Progesterone

Bile salt for fat digestion
Cholic Acid

Cell membrane component
Cholesterol

are not specified, are bonded to the carbon atoms as is necessary to satisfy the octet
rule. This method is illustrated below with estradiol, a female sex hormone. The figure
on the left includes all the atoms in the molecule, the one on the right gives the skele-
tal representation.

Estradiol $C_{18}H_{24}O_2$

The structures in Figure 11.12 indicate that some very subtle molecular differences
can result in profoundly altered properties. For example, maleness and femaleness rest
on just a few carefully-placed atoms on a steroid nucleus!

11.8	***Your Turn***

Carefully examine the structural formulas given in Figure 11.12. Identify the similarities and differences in the structures of the following:

a. the male and female sex hormones
b. cholesterol and cholic acid

In this chapter we will concentrate on only a small number of the many steroid compounds. We begin with cholesterol, the most abundant steroid in the body and probably the best known. The average-sized adult has about half a pound of cholesterol in his or her body! Cholesterol is a starting point for the production of steroid-related hormones and a major component of cell membranes. Because their shape is relatively long, flat, and rigid, cholesterol molecules help to enhance the firmness of cell membranes. Although cholesterol is essential for human life, there are concerns that too much of the compound in the blood can lead to the build-up of plaque, fatty deposits in the blood vessels. This plaque restricts blood flow and can lead to strokes or heart attacks. Therefore, people are advised to regulate their dietary intake of cholesterol, which is found in milk, butter, cheese, egg yolks, and other foods rich in animal fats. But one must keep in mind that some "cholesterol-free" foods can nevertheless contribute to the build-up of cholesterol in the body. It is synthesized there from fatty acids of animal or vegetable origin. A diet rich in "saturated fats" (those without double bonds between carbon atoms) is particularly likely to lead to elevated serum cholesterol.

The role of cholesterol in the body is relatively passive, but steroid hormones are involved in a tremendous range of physiologically vital processes, including such popular pastimes as digestion and reproduction. Because of the importance of these functions, medicinal chemists have, over the past 50 years, synthesized many derivatives of naturally occurring steroid hormones. These drugs, developed to mimic or inhibit the activities of the hormones in the body, have been variously described as "miracle drugs," "killer compounds," or "sleazy therapeutic agents." Perhaps more than any other type of pharmaceutical, steroid-related drugs are involved with social and ethical issues. These issues include birth control, abortion, diet, body-building, drug abuse, and drug testing. We begin by looking at drugs related to sex hormones.

■ *"The Pill"*

Sex hormones are the chemical agents that determine the secondary sex characteristics of individuals. Female sex hormones are classified as **estrogens;** male sex hormones as **androgens.** You may be surprised to learn that males have female sex hormones, and there are male sex hormones in females. However, androgens predominate in males and estrogens in females.

Because of their importance, androgens and estrogens were the first steroidal hormones studied in great detail. When this work was just beginning, techniques for determining molecular structure were in their infancy. A sample of several milligrams of the pure substance was required—much more than is needed today. Because sex hormones occur only in very small quantities, heroic efforts were required to obtain sufficient amounts for the early chemical studies. For example, one ton of bull testicles was processed to yield just 5 milligrams of testosterone, and four tons of pig ovaries provided only 12 mg of estrone. Fortunately, improved technology and instrumentation allow modern chemists to determine molecular structures with samples weighing only a fraction of a milligram.

After the molecular structures of the sex hormones were determined in the 1930s, work could proceed on the synthesis of drugs of similar structure. These efforts ultimately led to the creation of "the Pill"—the oral contraceptive that has had such a

Figure 11.13
Molecular structures of progesterone
and norethynodrel.

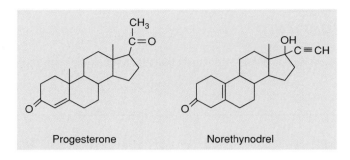

Progesterone   Norethynodrel

profound effect on modern society by launching the so-called "sexual revolution." As with aspirin, birth control drugs came about through molecular modifications, in this case, changing substituents on the steroid nucleus. The aim was to develop a drug that would mimic the action of progesterone on the complex female reproductive cycle. Interestingly, the initial motivation of the research was to improve fertility in women who found it difficult to conceive. Gregory Pincus and John Rock injected progesterone into patients in order to block ovulation and stimulate body changes related to pregnancy. Their hope was that when the therapy was discontinued, a kind of rebound would occur and fertility would increase. Such a response, now known as the "Rock rebound," does in fact take place.

Unfortunately, progesterone was expensive and not very effective when administered orally. It also caused some serious side effects in a small percentage of patients. Therefore, chemists working in a number of pharmaceutical firms set out to develop a synthetic analog for progesterone that could be taken orally, would reversibly suppress ovulation, and would have few side effects. The ultimate goal of these efforts soon became the inhibition of fertility, not its enhancement. In the mid-1950s, Frank Colton, a chemist at G. D. Searle, synthesized norethynodrel. The molecular structure of this compound (Figure 11.13) shows some subtle but significant differences from progesterone, notably the replacement of $-COCH_3$ on the D ring with $-OH$ and $-C\equiv CH$. As a consequence of these changes, the norethynodrel molecule is tightly held on a receptor site, which prevents its rapid breakdown by the liver and permits its oral administration. Norethynodrel became the active ingredient in Enovid, the first commercially available oral contraceptive, which was approved for sale in 1960.

Since the development of the original birth control pill, further molecular modifications (many of them minor) have led to decreased dosage and minimized side effects. A major contributor to these innovations has been Carl Djerassi, currently professor of chemistry at Stanford University and president of Zoecon Corporation. Djerassi, the author of hundreds of scientific papers and the holder of many patents for modified steroids, is also a novelist and poet. Recent research into an alternative birth control delivery system culminated in a plastic implant that releases a progesterone analog so slowly so that it can be effective for several years.

The mechanism for the action of steroid-based contraceptives is diagrammed in the simple schematic of Figure 11.14. In effect, the drug "fools" the female reproductive system by mimicking the action of progesterone in true pregnancy. When fertilization occurs, progesterone is released, carrying a number of chemical messages. Some of these help prepare the uterus for the implantation of the embryo. Others block the release of pituitary hormones that stimulate ovulation. The reason for this is clear: ovulation during pregnancy could lead to very serious complications. Birth control steroids, being progesterone-like molecules, send a chemical message much like progesterone. By thus simulating pregnancy, ovulation is inhibited. In effect, the message this time is not "Hey, Mom, I'm here!" but rather "Hey, you think I'm here, but I'm not!"

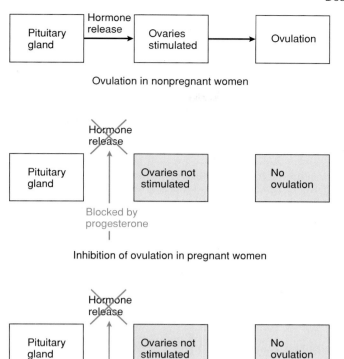

Figure 11.14

Action of a steroid contraceptive. (Reprinted with the permission of Macmillan College Publishing Company from *Drugs and the Human Body with Implications for Society*, Third Edition by Ken Liska. Copyright © 1990 by Macmillan College Publishing Company, Inc.)

Ovulation in nonpregnant women

Inhibition of ovulation in pregnant women

Inhibition of ovulation in nonpregnant women using a synthetic progestin

---

11.9 ■ *Consider This*

Oral contraceptives put control of fertility into the hands of women. What are the social and political implications of this development? What scientific and/or social reasons lie behind the fact that chemical control of reproduction was developed first for the female rather than the male reproductive system?

■ *RU–486: The "Morning-after Pill"*

Some controversy still surrounds the synthetic steroidal hormones that control fertility by inhibiting ovulation. But a far more controversial approach to birth control is a drug that can induce abortion with relative ease and safety—a "morning-after pill." The drug, currently available in France, Sweden, and the United Kingdom, is mifepristone. It is better known as RU–486, after its manufacturer, Roussel–Uclaf. Its invention was announced in 1982 by Etienne-Emil Baulieu, a French physician and researcher.

Comparison of the structural formula for RU–486 with that of progesterone shows that they are very similar, each having a steroid nucleus. The critical difference is the bulky benzene ring attached to the C steroid ring in RU–486. This bulky ring prevents changes that occur at the receptor site when progesterone is present. As noted above, when progesterone binds to the receptor, it initiates a chain of chemical events signaling the body to produce proteins needed in pregnancy. RU–486 is an **antagonist** for progesterone—it occupies the progesterone binding site, but it shows no activity. Thus, no pregnancy protein production signal is sent. Because progesterone activity is essential for implantation of the embryo in uterine cells, the developing embryo is spontaneously aborted.

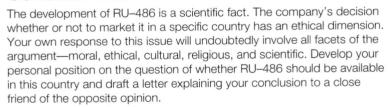

RU–486

RU–486 has been extensively tested and found to be 96% reliable. Moreover, it appears to have a relatively low incidence of serious side effects. Risk assessment studies suggest that the drug is probably the safest method to terminate a very early pregnancy. Nevertheless, initial opposition to the distribution and use of RU–486 was so strong that Roussel-Uclaf originally sought to suspend marketing the drug. But the announcement of that decision was opposed by a petition signed by 1000 leading physicians attending a world congress of gynecology and obstetrics. Claude Evin, the French Minister of Health, called the drug "the moral property of women, not just the property of the drug company" and ordered RU–486 to be put back on the market. Since then, over 100,000 French women have used the drug successfully.

The company has established five criteria, given below, that a country must meet before the company will seek approval of RU–486 in that country.

1. Abortion must be legal.
2. Abortion must be accepted by the public and the medical community.
3. A suitable synthetic prostaglandin must be available for use in the country. (RU–486 is given in conjunction with the prostaglandin for maximum efficacy and safety.)
4. Distribution must be strictly controlled.
5. A patient must agree in writing that if induced abortion fails, she will proceed with a surgical abortion.

As this book went to press, plans were being made to test RU–486 in the United States, and antiabortion forces were organizing their opposition to the testing.

11.10	*Consider This*
■	The development of RU–486 is a scientific fact. The company's decision whether or not to market it in a specific country has an ethical dimension. Your own response to this issue will undoubtedly involve all facets of the argument—moral, ethical, cultural, religious, and scientific. Develop your personal position on the question of whether RU–486 should be available in this country and draft a letter explaining your conclusion to a close friend of the opposite opinion.

■ *Anabolic Steroids: What Price Glory?*

Like birth control drugs, anabolic ("building up") steroids are controversial. Moreover, again like their contraceptive chemical cousins, anabolic steroids were created for quite a different purpose than their ultimate use. These steroids were developed initially to help patients suffering from wasting illnesses to regain muscle tissue. Ironically, their use has now become perverted by the strong who seek to become even stronger.

It has long been known that testosterone promotes muscle growth as well as the development of male secondary sexual characteristics. Drug companies sought to pursue this avenue to produce a testosterone-like drug that would stimulate muscle growth in debilitated patients, such as those recovering from long-term illness. The intent was to modify the testosterone molecule in such a way that its analog would have the desired effects on muscle development without serious negative side effects. Anabolic steroids were the result.

Figure 11.15
Molecular structures of testosterone
and two anabolic steroids.

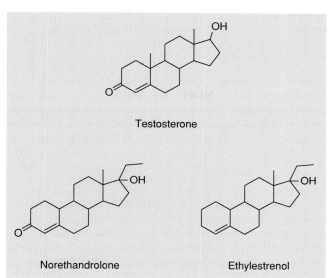

Testosterone

Norethandrolone   Ethylestrenol

This research project, as any involving sex hormones, required the use of a suitable animal model to evaluate the effectiveness and safety of the drugs. Ethical considerations and public opinion preclude the use of human subjects for testing in such unpredictable circumstances. Therefore, castrated rats were used to test anabolic steroids. Various trials compared prostate gland weight with the weight of an isolated abdominal muscle. The idea was to develop a drug that increased muscle mass (an anabolic effect) without increasing prostate mass (an undesirable side effect). Side effects such as this are said to be androgenic. They result from changes in the level of sex hormones, and they often accentuate female characteristics in males and male characteristics in females.

Several drug companies eventually succeeded in greatly reducing the androgenic effects of synthetic steroids, while not affecting their desired anabolic effects. The structural formulas of the two most potent anabolic steroids, norethandrolone and ethylestrenol, are given in Figure 11.15, along with that of testosterone for comparison. Note their very close similarities.

In other synthetic anabolic steroids, the molecular shape has been altered by adding substituents to the A ring. These substituents interfere with the fit of the molecule on the androgen activity receptor, but they do not impair anabolic activity. Two examples are stanozolol and oxymetholone.

Stanozol   Oxymetholone

Research efforts to properly balance anabolic and androgenic effects in synthetic anabolic steroids have not been completely successful. Unfortunately, this serious drawback has not precluded the widespread use of these drugs. A vast new market has sprung up on the world's playing fields and athletic training facilities, even though the drugs can be obtained legally only by prescription. Because anabolic steroids increase muscle mass, they appeal to some athletes who compete in strength-related sports such as football, weight lifting, and certain track and field events. The desire to win has led athletes to take anabolic steroids in order to gain a purported competitive edge. And the problem seems to be pervasive. It has been estimated that half of recent Olympic athletes, in sports ranging from weight lifting to figure skating, have used steroids at some time in their careers. Annual sales of illegal steroids to athletes are estimated to be in excess of 200 million dollars—in spite of the fact that legitimate experts in exercise physiology and related fields hold opposing opinions about the merits of steroid use to enhance athletic performance.

On the other hand, it is well established that using large doses of anabolic steroids over time can cause a variety of undesirable side effects. In males, these include shrinking testes, difficulty in urination, impotence, fluid retention, baldness, high blood pressure, and heart attack. Women suffer from masculinization in which female secondary characteristics are lost. Both sexes show increased aggressiveness, unpredictable periods of violent mood changes, and other behavior disorders. Heavy users often compound the problem of drug abuse by using more than one anabolic steroid at a time—sometimes in untested combinations. Many abusers seem to be convinced that "more is better." For example, steroids that are prescribed in legitimate therapeutic doses of a few milligrams have been taken in doses twenty times larger, in spite of the fact that such a large overdose may be lethal. Lyle Alzado, the NFL football star who died in May 1992, attributed his brain cancer to consuming $20,000–30,000 worth of steroids per year. Although Alzado championed a national campaign against steroid abuse, physicians have concluded that there is no evidence linking his use of these drugs and his cancer. Even without this connection, the physiological effects of anabolic steroid abuse are bad enough.

One of the most challenging competitions in sports is that between athletes who use performance-enhancing illegal drugs and the chemists who test for them. Very sophisticated chemical separation and analytical techniques have been developed to detect banned substances in blood and urine samples at the level of parts per billion. Detection of synthetic steroids is difficult because they are generally used in relatively small amounts. But the real detection problem arises because they are chemically very similar to compounds that occur normally in the body. The success of synthetic chemists now becomes the analytical chemists' burden.

Athletes who use illegal steroids have resorted to many strategies in order to avoid detection. These range from simple substitution of a "clean" urine sample for their own, to the rather extreme practice of draining their bladder and then using a catheter to fill the bladder with a sample of "pure" urine just before the drug test. Methods for steroid testing must take into consideration the fact that the fat-soluble steroids take some time to completely clear the body. Because of this time lag, elaborate schemes have been developed for tapering off illegal drugs in order to get below allowable limits just before competitions. Not surprisingly, some coaches and athletes with very little previous curiosity of medicinal chemistry now make it a significant area of interest—at least the part that applies to steroids.

11.11 | *Consider This*

In a recent survey of world-class athletes, 50% said they would take a drug that would enable them to win an Olympic gold medal, even though it would probably cause their death within ten years. Assume you are the editor of a sports magazine for teenage readers and write an editorial on this subject.

11.12 | *Consider This*

Before the unification of Germany, trainers in the East German Olympic program administered carefully monitored doses of anabolic steroids to athletes. It appears that under such controlled conditions, many of these drugs can be used quite safely. Should such a program be permitted and/or established for all Olympic athletes? As an Olympic judge, make a case either supporting or opposing a carefully monitored steroid supplement program.

■ *The Thalidomide Story*

The type of drug testing mentioned in the previous paragraph refers to determining the presence or absence of an illegal drug in a biological fluid and, if it is present, measuring its concentration. Another very different and far broader testing program is required of all new drugs. In accordance with the prevailing laws and regulations, drugs are subjected to an intensive, extensive, and expensive screening process before they can be approved for sale and public use. Ultimately, the question to be answered is, "Is the drug safe to use?" Considering the variability within target populations and the need to minimize unwanted side effects, it is somewhat remarkable that any drug ever receives approval. However, if an error must be made, it seems preferable that it be made on the side of rejecting rather than approving a drug that does not fully meet requirements after a thorough program of testing. Such was not the case for the drug thalidomide.

In 1956, thalidomide was put on the European market without sufficient screening because of an erroneous conclusion reached on the basis of incomplete testing data. The results were tragic. Discovered and developed by a small German drug company, Chemie Grunenthal, thalidomide was a by-product of research aimed at developing new antibiotics. A medicinal chemist at Chemie Grunenthal recognized thalidomide as an analog of a drug that had recently been put into use as a sedative. Testing of thalidomide on four species—mice, rats, guinea pigs, and rabbits—led to the conclusion that the drug was a remarkably safe sedative. Unfortunately, testing was not done to determine if thalidomide was a **teratogen,** that is, whether it could damage a developing embryo sufficiently to cause birth defects. Assuming the new drug to be safe, Chemie Grunenthal approached several companies who were interested in gaining a greater share of the sedative market. At least one United States drug company rejected the compound as worthless and possibly unsafe after carrying out its own testing of the drug. However, several pharmaceutical firms accepted Chemie Grunenthal's offer and marketed thalidomide in different parts of the world.

Soon after thalidomide was introduced, several physicians reported cases of nerve damage in patients who had taken the drug, but these reports were largely ignored. Five years later, a German pediatrician reported a large increase in the number of infants suffering from phocomelia. In this condition, the development of the bones in the arms and legs is severely arrested, producing flipper-like limbs, badly deformed limbs, or no limbs at all. Because phocomelia is one of the rarest birth defects known, the sudden increase in its occurrence caused alarm. Subsequently, the outbreak of phocomelia was traced to thalidomide taken during the first three months of pregnancy by mothers who used the drug to control morning sickness and nausea. All together, some 9000 deformed infants, known as "thalidomide babies," were born worldwide.

News of the thalidomide debacle brought expressions of grief and outrage from people around the world. Investigations indicated that greed plus inept, inadequate, and fraudulent testing were responsible for the disaster. Exceptional bad luck was also involved in that humans are more sensitive to the drug's teratogenic effects than are most test animals.

Thalidomide was not sold in the United States because Dr. Frances Kelsey, a pharmacologist with the Food and Drug Administration, had been unconvinced by the limited safety data supplied by the manufacturers. Thus, because of scientific integrity and the insistence on good testing procedures, relatively few thalidomide babies were born in the United States. Although the thalidomide story is probably the darkest chapter in the history of drug design, it was responsible for the establishment of greatly improved drug testing procedures throughout the world. Teratogenicity testing is now a standard part of new drug approval procedures in most western countries. The FDA testing procedures, considered slow and overly methodical by some, have been widely adopted.

Figure 11.16

Schematic of the drug approval process in the United States.

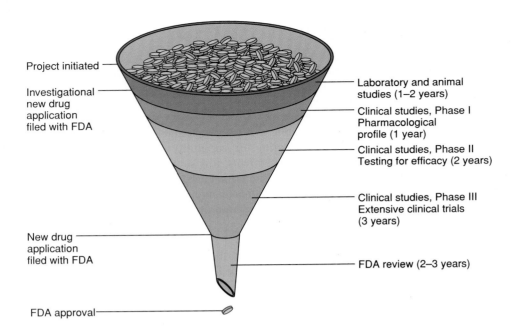

Project initiated

Investigational new drug application filed with FDA

New drug application filed with FDA

FDA approval

Laboratory and animal studies (1–2 years)

Clinical studies, Phase I Pharmacological profile (1 year)

Clinical studies, Phase II Testing for efficacy (2 years)

Clinical studies, Phase III Extensive clinical trials (3 years)

FDA review (2–3 years)

■ Drug Testing and Approval

All proposed new drugs, be they extracted from natural materials or synthesized in the laboratory, are subjected to exacting series of tests before they obtain FDA approval. These steps are summarized in Figure 11.16.

From discovery to approval, the development of a new drug takes, on average, nearly twelve years and more than 200 million dollars—over twice the cost of a decade ago. The expenses are principally for the various stages of drug testing, probably the most complicated and thorough pre-marketing process ever developed for any product. Although the number of pills getting through the funnel of Figure 11.16 gets progressively smaller with time, the diagram does not begin to convey the high mortality rate of proposed drugs. Currently, the odds of getting a candidate drug from identification to approval are 1 in 10,000. For every 10,000 trial compounds that begin the process, 20 make it to the level of animal studies, half that many get clearance for use in clinical testing with humans, and finally only one gets FDA approval.

Examples already encountered in this chapter have suggested the long process of chemical hide-and-seek that often precedes the identification of a compound as possibly having therapeutic properties. Once the promising candidates have been identified, they are subjected to *in vitro* studies. Simultaneously, a wide range of activity is undertaken by the pharmaceutical company. Chemists and chemical engineers investigate whether the compound can be produced in high volume with consistent quality control. Pharmacists carry out studies of the most effective way to formulate the drug for administration—as capsules, pills, injection, syrup, or perhaps something more unusual such as a nasal spray, skin patch, or implant. Stability and shelf-life are investigated. Economists, accountants, patent attorneys, and market analysts conduct research on the likelihood of deriving a profit from the product. A fair, responsible price must be established that allows the corporation to recapture the extensive development costs while keeping the drug affordable.

Only a small fraction of compounds survive this scrutiny to move on to animal testing. These *in vivo* tests are designed to determine the drug's efficacy, safety, dosage, and side effects. It is typically at this stage that pharmacologists determine the drug's mode of action, its metabolic fate in the test animals, and its rate of absorption and excretion. The tests are carefully controlled, requiring the collection of very specific kinds of data. For example, drugs are evaluated for their short- and long-term effects on particular organs (such as the liver or kidneys) and on more general systems (such as the nervous or reproductive system). Perhaps the most controversial toxicity testing involves the determination of LD_{50}, the lethal dose for 50% of the test animals.

11.13 *Consider This*

◼ Animal rights groups often target the LD_{50} requirement as an example of callous indifference to animal welfare. Others argue that such tests are essential to ensure drug safety and effectiveness. Take one of these positions and defend it in a statement that could be used in a public information campaign.

Results of animal tests must be submitted to the FDA for evaluation before permission is granted to proceed to the next stage—clinical testing of the drug on humans. In addition, approval must be obtained from local agencies and authorities such as a hospital's ethics panel or medical board. Typically, clinical studies involve the three phases identified in Figure 11.16: developing a pharmacological profile, testing the efficacy of the drug, and carrying out the actual clinical tests. The entire process often requires six years or more.

The clinical trials may involve only a few patients or several hundred. A large pool will more likely include a wide range of subjects. This is desirable because the drug in question may have markedly different effects on the young and the old; men and women; pregnant or lactating women; infants, nursing infants, and unborn infants; and persons suffering from diabetes, poor circulation, kidney problems, high blood pressure, heart conditions, and a host of other disabilities. Typically, double blind studies are carried out in order to obtain unbiased results. In this methodology, neither the patient nor the physician knows which patients are receiving the drug and which are receiving a placebo, an inactive imitation made to look like the "real thing."

11.14 *Consider This*

◼ A friend of yours is dying from AIDS. He is in a hospital where a double blind study of a new anti-AIDS drug is being tested. By the nature of the test, he may or may not get the drug. Draft a letter to the director of the hospital in which you express your opinion on the research methodology involved.

Once clinical trials have been completed successfully—typically by only 10 drugs out of an original pool of 10,000 compounds—the test data are submitted to the FDA. Upon review, the Agency may require the repetition of experiments or the inclusion of new ones, thus adding years to the approval process. Of the drugs submitted to clinical testing, only about one in ten is finally approved. Once approval is granted, the drug can be sold in the United States. Nevertheless, it still remains under scrutiny, monitored through reports from physicians. Drugs are removed from the market if serious problems occur. Some side effects show up only when large numbers of users are involved. For example, benoxaprofin, an anti-arthritic drug, was withdrawn from the market because of severe side effects that occurred with an incidence of 1 in 8400 patients (0.012%). It is estimated that to ensure detection of side effects at this level of incidence, nearly 30,000 people would have had to receive the drug.

11.15 **Consider This**

Streptokinase, a medication that dramatically increases the likelihood of surviving a heart attack, costs $700 per dose. Who should pay for this life-prolonging drug—individuals, medical insurance, pharmaceutical companies, the government? Take a stand and defend your position.

11.16 **Consider This**

Deciding to take a specific drug involves a personal risk/benefit analysis. Describe the toxicity level you would accept for the benefit of curing the following: chicken pox, polio, typhoid, inoperable cancer.

■ Conclusion

The molecular manipulations of chemists have created a vast new pharmacopeia of wonder drugs that have significantly increased the number and quality of our days. Today, the great majority of bacterial infections are quite easily controlled. Once-dreaded killers such as typhoid, cholera, tuberculosis, and pneumonia have been largely eliminated—at least in wealthy, industrialized societies. But after reading this chapter, you should be well aware that no drug can be completely safe and that almost any drug can be misused. Taking a medication is making a conscious choice between the benefits derived from the drug and the risks associated with its side effects and limits of safety. Because most drugs have very wide, carefully established margins of safety, their benefits far outweigh the risks. For some drugs, however, the trade-off between effectiveness and safety involves a different balance. A drug with severe side effects may be the only treatment available for a life-threatening disease. Someone suffering from AIDS or advanced, inoperable cancer will understandably have a different perspective on risks and benefits than a person with a severe cold. When there is nothing to lose, one is willing to take great chances, including an imperfectly tested drug. And the impersonal anonymity of averages take on new meaning at the bedside of a loved one. When chemistry is applied to medicine, science must be guided by morality, and reason must be tempered with compassion.

■ References and Resources

Cowart, V. S. "Dietary Supplements: Alternatives to Anabolic Steroids?" *The Physician and Sportsmedicine* **20,** March 1992: 189–98.

Griffith, H. W. *The Complete Guide to Prescription and Non-Prescription Drugs.* Tucson: HP Books, 1983.

Hanson, D. "Pharmaceutical Industry Optimistic About Improvements at FDA." *Chemical & Engineering News,* Jan. 27, 1992: 28–29.

Sneader, W. *Drug Discovery: The Evolution of Modern Medicines.* New York: Wiley, 1985.

Stinson, S. C. "Chiral Drugs." *Chemical & Engineering News,* Sept. 28, 1992: 46–79.

Weissmann, G. "Aspirin." *Scientific American,* Jan. 1991: 84–90.

■ Experiments and Investigations

14. Properties of Analgesic Drugs

15. Synthesis of Aspirin

■ *Exercises*

1. The recommended dose of the miracle drug described early in the chapter is 0.650 grams (2 tablets), whereas the lethal dose is 20–30 grams. Calculate the number of tablets that would have to be taken at one time to reach the estimated lethal limit.

2. Draw Lewis structures and determine the number and type of bonds (single, double, triple) used by each carbon in the following molecules.

   a. H_3CCN (acetonitrile)

   b. $H_2NC(O)NH_2$ (urea)

   c. C_6H_5COOH (benzoic acid)

   (See Chapter 2 if you have forgotten how to write Lewis structures.)

3. Write condensed structural formulas for the three isomers of pentane you drew in 11.4 Your Turn.

*4. Before the structure of benzene (shown in Figure 11.3) was determined, there was a great deal of controversy about how the atoms in a compound with this formula could be arranged. Draw three possible isomers.

*5. Using equation 11.2, calculate the minimum number of grams of salicylic acid needed to yield the aspirin in 100 aspirin tablets, each containing 0.325 g of the drug.

6. If aspirin is a specific chemical compound, what, if anything, justifies claims for the superiority of one brand of aspirin tablets over another, and differences in price?

7. Consider the formulas given for the analgesics in Figure 11.5. Identify the portions of each molecule you would expect to promote solubility in polar solvents and in nonpolar solvents.

*8. From the differences in the molecular structures of the analgesics in Figure 11.5 and the differences in their physiological activity, speculate on the possible relationships between their modes of activity and the presence of specific functional groups.

9. From the molecular structure of acetaminophen given in Figure 11.5, propose a possible equation representing its synthesis. (Hint: See the synthesis of nylon in Chapter 10.)

10. Sulfanilamide is the simplest of the class of antibiotics known as sulfa drugs. It appears to act against bacteria by replacing para-aminobenzoic acid, an essential nutrient for bacteria. Account for the substitution of the drug for the nutrient on the basis of their molecular structures.

Sulfanilamide

Para-aminobenzoic acid

11. Which of the compounds below can exist in optically active isomers?

   a.

   b.

   c.

12. As mentioned in the text, testosterone and estrone were first isolated from animal tissue. One ton of bulls' testicles were needed to obtain 5 mg of testosterone and four tons of pig ovaries were processed to yield 12 mg of estrone. Assuming complete isolation of the hormones was achieved, calculate the mass percentage of the two steroids in the original tissue. Explain why the calculated result is very likely incorrect.

13. In 1973 diethylstilbestrol (DES) was approved to induce abortions. Identify the similarities and differences between the structure of the DES molecule, shown below, and estradiol.

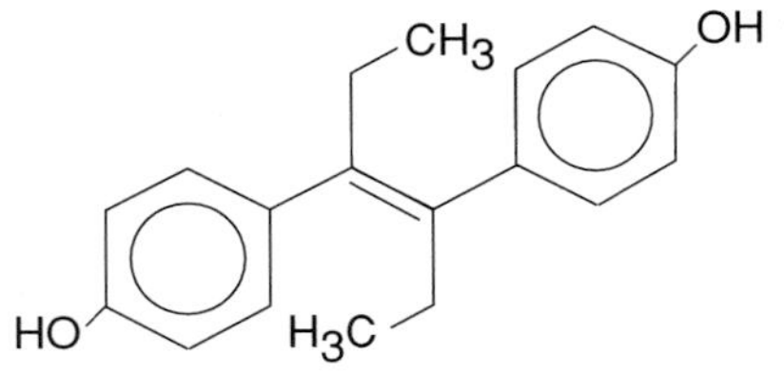

14. A new synthesis of ibuprofen has been reported (*Chemical & Engineering News*, Feb. 8, 1993) that involves three steps rather than the six steps used previously. If each step in these two syntheses gives a 75% yield, compare overall percentage yield of ibuprofen from each of the two processes.

15. A new development in contraception, called Norplant®, involves surgically implanting six matchstick-size capsules in a woman's arm. These capsules slowly release a low dosage of levonorgestrel, a synthetic hormone, that can prevent pregnancy for up to five years. Make lists of the advantages and disadvantages of such a contraception option.

16. Identify the molecular similarities and differences between testosterone and the two anabolic steroids depicted in Figure 11.15.

17. Estrogens, along with other drugs, are often given to men suffering from testicular cancer. On the basis of your knowledge of the effects of estrogen, speculate on why this therapy is effective.

18. Why might the teratogenic effects of thalidomide not have shown up in the test animals used even if they had been sought? What implications does this have for drug testing in general?

19. Potential treatments for fatal diseases such as AIDS and cancer have often been denied to people suffering from these diseases if they are not part of controlled studies. Discuss the possible reasons for such action and the pros and cons of such regulations.

20. The drug-testing laws in some other nations are not as stringent as those in the United States. As a result, some American citizens purchase drugs that are illegally imported from countries where they are approved for use. Under what circumstances, if any, would you resort to this practice?

12

Nutrition: Food for Thought

"I'm hungry." (English); *"Tengo hambre."* (Spanish); *"Ana joann."* (Arabic);
"Ich habe hunger." (German); *"Mimi taka kula."* (Swahili);
"Bae gob pub ni da." (Korean); *"Muje bhookh lagi he."* (Hindi);
"J'ai faim." (French); *"Aniguwaan gookoonayaa."* (Somali)

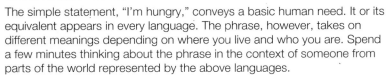

12.1   Consider This

The simple statement, "I'm hungry," conveys a basic human need. It or its equivalent appears in every language. The phrase, however, takes on different meanings depending on where you live and who you are. Spend a few minutes thinking about the phrase in the context of someone from parts of the world represented by the above languages.

a. Describe what *you* mean when you say "I'm hungry" and what you expect in order to satisfy that need.

b. Next, put yourself in the place of someone from France, India, Korea, and Somalia. Describe what you think that person means by saying "I'm hungry" and how he or she expects to satisfy it.

The exercise you have just completed very likely generated a range of responses about the meaning of the phrase, "I'm hungry." The responses vary by gender, age, and locale. They are obviously very individualistic. What could be more personal than taste for food? But for millions, "I'm hungry" is not a matter of taste; it is a matter of survival. Thus, what at first seems individual and personal has great global consequences. This is the message conveyed by the two quotations that follow.

> No contemporary problem compares in gravity to the human devastation caused by persisting hunger and malnutrition.[1]

> Americans are zany about food and diet. No other country gorges itself on junk food the way we do, and no other country has as many "experts" on health diets. We have become more concerned about what we should not eat than what we should.[2]

In many ways, these two statements summarize the fundamental issue of world nutrition. Above all, it is nutrition that differentiates between the "haves" and the "have nots." "I am hungry" has very different meanings in a developing, cash-poor country and a developed nation that has sufficient financial means to feed its citizens. On one hand is the grim, persistent specter of massive hunger and starvation. In stark contrast is the overabundance of food and the wasteful (as well as "waistfull") eating habits of many Americans and those in other developed countries. This chapter is devoted to examining this central issue and the wide range of problems associated with nutrition and the food we eat.

■ Chapter Overview

In this chapter we approach nutrition from both personal and global perspectives. We begin at the latter level, but soon shift to an examination of nutritional information on cereal boxes. This prompts a simple-minded question: "Why eat?" The answer leads to a general consideration of the three main classes of food compounds—carbohydrates, fats, and proteins. We identify their sources and consider

[1.] *1990 State of Food and Agriculture.* Food and Agricultural Organization of the United Nations (FAO).

[2.] Dr. C. Everett Koop (former Surgeon General of the United States). *The Memoirs of America's Family Doctor.* New York: Random House, 1991, pp. 293–94.

the current dietary recommendations for the three groups, then examine each of these macronutrients at some depth. We devote particular attention to molecular structure and its relationship to properties and functions. In the process, you will learn about specific topics such as lactose intolerance, saturated and unsaturated fats and oils, cholesterol and heart disease, essential amino acids, and phenylketonuria.

Food is, of course, the source of the energy that powers our bodies and our brains. Therefore, we consider the caloric content of various foods, the recommended food energy intake for men and women of various ages and weights, and the energy expenditures associated with a number of activities. Data cited in the text indicate that the energy supplies available per person per day vary widely across the planet. But calories are not enough to assure a balanced diet. That also requires the correct concentrations of a wide array of vitamins and minerals. Therefore, we include a section of the role of a number of these micronutrients and the hazards of an insufficient supply. Methods of food preservation, especially by gamma irradiation, brings the focus back to broader issues, and the chapter ends with an overview of the problems associated with efforts to feed our fellow human beings all over the globe.

The chapter is long and contains a wide range of information, but our goals are straightforward: (1) To help you see the connection between chemistry and nutrition by applying some chemical principles in the context of the composition and reactions of foodstuffs; (2) To provide information that you can use to make daily choices about personal nutrition and health in terms of the foods you eat; (3) To analyze a number of nutrition-related controversies, some of which have appeared in the popular press; and (4) To follow the thread that connects individual concerns about personal nutrition to the nutritional needs of a hungry planet.

■ *Famine and Feast*

In late 1992, the attention of the world was riveted on the plight of Somalia. The very name brings haunting images of famine, human suffering, and death on a massive scale: 350,000 dead during two years and nearly two million at the brink of starvation. Unfortunately, Somalia is not alone in this regard. During the past decade, Ethiopia and Sudan have also vividly exemplified the fact that for many developing nations, hunger, undernourishment, malnutrition, and starvation are never far away. Nor are these conditions restricted to sub-Saharan Africa; other regions of the world are also chronically plagued by them. There are an estimated half billion hungry people in the world, and four million with chronic malnutrition. For example, it is estimated that 40% of Brazil's population is malnourished and as much as 80% of rural Mexicans suffer from malnutrition. The meaning of this commonly used term is important. **Malnutrition** is caused by a diet lacking in the proper mix of nutrients, even though enough calories are eaten daily. It contrasts with **undernourishment.** People who are undernourished do not ingest a sufficient number of calories daily, regardless of what they eat. According to the Food and Agricultural Organization of the United Nations (FAO), there are 98 developing countries whose populations are seriously undernourished. The diagram in Figure 12.1 dramatically illustrates the persistent undernourishment found in developing regions of the world over nearly two decades.

The percentages cited in Figure 12.1 are the percentage of the population whose daily food supply is inadequate to meet basic nutritional requirements. Translating these sterile statistics into terms of human misery means that nearly one in three people (32%) in Africa is undernourished, one in five (22%) in the Far East, and nearly one in seven (14%) in Latin America. From such data, the 1990 FAO report draws the bleak conclusion quoted above. That strong statement calls attention to the magnitude and tenacity of hunger in the world.

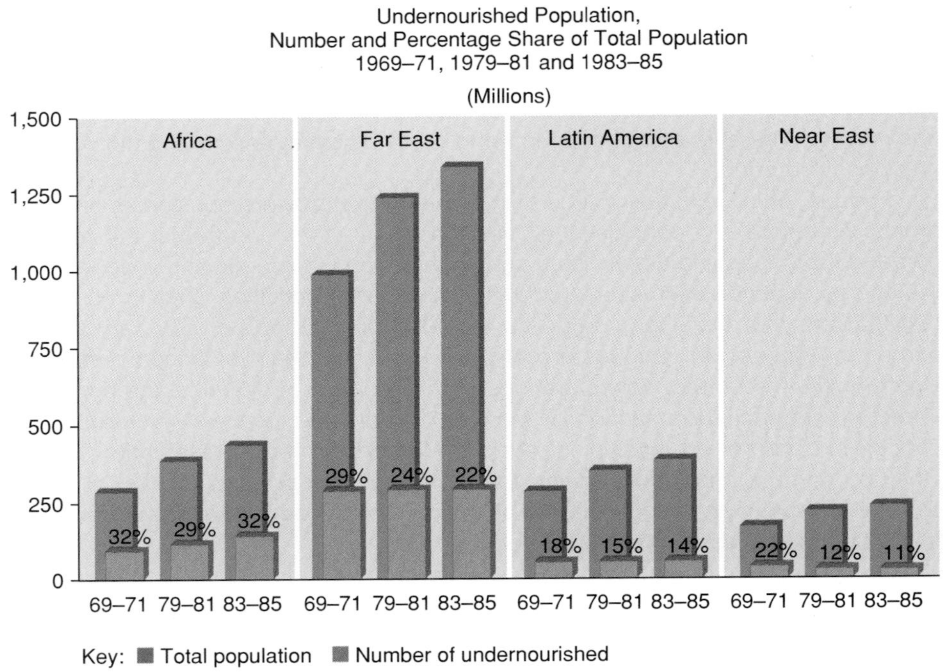

Figure 12.1
Percentage of undernourished populations. (Source: Data from *The State of Food and Agriculture: 1990*. Rome: Food and Agriculture Organization of the United Nations, 1991.)

Undernourished Population,
Number and Percentage Share of Total Population
1969–71, 1979–81 and 1983–85
(Millions)

Key: ■ Total population ■ Number of undernourished

Contrast this grave accounting with the dietary circumstances of most people in the United States. Although there are malnourished and undernourished people in this country, hunger is typically not a consideration in a nation where 27% of the population is up to 40% over their ideal weights, each consuming nearly 1000 excess Calories per day. One result of such excess is that about 50% of adult women and 25% of men annually attempt to lose weight by dieting, generally with very limited long-term success. No wonder that the former Surgeon General of the United States has expressed his concern.

12.2 ■ *Consider This*

In late 1992, American troops were sent to Somalia to help maintain peace and assure that food supplies would be distributed to the starving population. How were the vital economic or security interests of the United States involved in Somalia? What was the justification of this military intervention? Read several articles and editorials on the subject and report on the positions taken.

■ *Dietary Research at the Breakfast Table*

To put nutrition into a personal perspective, you might well ask "What do I eat?" and "What should I eat?" Suppose you decide to do a little research at the breakfast table. You begin by reading the nutritional information listed on the packages of cereals you have on hand, paying particular attention to the product you are eating. Information of this sort is reproduced in 12.3 Consider This.

12.3   *Consider This*

■

Listed here is nutritional information from four popular cereals, disguised by pseudonyms:

Cereal Brands: (1 oz. serving)	"Ko-ko Krunchies"	"Vita-Max"	"Health Nuts"	"Oaties"
Calories	110	100	107	110
Protein, g	1	3	3	4
Carbohydrates, g	25	22	22	20
Fats, total, g	1	1	2	2
Fats, unsaturated, g	—	—	2	—
Fats, saturated, g	—	—	0	—
Cholesterol, g	0	0	0	0
Sodium, mg	180	200	190	290
Potassium, mg	55	105	83	105
Percent of U.S. Recommended Daily Requirements (RDA)				
Protein	<2	4	4	6
Vitamin A	<2	100	10	25
Vitamin C	25	100	0	25
Thiamin	25	100	2	25
Riboflavin	25	100	2	25
Niacin	25	100	2	25
Calcium	4	20	0	4
Iron	25	100	7	45
Vitamin D	<2	10	10	10
Vitamin B_6	25	100	20	25
Folic Acid	25	100	20	25
Phosphorus	4	20	7	10
Magnesium	<2	8	5	10
Zinc	<2	100	20	6
Copper	<2	6	3	6

Note: *The Calories listed in this table are kilocalories.* (See Chapter 4.)

By comparing information from various brands, you can form opinions about the claims made by the cereal manufacturers. Consider the following.

a. Notice that both the nutritional information and the percent U.S. RDA are given for a 1-oz serving. Estimate how many ounces constitute a normal serving for you? For comparison, single-serving boxes contain 0.75 ounce of cereal.

b. Based on your normal serving size, how many Calories-worth would you consume of an "adult" cereal such as "Vita-Max" compared to a "kids" cereal like "Ko-ko Krunchies"? Remember this does not include the Calories you get from milk used with the cereal.

c. Eating too much sodium can result in water retention and high blood pressure (hypertension). If you must avoid excess sodium for health reasons, which of these four cereals should you not eat?

d. "Health Nuts" is usually marketed as being the choice of a health-conscious population. No such claims are made for "Ko-ko Krunchies". Compare the percent U.S. RDA for these two cereals. Make a case either supporting or refuting the health claims for "Health Nuts".

e. Now that you have considered all the cereal data, which of these cereals would you choose to eat and why?

If you actually completed this exercise at the breakfast table, your own cereal probably wound up as a rather soggy lump in the bottom of the bowl. But, you learned quite a bit about cereals and about reading labels. Perhaps you have also been convinced of another of Dr. Koop's observations in his 1988 Surgeon General's report: "Your choice of diet can influence your long-term health prospects more than any other action you might take." If indeed "you are what you eat," your diet demands even more attention.

■ *Why Eat?*

Our ancient ancestors had a rather unattractive diet, restricted to what they got from foraging and hunting. However, anthropological research has suggested that it was a reasonably balanced diet. The development of agriculture, starting about 10,000 years ago, provided a more stable supply of certain foods. Ironically, this sometimes led to unbalanced diets, because various cultures came to depend heavily on locally-produced foods, some of which were lacking in certain nutritional components. Now, the advances in agriculture coupled with modern food distribution and processing methods have significantly extended the range of our food choices, especially in developed nations, but not necessarily in healthful ways.

Depending on circumstances, the reasons for eating range from satisfying a basic need to celebrating a memorable social occasion. But the real reason we eat is because food provides the four fundamental types of materials required to keep our cells functioning. The four types of materials are water, energy sources, raw materials, and metabolic regulators. Water serves as both a reactant and a product in metabolic reactions, and as a coolant and thermal regulator (see Chapter 5). Sources of energy are required to provide the energy needed for processes as diverse as muscle action (especially heart contractions), brain and nerve impulses, and for moving and arranging molecules and ions in suitable ways at appropriate times and places. In fact, most of the energy provided by the food we eat goes to maintain differences in ionic concentrations in spite of the natural tendency for diffusion to occur from regions of higher to lower concentration. Such concentration differences are essential for nerve action and metabolic processes. Raw materials are needed for the syntheses of new bone, blood, enzymes, muscles, hair, and for the replacement and repair of cellular materials. Chemical regulators such as enzymes and hormones control the overall metabolic processes associated with nutrition.

12.4   *The Sceptical Chymist*

During a lifetime, you will eat a truly prodigious amount of food—about 700 times your body weight! Do a calculation to check the validity of this assertion.

Soln: To carry out the calculation, we must first estimate the weight of food an individual eats daily. This estimation presents a real difficulty because of the wide variations in amounts and types of foods eaten. At best, we would have to make a guess that would likely be inaccurate. Another way to approach the problem is to work backwards and see if the estimate of daily food intake is reasonable. Let's assume the individual is a 50 kg (110 lb) female. Therefore, the amount of food consumed in a lifetime is 700 times the body weight, or 700×50 kg $= 35,000$ kg (77,000 lb). Assuming a 75 year life span, the weight of food eaten daily is: 35,000 kg/75 yr $\times$ 1 yr/365 days $= 1.3$ kg/day. This value (about 2.8 pounds per day) may seem somewhat high, but most foods have a high water content, frequently greater than 50%.

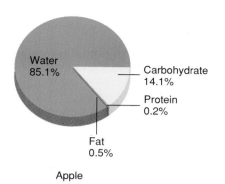

Apple

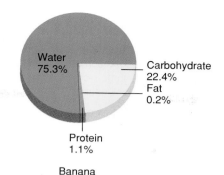

Banana

Figure 12.2

Percentage of water, carbohydrates, proteins, and fats in several foods. (From "Postharvest Physiology and Biochemistry of Fruits and Vegetables," by N. Haard, *Journal of Chemical Education,* 1984, 61, 277–83. Copyright © 1984 Division of Chemical Education, American Chemical Society. Reprinted by permission.)

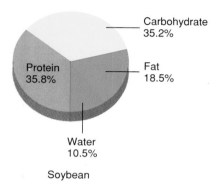

Soybean

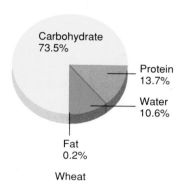

Wheat

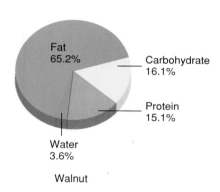

Walnut

It goes without saying that eating does not in itself guarantee good health or even good nutrition. It is possible to consume food regularly, even to the point of being overweight, and still be malnourished. That breakfast cereal will again provide evidence. It is obvious from the label (and the table) that one bowlful of cereal is not nearly enough to supply the body's daily energy needs. Nor does it meet the recommended daily dietary allowances (RDAs) of vitamins and minerals. To compensate for such shortfalls, you should eat foods during the remainder of the day that provide appropriate nourishment to fill in the remaining gaps of nutrients. Nutrients are substances of major dietary consequence categorized into three main types: (1) water; (2) macronutrients—fats, carbohydrates, and proteins; and (3) micronutrients—vitamins and minerals. To be sure, water is not a source of energy; it cannot be "burned" in the body or elsewhere. However, water is essential for biochemical processes. Our bodies are about 60% H_2O. Similarly, micronutrients do not provide a significant number of calories, but they too are necessary for life.

Although water content is not typically listed on food labels, information about macronutrients and micronutrients is included. Figure 12.2 indicates the percentage of water and the three macronutrients in a variety of foods. Notice their considerable

12.5 | Your Turn

One of the authors of this textbook once taught a student who subsisted for one semester on nothing but the cereal we have called "Oaties." Assume his daily food energy intake was 2500 Cal. Consult the data in 12.3 Consider This and answer the following questions.

a. How many one-ounce servings of "Oaties" did the student need to eat each day to meet his energy requirements?
b. How many grams (or milligrams) of the various food components listed in the table did the student obtain each day from his cereal diet?
c. What percentage of the RDA values for the various vitamins and minerals did the "Oaties" yield?
d. Provide a critique of this rather peculiar diet, identifying its strengths and weaknesses.

variation among the different foods. Carbohydrates, fats, and proteins are listed because of their special importance in nutrition. Indeed, they are so important that a good deal of this chapter will be devoted to these fundamental food stuffs.

■ The Macronutrients: Carbohydrates, Fats, and Proteins

Figure 12.3 illustrates the composition of the human body. From the data given we can calculate that if you weigh 150 pounds, water represents 90 pounds of your body weight (150 lb × 60 lb water/100 lb body), fat accounts for 30 pounds, and the remaining 30 pounds consist of various carbohydrates and proteins, along with calcium and phosphorus in the bones. Surprisingly, the other minerals and the vitamins make up less than one pound. This indicates that a little bit of each of them goes an awfully

Figure 12.3
Composition of the human body.

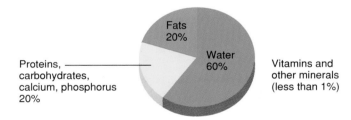

long way! Perhaps even more surprising is that of the nearly ninety naturally-occurring elements, only eleven of them make up over 99% of the mass of your body! These elements and their mass percentages (in descending order) and their relative abundances are given in Table 12.1.

The eleven elements of Table 12.1, highlighted in the skeleton periodic table, are vital to our well-being. But why is it that the most abundant elements—oxygen, carbon, hydrogen, and nitrogen—have such preeminence in human health? Oxygen and hydrogen are, of course, the elements of water. Moreover, along with carbon they constitute all carbohydrates and all fats. Finally, these three elements plus nitrogen are found in all proteins. Thus, nature uses very simple units, atoms of oxygen, carbon, hydrogen, and nitrogen, in a myriad of elegantly functional combinations to produce water, carbohydrates, fats, and proteins—all necessary parts of a healthy body and a healthy diet.

■ Table 12.1	Major Elements of the Human Body		
Element	Symbol	Grams/100 g Body Weight	Relative Abundance in Atoms/million Atoms in the Body
Oxygen	O	64.6	255,000
Carbon	C	18.0	94,500
Hydrogen	H	10.0	630,000
Nitrogen	N	3.1	13,500
Calcium	Ca	1.9	3100
Phosphorus	P	1.1	2200
Chlorine	Cl	0.40	570
Potassium	K	0.36	260
Sulfur	S	0.25	490
Sodium	Na	0.11	410
Magnesium	Mg	0.03	130

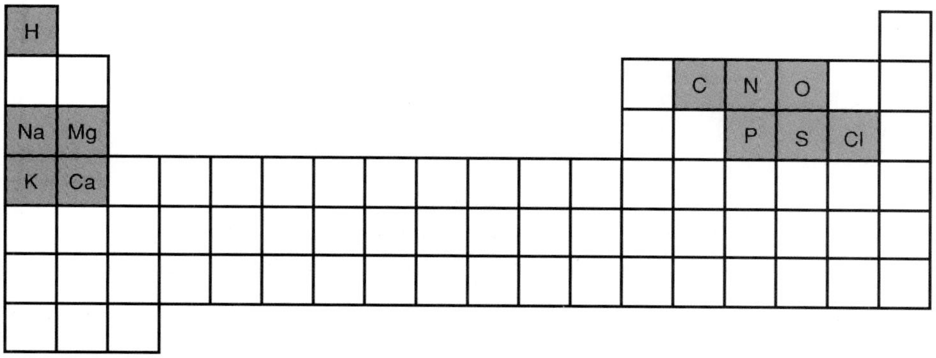

The current recommendations for a healthy diet, approved in 1991, are based on a pyramid relationship among the various foods we should eat. The food pyramid shown in Figure 12.4 incorporates the traditional basic four food groups of the 1958 pie chart, but with much different emphases. In particular, the pyramid calls for eating habits that increase the proportions of foods at the base of the pyramid and decrease those near or at the top. We are now urged to eat relatively more bread, cereal, rice, and pasta (the base of pyramid), fruits, and vegetables. Simultaneously, we are asked to significantly reduce the percentage of fats, oils, and sweets in our daily diet. This approach is a major departure from the 1958 dietary guidelines that placed a heavy dependence on red meats and fats.

While there is some controversy concerning the ideal balance of the macronutrients and the best sources of these compounds, there is no question that a healthy diet requires carbohydrates, fats, and proteins. We therefore turn to a consideration of each

Figure 12.4
The food pyramid and the basic four
food groups.

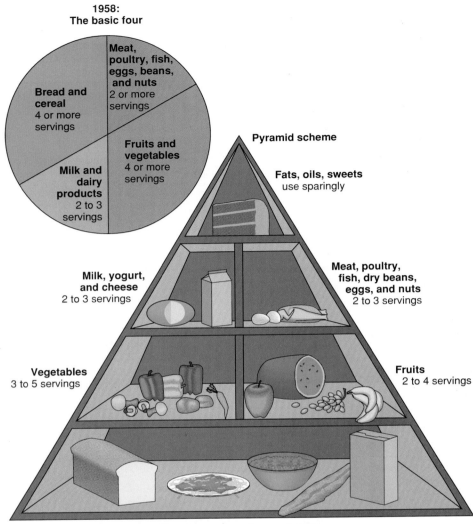

**1958:
The basic four**

Bread and
cereal
4 or more
servings

Meat,
poultry, fish,
eggs, beans,
and nuts
2 or more
servings

Fruits and
vegetables
4 or more
servings

Milk and
dairy
products
2 to 3
servings

Pyramid scheme

Fats, oils, sweets
use sparingly

**Milk, yogurt,
and cheese**
2 to 3 servings

**Meat, poultry,
fish, dry beans,
eggs, and nuts**
2 to 3 servings

Vegetables
3 to 5 servings

Fruits
2 to 4 servings

Bread, cereal, grains, and pasta
6 to 11 servings

12.6

■

Consider This

Politics entered the picture when the United States Department of
Agriculture (USDA) initially released the food pyramid. The meat and dairy
industries pressured the USDA into delaying public dissemination of the
pyramid. Look at the 1958 basic-four pie chart and the new pyramid
scheme. Although both charts list approximately the same number of
servings for meat and dairy products, the serving sizes on the pyramid
scheme are smaller than those on the pie chart. There are nutritionists
who claim that the USDA, which is obligated to both promote and police
agricultural products, has a conflict of interest between this responsibility
and its other obligation to monitor the health of Americans.

Suppose you are an American cattle rancher. How would you
reconcile your concern for your livelihood and your interest in preserving
your health? In other words, would you support or oppose public
dissemination of the food pyramid? Draft a letter to the Secretary of
Agriculture, the chief of the USDA, stating and defending your position on
this issue.

of these macronutrients—their chemical composition, molecular structure, and properties. You will encounter a good deal of chemistry, but most of the principles involved have already been considered in previous chapters.

■ *Carbohydrates—Simple and Complex*

Carbohydrates are compounds containing carbon, hydrogen, and oxygen, the latter two in the same 2:1 atom ratio found in water. Glucose, for example, has the formula $C_6H_{12}O_6$. The name "carbohydrate" (literally "carbon plus water") developed from the early idea that these compounds were "hydrates of carbon." This would imply the formula $C_6(H_2O)_6$ for glucose, with the hydrogen and oxygen atoms bound to each other as in water. It has since been well established that carbohydrates are not hydrates of carbon, but rather compounds with a carbon-chain "backbone" linked to the requisite number of oxygen and hydrogen atoms.

Carbohydrates come in a variety of forms. There are simple ones such as glucose or fructose made up of single molecules (*mono*saccharides-"one sugar"). There are also *di*saccharides ("two sugars"), formed by joining two monosaccharide units. Sucrose (table sugar) and lactose, found in milk, are disaccharides. In a disaccharide, the monosaccharide units are joined by a C–O–C bond created when a water molecule is formed and split out from between them. These structural principles and the shorthand used to represent the molecules are illustrated below. In the shorthand version, only those –OH groups directly involved in the reaction are shown. Remember that a carbon atom is assumed to be present at each corner of the molecular ring.

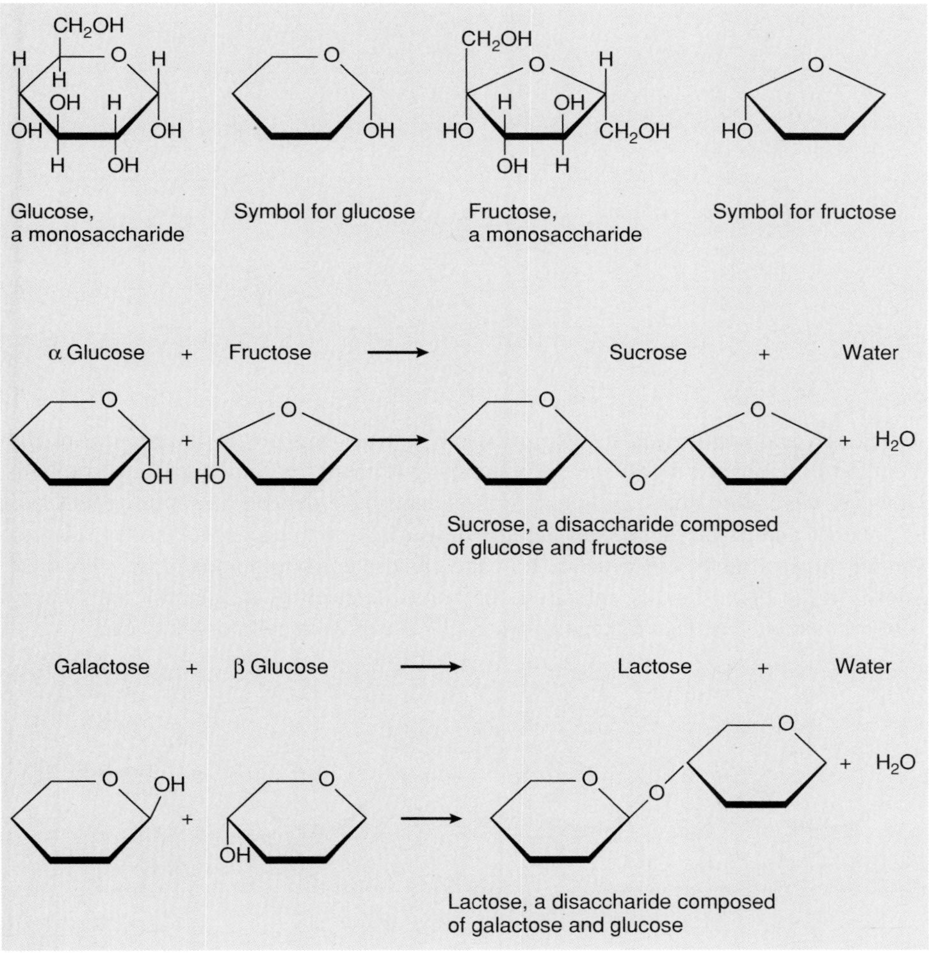

Glucose, a monosaccharide

Symbol for glucose

Fructose, a monosaccharide

Symbol for fructose

α Glucose + Fructose ⟶ Sucrose + Water

Sucrose, a disaccharide composed of glucose and fructose

Galactose + β Glucose ⟶ Lactose + Water

Lactose, a disaccharide composed of galactose and glucose

The linking of small units called monomers to produce a longer, more complex molecule is called polymerization (see Chapter 10). In the case of carbohydrates, the monomers are monosaccharides. If sufficient monosaccharide units are joined, a *poly*saccharide ("many sugars") forms. Polysaccharides are also known as complex carbohydrates. Starch, glycogen, and cellulose are polysaccharides in which the constituent molecules are made of several thousand glucose units linked in long chains.

Starch

A polysaccharide that is similar to starch is glycogen, a compound vital to nutrition because it is how the body stores excess glucose in muscles and the liver. Glycogen is also made exclusively from glucose units, testimony to the body's ability to convert most other monosaccharides and disaccharides to glucose. The molecular structure of glycogen is like that of starch except that the polysaccharide chains are longer and more branched in glycogen.

Glycogen

Starch and cellulose differ significantly in what happens to them nutritionally. You and I are able to digest starch by breaking it down into individual glucose units. However we cannot digest cellulose. Consequently, we depend on starchy foods such as potatoes and pasta, rather than literally devouring toothpicks or textbooks. The reason for this is a subtle difference in how the glucose units are joined in starch and cellulose. In the *alpha* linkage in starch, the bonds connecting the glucose units have a bent orientation. The *beta* linkage between glucose units in cellulose is linear.

α - linkage   β - linkage

Lactose Intolerance

An interesting metabolic anomaly involving carbohydrates is **lactose intolerance**, a condition affecting up to 80% of the world's population. Most Northern Europeans, Scandinavians, and people of that ethnic background do not have the condition. They are lactose tolerant; most others are not. Lactose intolerance is characterized by diarrhea and excess gas after eating milk, cheese, ice cream, or other dairy products. This results from the inability to break down lactose (milk sugar) into its two component monosaccharides, glucose and galactose. It is a condition caused by a lack of or a low concentration of lactase, the specific enzyme that catalyzes this breakdown. In lactose-intolerant individuals the lactose is instead fermented by intestinal bacteria, generating carbon dioxide and hydrogen gases, as well as lactic acid, in this case the principal source of diarrhea.

Given milk's importance for growing bones and teeth, it is significant that infants generally can produce sufficient lactase. But, as we age, this production decreases. By adulthood, most people in the world are lactase deficient and do not have enough of it to accommodate a diet heavy in dairy products. However, most adults have enough lactase to be able to digest modest amounts of dairy products, perhaps the equivalent of a glass of milk per meal, without experiencing difficulties.

Our enzymes, and the enzymes of other mammals, are unable to catalyze the breaking of beta linkages. However, cows, goats, sheep, and other ruminants manage to do it with a little help. Their digestive tracts contain bacteria that are able to break down cellulose into glucose monomers. The animals' own metabolic system then takes over. Similarly, the fact that termites contain cellulose-hungry bacteria means that wooden structures are sometimes at risk.

■ *The Fat Family*

Fats are the second type of macronutrient made exclusively from carbon, hydrogen, and oxygen. To the American public, "fats" are some sort of material that we are supposed to limit in our diets. Fats currently make up about 40% of the calories in the average American diet, even though healthcare specialists recommend it to be 30% or less. Although we may not have a precise definition for "fats," from experience we know some of their properties: a greasy texture, a slippery feeling; and an insolubility in water.

From a chemical and nutritional standpoint, fats are classified as **triglycerides,** compounds made from fatty acids and glycerol, $C_3H_8O_3$. Fats are triglycerides that are solids at room temperature. Triglycerides that are liquid at room temperature are called oils. Whether solid or liquid, triglycerides are a major portion of a broader class of compounds called **lipids.** In addition to fats and oils, lipids also include steroids such as cholesterol (see Chapter 11), and complex compounds like lipoproteins.

Most of the lipids in the body and in foods are triglycerides formed by the following reaction:

$$3 \text{ Fatty acids} + \text{Glycerol} \rightarrow \text{Triglyceride} + 3 \text{ Water} \qquad (12.1)$$

Note that in this reaction, as with polysaccharide formation, smaller units are joined to form a more complex unit (a triglyceride) by splitting out water. The reaction is an example of esterification, a process described in Chapter 10 as leading to the formation of polyesters. In order to see the connection, we need to consider the molecular structures of glycerol and fatty acids.

The molecular structure of glycerol (commonly called glycerine) determines its chemical properties and its dietary function. The figure below indicates that a molecule of this compound includes three –OH groups. An –OH group is an example of a functional group that characterizes a class of compounds. (A number of functional groups are introduced in Chapters 10 and 11.) In an organic compound, an –OH group attached to a carbon atom indicates an alcohol. Thus, glycerol has chemical properties characteristic of alcohols.

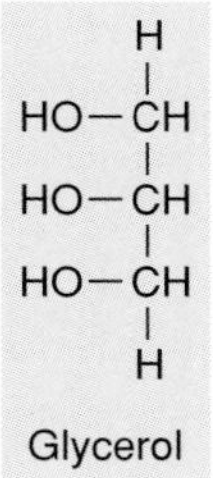

```
              H
              |
      HO — CH
              |
      HO — CH
              |
      HO — CH
              |
              H

      Glycerol
```

Fatty acids are a class of compounds characterized by two parts: a long hydrocarbon chain containing an odd number of carbon atoms (generally 11–23), plus a carboxylic acid group (–COOH) at the end of the long chain. This functional group is what puts the *acid* in fatty acids because it can provide a hydrogen ion (H^+). Stearic acid, $C_{17}H_{35}COOH$, is a fatty acid found in animal fats; its structural formula and condensed molecular formula are given below.

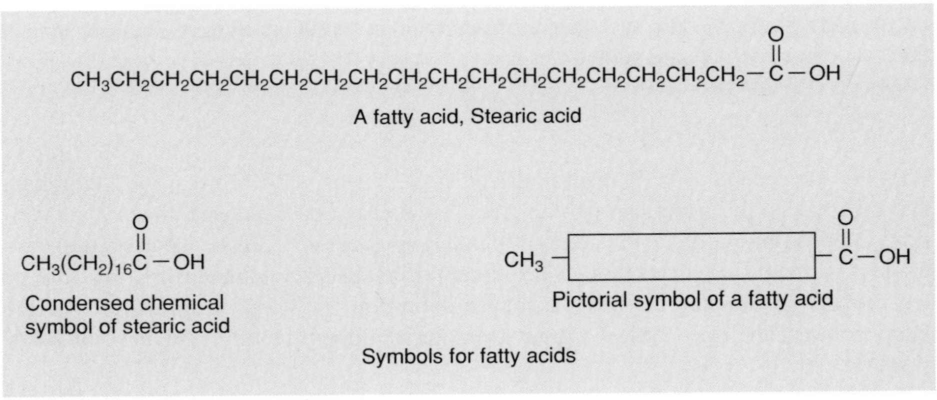

$$CH_3CH_2CH_2CH_2CH_2CH_2CH_2CH_2CH_2CH_2CH_2CH_2CH_2CH_2CH_2CH_2CH_2 - \overset{\displaystyle O}{\overset{\|}{C}} - OH$$

A fatty acid, Stearic acid

$$CH_3(CH_2)_{16}\overset{\displaystyle O}{\overset{\|}{C}} - OH$$

Condensed chemical
symbol of stearic acid

$$CH_3 - \boxed{} - \overset{\displaystyle O}{\overset{\|}{C}} - OH$$

Pictorial symbol of a fatty acid

Symbols for fatty acids

The figure also includes a third way of representing the structure. The box signifies the $(CH_2)_{16}$ chain. We will use variations of this symbolism below.

Triglycerides form via a process in which three fatty acid molecules react with the alcohol groups of glycerol. Because a glycerol molecule has three alcohol groups, each $-OH$ requires a fatty acid molecule to react with it. The reaction of an $-OH$ group and a $-COOH$ group results in the elimination of an H_2O molecule and the formation of an ester linkage. Thus, one glycerol molecule and three fatty acid molecules yield a *tri*ester, commonly called a *tri*glyceride. This is the reaction represented by equation 12.1.

The particular triglyceride formed depends on the fatty acids used. In some cases, only a single fatty acid, for example stearic acid, is involved, and all three hydrocarbon chains are identical. In some fats, three different fatty acids are represented. Such a triglyceride is pictured below. Note that the boxes representing the hydrocarbon chain include what looks like zero, one, or two equal signs (=). These indicate double bonds. Stearic acid contains no double bonds, oleic acid contains one double bond per molecule, and linoleic acid contains two.

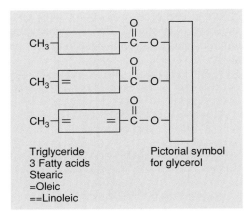

Triglyceride
3 Fatty acids
Stearic
=Oleic
==Linoleic

Pictorial symbol
for glycerol

Now that you know the structural formula of a triglyceride, let's look more closely at its formation.

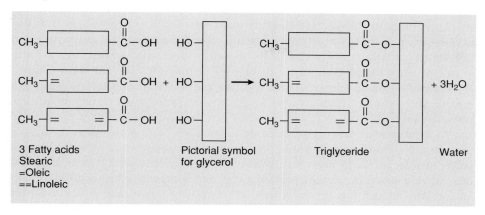

3 Fatty acids
Stearic
=Oleic
==Linoleic

Pictorial symbol for glycerol

Triglyceride

Water

Regardless of which fatty acids are used, the major point is that the overwhelming majority of fatty acids in the body are transported and stored in the form of triglycerides, accounting for almost 95% of dietary fat.

■ *Saturated and Unsaturated Fat*

It may not be obvious to you, but we have already introduced saturated and unsaturated fats. Remember that in the box representing the hydrocarbon chain of fatty acids, we enclosed zero, one, or two equal signs indicating double bonds. If the chain contains only single bonds between the carbon atoms, C—C, and no double bonds, the fatty acid is said to be **saturated**. This is the case with stearic acid. If, however, there is one or more double bond between carbon atoms, C=C, then the fatty acid is **unsaturated**. Oleic acid, with one double bond per molecule, is classified as *mono*unsaturated. Those fatty acids containing more than one C=C double bond per molecule are called **polyunsaturated.** Linoleic acid, which contains two double bonds per molecule, and linolenic acid with three double bonds per molecule are polyunsaturated.

$$CH_3(CH_2)_{16}COOH$$

Stearic acid, Saturated fatty acid

$$CH_3-(CH_2)_7-CH=CH-(CH_2)_7-COOH$$

Oleic acid, Monounsaturated fatty acid

$$CH_3-(CH_2)_4-CH=CH-CH_2-CH=CH-(CH_2)_7-COOH$$

Linoleic acid, Polyunsaturated fatty acid

$$CH_3-CH_2-CH=CH-CH_2-CH=CH-CH_2-CH=CH-(CH_2)_7-COOH$$

Linolenic acid, Polyunsaturated fatty acid

Our bodies can chemically make all of the required fatty acids from the foods we eat, except linoleic and linolenic acids; it cannot make these two. This inability requires that these two acids, called **essential fatty acids,** must be contained in the foods we eat. Fortunately, linoleic and linolenic acids are present in many foods including plant oils, fish, and leafy vegetables. A healthy diet should consist of polyunsaturated fats, certainly linolenic and linoleic acids, as well as some saturated fats. Both are necessary because of the wide range of biological functions carried out by lipids. Obviously fats are an energy source, and they provide insulation for the body. But triglycerides and other lipids are also the primary components of cell membranes and nerve sheaths.

As mentioned earlier, triglycerides are either solids (fats) or liquids (oils) at room temperature. Within a given family, for example the saturated fatty acids, melting points increase as the number of carbon atoms per molecule and the molecular mass increase. Moreover, the melting point of a triglyceride depends on its degree of unsaturation. Unsaturated fats generally have lower melting points than saturated fats of similar molecular weight. Therefore, unsaturated fats are typically liquids at room or body temperatures, whereas saturated fats are solids or semi-solids (see Table 12.2). In general, animal fats are more saturated than vegetable oils with the same number of

■ Table 12.2	Melting Points of Some Fatty Acids	
Name	**Carbon Atoms per Molecule**	**Melting Point, °C**
Saturated fatty acids		
Capric	10	32
Lauric	12	44
Myristic	14	54
Palmitic	16	63
Stearic	18	70
Unsaturated fatty acids		
Palmitoleic (1 double bond/molecule)	16	0
Oleic (1 double bond/molecule)	18	16
Linoleic (2 double bonds/molecule)	18	5
Linolenic (3 double bonds/molecule)	18	−11

carbon atoms, although some differences exist among oils. For example, palm oil is much more saturated than corn oil or sunflower oil. Ironically, coconut oil, used as a cream substitute in non-dairy creamers, is more saturated than the cream it replaces. In fact, coconut oil contains even more unsaturated fats than does lard.

The fat composition in various common foods is noted in Figure 12.2. From Figure 12.5 we can understand why lard (41% saturated fat) is a solid, chicken fat (31% saturated fat) is a bit softer, but still a solid, and safflower oil (78% *un*saturated fat) is a liquid. Nutritionists recommend that saturated fats make up no more than 30% of our total fat intake.

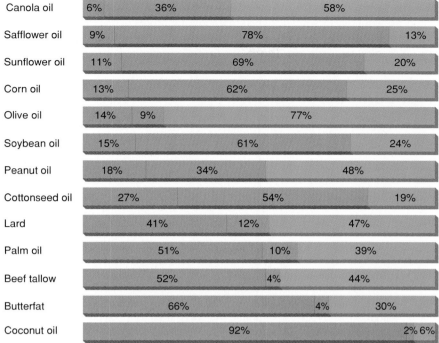

Dietary oil/fat

Canola oil	6% 36%	58%
Safflower oil	9% 78%	13%
Sunflower oil	11% 69%	20%
Corn oil	13% 62%	25%
Olive oil	14% 9% 77%	
Soybean oil	15% 61%	24%
Peanut oil	18% 34%	48%
Cottonseed oil	27% 54%	19%
Lard	41% 12%	47%
Palm oil	51% 10%	39%
Beef tallow	52% 4%	44%
Butterfat	66% 4%	30%
Coconut oil	92%	2% 6%

Key: ■ Saturated fat   ■ Polyunsaturated fat   ■ Monounsaturated fat

Figure 12.5

Saturated and unsaturated fats. (Source: *Food Technology*, April 1989.)

A wide variety of foods contain partially hydrogenated triglycerides. These are oils that have been carefully reacted with hydrogen, H_2. The hydrogen adds to the double bonds, converting some, but not all, of the C=C into C–C bonds.

$$CH_3 = = = C-OH + 2H_2 \longrightarrow CH_3 = C-OH$$

$$CH_3 = = = C-OH + H_2 \longrightarrow CH_3 = = C-OH$$

After partial hydrogenation, the percentage of unsaturated sites in the oil decreases as it becomes more saturated (less C=C, more C–C). Because of these changes in molecular structure, the oil becomes a semi-solid, hard enough to hold together, yet soft enough to spread or mold easily. Margarine, cookies, candybars, and peanut butter typically contain partially hydrogenated oils and therefore have such a blend of saturated and unsaturated fats. For example, a commercial soft-spread margarine contains 10 g of fat per tablespoon, 5 g of it polyunsaturated and 1 g saturated fat. The balance represents 4 g of monounsaturated fat. Because it is of vegetable origin, margarine contains no cholesterol. By comparison, a tablespoon of butter contains 12 grams of fat. But only 0.4 g of the fat is polyunsaturated and 7.2 g are saturated fat. Butter is an animal-derived substance and, consequently, it also contains cholesterol, 31 mg of it per tablespoon. New margarines are designed to be low-fat products that taste good, spread easily, and behave like butter in cooking and baking. The aim is to achieve all these qualities while still having the majority of the fat polyunsaturated or monounsaturated.

12.7 *Consider This*

Listed here is the fat content for Crisco®, a partially hydrogenated shortening, and the soft, tub margarine Promise®.

	Crisco®	Promise®
Total fat	12 g	10 g
Polyunsaturated	3	5
Saturated	3	1

As promised, Promise® has a lower total fat content. Notice that the sum of the saturated and polyunsaturated fats does not add up to the total fat value. The difference is the amount of monounsaturated fats. Calculate the number of grams of monounsaturated fats for the two substances given here.

Conduct a mini-survey of margarine products on the market in your area. List the total fat content and the content of polyunsaturated, saturated, and monounsaturated fats for at least three different margarines. One of the samples should be a soft margarine packaged in a tub; the others in stick form. Be sure to include one example of "light" margarine in your survey.

a. How does Crisco® compare to the regular stick, regular tub, and light margarines in terms of total fat and various types of fats?

b. What are the major differences in total fat and types of fats between the following: regular stick and light margarines; regular stick and tub margarine; and regular tub and light stick margarine?

c. Which margarine would you choose based on its fat content and why?

"Are you a no-cholesterol doctor or are you one of those no-cholesterol-is-all-bosh doctors?"
Drawing by Stan Hunt; © 1959, 1987. The New Yorker Magazine, Inc.

▪ *Controversial Cholesterol in the Diet and the Blood*

At room temperature, cholesterol is a white solid that dissolves in oil or fat, but not in water. Cholesterol is also a loaded word that has drawn heavy media attention, but not necessarily total accuracy or objectivity. The attention arises because of the apparent connection between serum cholesterol (cholesterol in the blood) and cardiovascular disease. Over the past decade, several national reports based on research and issued by medical professionals and public health officials have reversed positions on the matter.

In 1980, the Food and Nutrition Board of the National Academy of Science issued a controversial report, "Toward Healthful Diets." The report cited the inconsistency of data collected in medical studies done up to that time. A direct causal connection had not been unambiguously established between dietary cholesterol and atherosclerosis, the thickening of the arterial walls. Therefore, the report concluded that healthy adults did not have to reduce dietary cholesterol. However, by the end of the 1980s, a reversal of this position had occurred, based on the aggregate data accumulated from a wide variety of new and continuing studies. These studies led to the conclusion that high serum cholesterol levels *appear* to be a predictor of the potential for a stroke or heart attack. Although a direct linkage has not yet been shown irrefutably, the data seem compelling. Waxy deposits of excess cholesterol cause arteries to narrow and harden, which may elevate blood pressure and increase the risk of heart disease (Figure 12.6).

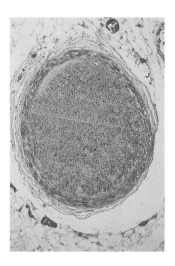

Figure 12.6
Cross sections of a healthy artery (left) and an artery clogged with atherosclerotic plaque (right).

■ *Table 12.3*	*Cholesterol Content of Various Foods*		
Food	**Cholesterol (mg)**	**Food**	**Cholesterol (mg)**
Fruit and vegetable	0	Milk, whole (8 oz)	33
Pork chop (3 oz)	83	Low-fat milk (8 oz)	22
Chicken, skinless white meat (3 oz)	71	Low-fat yogurt (8 oz)	14
Steak (3 oz)	70	American processed cheese (1 oz)	27
Shrimp (3 oz)	166	Cheddar cheese (1 oz)	30
Hot dog (3 oz)	43	Ice cream (4 oz)	29
Egg yolk	213	Butter (1 tbsp)	33

While there is general agreement that elevated blood cholesterol levels are associated with atherosclerosis, there is no consensus on what constitutes dangerously high concentrations. Many medical researchers and the American Heart Association consider serum cholesterol values above 200 mg cholesterol per 100 mL of blood as the critical point for medical intervention. Other investigators are more generous, citing values over 240 mg/100 mL as the threshold for such action.

What that action should be is not always obvious. The issue is one of whether to avoid foods containing fats, foods containing cholesterol, or foods containing both fats and cholesterol. To keep the record straight, it must be pointed out that serum cholesterol comes from two sources. One source is cholesterol in the foods we eat. The second is the cholesterol manufactured by our bodies. We can control cholesterol from the first source, our diet, by cutting down on eating foods high in cholesterol. Such foods include fatty red meats, cream, butter, and egg yolks, which contain a whopping 213 mg of cholesterol per yolk. The American Heart Association suggests no more than four eggs per week. In contrast, egg whites contain no cholesterol, nor do vegetables, fruits, and oils (see Table 12.3).

The principal source of serum cholesterol is not that ingested directly, but rather that made by the body. It is synthesized in the liver to sustain the minimum level required for use in cell membranes and to produce estrogen, testosterone, and other steroid hormones. Herein lies the crux of the cholesterol/fats controversy. The liver produces cholesterol principally from dietary saturated fats. Therefore, although cutting down on dietary cholesterol is important, apparently it is even more significant for controlling cholesterol to reduce the amounts of saturated fats in the diet. Eating foods low in cholesterol but high in saturated fat simply provides the liver with the starting materials to make more serum cholesterol.

Cholesterol produced by the body is difficult to control because diet is only one among several factors influencing the rate of its synthesis. Perhaps the most important is genetic makeup. This may explain why some people seem to eat fatty foods without suffering from heart disease, while others who scrupulously watch their diets are afflicted with it. Because it's impossible to pick your biological parents, you are encouraged to do what you can to reduce the level of serum cholesterol and the chance of cardiovascular disease. Exercising regularly, reducing stress, and eating certain types of dietary fiber appear to be effective.

■ Lipoproteins: Good and Bad

Cholesterol and triglycerides are not water soluble, so to be transported in blood, which is an aqueous medium, they must be linked to a carrier. Certain proteins in serum, called **lipoproteins,** combine with cholesterol or triglycerides, and carry them through the bloodstream. As the name suggests, lipoproteins are composed of a lipid part and a protein part. Two types of lipoproteins, high-density lipoproteins (HDL) and low-density lipoproteins (LDL), are associated with cholesterol transport and heart-disease risk. The difference in their densities is related to their relative amounts of protein and lipid: high-density lipoproteins contain a higher proportion of protein to cholesterol or triglycerides. Low-density lipoproteins contain a lower proportion. HDLs are the "good kind" associated with a lower risk of heart disease because of their efficient transport of cholesterol in serum. On the other hand, elevated serum levels of LDLs imply an increased risk of heart disease. Normal HDL levels (mg/100 mL blood) are 45–50 for men and 50–60 for women. Blood LDL level should be less than 130 mg/100mL. Blood tests measuring HDL/LDL ratios are used to forecast potential heart disease problems (see Figure 12.7).

Figure 12.7

Blood analysis report form.

CHEMISTRY 1	CHART COPY		LAB-13
REPORTED BY:		DATE COMPLETED:	

✓	TEST	RV	RESULT
	GLUCOSE (FASTING)	60–110	mg/dl
	2 HR. P.P. GLUCOSE		mg/dl
	BUN	8–20	mg/dl
	CREATININE	0.8–2	mg/dl
	URIC ACID	2.6–7.5	mg/dl
	CHOLESTEROL	160–230	mg/dl
	HDL-CHOL		mg/dl
	TOTAL PROTEIN	6–8	gm/dl
	ALBUMIN	3.2–5.6	gm/dl
	GLOBULIN	1.5–3.0	gm/dl
	A/G RATIO		
	BILIRUBIN T.	0.2–1.0	mg/dl
	BILIRUBIN		
	SGOT	11–35	IU
	LDH	26–186	IU
	CPK	16–164	IU
	CPK/MB	0–25	IU
	CALCIUM	8.5–11	mg/dl
	PHOSPHORUS	2.5–4.5	mg/dl
	ALK. PHOS.	10–85	IU
	ACID PHOS.	0–1	UNIT
	SODIUM	135–147	meq/L
	POTASSIUM	3.8–5.3	meq/L
	CHLORIDE	96–106	meq/L
	ACETONE	NEG.	
	MAGNESIUM	1.9–2.5M	mg/dl
	OSMOL SR.	275–295	mos/kg
	OSMOL UR.	400–800	mos/kg
	SGPT	6–36	IU

✓	GLUCOSE TOLERANCE	BLOOD mg/dl	URINE	
			GLUCOSE	ACETONE
	FASTING			
	1/2 HR.			
	1 HR.			
	2 HR.			
	3 HR.			
	4 HR.			
	5 HR.			

DEPARTMENT OF PATHOLOGY
MIDTOWN MEMORIAL HOSPITAL
P.O. Box 27
Midtown, CA. 95492 • 208-443-9425
ACCESSION #
□ P.A.T. □ STAT □ ROUTINE □ PRE-OP □ SCHEDULED
A.M. P.M.
DRAWN BY:
ON
SPECIAL INSTRUCTIONS
DATE ORDERED

Coronary disease specialists have suggested that the cholesterol-to-HDL ratio is a crucial factor. The cholesterol/HDL ratio should be less than 4.5. A ratio greater than 4.5 is typical among those who develop heart disease. The good news is that it has been demonstrated that decreasing LDL serum levels lowers serum cholesterol concentration. Because saturated fats raise serum LDL levels, shifting to a diet with foods rich in mono- and polyunsaturated fats reduces LDL levels as well as serum cholesterol. Regular exercise and loss of excess weight also have been shown to increase HDL levels. Thus exercise, weight control, and smoking cessation all reduce the risk of atherosclerosis. It makes little sense from a health standpoint to eat low-cholesterol foods while continuing to smoke. Dr. Richard Peto, an Oxford University epidemiologist, puts this into context in rather blunt language: "You can't offer eternal life to old people. But what you can do is to avoid death in middle age. At the moment, about a third of all Americans die in middle age, and that isn't necessary. About half of those premature deaths could be avoided if people took smoking, blood cholesterol, and blood pressure more seriously."[3]

12.8   Consider This

Medical diagnosis by amateurs can be dangerous. Indeed, it is often uncertain when done by doctors. Nevertheless, try your hand at evaluating the following blood test results for a 58-year-old man, weighing 175 pounds, at a height 5 ft 10 in.

Cholesterol	250 mg/100 mL
HDL	50 mg/100 mL
LDL	180 mg/100 mL

a. Using the information provided in this chapter, assess the risk of coronary artery disease for this individual.
b. What three changes would you suggest this person make in his lifestyle to lower his serum cholesterol level?

■ Proteins: Our Major Structural Building Blocks

Thus far we have managed to describe two macronutrients in terms of just three elements: carbon, hydrogen, and oxygen. **Proteins,** the third macronutrient, are present in every body cell. They are needed for cholesterol transport, metabolism regulation, and many other functions. If you studied Chapter 11, you already know that proteins are polymers. But, unlike starch or cellulose, which contain only glucose units, proteins are polymers made up of various amino acids as the monomers. **Amino acids** are compounds that contain nitrogen in addition to carbon, hydrogen, and oxygen. A few amino acids also contain sulfur.

Protein formation, like that of starch, is a condensation polymerization reaction in which water is eliminated. As the amino acid units link together in a growing chain, one water molecule is split off for each new amino acid added. When two amino acid units join, a *di*peptide is formed. When three link together, a *tri*peptide results.

Amino Acid 1 + Amino Acid 2 → Dipeptide + Water

Dipeptide + Amino Acid 3 → Tripeptide + Water

Eventually, enough amino acid units are incorporated into the chain so that a large polymer, a protein, is produced.

Many amino acids → Protein + Water

The length of the polymer chain can vary from relatively short to very long indeed. For example, there are 51 amino acid monomer units in the metabolic hormone insulin

---

[3.] "Lipoproteins: Good and Bad." Excerpted with permission from the December 1989 issue of the *Harvard Health Letter,* 1989 President and Fellows of Harvard College.

and 574 amino acids in hemoglobin, the blood protein involved in oxygen transport. The particular protein formed depends on which amino acids are in it as well as their sequence in the polymer chain. Assembling the correct amino acid sequence is like putting letters in a word; if they are in a different order, a completely new meaning results. For example, the sequencing of the three letters *a, e,* and *t* forms the words *eat, ate,* or *tea,* as well as some combinations that not recognizable English words (*aet* or *tae*). Having a misplaced amino acid in a protein sequence can have serious, even lethal effects, as in sickle-cell anemia and other genetically-determined conditions.

All amino acids have four things attached to a particular carbon atom: 1) an acid group, $-COOH$; 2) an amine group, $-NH_2$; 3) a hydrogen atom, $-H$; and 4) either a hydrogen atom, or another carbon atom, generally designated as *R* to represent a carbon-containing side chain in the fourth position.

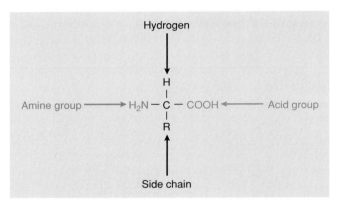

Variations in the side chain are what differentiates amino acids. In glycine, the simplest amino acid, the fourth position is occupied simply by a hydrogen atom. In alanine, R is a $-CH_3$ group, in aspartic acid (found in asparagus) it is $-CH_2COOH$, and in phenylalanine it is a ring-shaped group with the formula $-CH_2(C_5H_6)$.

Because all amino acids except glycine involve four different units bonded to a central carbon atom, they all exhibit optical isomerism (see Chapter 11). The naturally occurring amino acids that are incorporated into proteins are all in the left-handed isomeric form.

Using the structural formulas for glycine and alanine given above, we can demonstrate how they join to form a dipeptide. The acidic $-COOH$ group reacts with the basic $-NH_2$ group and a molecule of water is eliminated. In the process, the two amino acids become linked by a peptide bond (indicated in the box below).

(12.2)

Glycine + Alanine ⟶ Dipeptide + Water

Because each amino acid contains an amine group and an acid group, there are two ways the amino acids can join. Hence, two dipeptides are possible. We will illustrate the options with simple block diagrams. In the first case glycine acts as the acid and alanine as the amine:

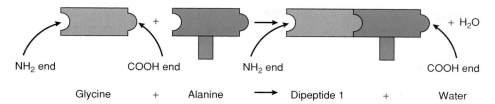

NH₂ end   COOH end   NH₂ end   COOH end

Glycine   +   Alanine   ⟶   Dipeptide 1   +   Water

In the second case, the amino acids reverse roles; alanine is the acid and glycine the amine.

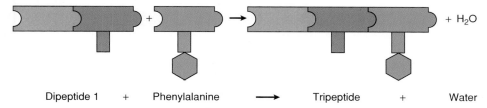

Alanine   +   Glycine   ⟶   Dipeptide 2   +   Water

You might be tempted to say that the two dipeptides are the same. But a closer examination of their molecular structures will indicate that in the first dipeptide, the unreacted amino group is on the glycine and the unreacted acid group is on the alanine. In the second, the $-NH_2$ is on the alanine and the $-COOH$ on the glycine. Thus, the two dipeptides are structural isomers (see Chapter 10).

Because any dipeptide has an uncombined amine group at one end and an uncombined acid group on the other, it can react with yet another amino acid to form two possible tripeptides. Pictured below is one of the tripeptides formed from what we have called dipeptide 1 and phenylalanine.

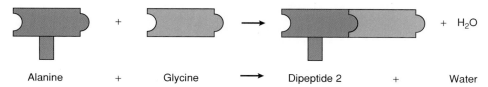

Dipeptide 1   +   Phenylalanine   ⟶   Tripeptide   +   Water

Alternatively, the acidic end of the phenylalanine molecule can react with the amine end of dipeptide.

If enough of the appropriate amino acid units are joined in the correct sequence, eventually a particular protein forms. The great majority of proteins are made from a pool of 20 different amino acids. Of these, nine are essential amino acids (Table 12.4). Like the two essential fatty acids mentioned earlier, these nine essential amino acids must be part of our diet. They cannot be made by the body from other foodstuffs.

■ *Table 12.4*	*The Essential Amino Acids*	
Histidine	Lysine	Threonine
Isoleucine	Methionine	Tryptophan
Leucine	Phenylalanine	Valine

A packet of the artificial sweetener, aspartame, with a warning to phenylketonurics.

■ *How (Synthetically) Sweet It Is*

Because of the great national preoccupation with excess calories and excess pounds, artificial sweeteners have become a multimillion dollar business. Gram for gram, these compounds are much sweeter than sugar, but they have little, if any, nutritive value. Hence, they are nonfattening. Currently the most widely used artificial sweetener is aspartame, the principal ingredient in NutraSweet® and Equal®. Somewhat surprisingly, it is related to proteins. Aspartame, is a dipeptide made from two amino acids— aspartic acid and a slightly modified phenylalanine.

Curiously, neither of the two amino acids alone tastes sweet. However, combined as aspartame, they are about 200 times sweeter than sucrose, even though aspartame contains no saccharides. This is evidence that sweetness is a property not restricted to sugars alone. Not surprisingly, it is related to molecular structure (see Chapter 11 for more on molecular form and function). The current thinking about what triggers a response from the sweetness receptors in our taste buds is a distinctive triangular arrangement of atoms within a molecule (Figure 12.8). Molecules that have this "sweetness triangle" taste sweet. Evidently, the triangular arrangement provides three sites in the molecule that anchor it to three key locations of a taste bud receptor.

Figure 12.8

The sweetness triangle. (Source: Reprinted with permission from *Chem Matters,* Vol. **6,** No. 1, February 1988, page 8. Copyright 1988 American Chemical Society, Washington D.C.)

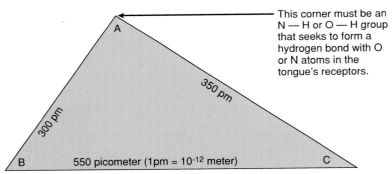

This corner must be an N — H or O — H group that seeks to form a hydrogen bond with O or N atoms in the tongue's receptors.

Corner B needs to be a basic atom (O or N) that can attract N — H or O — H centers in the receptor.

Corner C should be any group that repels water, such as CH_2, CH_3, or C_6H_5.

The Food and Drug Administration approved aspartame for consumer use in 1981. Since that time, the compound has been used by millions of people worldwide. A relatively few cases of adverse side effects have been attributed to aspartame. But exhaustive reviews of these cases have failed to show unequivocally a direct connection between the symptoms and the sweetener. Aspartame continues to be safe when used in moderation for the vast majority of consumers. There is, however, one group of people who definitely should not use aspartame. The warning on packets of artificial sweeteners and products containing aspartame is explicit: "Phenylketonurics: Contains phenylalanine." The paragraph in the margin gives more information on this classic case where "one man's meat is another man's poison."

<table>
<tr><td>**12.9**</td><td>### *Consider This*</td></tr>
</table>

We don't need sugar or other sweeteners to have a healthy diet. Nevertheless, the per capita national consumption of sugar and other sweeteners is enormous. (A 12-oz can of cola contains the equivalent of four tablespoons of sucrose.) Sweeteners are very profitable because of this heavy consumer demand, and food companies compete to develop new sweeteners for such a lucrative market. Examine how pervasive the use of sweeteners is by doing the following:

a. Review your diet and list all foods that you eat in a typical three-day period that contain sugar or a sugar substitute.
b. From labels and tables giving food composition, estimate the total mass of sugar and other sweeteners you consume in an average day.
c. Decide which foods would still be palatable without any sweetener.

■ *Enough Protein: The Complete Story*

Dietary protein requirements are generally expressed in terms of grams of protein per kilogram of body weight per day. This need varies with age, size, and energy demand. Infants require 1.8 grams/kg/day, middle school children about 1.0 g/kg/day, and adults 0.8 g/kg/day. Therefore, a 20 lb infant (9 kg) needs 16 grams of protein daily, while a 165 lb (75 kg) college student requires 60 grams each day.

Normally, the body does not store protein. Because proteins are the major source of nitrogen in the body, they are constantly being broken down and reconstructed. Evidence for this is that a healthy adult on a balanced diet will be in nitrogen balance and will excrete as much nitrogen (primarily as urea in the urine) as she or he ingests. Growing children, pregnant women, and persons recovering from long-term wasting illnesses have a positive nitrogen balance. These individuals excrete less nitrogen than they take in because they are making supplemental proteins. A negative nitrogen balance exists when the opposite condition occurs, as in starvation. Insufficient protein is being synthesized, perhaps because essential amino acids are lacking in the diet. In cases of starvation, muscle is metabolized as an energy source.

To ensure against malnutrition, it is important not only to have enough dietary protein, but the proper quality of protein as well. Meat (beef, fish, poultry) contains all of the essential amino acids in approximately the proportions found in the human body. Therefore, meat is termed a *complete* protein. However, most of the people in the world depend on grains, not meat, as their major source of proteins. In regions of Southeast Asia, large portions of the population survive on a diet consisting largely of rice and soy products without suffering from malnutrition. However, in Mexico and Latin America, where corn is the main dietary staple, malnutrition exists. These two cases point out that although adequate quantities of protein are available in each case, the quality of the protein differs. Rice and soy products combine to make a complete protein diet. Corn, however, is too low in tryptophan, an essential amino acid; thus, corn is called an *incomplete* protein. Obviously, a diet made up exclusively of incomplete protein can be hazardous to your health. Protein synthesis requires that all of the essential amino acids be present. Any essential amino acid in short supply limits the production of proteins that rely on that amino acid as part of the proper sequence of amino acid units in building the protein chain.

We avoid such problems by eating proteins from a variety of sources at the same meal so that adequate amounts of the essential amino acids are obtained. For example, when we eat a peanut butter sandwich, we are using a principle nutritionists call *complementarity,* combining foods that complement each others' essential amino acids to collectively supply the amounts needed. Bread is deficient in lysine and isoleucine;

Phenylketonuria

Phenylalanine is an essential amino acid, principally because it is converted to tyrosine, another amino acid. Individuals with **phenylketonuria**, a genetically transmitted disease, lack the enzyme (phenylalanine hydroxylase) that catalyzes this conversion. Consequently, in those individuals (phenylketonurics), the conversion of dietary phenylalanine to tyrosine is blocked and the phenylalanine concentration rises. To compensate for the increased phenylalanine, the body resorts to converting it to phenylpyruvic acid and excreting large quantities of this acid in the urine. Phenylpyruvic acid is termed a "keto" acid because of its molecular structure; hence the disease is called phenylketonuria or PKU for short. Excess phenylpyruvic acid causes severe mental retardation. Therefore, the urine of all newborn babies is tested for phenylpyruvic acid, using special test paper placed in the diaper. Infants diagnosed with PKU must be placed on a diet with limited phenylalanine. This means avoiding excess phenylalanine from milk, meats, and other sources rich in protein. Commercial food products are available for such diets, their composition adjusted to the age of the user. Because phenylalanine is an essential amino acid, a minimum amount of it must still be available, even for phenylketonurics. Supplemental tyrosine may also be needed due to the lack of normal conversion of phenylalanine to tyrosine. A phenylalanine-restricted diet is recommended for phenylketonurics at least through adolescence. Adult phenylketonurics must also limit their phenylalanine intake, and hence curtail their use of aspartame.

these amino acids in peanut butter overcome the deficiency. Peanut butter is low in methionine; bread provides this component. Protein complementarity is used widely in many developing nations where meat is not the major protein source. As we have just seen, soy foods are eaten with rice in parts of Southeast Asia and Japan. In Latin America, beans are used to complement corn tortillas. People in the Middle East combine bulgur wheat with chickpeas or eat hummus, a sauce of sesame seeds and chickpeas, with pita bread. In India, lentils and yogurt are eaten with unleavened bread.

Livestock, especially beef cattle, are fed a variety of grains to provide a diet of complete proteins necessary to produce meat. However, we should keep in mind that there is a loss of efficiency with each step during energy transfer, whether from electrical power plants (as we saw in Chapter 4) or from cells during metabolism. Cattle are notoriously inefficient in converting the energy in their feed into meat on the hoof. It takes about seven pounds of grain to produce one pound of beef. Put into human terms, the 1.75 pounds of grain used to produce a "quarter-pounder" can provide two days of food for a person in a poorly-developed country. Other livestock animals are a bit more efficient than cattle in converting grain to meat. Hogs require 6 pounds of grain per pound of meat, turkeys need 4, and chickens even less, only 3 pounds.

It is much more efficient to get the energy directly from grains, rather than through secondary or tertiary sources still further along in the food chain. Most of the world's people use grains as their primary protein source. In addition to being more energy efficient, this type of diet also is less likely to bring on long-range health problems associated with a dependency on red meat as a principal protein source. But no matter what the source, proteins provide only about 10% of the energy our bodies require. Carbohydrates and fats are the principal energy sources, a topic we now consider.

■ *Energy: The Driving Force*

One of the main reasons we must eat is to obtain energy to run the complex chemical, mechanical, and electrical system called the human body. The energy in our foods is energy we obtain indirectly. Initially, the energy arrives in the form of sunlight which is absorbed by green plants during photosynthesis. You have already encountered photosynthesis in Chapters 3, 4, and 9. Under the influence of a catalyst called chlorophyll, carbon dioxide and water are combined to form glucose. In the process, the Sun's energy is "stored" in chemical bonds of the sugar.

$$\text{Energy (from sunshine)} + 6\ CO_2 + 6\ H_2O \rightarrow C_6H_{12}O_6 + 6\ O_2$$

During metabolism the photosynthetic process is reversed, the food is broken down into simpler substances, and the stored energy is released.

$$C_6H_{12}O_6 + 6\ O_2 \rightarrow 6\ CO_2 + 6\ H_2O + \text{Energy (from metabolism)}$$

These two relationships are shown in Figure 12.9.

Figure 12.9

Energy from photosynthesis and metabolism. (From Curtis T. Sears, Jr., and Conrad L. Stanitski, *Chemistry for Health-Related Sciences: Concepts and Correlations,* 1976, p. 315. Reprinted by permission of Prentice Hall, Englewood Cliffs, New Jersey.)

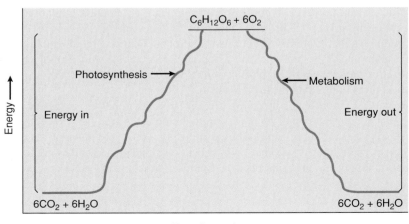

To derive energy, we eat green plants, their fruits, or the meat from fish, fowl, or hoofed animals that have eaten the green plants. We get this energy secondhand from plants and fruits, at least thirdhand in the case of meats. Whatever their sources, foods are mixtures containing a dazzling array of compounds. Thus, whether grown wild, by cultivation, or "organically," *all foods contain chemicals.* This rather basic fact refutes the people, especially food faddists, who claim to want foods or to supply foods "without chemicals," an obvious contradiction in terms.

The energy derived from foods has been aptly described as "the fire of life." During digestion, the energy from foods is extracted from the bonds holding atoms together in the compounds making up foods. Complex molecules are broken down into simpler structures, which are then available to be reformed into other complex substances. Bonds are broken, an endothermic (energy-requiring) process, while other new bonds are made, which is an exothermic (energy-releasing) event. However, we need to keep in mind that we cannot escape the laws of thermodynamics (see Chapter 4), even during cellular metabolism. The first law specifies that energy is neither created nor destroyed, but only transformed and transferred. And the second law predicts that the efficiency of energy transfer decreases with each step in the food chain, as energy becomes more degraded into heat, and disorder and entropy increase.

Carbohydrates and fats are the major energy sources for the body. When metabolized, they provide 4 kcal/g of carbohydrate and 9 kcal/g of fat, respectively. Thus, gram-for-gram, fats provide over twice as much energy as carbohydrates. In general, a balanced diet provides about 60% or more of its calories from carbohydrates, no more

---

12.10

The Sceptical Chymist

We have just reported that gram-for-gram, fats yield over twice as much energy as carbohydrates. Use your knowledge of the chemical composition of these two classes of compounds to account for this difference.

Soln. Fats, made up of many CH_2 units per molecule, have high percentages of carbon and hydrogen. Therefore, they closely resemble hydrocarbons, which are excellent fuels. Carbohydrates contain a relatively higher percentage of oxygen than do fats. We say that carbohydrates are "more oxygenated." To see this difference, consider two compounds, each containing 12 carbon atoms per molecule: one is lauric acid, $C_{12}H_{24}O_2$, a saturated fatty acid; the other is sucrose (table sugar), $C_{12}H_{22}O_{11}$. When metabolized (burned), each is converted to carbon dioxide and water.

For lauric acid:

$$C_{12}H_{24}O_2 + 17\ O_2 \rightarrow 12\ CO_2 + 12\ H_2O$$

It can be determined experimentally that 36.8 kJ of energy are released per gram of lauric acid metabolized. This amounts to 36.8 kJ/g × 1 kcal/4.18 kJ = 8.8 kcal/g of lauric acid.

For sucrose:

$$C_{12}H_{22}O_{11} + 12\ O_2 \rightarrow 12\ CO_2 + 11\ H_2O$$

The metabolism of one gram of sucrose generates 16.4 kJ/g corresponding to 16.4 kJ/g × 1 kcal/4.18 kJ = 3.9 kcal/g.

Notice from the balanced equations that it takes more oxygen to burn lauric acid than to burn sucrose. This is because sucrose already contains more oxygen to start with. It is already partially oxidized and thus "closer" to the end products, CO_2 and H_2O. This helps explain why burning a gram of fat produces around 9 kcal of energy while a gram of carbohydrate provides only about 4 kcal when metabolized.

than 30% from fats, and the remainder from proteins. Proteins, which also furnish 4 kcal of energy per gram when metabolized, are normally used to make structural units such as skin, muscles, tendons, and ligaments.

All energy changes, whether in a laboratory or in your body, are subject to the law of conservation of energy. This accounting requires that a balance exist between the energy taken in with that used and/or stored:

$$\text{Energy in (from food)} = \text{Energy used (basic body functions plus activity)}$$
$$+ \text{Energy stored (fat, glycogen)}$$

In more personal terms, this equation indicates that if less energy is used than is taken in, the excess energy is stored, typically as fat. Putting it more crassly, "those who indulge, bulge." Regular exercise, along with eating less, increases the amount of energy used and decreases what is stored.

The energy available to the body from foods is used to drive the chemical reactions in nonspontaneous processes such as the contraction of heart muscles and the transmission of nerve impulses (see Chapter 4 for more about non-spontaneous processes). In short, spontaneous processes or reactions that release energy allow nonspontaneous processes to occur by furnishing the energy necessary for them to happen. As an analogy, consider an auto battery. The battery produces electrical energy spontaneously because of chemical reactions in the battery. These spontaneous processes provide energy that can be used to power nonspontaneous processes, for example, starting the car or making the car tape player, headlights, and horn work.

In addition to having sufficient energy available, the body must also have some way of regulating the rate at which the energy is delivered to avoid wild temperature fluctuations or too rapid release of energy. Think about it in terms of an automobile, its ignition system, and the gasoline in its fuel tank. If a lighted match were dropped into the fuel tank, the gasoline would ignite, burning all the gasoline as well as the car. The temperature would rise dramatically, too. Igniting all the gasoline at once is a rather drastic means to move a car. However, under normal operating conditions, just enough fuel is delivered to the ignition system to supply the automobile with the energy it needs without raising the temperature of the car and its occupants beyond reason. By releasing a little energy at a time, rather than all of it at once, the efficiency of the process is enhanced. So it is with the body. Enzymes, enzyme regulators, and hormones mediate the rates of metabolic reactions so that the energy is released incrementally, slowly, and as needed.

From Chapter 4, you are well aware that energy, the "capacity to do work," is frequently expressed in joules or calories. However, by tradition, nutritionists use a version of the calorie, the Calorie (capital C), for describing dietary energy values. Sometimes referred to as the "big calorie," the nutritionists' Calorie is simply a kilocalorie: 1 Calorie (Cal) = 1 kilocalorie (kcal) = 1×10^3 calories (cal). Therefore, numerical tables that state how much energy we get from our dining tables are stated in Calories, for example, "eating one chocolate chip cookie provides 50 Calories (kcal)."

You have already learned that fats yield about 9 Cal/g when metabolized, and carbohydrates and proteins yield 4 Cal/g. Therefore, if you know the composition of a particular food, you can calculate the number of Calories it provides. The process is illustrated in 12.11 The Sceptical Chymist and 12.12 Your Turn.

It is instructive that only 1% of calories in uncooked potatoes comes from fats. The additional "fat" calories calculated in 12.12 Your Turn come from the cooking process during which the fresh, white potato slices get converted into crisp, golden, fat-laden chips. Calories obtained in this way, from high-fat foods, sweets, or cola beverages, are actually "empty calories." They are high in energy content (the number of calories), but lacking in the other basic nutrients needed for a well-balanced

12.11 *The Sceptical Chymist*

A brand of turkey bologna proudly advertises that it provides 60 Cal per one ounce slice, 82% of which is fat free, by weight. Is this product really "low cal" and "fat free"?

Soln. First of all, it might be good to recognize that bologna that is 82% fat free still contains 18% fat by weight. A 1-ounce slice of this bologna weighs 28 grams, five of which are fat (28 g bologna × 18 g fat/100 g bologna = 5 g fat). Consequently, fat makes up 75% of the calories obtained from one slice of the bologna: 5 g fat × 9 Cal/g fat = 45 Cal; 45 Cal from fat/60 Cal total × 100 = 75% calories from fat. It seems that the advertising is "bologna" as well. Similar calculations can be done to determine the percentage of calories derived from fats in lean, chopped meat (61%), "lite" hot dogs (76%), or other foods. To put things in perspective, remember that the dietary recommendation is that no more than 30% of calories should come from fat.

12.12 *Your Turn*

Potato chips are 65% fat free, meaning that 35% of their weight comes from fat, a fact that is not readily apparent from advertisements for potato chips. Because, "no one can eat just one" or hardly even limit themselves to the one ounce portion in a small bag, potato chips can be a substantial source of dietary fat. From the nutritional information given below calculate the percentage of calories in the chips from fat, carbohydrate, and protein. Keep in mind the following relationships: 9 Cal/g fat; 4 Cal/g carbohydrate; and 4 Cal/g protein.

Nutritional Data per One-Ounce Serving

Weight	28 g
Calories	150
Protein	1 g
Carbohydrate	15 g
Fat	10 g
Cholesterol	0 mg
Potassium	390 mg
Sodium	170 mg

Ans. 58% from fat, 39% from carbohydrate, 3% from protein

diet. Thus, a person can be simultaneously overfed with excess calories but malnourished through an incomplete diet. A diet with high **nutrient density** is rich in nutrients, compared to the number of calories eaten and hence is much more beneficial.

■ *Energy: How Much Is Needed?*

The question is simple: "How much energy do we need daily?" The answer is "It depends." The number of calories you require each day depends on your level of exercise or activity, the state of your health, your gender, age, body size, and a few other factors. Table 12.5 summarizes the recommended food energy intakes for Americans.

Some generalizations can be drawn from Table 12.5. Most men require more calories per day than women do because of differences in body mass. For those in the traditional college/university age range, the daily energy requirement is about 2200

Table 12.5	Recommended Energy Intake (United States)				
Age (yrs)	Avg. weight (kg)	Avg. weight (lb)	Avg. height (in)	Avg. Cal/kg	Avg. Cal/day
0.5–1.0	9	20	28	98	850
4–6	20	44	44	90	1800
7–10	28	62	52	70	2000
Males					
15–18	66	145	69	45	3000
19–24	72	160	70	40	2900
25–50	79	174	70	37	2900
51+	77	170	68	30	2300
Females					
15–18	55	120	64	40	2200
19–24	58	128	65	38	2200
25–50	63	138	64	36	2200
51+	65	143	63	30	1900

Cal for women and 2900 Cal for men. Generally, fewer calories are needed daily by children, because of their smaller mass, and by older people, due to typically lower activity levels. But note that caloric requirements per kilogram of body weight are a good deal higher for growing children than for adults. Most American adults could easily survive a decrease of 300 Cal per day. However, a decrease of 300 Cal per day for a child or infant would be proportionally much more serious than for an adult. This inequality is significant in the disproportionate mortality rates among infants and young children in Somalia and other famine-stricken countries. Since 1988 in Somalia, the average caloric intake dropped drastically to a reported 200 calories daily, far below the required minimum.

The minimum amount of energy required daily, the **basal metabolism rate** or **BMR** is the amount necessary to support basic body functions—to keep the heart beating, the lungs inhaling and exhaling, the brain functioning, the blood circulating, all major organs working, and body temperature maintained at 37°C. This amounts to approximately one Calorie per kilogram (2.2 lb) of body weight per hour, varying with size and age. The actual value of the BMR is determined experimentally in a resting state. The quantity of energy used in digestion and metabolism is eliminated from the BMR value by having the subject fast for 12 hours before the measurement is made.

Using the approximation, we can calculate the BMR of the heaviest of the authors of this book, who weighs in at 85 kilograms. His basal metabolism rate over a 24 hour day is calculated as follows:

$$85 \text{ kg} \times 1 \text{ Cal/kg} \times 24 \text{ hrs/day} = 2040 \text{ Cal/day}$$

By contrast, the lightest member of the writing team, a 50 kg woman, has a BMR of 1200 Cal/day. Applying the relationship to a 20 kg child (44 lb) yields a BMR of only 480 Cal/day. However, children have higher BMRs than adults, so the 480 Cal/day is only approximate, and very likely low.

You will note that the basal metabolism accounts for a surprisingly large proportion, about 70%, of the energy used in a typical day. Obviously, additional energy is required beyond the BMR value for any physical activity or exercise. Extensive tabulations have been made of energy expenditures for a wide range of activities. Some are listed in Table 12.6.

An alternative way of expressing these energy expenditures is in terms of their dietary equivalents, that is, how much exercise is required to "work off" a cookie or a serving of ice cream (Table 12.7).

■ *Table 12.6*	*Energy Expenditures (Cal/min) for Various Activities in Relation to Body Weight (lbs)*		
Activity	**125 lb Cal/min**	**150 lb Cal/min**	**175 lb Cal/min**
Aerobics (vigorous)	7.8	9.3	10.9
Bicycling (15 mph)	6.1	7.4	8.6
Golf (carry clubs)	5.6	6.8	7.9
Jogging (5 mph)	7.6	9.2	10.7
Studying	1.4	1.7	1.9
Swimming (20 yds/min)	4.0	4.8	5.6
Tennis (novice)	4.0	4.8	5.6
Walking (3.5 mph)	4.4	5.2	6.1

■ *Table 12.7*	*How Much Exercise Must I Do If I Eat This Cookie? Calories and Minutes of Exercise for a 150-Pound Person*		
Food	**Calories**	**Walk (min)**	**Run (min)**
Apple	125	24	8
Beer, 8 oz (regular)	100	19	6
Chocolate chip cookie	50	10	3
Hamburger	350	67	20
Ice cream, 4 oz	175	34	10
Pizza, cheese, 1 slice	180	35	10
Potato chips, 1 oz	108	21	6

12.13 *Your Turn*

A 150 lb person consumes a meal consisting of 2 hamburgers, 3 oz of potato chips, 8 oz of ice cream, and a 12-oz beer. Calculate the number of Calories in the meal and the number of minutes the person would have to run to "work off" the meal.

Ans. 1524 Cal, 87 min (from Table 12.7) or 166 min jogging (from Table 12.6).

12.14 *Consider This*

First calculate your BMR. Then select from Table 12.6 the activities you do in a typical day. Calculate the supplemental energy you need for these activities. Then add your BMR and your supplemental energy needs to determine the total number of Calories you require per day. How does this result compare with the recommended energy intake for your age and gender (see Table 12.5)?

As we consider how many dietary Calories we take in daily to meet our BMR and activities, it is instructive to compare that value with the energy available from diets in various regions of the world. Table 12.8 provides these data as nutritional Calories (kcal) per person per day or Dietary Energy Supplies (DES values).

As expected, in each time interval, the average DES value for developed nations is greater than that for developing countries, currently by about 40%. From 1986 to

■ *Table 12.8*	Dietary Energy Supplies (DES) in Calories per person per day		
Region	1969–71	1979–81	1986–88
Developed Countries	3186	3300	3389
North America	3371	3487	3626
Western Europe	3233	3371	3445
Developing Countries	2158	2317	2352
Africa	2046	2148	2119
Latin America	2514	2675	2732
Near East	2399	2794	2914
Far East	2049	2185	2220
World	2435	2587	2671

(Data taken from *The State of Food and Agriculture:1990*. Rome: Food and Agriculture Organization of the United Nations, 1991, Table 1.7, p. 31.)

1988, North Americans had a daily Dietary Energy Supply nearly 71% greater than individuals living in Africa, 63% greater than people in the Far East, and 33% greater than Latin Americans. In fact, the DES for North Americans is the highest in the world, exceeding those of other developed nations.

■ *Vitamins and Minerals: The Other Essentials*

It is essential to take in an adequate number of calories daily. However, calories alone are not enough. Over the ages, human beings have learned that if certain parts of the diet were lacking, illness occurred. Thus, a link between nutrition and health was created—the hard way. Through such experiences, a list of essential foods developed, more by accident than design, and with little understanding of why certain items were necessary. As the study of nutrition developed more fully, the chemical basis for sound dietary habits was established. By the turn of this century, the necessity of carbohydrates, proteins, and fats was clearly established. Additionally, the use of iron to treat anemia and iodine for goiter had shown the dietary importance of these two elements. However, it was not until the early part of this century that it was realized that good health required substances we now categorize as the micronutrients—the vitamins and minerals.

Vitamins were discovered during this century and shown to be absolutely indispensable. Vitamin B_1 (thiamine) was the first to be identified. It was designated B_1 because that was the label on the test tube in which it was collected. It was called a *vitamine* because it is *vital* for life and is chemically an *amine*. The *e* has long since been dropped because not all vitamins are amines. Many additional vitamins have been isolated and the daily requirements of the vitamins have been assessed. Today **vitamins** are defined by their properties: they are essential in the diet, although required in very small amounts; they all are organic molecules; and they generally are not used as a source of energy, although they help break down macronutrients.

Vitamins are typically classified by their solubilities. Because of their molecular structures, vitamins either dissolve in water or they do not, which is a very important distinction. Those that are not water soluble dissolve in fat. There are four fat-soluble vitamins: A, D, E, and K. Their solubility is a consequence of the fact that they are non-polar molecules. Their molecular structures are similar to a hydrocarbon with many CH_2 units. All other vitamins are water soluble because their molecules contain several –OH groups. These hydrogen bond to water molecules, thus bringing the vitamins into aqueous solution (see Chapter 5).

Vitamin A, a fat-soluble vitamin   Vitamin C, a water-soluble vitamin

These solubility differences have significant health implications. Because of their fat solubility, vitamins A, D, E, and K can be stored in the body in cells rich in lipids, to be made available as needed. This means that the fat-soluble vitamins do not have to be taken daily, only periodically. It also means that fat-soluble vitamins may build up to toxic levels if taken far in excess of normal requirements. By contrast, water-soluble vitamins cannot be stored. Any excess is generally excreted in the urine. Thus, water-soluble vitamins need to be supplied more frequently and in small doses. Under atypical circumstances, water-soluble vitamins, when taken in megadoses, can collect in excess, and reach toxic levels. For example, there are reports of toxic side effects with vitamin B_6 supplements taken at levels 1000 times than the recommended dosage to alleviate symptoms of pre-menstrual syndrome (PMS). Therefore, an excess of vitamins, whether fat soluble or water soluble, should be avoided. Taking vitamin supplements is generally not necessary if you eat a well-balanced diet. Such a diet provides all the necessary vitamins in the required amounts.

12.15 *The Sceptical Chymist*

Arguably, Linus Pauling has contributed more to chemistry than any chemist of the twentieth century. He has been awarded both the Nobel Prize in Chemistry and the Nobel Peace Prize. Yet he is probably best known for his advocacy of massive doses of vitamin C as a way to avoid common colds and possibly to reduce the risk of certain cancers. To say the least, Pauling's claims for vitamin C have been controversial. They have attracted many advocates and many skeptics. Prepare a research paper in which you critically analyze Pauling's claims and the responses to them.

■ *Some Selected Vitamins and Minerals*

A major function of vitamins, especially the B family of vitamins, is to act as coenzymes. **Coenzymes** are generally small molecules working in conjunction with enzymes to enhance the enzymes' activity. **Niacin,** in particular, is an extremely important coenzyme in energy transfer during glucose and fat metabolism. The synthesis of niacin requires tryptophan, thus making tryptophan an essential amino acid. A diet deficient in tryptophan may lead to niacin deficiency. Such a deficiency causes pellagra, a condition involving a darkening and flaking of the skin as well as behavioral aberrations.

Vitamin C, also called ascorbic acid, must be supplied in the diet. Although other animals can synthesize vitamin C from glucose, humans cannot because we lack an enzyme necessary to convert glucose to ascorbic acid. Citrus fruits and green vegetables are the common sources of vitamin C. Ascorbic acid is necessary to prevent scurvy, a disease in which collagen, an important structural protein, is broken down. The link between vitamin C and scurvy was discovered over 200 years ago when it was found that feeding British sailors limes or lime juice on long sea voyages prevented scurvy. It also led to British sailors being called "limeys." Vitamin C is also necessary for the uptake, use, and storage of iron, important in the prevention of anemia. We will look at this aspect shortly.

Vitamin E is important in the maintenance of cell membranes and as protection against high concentrations of oxygen, as in the lungs. In general, vitamin E is so widespread in foods that it is difficult to create a vitamin E-deficient diet. People who eat very little fat may need vitamin E supplements. Vitamin E deficiency in humans has been linked with nocturnal cramping in the calves and fibrocystic breast disease.

An adequate supply of **minerals** is also essential to the body. Table 12.1 lists calcium, phosphorus, chlorine, potassium, sulfur, sodium, and magnesium among the major elements of the body. Not nearly as abundant as carbon, nitrogen, hydrogen, or oxygen, these seven elements are nevertheless vital for good health. Also called **macrominerals,** they play very important roles in metabolism. However, there are other minerals, the **microminerals,** that are also essential, but occur in the body in lesser amounts than the macrominerals. The recognized microminerals include iron, manganese, copper, and iodine. Other elements are present in the body at even lower concentration. These include zinc (for basal metabolism), fluorine (in bones and teeth), and cobalt (one atom per vitamin B_{12} molecule). Even arsenic, which is generally classified as toxic, is needed in trace amounts.

These various essential elements are identified in the skeleton periodic table that appears as Figure 12.10. The metals exist in the body as cations, for example, Ca^{2+} (calcium), K^+ (potassium), Fe^{2+} (iron), and Na^+ (sodium). The nonmetals typically are present as anions, thus chlorine is found as Cl^- and phosphorus appears in the phosphate ion, PO_4^{3-}.

Calcium, along with phosphorus and lesser amounts of fluorine, forms bones and teeth. You also need adequate calcium for use in blood clotting, muscle contraction, and transmission of nerve impulses.

Sodium is a vital micronutrient, but it is not needed in the high quantities found in most American diets, which often include excessive amounts of salt (sodium chloride). Physicians recommend ingesting about 1.2 grams of sodium per day. This corresponds to 3 grams of salt and is twice the estimated minimum requirement. Most Americans exceed the recommended daily sodium intake, sometimes by three to four times. The major culprits are processed foods and "fast foods" which are very heavily salted for flavor. For example, a typical "double burger" contains nearly one gram of sodium,

Figure 12.10

Periodic table indicating macrominerals, microminerals, and some trace minerals necessary for human life.

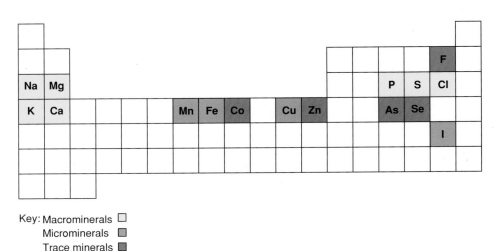

Key: Macrominerals □
     Microminerals ▨
     Trace minerals ▪

and that's before the accompanying french fries are salted! The very real concern with excess dietary sodium is its correlation with high blood pressure.

You might recall that **potassium** is included in cereal nutrition information data and is also listed in Table 12.1. Even though less abundant in the body than its sister element, sodium, potassium is essential to intra-cellular enzyme activity and the transmission of nerve impulses. Peaches, watermelon, bananas, and potatoes are rich sources of potassium and can be eaten to meet the estimated 2 g (2000 mg) daily requirement. Chemically, potassium is similar to sodium because they both have one electron in their outer level as neutral atoms and an octet of outer electrons as ions. Potassium is the chief cation within cells, more abundant in them than sodium, but in much lower concentration outside of cells. The regulation of the cellular concentration of K^+ relative to Na^+ is especially important for the proper rhythmic beating of the heart. Individuals who are on diuretics commonly take potassium supplements to replace potassium excreted in the urine. However, such supplements should be taken only under a physician's direction, because of the potential danger that they could dramatically alter the potassium/sodium balance and lead to cardiac complications.

Iodine is essential because it helps to form the hormones of the thyroid gland, such as thyroxine, which regulate the basal metabolism rate. Iodine deficiency causes goiter, an enlargement of the thyroid gland (Latin *guttur,* throat). Curiously, moderate goiter was considered to be fashionable and a mark of beauty during the late Renaissance. The condition does not look particularly fashionable in Figure 12.11. In order to prevent goiter, adequate amounts of iodine can be obtained from seafood or from iodized salt, normal table salt (NaCl) to which 0.02% of potassium iodide (KI) has been added.

Figure 12.11
A woman in Bangladesh suffering from a severe goiter.

Popeye, the nautical cartoon hero, gets his mythical super strength from spinach, a vegetable rich in **iron,** another essential micronutrient. Without iron, we would die. In the body, iron is part of a molecular unit known as heme, which is critical to the transport of oxygen by hemoglobin and to the temporary storage of oxygen in myoglobin, especially in heart muscles. Insufficient iron in the diet causes iron-deficiency anemia, a condition in which the red blood cells are low in hemoglobin and correspondingly carry a decreased oxygen supply. Fatigue, listlessness, and decreased resistance to infection are symptoms of this condition. Iron-deficiency anemia is a major problem in developed as well as developing nations. Iron deficiencies have been estimated as high as 20% in the United States, particularly among post-puberty women. To be used by the body, iron must be absorbed as Fe^{2+} ions, not simply as elemental iron. Iron-fortified foods, such as breads and cereals to which iron compounds have been added, or iron supplements, are sometimes used to help offset iron deficiency. Iron in foods tends to be better absorbed by the body than that ingested in pill form. However, caution must be used because excess iron can be toxic.

"OUCH!...THE FDA SURE ISN'T PULLING ANY PUNCHES THESE DAYS!!"

(Reprinted with special permission of King Features Syndicate.)

■ *Food Preservation*

It is not enough for a society to have an adequate initial supply of food. That food must stay uncontaminated until ready to be eaten. For centuries people have used a variety of methods to preserve foods. Before modern refrigeration, the traditional methods of salting or storing foods in concentrated sugar syrups were used. Both of these methods rely on having the concentration of solute (salt or sugar) in vast excess of that in any harmful organisms. Under these conditions, osmosis causes water to diffuse from the cells into the surrounding solution. As this occurs, the cell membranes are ruptured and the organisms killed. Heat is also used to kill microorganisms, as in home canning or pasteurization. Refrigeration retards, but does not ultimately prevent, spoilage.

In addition to these preservation methods, substances are added to foods in order to reduce spoilage and to extend their useful shelf-life. Anti-oxidants are one type of such additives used to prevent packaged, processed foods from becoming rancid due to oxidation of the oil or fat. If you examine the label of any such processed food (dry cereals, potato chips, and various other snacks), you are likely to see the letters BHT or BHA. These stand for butylated hydroxytoluene (BHT) and butylated hydroxyanisole (BHA), the two most common anti-oxidants added to such foods.

BHT   BHA

These compounds act as anti-oxidants by preventing a build-up of free radicals, which are molecular fragments formed when fats or oils react with oxygen from the air in the food package.

Fat + oxygen → Free radical·

A free radical has an unpaired electron, designated by the dot. This makes such species very reactive. (Recall that the role of free radicals in ozone depletion was men-

Figure 12.12
International symbol for irradiated food.

Food irradiation logo

tioned in Chapter 2.) By scavenging the unpaired electron from the free radical, BHT or BHA forms a stable radical species that prevents further oxidation of the fat, thus preventing rancidity.

A modern method of food preservation is by radiation. Of course, food is irradiated in a number of different circumstances, but not usually to preserve it. It's a case of what part of the electromagnetic spectrum is used. Needless to say, radiation is essential for the production of food. Visible light from the sun drives photosynthesis and gives us fruits and vegetables. We use longer wavelength infrared radiation from a stove, or still longer wavelength microwaves to cook food or warm up leftovers. However, irradiating food to preserve it is a different matter. That process uses short-wavelength, high-energy gamma radiation to kill microorganisms.

Classified as a food additive by Congress in 1958 and approved by the FDA in 1963, the preservation of food by irradiation is controversial. The method is used in over 30 countries around the world and in preserving food for astronauts to take with them as they circle the world. It has the enthusiastic endorsement of the Food and Agricultural Organization of the United Nations. Irradiated foods even have their own international symbol (Figure 12.12). So, why the controversy?

Opponents of irradiated foods question whether they are safe to eat. In addition, they raise concerns about the safety of the process, which uses cobalt-60 or cesium-137 sources. The effectiveness as well as the need for this technology are also questioned. Finally, there is the issue of the proliferation of radioactive materials to be used for questionable, possibly unnecessary commercial applications. But, there is also some common ground. Opponents and proponents agree that irradiating food does not make it radioactive, other than the initial normal background radiation it contains (see Chapter 8).

The most serious charge brought by critics concerns the formation of radiolytic products generated by the breaking of chemical bonds by gamma radiation. Recall that with its shorter wavelength, gamma radiation is much more energetic than microwave, infrared, or even ultraviolet radiation. Opponents of preservation by irradiation claim that the radiolytic products are unique, uncommon, and untested. Proponents say that such products are neither unique nor uncommon; some are produced by normal cooking. They also point out that radiolytic products are present in very small amounts, far less than the 3–4 grams per day we consume of other food additives which also have not been tested and are assumed to be safe. The radiolytic products are even in smaller concentrations than the 500 micrograms of natural cancer-causing materials in a cup of coffee or the nearly 200 micrograms of such carcinogens in a slice of bread. Irradiated foods fed to test animals have measured the efficacy of this preservation procedure. What has not been done is to isolate and test all of the radiolytic products, a process that some scientists feel is impossible to do with a sufficient level of certainty about the data that would be gathered.

Figure 12.13

An illustration of the effectiveness of gamma radiation in the preservation of strawberries. Those on the right were irradiated; those on the left were not. The photograph, taken after 15 days of storage, shows that the irradiated berries remained firm, fresh, and free of mold.

NON - IRRADIATED -   IRRADIATED - (0.2 M RAD)

STRAWBERRIES -
15 DAYS STORAGE 38°F (4°C)

In January 1992, strawberries preserved by irradiation went on the market, a product of the first large-scale commercial food irradiation facility in the United States (Figure 12.13). Fruit growers had been seeking an alternate large-scale method to kill insects in fruits since the use of ethylene dibromide was banned in 1984. Irradiation of seafood, pork, and poultry have also been approved by the FDA. It is estimated that as much as 40% of chickens sold in the United States are contaminated with salmonella bacteria. Food contaminated with this organism has been linked to 4000 deaths annually in this country alone. Irradiation of chickens would reduce the threat of accidental poisoning from salmonella. However, the major American chicken producers have said that they will not irradiate chickens to preserve them. Apparently they feel that the risks (or costs) do not yet outweigh the benefits.

12.16 ■ *Consider This*

As you have just read, emotions run high when people attempt to assess the risks and benefits of food preservation by irradiation. To evaluate the situation, find two newspaper or magazine articles on food irradiation—one supporting it and one opposing it. List the arguments cited for and against this method of food preservation. Then state your position on the issue, giving your reasons.

■ *Feeding the World*

Food preservation is important, but first there must be food to preserve. A nation can obtain its food only by growing it or importing it. These two sources lie at the very core of the world's food problem. Some areas of the world simply lack enough usable land, adequate soil, water, seeds, money, or mechanization. Prolonged droughts limit water while episodic floods wash away crops. Poor soils cause low crop yields and fertilizers may not be available. Beasts of burden, to say nothing of tractors, may be too expensive for the farmer. In many developing nations, arable land (that suitable for cultivation) is owned by only a very small percentage of the population who control what is grown and where the agricultural products are sent.

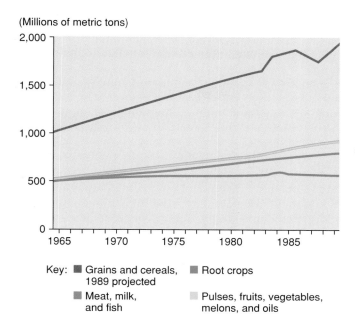

(Millions of metric tons)

Figure 12.14

World production of selected food crops. (Data from the Food and Agriculture Organization of the United Nations.)

Key:
■ Grains and cereals, 1989 projected
■ Meat, milk, and fish
■ Root crops
□ Pulses, fruits, vegetables, melons, and oils

Nations that need to import food must be able to pay for it; they must have products to sell. Some developing countries grow adequate amounts of grain to be self-sufficient, but rather than use the grain for domestic human consumption, it is fed to cattle that are exported to richer nations. Profits earned from exporting cattle or other products are generally reinvested to enlarge the export-product base, rather than to improve agriculture or to ensure sufficient grain to feed their people. This, then, becomes a vicious cycle in which government funds are used to create an exportable product, but not necessarily to provide an adequate food supply. There are economists who feel that such funds would be better used to raise grain and other basic foodstuffs needed to feed the domestic population, rather than to contribute to an economy based on international trade.

The worldwide production of grains and cereal crops is commonly used as an indicator of the world's food supply and the ability of the world to feed itself adequately. Figure 12.14 indicates that global grain and cereal production has grown rapidly over the past 25 years. But so did the world's population, and it is the per capita food supply in a particular region that determines feast or famine.

The need to feed the peoples of the world is infused with a bitter irony. On the one hand, the quantity of food currently produced on the planet is sufficient to feed its entire population. Indeed, it has been estimated that at current levels of production, there is enough food to provide for 6.1 billion people—the projected world population at the turn of the twenty-first century. Nevertheless, today over 500 million of our fellow human beings are hungry and undernourished, more than at any other time in human history. Clearly, the world's food supply is not equally distributed among its inhabitants. Piles of corn and wheat rot in the American midwest or on docks around the world while half a billion men, women, and children go to bed hungry.

There are many reasons for this inequity, reasons that are economic, political, and social, as well as agricultural. Some have to do with geography. Because of marginal soils, inadequate water, or limited arable land, some countries simply cannot produce enough food for their people, especially if the population is burgeoning. For others, there is insufficient money to support agriculture adequately or to buy supplemental food imports. And, as the civil strife in Somalia painfully illustrates, in some countries political and military actions block the flow of food and agricultural products.

The disparity is particularly great between the developed and the developing nations. The developed nations, including the United States and Canada, produce nearly 50% of the world's food, but have only 20% of its population. Figure 12.15 illustrates

a. Developing regions

Index numbers
(1965 = 100)

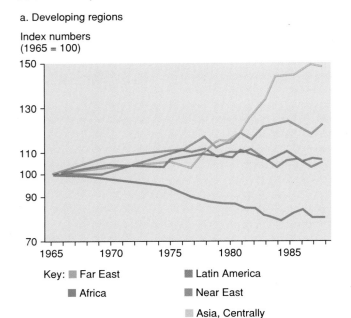

Key: ■ Far East   ■ Latin America

       ■ Africa   ■ Near East

       ■ Asia, Centrally
           Planned Economies

b. Developed regions

Index numbers
(1965 = 100)

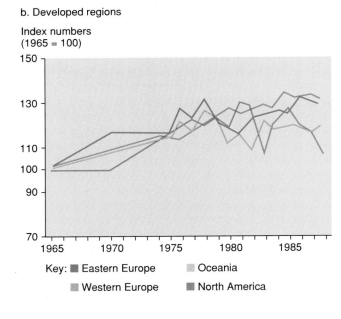

Key: ■ Eastern Europe   ■ Oceania

       ■ Western Europe   ■ North America

Figure 12.15

Per capita food production in various regions. (Sources: For 1965 and 1970, United Nations Food and Agriculture (FAO), *1987 Country Tables* (FAO, Rome 1987), pp. 312–36; for 1975–76, FAO, *1986 Production Yearbook* (FAO, Rome, 1987), p. 48; for 1977–88, FAO, *FAO Quarterly Bulletin of Statistics,* Vol. 2 (1989), pp. 17–18.)

the variability in per capita food production among developed and developing countries over the period 1965–1988. The 1965 per capita food production is used as an index for comparison, so 1965 represents 100%. Subsequent increases or decreases are evaluated relative to that index.

Two trends are evident from Figure 12.15. Overall, per capita food production in developed countries has increased more rapidly during the nearly 25-year span than it has in developing countries. Even in the developing countries, food production has generally kept pace with the growing populations, and in the case of Asia, it has significantly exceeded it. In Africa, however, there has been a long-term, continuing decline in per capita food production.

Perhaps the dramatic increase of food production in developing Asian countries can hold some promise for the plight of Africa. Certainly one reason for the increase in Asian crop yields has been greater use of fertilizers and pesticides. The application of both has often been criticized as being harmful to the environment, but the fact remains that millions have been saved from starvation thanks to fertilizers and pesticides.

An even more striking and generally less controversial contribution to world agriculture has been the **Green Revolution.** Fundamental to this enterprise has been the development of high-yield grains, principally wheat and rice, suited to particular regions. These new varieties mature faster, permitting more harvests per year, so that the same amount of cultivated land can produce more crops. But the Green Revolution is neither a panacea nor the ultimate answer. It has not been without costs, and it is not universally applicable. It has worked best in areas where money is available for supplemental fertilizers such as ammonia or nitrates, where water is abundant, and where technological understanding and application exist.

Genetic engineering and other applications of biotechnology now hold out promise for a second Green Revolution. Dr. Norman Borlaug, the Nobel Prize winner who developed the high-yielding wheats that spawned the first Green Revolution, sees such gains coming first in animal science and microbiology and only later in traditional field crops. In his opinion, "It will take considerably longer to develop biotechnological research techniques that will dramatically improve the production of our major crop species."

12.17	**Consider This**
■	It has been argued that sending food to countries facing famine is at best a temporary stop-gap solution. Suggest a strategy that will be effective over the long term in solving the problems of world hunger. Your plan should include specific recommendations for organization, administration, timing, and funding.

■ Conclusion

This chapter began by asking what it means to say "I'm hungry" in a number of languages. An American college student may expect a hamburger and French fries in return. A Japanese business executive may be eagerly anticipating an elegant array of sushi—raw fish and rice. A child in Somalia would be grateful for anything edible. Our tastes may vary, but our biological needs are much the same. We all need carbohydrates as our primary energy source; fats for cell walls, synthesis, and lubrication; proteins to build muscle and create the enzymes that catalyze the wonderful chemistry of life; and vitamins and minerals that help make that chemistry happen. People with too much to eat, like most Americans, are preoccupied with food. So are the hungry and the starving; they think of little else. It is a truism that if all individual dietary needs could be met, problems of global nutrition would also be solved. How to accomplish that end, when people are starving in the midst of planetary plenty, remains one of the great challenges of our time. Chemistry is only part of the solution.

■ References and Resources

Food and Agriculture. A Scientific American Book. San Francisco: Freeman, 1976.

Mela, D. J. "Nutritional Implications of Fat Substitutes." *Journal of the American Diabetes Association* **92,** (1992): 472–76.

Shapiro, L. "The Fat That's Good for You." *Newsweek*, Jan. 6, 1992: 48.

The State of Food and Agriculture: 1990. Rome: Food and Agriculture Organization of the United Nations, 1991.

Thayer, A. M. "Food Additives." *Chemical & Engineering News*, June 15, 1992: 26 ff.

World Resources Institute. *World Resources: 1990–91.* New York: Oxford University Press, 1990.

■ Experiments and Investigations

18. The Worrisome Three, Part I

19. The Worrisome Three, Part II

20. Vitamin C Content in Foods

23. Determination of Iron Content in Foods

■ Exercises

1. According to the text, 27% of United States citizens consume 1000 excess Calories per day. Taking the current population to be 250×10^6, calculate the number of excess Calories consumed daily. If the minimum daily requirement is 2000 Calories, how many extra people could be fed by this excess?

2. Use Figure 12.2 to determine which of the foods listed would be the best source of carbohydrates or protein and which should be avoided if one wished to control the intake of fat.

3. Examine Table 12.1 and explain why hydrogen ranks first in atomic abundance in the human body, but third (behind oxygen and carbon) in terms of mass percent.

4. Account for the differences in human ability to digest starch and cellulose on the basis of the lock-and-key model (Chapter 11).

5. Explain the fact that cellulose, starch (and proteins) are considered polymers whereas fats are not, even though they are all made from smaller molecules by eliminating a molecule of water.

6. At one time soaps were made from fats by boiling fats in a sodium hydroxide solution. Write a chemical equation for such a reaction, identify the portion of the initial fat that functions in the soap and explain how it does so. (See Figure 5.14 and the accompanying discussion.)

*7. Oleic, linoleic, and linolenic acid each contain 18 carbon atoms per molecule. From the formulas given in the chapter, determine to what extent the increasing degree of unsaturation affects the percent of carbon present.

8. Saturated fats generally have higher melting points than unsaturated fats of comparable molar mass as indicated by the data in Table 12.2. Account for this observation in terms of the structures of these molecules. (See the discussion of HDPE and LDPE in Chapter 10.)

9. Explain why protein molecules can exhibit optical activity.

*10. Determine how many tripeptides can be formed by the combination of three different amino acids if each amino acid can be used

   a. only once.

   b. more than once.

11. The 20 common amino acids can be arranged in a mind-boggling number of ways to form different proteins. If a protein consists of n amino acid units, drawn from a pool of 20 different amino acids, the number of possible different molecular arrangements is 20^n. The hormone insulin is a small protein made up of 51 amino acid units. How many different proteins with this number of amino acids might exist? What are the implications of your answer?

*12. The human body is approximately 60% water, and one half of the dry portion is protein. Calculate the mass of protein in a 150 lb (68 kg) human and determine the approximate number of amino acids needed to make that protein. (Assume the average amino acid has a molar mass of 120 g.)

13. Suggest an explanation for the decrease in daily protein requirements (g/kg of body weight) with increasing age.

14. How long would you have to study to burn up the 150 Calories contributed by a bag of potato chips? How long would you have to jog to accomplish the same goal?

15. Briefly explain why the DES values listed in Table 12.8 do not present the full picture of the nutrition picture in places like Somalia.

16. Vitamins are specific chemical compounds, whether they occur naturally or are produced synthetically. Yet vitamins from certain natural sources, for example vitamin C from rose hips, command a premium price in health stores. Explain whether or not there is a justification for this pricing practice.

*17. The recommended daily intake of sodium is 1.2 g and that of potassium is 2.0 g. Determine the number of Na^+ and K^+ ions in these masses.

*18. Calculate the number of milligrams of potassium iodide present in one pound (454 g) of table salt to which 0.02% of KI (by mass) has been added? How many milligrams of iodine does this mass of KI provide?

19. Some iron-fortified cereals contain iron in its elemental form. (You can attract the iron filings from a water-slurry of the cereal with a magnet.) From what you have read about the ionic form of iron used by the body, comment on the claim that these cereals provide 100% of the body's daily iron requirement.

20. The per capita direct consumption of grain as human food is about the same in the United States, India, and Mexico. However, the total per capita use of grain in the United States is several times that for the other two countries. Explain this apparent inconsistency.

21. Estimate your beef consumption for one year. Using the information in the text, calculate the number of pounds of grain needed to produce this quantity of beef. If this grain were used to feed a person in a poorly developed country, how many days of food would it provide?

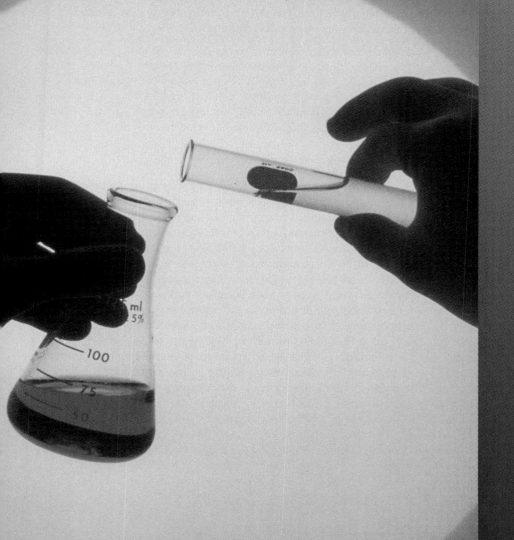

13

The
Chemistry of
Tomorrow

■

It is the year 2030. You are a technical representative for a computer company located near Palo Alto, California, in what was called "Silicon Valley" back in the 1990s. Now it's called "Germanium Arsenide Valley." You receive an urgent message on your microwave electronic mail system that a client in Los Angeles needs a special integrated molecular circuit component for a high speed superconducting supercomputer. Although science and technology have made great strides since the primitive days of the late twentieth century, there is still no way to fax computer parts and the technicians to install them. You have no alternative but to go to L.A. and make the repairs.

Fortunately, it's no big deal. You grab the part and rush down to the station just in time to catch the levitated train from San Francisco. It appears suddenly and soundlessly, riding on a magnetic field, its sleek composite skin shining in the Sun. You enter the car, settle into your individually climate-controlled chair with its personal stereo and television system, and prepare to enjoy the trip. The only problem is that at 300 miles per hour, the train is moving too fast to permit you a good view of the scenery. Instead you decide to call some of your clients in Boston, Kobe, and St. Petersburg. In fact, you manage to accomplish quite a bit during the hour and half you have on this 400 mile ride. As you get off the train in Los Angeles you notice that the air quality is really very good. They finally seem to have licked the smog problem. With any luck you should be able to finish the job, catch the 4:00 p.m. northbound train, and be back for your daughter's birthday party at 6:00. Now if you could only find a taxi! The shift to electrically powered ceramic engines hasn't made cabs any more plentiful or the drivers any more polite!

Fantasy? Perhaps so, but much of it may come true. All of the elements in this scenario are being seriously considered by scientists and engineers. And nowhere are things changing more rapidly than in materials science and in the fantastic forms of matter created by modern chemistry. Mary Good of Allied-Signal Chemical Company, a former president of the American Chemical Society, has summarized it thus: "Ultimately, materials development will determine our standard of living."

■ *Chapter Overview*

In this chapter we look to the future and focus on areas of innovation where chemistry will undoubtedly transform human life through the transformation of matter. There are dozens of possible examples, but we restrict ourselves to only three. We first return to the superconductors that introduced the chapter and explore these revolutionary devices. We briefly consider the phenomenon of superconductivity and some of its applications in science and medicine. Because the recent discovery of high-temperature superconductors may be of revolutionary significance, we devote several pages to the structure and properties of these interesting materials and speculate about some of their possible uses.

Next we turn to catalysts, the marvelous materials that make possible the production of many other new forms of matter. Among the topics considered are the mode of action of catalysts, some specific examples, and a variety of applications including manufacturing, environmental clean-up, and energy production.

Biological catalysts or enzymes provide a link to the final example of the chemistry of the future—the way in which chemistry can be used to manipulate the molecules of life. After a brief review of protein structure we consider the mechanism by which enzymes influence reaction rates and pathways. A discussion of enzyme synthesis soon leads to DNA and the molecular basis of genetics. This, in turn, prompts a final section on the risks and benefits of genetic engineering.

■ *Supercold Superconductors*

The levitated train has become sort of a contemporary magic carpet. The idea of tons of train and passengers floating on a magnetic field and propelled at high speeds by a modest push sounds very appealing to someone stuck in bumper-to-bumper traffic, or fogged-in at an airport. Engineers began thinking seriously about the possibility in 1987, when two American scientists, Paul Chu of the University of Houston and M.K. Wu, Jr., of the University of Alabama announced a remarkable scientific discovery. They had succeeded in making a ceramic that exhibited superconducting properties at temperatures as high as −196°C. Although −196°C does not sound particularly high (it corresponds to −385°F), it is the temperature of boiling liquid nitrogen. Just why that is a breakthrough requires a little explaining and a little history.

Around the turn of this century, the Dutch physicist, Heike Kamerlingh Onnes (1853–1926), first liquified helium. Under ordinary circumstances helium is a gas made up of individual He atoms. The attractive forces between these small, compact atoms are so weak that they have very little tendency to approach each other closely enough to form a liquid. Instead, they keep moving in the chaotic pattern characteristic of gases. However, as gaseous helium is cooled, the motion slows, and as the pressure on the gas is increased, the atoms are squeezed closer together. When the temperature is low enough and the pressure high enough, gaseous helium is converted into a liquid.

Liquid helium has a boiling point of −269°C at a pressure of one atmosphere. This temperature is very close to the coldest temperature possible—**absolute zero** or **−273°C**. At −273°C the random thermal motion of atoms and molecules ceases, and solids become perfectly ordered. This temperature becomes the zero of the Kelvin or absolute scale. This scale is related to the Celsius scale by the following equation.

$$T \text{ (in Kelvins)} = t \text{ (in degrees Celsius)} + 273$$

Thus, the boiling point of helium in Kelvins is −269 + 273 or 4 K. All temperatures on the Kelvin scale are, of course, positive.

13.1	*Your Turn*
■	Express the following temperatures, given in degrees Celsius, in terms of the Kelvin or absolute scale.

    a. the boiling point of water: 100°C
    b. the temperature of the surface of the Sun: 6000°C
    c. the maximum operating temperature of a new superconductor: −163°C

Ans. **a.** 373 K

Once liquid helium became available, Kamerlingh Onnes and others began to observe and measure the properties of a variety of materials at these low temperatures. In 1911, the Dutch scientist was studying the way in which the electrical conductivity of mercury varies with temperature. As is the case with most metals, mercury becomes a better conductor as the temperature decreases. Metals can be regarded as positive ions in a sea of electrons. When an electric field is applied across the metal, the electrons move, transporting the current. As the metal gets cooler, the ions vibrate less, causing less interference for the migrating electrons. What was unexpected, though, was the large, abrupt increase in the conductivity of mercury that occurred at about 4 K. The metal appeared to exhibit zero resistance to the flow of electrical current. The first **superconductor** had been discovered. Since that time, superconductivity has been observed in the 26 elements highlighted in Figure 13.1. In none of these cases does the phenomenon occur at temperatures above 10 K.

Figure 13.1
Skeleton periodic table with superconducting elements highlighted.

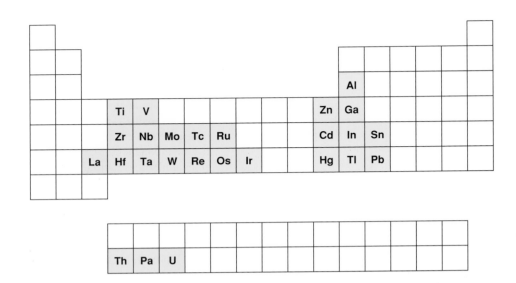

But superconductivity is much more than a fascinating (and not fully understood) scientific curiosity. The phenomenon is of tremendous potential utility because so much of modern technology involves the passage of electrical current. Electric generators, motors, electromagnets, and transmission lines use millions of miles of copper wire—a good but not perfect conductor. If the wire could be replaced by a superconductor, the savings in energy would be monumental and undreamed of devices might be possible.

13.2   Consider This

Examine Figure 13.1 and suggest some chemical generalizations about the elements that can act as semiconductors.

■ NMR and MRI

The first large-scale practical application of superconductivity has been in superconducting magnets for scientific research and medical diagnosis. Most of these applications relate to the phenomenon of **nuclear magnetic resonance (NMR).** The nucleus of an atom of ordinary hydrogen atom can be regarded as similar to a spinning top. In the presence of a magnetic field, the energy associated with a clockwise spin is not identical to that associated with a counterclockwise spin. The energy difference is very small, corresponding to the energy of a photon in the radio wave region of the spectrum. When the nucleus absorbs a photon of the appropriate energy, it flips from one spin state to the other. It turns out that the specific amount of energy necessary to cause this flip depends on the immediate electromagnetic environment of the atom involved. This environment, in turn, depends on the structure of the molecule. The NMR spectrum of a compound shows different energy absorption peaks for different hydrogen atoms within the molecule. If the radio frequency and the magnetic field are in the proper relationship, the nuclei will change their spins and a signal will be recorded. From this spectrum, chemists can often determine the arrangement of the atoms. The nuclei of a number of other isotopes, including carbon-13, nitrogen-14, and phosphorus-31, can also exhibit nuclear magnetic resonance and yield important structural information.

An NMR spectrometer consists of a powerful magnet, a radio transmitter, and a detector of radio waves. The sample is placed between the poles of the magnet, the transmitter is turned on, and the spectrum is measured. Some NMR spectrometers have permanent magnets, others have electromagnets, but the most powerful have superconducting magnets made of niobium-tantalum alloys or a compound of niobium and germanium with the formula Nb_3Ge. The principle behind a superconducting

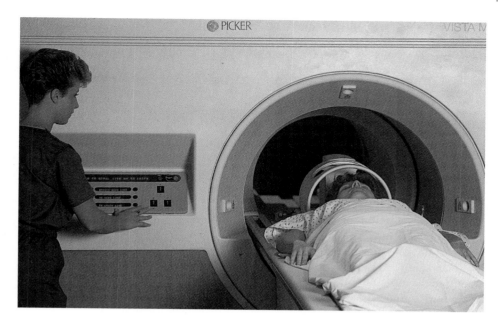

Figure 13.2
A patient in a magnetic resonance imaging (MRI) apparatus.

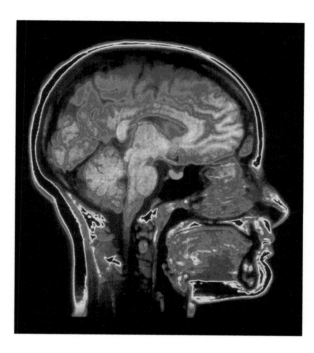

Figure 13.3
A color-enhanced MRI scan of a normal human brain. (The small blue object is the pituitary gland.)

magnet is similar to that associated with an electromagnet. A nail wrapped with wire connected to a source of electric current is an example of a simple electromagnet. As long as the electrons keep moving through the wire, they will induce a magnetic field in the iron core. But when a current of electrons is introduced into a superconductor, they keep flowing indefinitely, unimpeded by resistance. The result is the build-up of a strong magnetic field.

Recently, this sort of equipment has been scaled up so that a human being can be placed between the poles of the magnet (Figure 13.2). The technique is called **magnetic resonance imaging** or **MRI.** (Presumably "nuclear" was dropped because someone decided the average citizen was too afraid of that word to ever agree to the procedure.) An MRI instrument can detect differences in the energy absorption by hydrogen nuclei in various sorts of cells. In particular, the technique can be used to locate and identify tumors (Figure 13.3). MRI is thus a valuable (and expensive) diagnostic tool in modern medicine.

■ *High-Temperature Superconductors*

The superconducting devices just described have all suffered from one practical problem: they can only operate near absolute zero. Liquid helium is an appropriate coolant, but it is expensive and inconvenient. Therefore, the announcement of the synthesis of a substance that superconducted at around 80 K was greeted with enthusiasm. Liquid helium could now be replaced with the more plentiful, higher boiling liquid nitrogen, the substance frequently demonstrated as "liquid air."

Nitrogen boils at a higher temperature than helium because the attractive forces between N_2 molecules are greater than those between He atoms. It takes more energy and a higher temperature to overcome them and vaporize the liquid. Looking at it from the opposite direction, nitrogen is easier to condense than helium. The explanation for this is that an N_2 molecule is considerably larger than an He atom. Its electrons are farther from the nuclei and less tightly held. This means that the 14 electrons in N_2 can be more readily distorted than the two electrons in He. As the electrons shift around, one part of the molecule can temporarily experience a build-up of negative charge, while another part is temporarily positive. The positive part of one N_2 molecule is attracted to the negative part of another, giving rise to what are called **dispersion forces.** In a group of similar elements or compounds, the dispersion forces become stronger as the sizes of the atoms or molecules increase. This explains why the boiling points of the rare gases increase down the far right column of the periodic table, from helium through radon (Rn). The same trend accounts for the fact that fluorine and chlorine are gases at room temperature, bromine is a liquid, and iodine is a solid.

13.3	**Your Turn**
■	List the following compounds in predicted order of increasing boiling points.

a. C_8H_{18} **b.** CH_4 **c.** C_4H_{10}

The newly discovered substances that superconduct at liquid nitrogen temperatures are ceramics, which are made by grinding and heating certain metallic oxides and carbonates. Perhaps the most thoroughly investigated have been the 1-2-3 superconductors, so called because of the formula $YBa_2Cu_3O_{7-x}$ (Figure 13.4). The three metallic elements involved are yttrium (Y), barium (Ba), and copper (Cu), and their molar ratio is 1:2:3. The formula is unlike any you have seen in this text because the amount of oxygen is not fixed. The x in the subscript, $7-x$, is approximately 0.1, but it can vary somewhat. Figure 13.4 shows the way in which the atoms are arranged in a crystal of this superconducting compound.

■ *Superconductors for a Super Tomorrow*

Some of the early optimism over the possible practical applications of high temperature superconductors has been tempered by manufacturing problems. The 1-2-3 superconducting compounds are easy to make, but quite expensive to buy. The 1992 Aldrich Chemical catalog lists a price of $323.40 for 250 g (about half a pound). The greatest problem appears to be in fabricating the oxide into the forms that would be most useful—especially wires and cables. Like most ceramics, these compounds are brittle and break easily. There has been some success in forming thin superconducting films on metal wires or tapes, but much work remains to be done before the great potential of high-temperature superconducting can be realized.

The ultimate goal is to make a superconductor that will work at room temperature. All superconductors appear to have a **critical temperature** above which they lose their remarkable conductivity and become electrical insulators. The highest critical

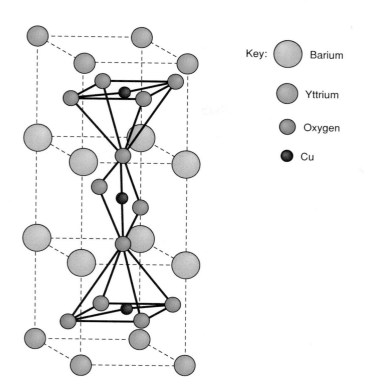

Key:
○ Barium
○ Yttrium
● Oxygen
● Cu

Figure 13.4
Crystal structure of $YBa_2Cu_3O_{7-x}$, a 1-2-3 superconductor.

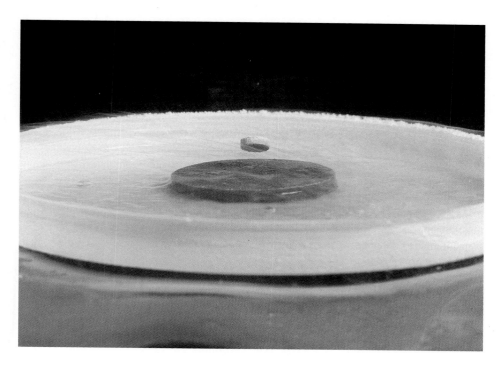

Figure 13.5
Photograph of Meissner Effect. The small magnet is floating over a superconductor cooled with liquid nitrogen.

temperature achieved thus far is about 110 K. As long as a refrigerant such as liquid nitrogen is required, some of the more inviting applications remain expensive and elusive. The magnetically levitated train is an example. A magnet brought near a superconductor will induce a moving current and an opposing magnetic field in the surface of the superconductor. The magnet will therefore be repelled and float above the superconductor in what is called the Meissner Effect. This unusual property is pictured in Figure 13.5.

Figure 13.6

A magnetically levitated train being developed and tested in Japan.

To reduce the Meissner Effect to practice requires the use of superconductors as either the rails or the wheels of a transportation system. Given current superconductors, a complicated low-temperature cooling system is also needed. In spite of these difficulties, the Japanese have tested a magnetically levitated train, using helium-cooled superconductors (Figure 13.6).

Low critical temperatures also bedevil attempts to use the current crop of superconductors as transmission lines for electric power. Another complication is the fact that superconductivity disappears at certain levels of current flow or magnetic field. Unfortunately, most of the proposed uses of superconductors involve high current flows and/or high magnetic fields. Thus, the system fails when it performs its intended function. Nevertheless, a major revolution could be achieved by resistance-free long-distance power transmission lines, superconducting coils for long-term storage of electrical energy, highly efficient electrical generators and motors, superfast computers, and magnetic levitation. As a consequence of these potential applications, a good deal of research and development is being carried out. In the United States, there are at least two major collaborative projects involving industry, academic institutions, and government laboratories. But Japan remains the leader in superconductivity research. Japanese investigators have filed more patent applications in this area than the rest of the world combined. Perhaps our starting scenario would be more likely to come true if the cities involved were Tokyo and Osaka. For the present, it seems most likely that the first applications of high temperature superconductors will be in NMR and MRI instruments, in detectors of infrared, microwave, and other forms of radiation, and in special research installations.

Probably the most discussed research application of superconducting technology is the superconducting super collider or SSC. This is a proposed multi-billion dollar accelerator that would fire nuclear particles at each other with fantastic speeds and energies. From the impacts and the fragments formed, physicists hope to learn fundamental facts about the structure and ultimate nature of matter. The SSC is being built in Texas, but the massive expenditure of federal funds for the program remains controversial. Not all scientists think this a wise use of limited national financial resources, because such big ticket items of big science significantly limit the funding available for more modest research projects. The next activity gives you an opportunity to study the issue and express your opinion.

13.4 **Consider This**

The superconducting super collider requires superconducting magnets capable of generating very high magnetic fields. Assuming for the sake of argument that the SSC is an important national priority, should the design and construction of the SSC continue using the currently available helium-cooled superconductor technology or should it be deferred until the use of high temperature superconductors becomes feasible? List the questions you would like answered and try to find the answers. Then draft a letter to the President's advisor for Science and Technology expressing your opinions and citing your reasons.

13.5 **Consider This**

This unit started with a fantasy. Now it's your turn to construct one. What applications of superconductors not mentioned above can you devise for entertainment, amusement, or any other purpose?

■ *Catalysts: Agents of Change*

You are literally surrounded by substances that do not occur in nature. You are probably wearing clothing woven from fibers that contain at least some polymeric materials produced in a chemical plant rather than a botanical plant. Your food comes wrapped in transparent film made from petroleum. The transistors in your portable tape player and the tapes you play in it would not exist without the science of chemistry. The drugs and medicines that control disease and prolong life are largely the products of chemical synthesis. Without synthetic materials, your life would be very different, and very likely a good deal shorter. And it seems inevitable that in the future you will encounter many materials that are, at present, only an idea in the mind of a chemist.

The unnatural products that have become indispensable for our modern lifestyle are formed in chemical reactions first carried out in laboratories and eventually in factories. In a very fundamental sense, reactions are what chemistry is all about. Therefore, it is appropriate that the Chinese characters representing "chemistry" are literally translated as "the study of change." At the heart of the discipline is the transformation of materials and properties—the reorganization of matter. In this book you have encountered dozens of chemical reactions, but only a minuscule fraction of those that are known. On several occasions we have mentioned the role played by **catalysts** in promoting certain reactions: the breakdown of large hydrocarbon molecules to smaller ones in the catalytic cracking of crude oil, the conversion of coal and steam into gasoline, and the polymerization of small molecules into large chains. In every case, the catalyst is involved in the reaction, usually to speed it up or to influence the products formed. But the catalyst is always regenerated and does not undergo permanent change. In short, a catalyst provides the chemist with a way of promoting and controlling natural change.

Note the emphasis on *natural* change. No catalyst in the world can cause a reaction to occur that violates the laws of thermodynamics. Thus, carbon dioxide and water cannot spontaneously combine to form glucose and oxygen without the absorption of energy. Even if energy is available, such a transformation is highly improbable. You are likely to be disappointed if you attempt to sweeten your soda water by simply exposing it to sunlight. On the other hand, given the right catalyst (a compound called chlorophyll) and the right sort of chemical plant (a green one!) such a transformation is not only possible, it is commonplace.

Chinese characters for *huā-xue*, "the study of change."

Biological catalysts or **enzymes** abound in nature, and the chemist attempts to mimic nature by designing catalysts to promote reactions that would otherwise occur slowly or with great difficulty. Obviously, the rate of reaction is of economic significance to a corporation set up to produce ammonia from nitrogen and hydrogen or plastic bags from ethylene. The most direct way to increase the speed of a reaction is to increase the temperature of the reaction mixture. The added energy helps the reactants overcome the activation energy barrier (Chapter 4). But increasing the temperature increases fuel demands and consequently raises costs and environmental deterioration. Moreover, many reactions are more likely to yield unwanted side-products at elevated temperatures. The yield of a desired product formed in a given amount of time can even drop as the reaction temperature increases, especially for reactions that give off energy.

Catalysts make it possible to increase the rate of reaction without increasing temperature. They typically do this by providing alternate reaction pathways with lower activation energies. Some catalysts can speed up reactions by as much as 10 billion times. Furthermore, catalysts can pick out one reaction, and hence one product, out of many possibilities. Given this power to influence the rate and direction of chemical change, it is not surprising that catalysts play an essential role in almost all industrial chemical processes. Many of the manufactured products we take for granted would simply not exist without catalysts. It has been estimated that 20% of the gross national product of the United States is, in one way or another, accounted for by catalysts. No wonder that the metaphor has been borrowed and the word "catalyst" applied to any individual or agent that promotes social, political, or economic change.

In the pages that follow, we consider three of the most important areas where catalysts have been employed—in creating new materials, in protecting the environment, and in providing more convenient or more efficient energy sources. You will soon see that there is, inevitably, a good deal of overlap among these categories.

13.6   *The Sceptical Chymist*

Occasionally chemistry textbooks define a catalyst as "a substance that influences the rate of a chemical reaction without taking part in the reaction." Critically analyze this definition and point out why it must be incorrect.

■ *Creative Catalysis*

Many of the materials you use every day owe their existence to catalysts. This is especially true of polymers (Chapter 10). Every year, 2 million tons of high-density polyethylene are used to mold vast numbers of bottles and other objects. The long molecular chains of this polymer are formed when individual molecules of ethylene ($CH_2=CH_2$) combine under the influence of a catalyst. The most commonly used catalysts for this reaction are chromium (Cr) and titanium (Ti). These metals are used in the solid state and ethylene is a gas. Thus the reaction involves two different physical phases and is an example of **heterogeneous catalysis.**

In heterogeneous catalysis, the reaction occurs on the surface of the catalyst. Therefore, the catalyst is finely divided in order to increase its surface area. It is not uncommon for a single gram of a catalytic powder to have a total surface area equal to that of a tennis court. The starting materials for a particular synthesis are absorbed on this surface and held in close proximity to enhance their likelihood of reaction. Often the interaction between the catalyst and the absorbed molecules helps promote the breaking of bonds and the forming of new ones. In the case of ethylene, the double bond opens and the electrons previously involved in the double bond become available to form bonds with other $CH_2=CH_2$ molecules, thus creating the polymer.

Some catalysts improve on nature by providing cheaper alternative routes to naturally occurring materials. Acetic acid, CH_3COOH, is formed by the fermentation of sugars in grains, apples, grapes, and many others plant products. This is the source of the wide range of vinegars available in specialty food shops. But this reaction, catalyzed by naturally occurring enzymes, is relatively expensive and it cannot produce sufficient quantities of acetic acid for its many industrial uses. Each year, almost one billion pounds of acetic acid are used in the manufacture of vinyl acetate (for coatings) and polyvinyl alcohol polymers. Much of it is made by the direct combination of methyl alcohol and carbon monoxide.

$$CH_3OH + CO \rightarrow CH_3COOH$$
$$\text{methyl alcohol} \qquad \text{acetic acid}$$

When this reaction is catalyzed by the rhodium dicarbonyl diiodide ion, $[Rh(CO)_2I_2]^-$, it produces acetic acid in high yield and excellent purity. The catalytic ion is actually dissolved in the reaction mixture, so the reaction occurs in solution, not on a surface. This is an example of **homogeneous catalysis.** The starting materials, CH_3OH and CO, are in turn produced from coal and water by the action of a catalytic mixture including the oxides of zinc, copper, and aluminum.

All of the catalysts identified in these examples involve metallic elements that are found in the middle of the periodic table. Chromium, titanium, rhodium, zinc, and copper are **transition metals.** There are a total of 30 transition elements in 3rd through the 12th column of the periodic table (Figure 13.7). Included are some of the most common structural metals—iron (Fe), nickel (Ni), tungsten (W)—and some of the most precious—silver (Ag), gold (Au), and platinum (Pt). Many of the transition metals have catalytic properties, and the most versatile and effective are often the rarest. Thus rhodium, platinum, ruthenium (Ru), palladium (Pd), and iridium (Ir) can be used to catalyze many reactions. The fact that catalysts are not consumed in a reaction means that a relatively small quantity of these rare and expensive substances can produce many times their mass of product. Moreover, as we have seen, a tiny amount of platinum powder can have a huge catalytic surface.

In spite of the phenomenal success of catalytic chemistry, or perhaps because of it, a good deal of research is being conducted to create new catalysts that will be less costly and even more effective. We can be sure that some of the most important new materials of the twenty-first century will be catalysts. An area that seems especially promising is that of **metallic clusters.** Chemists have recently synthesized ions and molecules containing clusters of transition metal atoms. One of the largest is a compound of platinum having the formula $[Pt_{38}(CO)_{44}]^{2-}$. Other clusters consist only of the metal atoms. These relatively large molecules would seem to represent the ultimate in a finely divided catalyst. Therefore, there are high hopes that at least some of these clusters will constitute a new generation of super catalysts.

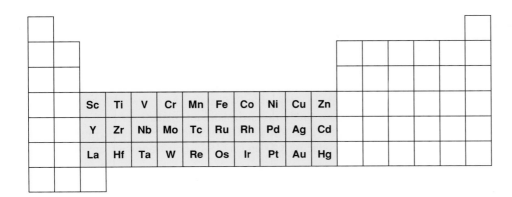

Figure 13.7

Periodic table with the transition elements highlighted.

For chemists, at least, the most exciting new cluster molecule contains no metal at all. It consists of 60 carbon atoms and bears the unlikely name **buckminsterfullerene.** The soccer ball structure of C_{60} has already been illustrated in Figure 10.10. We reintroduce this remarkable molecule here because it holds promise for some very interesting applications. Since the initial discovery in 1985, a number of fullerenes have been isolated, including a distorted sphere with the formula C_{70} and various tube-like molecules. Functional groups and atoms of other elements have been attached to the outside of the carbon framework, and metallic atoms have been trapped within the carbon cage. It has been suggested that some of these fullerene derivatives could be excellent catalysts. Others appear to have superconducting properties. And C_{60} molecules might serve as submicroscopic ball bearings. Whatever the ultimate applications, it is likely that "buckyballs" (as they are called) will be part of the Chemistry of Tomorrow.

Catalyzing a Cleaner Environment

Since 1976, every new car and truck sold in the United States has contained a catalyst designed to protect the environment (Figure 13.8). You already know from earlier chapters that an internal combustion engine releases a number of compounds that can contribute to atmospheric pollution, especially carbon monoxide, oxides of nitrogen, and unburned hydrocarbon fragments. The function of automotive catalytic converters is to transform these compounds into less harmful ones.

Figure 13.8

Cut-away photo of a car catalytic converter.

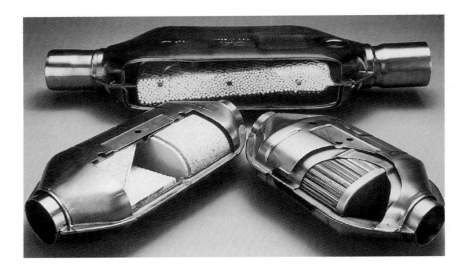

The idea is to take a slow but spontaneous reaction and speed it up. A case in point is provided by the conversion of carbon monoxide to carbon dioxide.

$$2\,CO(g) + O_2(g) \rightarrow 2\,CO_2(g) \tag{13.1}$$

Carbon monoxide is formed in the engine when there is insufficient oxygen present or insufficient burning time to completely convert the carbon from the gasoline into carbon dioxide. However, the reaction represented by equation 13.1 will occur if both carbon monoxide and oxygen are present and a little push is provided. In a catalytic converter the push comes from finely divided platinum, palladium, or rhodium on a ceramic support. To be sure, the CO_2 formed is a potential contributor to global warming, but at least the smog-producing CO and hydrocarbon fragments are eliminated from the exhaust.

Catalytic converters are less successful at removing NO and NO_2. You have learned that nitric oxide is formed by the direct combination of nitrogen and oxygen in the cylinders, where burning gasoline generates temperatures as high as $2500°C$.

$$N_2(g) + O_2(g) \rightarrow 2\,NO(g)$$

As the exhaust gases cool, there is a tendency for this reaction to reverse itself, converting the potentially harmful NO to its constituent elements.

$$2\,NO(g) \rightarrow N_2(g) + O_2(g) \qquad (13.2)$$

Although the dissociation reaction is slow, it is a natural candidate for catalytic acceleration. Unfortunately, reactions 13.1 and 13.2 can interfere with each other, because the former represents an oxidation and the latter a reduction. Moreover, reaction 13.1 generates heat, which inhibits 13.2. One solution has been to separate the catalytic converter into two stages, but even so, there is still need for additional research and the development of more effective catalysts.

In addition to reducing the emission of smog-causing compounds, the introduction of automotive catalytic converters had another environmentally beneficial side effect: it was a major factor in the removal of lead from gasoline. You recall from Chapter 4 that tetraethyl lead was, for many years, added to gasoline to increase its octane rating. The undesirable side of that coin was the release of vast quantities of lead. Even the glaciers of Greenland were not spared. Between 1940 and 1980, the level of lead measured in the snow there doubled. But the greater concern was in urban areas, where the possibility of increased frequency of cumulative lead poisoning alarmed some public health officials. Perhaps a greater stimulus to the drastic reduction of tetraethyl lead use was the fact that lead poisons platinum as well as people. Several tanks of leaded gasoline can cause a major reduction in the effectiveness of a catalytic converter, and consequently all cars and trucks sold since 1976 have been designed to run on unleaded gasoline. The quantity of leaded gasoline being sold has been reduced drastically, and will continue to decline in the future. The environmental effects are already noteworthy: the concentration of lead in the air fell 88% between 1978 and 1988. Thus, everyone seems to have profited from the decision to eliminate tetraethyl lead from gasoline. This includes the manufacturers of catalysts, because the new gasoline formulation has required the conversion of crude oil into new mixtures of hydrocarbons.

13.7 | **Consider This**

∎

Decisions such as the one requiring all new cars to operate on unleaded gasoline are driven by a complex array of forces and factors. Using this particular example, design a schematic flow chart representing the process that starts with a commitment to reduce air pollution and culminates in regulatory legislation. Include the input of government, industry, and the public; and address political, economic, and technical issues.

There is no doubt that efforts to improve environmental quality will continue to demand innovations in catalysis. For example, acid rain could be considerably reduced if new methods were developed to remove sulfur dioxide from power plant smoke stacks. Other efforts are being directed toward catalysts for the conversion of toxic wastes into less harmful forms. For example, cyanide can be destroyed by sunlight on a titanium dioxide surface. Progress has also been made by using the methods of molecular engineering (see below) to create new enzymes and new bacterial species that transform toxic substances into simpler, safer compounds.

Yet another strategy is to devise new catalytic procedures to replace environmentally detrimental manufacturing processes. The modifications can result in fuel savings associated with lower reaction temperatures, increased efficiency of conversion of reactants into products, or the elimination of unwanted by-products. New electrocatalytic technology has greatly reduced the serious contamination formerly associated with the chlor-alkali process. You will recall from Chapter 7 that mercury can escape from the cells in which salt brine solutions are electrolyzed to yield chlorine and sodium hydroxide. New cell designs eliminate mercury from the process, and thin catalytic films of ruthenium dioxide on the electrodes significantly increase efficiency and reduce maintenance.

■ *Catalysts and Energy*

Perhaps the greatest potential positive impact of catalysis on the environment is in the production of new and improved energy sources. Catalysts have long been part of petroleum refining. The cracking of hydrocarbon molecules containing 16 to 24 carbon atoms to species containing 7 to 9 carbon atoms is an important step in making gasoline. The catalysts used for this process are largely zeolites: substances containing aluminum, silicon, and oxygen atoms. Natural zeolites are related to mica and asbestos, and the ion exchangers you encountered in Chapter 5 are generally members of this same mineral group. Because solid zeolites contain holes or channels, they are referred to as "molecular sieves." Catalyzed reactions occur when molecules enter these channels. Because the size of the channel restricts the molecules that can enter it, it is possible to design zeolites with specific and selective catalytic properties. Indeed, most of the zeolites used in the petroleum industry today are artificially produced.

Heterogeneous metallic catalysts, including platinum, rhenium (Re), palladium, and iridium, are usually employed to rearrange or "reform" unbranched hydrocarbons into molecules that have a higher octane rating or that burn cleaner. Some of these changes in the composition of gasoline have been necessitated by the elimination of tetraethyl lead. Catalysts have also made possible the conversion of low grade crude oil to useful products. As high quality petroleum resources become further depleted, such transformations will become more important and more common.

Successful as chemists have been in developing new catalysts to make improved fuels, natural catalysts still serve as a model. A case in point is provided by the anaerobic bacteria that convert organic material into methane. To be sure, the CH_4 that is produced in rice paddies, marshes and swamps, cleared forests, and landfills can contribute to the greenhouse effect. But if this natural methane is trapped and burned (as it is in a limited number of facilities) it can serve as a source of relatively clean energy. An even better approach would be to learn the mechanism by which nature catalyzes the formation of methane and copy, adapt, or improve on it. It is known that nickel is involved in at least one of the enzymes, but the details remain to be discovered. With the appropriate catalysts it will become possible to convert biomass into convenient forms of energy. Fermentation, another naturally catalyzed process, already provides a method for producing ethyl alcohol from vegetable materials. A greater diversity of catalysts will provide a greater diversity of fuels, as well as biologically based intermediates for chemical syntheses.

The prospect of converting garbage to fuel is, of course, very attractive. But there is another great benefit associated with using biomass as a source of fuel. The growing plants remove from the atmosphere most of the CO_2 that is released by the burning fuels. Through the carbon cycle, the net increase in atmospheric CO_2 is kept to a minimum. Unfortunately, biomass can only supply a relatively small fraction of the Earth's energy needs. It has been estimated that one quarter of the planet's cropland would have to be removed from food and feed production in order to produce sufficient ethyl alcohol to meet 10% of the world's current energy demand.

In the long run, a more promising prospect than using green plants to feed our global energy hunger is to develop catalysts that emulate those plants in their ability to make fuel from carbon dioxide, water, and sunlight. Much time and effort have been devoted to studying photosynthesis, and the process is known in considerable detail. Because you have already encountered the topic elsewhere, a few major points should be sufficient here. An obvious first step is the absorption of radiant energy by the plant. This function is carried out by two closely related compounds: chlorophyll *a* and chlorophyll *b*. These green pigments absorb red light with a wavelength of about 700 nm at the low-energy end of the visible spectrum. It seems a little surprising that nature should use photons of such low energy to drive photosynthesis. One photon of red light does not have sufficient energy to split a water molecule into hydrogen and oxygen. That seems to contradict the message you read in Chapter 2. There we argued that the interaction between radiation and matter is an either/or situation—either a photon has enough energy to cause a chemical change or it doesn't, in which case nothing happens. But thanks to the elegant chemistry of photosynthesis, the energy of two photons is combined and the resulting energy is used to decompose water.

$$2 H_2O \rightarrow O_2 + 4 H^+ + 4 e^- \qquad (13.3)$$

The hydrogen ions and the four electrons are passed along a chain of intermediates in a process that ultimately results in the synthesis of glucose. The oxygen is released.

In order for us to chemically trap sunlight, it is not necessary to completely mimic photosynthesis and produce glucose. It would be sufficient to carry out a variant of reaction 13.3.

$$2 H_2O \rightarrow 2 H_2 + O_2$$

Efforts to find a catalyst for this reaction are a high priority, because hydrogen produced from water by solar energy is potentially an excellent fuel. It can be burned in power plants and cars, and used to generate electricity in fuel cells. As you already learned in Chapter 9, this could do much to solve our impending global fuel shortage and halt the growth of the greenhouse effect. And, in one of those happy coincidences that occur so frequently in this discipline, the chemistry involved is not only important, it is also fascinating. Currently, the most promising compound for capturing sunlight and using it to decompose water is ruthenium tris(bipyridine) chloride, $Ru(bpy)_3Cl_2$ (Figure 9.4). Although "rubippy" has the necessary electron transfer capabilities, it requires a catalyst such as finely divided platinum to optimize its efficiency.

■ The Chemistry of Life

Without a doubt, the human body is the world's most complicated chemical factory. Thousands of chemical reactions involving an even greater number of chemicals occur each second. Compounds are decomposed and others are synthesized; energy is used, released, and transformed; and information is transferred and processed. But in spite of the dazzling complexity of these processes, the last half century has seen a phenomenal increase in our knowledge of the chemistry of life. Indeed, in many respects, biology has become a chemical science. Today, the molecule and not the cell is the locus of much biological research.

The practical manifestations of this intellectual achievement are manifold. In 1900, life expectancy at birth in the United States was under 50 years, today it is over 75 (Figure 13.9). There are many reasons for this dramatic increase: better nutrition, improved sanitation, advances in public health, more accurate diagnoses, new medical

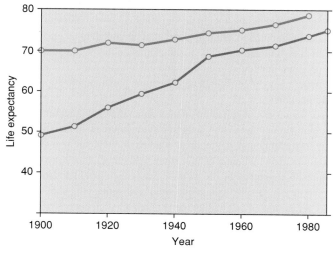

Figure 13.9

Life expectancy in the United States. (Reprinted by permission from *Opportunities in Chemistry: Today and Tomorrow* by George C. Pimental and Janice A. Coonrod. Courtesy of the National Academy Press, Washington, D.C. 1987.)

Key: ■ Life expectancy at age 45
■ Life expectancy at birth

procedures, and numerous medicines, drugs, and vaccines. Chemistry has contributed to all of these innovations, and it is an integral part of the latest revolution in health care—**biotechnology** and **molecular engineering.** There is little doubt that molecular engineering will profoundly alter human life in the twenty-first century.

Consider, for example, what scientists have learned about enzymes. We know that these biological catalysts are proteins—polymers built up of amino acids. One of the first proteins studied in structural detail was chymotrypsin, an enzyme that catalyzes the breakdown of other proteins in the small intestine. Each molecule of chymotrypsin consists of 243 amino acid residues and has a molecular mass of about 25,000. In descending order of abundance, it is made up of hydrogen, carbon, oxygen, nitrogen, and sulfur atoms.

If you studied Chapter 12, you already know that ordinary proteins are made up of various combinations of 20 different amino acids. All 20 share the common formula given below.

$$H_2N - \underset{\underset{R}{|}}{\overset{\overset{H}{|}}{C}} - COOH$$

You will recognize the functional groups introduced in Chapter 11. The amino ($-NH_2$) and acidic ($-COOH$) groups provide the name for this family of compounds. The individual members of the family differ in the identity of R, the group attached to the carbon atom that is situated between the amino and acid groups. A few examples are given in Chapter 12.

In that same chapter you read that the process by which amino acids polymerize to form proteins is an example of an acid/base reaction. The $-NH_2$ group has basic properties, and it reacts with $-COOH$ to eliminate an H_2O molecule and form what is called a peptide bond. This neat picture, summarized in equation 12.2 is a vast oversimplification of what actually happens when proteins are synthesized in a biological cell. There, other enzymes catalyze the reaction and, as we will soon see, nucleic acids direct the process with great specificity.

Again consider chymotrypsin and its 243 amino acid residues. The synthesis of a protein this size can be viewed as comparable to making a necklace by stringing 243 individual beads, selected from an assortment of 20 different kinds of beads. According to statistics, such a process could yield 20^{243} different possible combinations. Similarly, 20^{243} different protein molecules could be made from 243 amino acid residues. Expressed relative to the more familiar base 10, this number corresponds to 1.4×10^{316}—a number larger than the estimated number of atoms in the universe. Each member of this immense group of macromolecules would have its own **primary structure,** defined by the identity and sequence of the amino acids present. One specific primary structure is the biologically correct form of chymotrypsin with the desired enzymatic properties.

The primary structure of a protein determines the shape the molecule will assume under various conditions. Thanks to a technique known as X-ray diffraction, we have a good deal of information about the three-dimensional structure of these large molecules. The details of how a beam of X-rays bounces off the atoms in a crystal can be used to calculate the positions of the atoms. Such experiments show that the long protein chains are typically folded in complicated and quite compact structures. Thus, individual amino acids that are far apart in the primary structure of the protein may be close together in the folded form. Moreover, the shape of a particular protein molecule appears to be pretty much the same if variables such as temperature, pH, and solvent are kept constant.

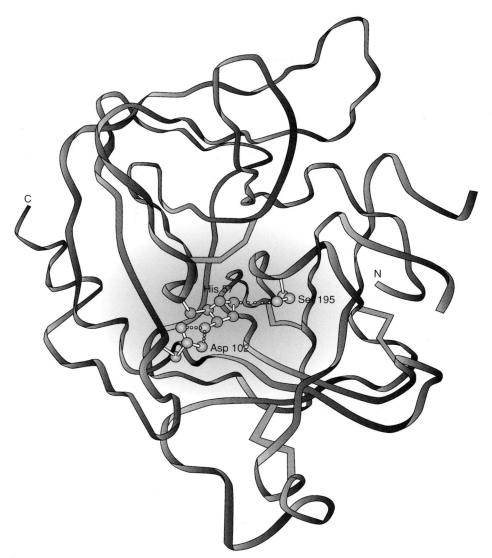

Figure 13.10
Tertiary structure of chymotrypsin.
The tape represents the polymerized
amino acid chain and the active site is
shown in color. (From B. S. Hartley
and D. M. Shotten, in P. D. Boyer,
(ed.), *The Enzymes,* 3d ed., Vol. 3.
Copyright © 1971 Academic Press,
Inc., Orlando, Florida. Reprinted by
permission.)

Evidence suggests that the three-dimensional conformation of a protein molecule
is stabilized by the interaction of various functional groups. Hydrogen bonds (see
Chapter 5) are particularly important in stabilizing structural subunits that occur in
many proteins. These include the helices and parallel chains that constitute the **secondary structure** of proteins. The overall shape of the molecule is termed its **tertiary
structure.** For example, Figure 13.10 depicts the tertiary structure of chymotrypsin.
The secondary structure is evident in the helical segments.

■ *How Enzymes Work*

The shape of a protein turns out to be closely related to its properties and functions.
Thus, the catalytic action of an enzyme such as chymotrypsin depends, in part, on the
way in which the chain of amino acids is folded. If an alteration in temperature or pH
changes the shape of the molecule, it loses its ability to catalyze the decomposition of
proteins. Such observations have led to greatly improved understanding of enzymatic
action. Every enzyme has a characteristic active site where the catalytic activity
occurs. The **substrate**—the molecules(s) whose reaction is catalyzed by the enzyme—
is held in close proximity to the active site. In some cases, the substrate forms chemical bonds to the active site, in others, it is temporarily attached to different parts of the
folded protein. The involvement of the active site in chymotrypsin promotes the

breaking of peptide bonds in the protein substrate. In certain other enzymes, the active site catalyzes the formation of bonds. In all instances, however, the orientation of the active site and the conformation of the rest of the molecule are of critical importance.

Improved understanding of the mechanism of enzymatic action has led to the design of new drugs (Chapter 11). It is known that some diseases can be attributed to elevated levels of hormones, neurotransmitters, or other biochemicals produced within the body. Enzymes catalyze the synthesis of these compounds. Therefore, a way to control such diseases is to slow down the rate of reaction by inhibiting the enzyme from carrying out its catalytic function. This can often be accomplished by introducing a drug that binds chemically to the active site of the enzyme. The discovery that aspirin inhibits an enzyme called cyclooxygenase has resulted in the design and synthesis of a number of other compounds that have similar inhibitory and analgesic properties. Other enzyme inhibitors are used to treat high blood pressure, asthma, and atherosclerosis.

■ Making Enzymes

Having learned much about how enzymes work and how to control their action, a next logical step is to actually produce enzymes. The easiest way to make enzymes is to let nature do it. Enzymes and their products were first obtained from bacteria, yeasts, and molds; and fermentation continues to be of great industrial and economic importance. Chapter 11 describes how penicillin was originally produced by the action of the enzymes from a certain strain of mold. Other naturally occurring enzymes are used to catalyze the conversion of carbohydrates, from renewable biomass, to a wide variety of compounds including ethanol, acetic acid, methane, and glycerol. These products have broad utility as fuels or as intermediates for chemical synthesis and manufacture. Today, new technologies are being developed to use natural catalysts even more efficiently.

Enzymes of bacterial origin have also been used to help solve environmental problems. Thus, one way to remove the residue of oil spills is to introduce bacteria that feed on the oil, breaking down the hydrocarbons into simpler and safer molecules. Other bacterial strains have been developed that consume potentially hazardous wastes such as phenol, C_6H_5OH, and trichloroethylene, C_2HCl_3.

It is even possible to synthesize enzymes and other proteins in the laboratory. The first successful effort was in 1968, when two groups of scientists, one at Rockefeller University and the other at the pharmaceutical firm of Merck, Sharpe and Dohme, independently prepared bovine ribonuclease A. Ribonuclease is an enzyme that catalyzes the cleavage of the genetic polymer ribonucleic acid (RNA). Once the 124 amino acid residues that make up bovine ribonuclease A were linked in the correct sequence, the resulting molecule possessed the catalytic activity of the naturally produced enzyme. This phenomenal feat required about 18 months of actual work, plus some 50 years of preliminary research. A cow does it in a minute or two.

The synthesis of bovine ribonuclease A was a great scientific achievement, richly deserving of the Nobel Prize that rewarded it. Although this research proved that the method worked, the direct laboratory or industrial synthesis of proteins is seldom carried out. There are easier ways of doing it. In fact, we can replace both the cow and the chemist with bacteria, thanks to recombinant DNA technologies.

■ Heredity and Chemistry

An important step in the long march that has lead to genetic engineering was the elucidation, in 1953, of the molecular structure of **deoxyribonucleic acid (DNA)** by James Watson and Francis Crick. DNA is the stuff from which genes are made, the molecular

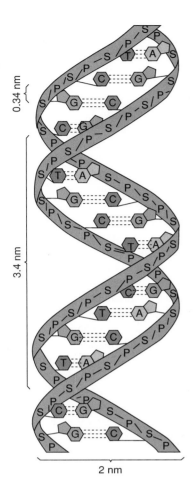

a.

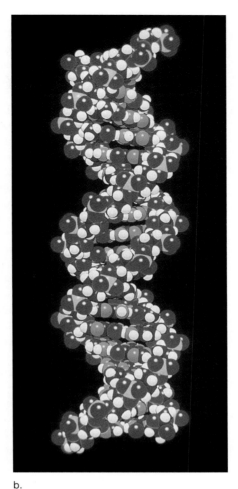

b.

Figure 13.11
The molecular structure of DNA.
a. A schematic representation in which P=phosphate group, S=sugar, A=adenine, C=cytosine, G=guanine, and T=thymine. (From Sylvia S. Mader, *Biology,* 3d ed. Copyright © 1990. Wm. C. Brown Communications, Inc., Dubuque, Iowa. All Rights Reserved. Reprinted by permission.) *b.* A space-filling model in which atoms are represented by spheres.

blueprint that stores the instructions for making every living thing, from slime mold to Albert Einstein. The molecule is an intertwined double helix consisting of a backbone of phosphoric acid groups to which are attached sugars in the form of deoxyribose rings. Bonded to each sugar residue is an amine base, a nitrogen-containing ring that, as the name suggests, has basic properties (see Figure 13.11).

Although the molecular mass of DNA may run into the millions, only four different bases are present: adenine (abbreviated A), thymine (T), cytosine (C), and guanine (G). In the center of the double helix, adenine and thymine always hydrogen bond to each other and cytosine and guanine are similarly coupled. These base pairs form the flat, parallel steps in the spiral staircase that is the nucleic acid molecule. Ten of these "steps" represent a 360° turn of the helix.

The two chains of the double helix are thus complementary; the bases of one matching the bases of the other in the A=T, C=G pairing described above. The consequences of this structure were immediately obvious to Watson and Crick who, in their historical first scientific paper on the subject, interjected perhaps an excess of British understatement: "It has not escaped our notice that the specific pairing we have postulated immediately suggests a possible copying mechanism for the genetic material."

By now, that copying mechanism has been thoroughly studied and the genetic code itself has been deciphered through an amazing piece of chemical cryptography. The code consists of units of three sequential bases called **codons.** Four different bases can be arranged in 4^3 or 64 different three-base combinations. It is as though one were making three-letter words out of A, T, C, and G. Of the 64 possible combinations, a few, like CAT, TAG, and ACT, make sense. Most are like GGG, TCC, or ATC, and do not have any meaning—at least not in English. But nature is clever at creating codes. Sixty-one of the 64 triplet codons correspond to specific amino acids. The identity and sequence of these codons are the instructions for synthesizing proteins. Thus, the codon sequence GTA in a DNA molecule signals that a molecule of the amino acid histidine should be incorporated into the protein, AAA codes for phenylalanine, and GGC stands for proline. Because there are 61 code "words" and only 20 amino acids, most amino acids are represented by more than one base triplet. In fact, three amino acids each correspond to six different codons. The three triplets that are not part of the amino acid code serve as signals to stop the synthesis of a protein.

It turns out to be relatively easy to determine experimentally the identity and order of bases in a segment of DNA. In fact, a massive research program is currently underway to determine the sequence of all the DNA in the **human genome.** This means all of the DNA in all 100,000 genes in 23 pairs of human chromosomes. To put this into personal perspective, you contain about 10 million million (10×10^{12}) nucleated cells. Within each of these cell nuclei is a complete set of the genetic information that makes you what you are—at least biologically. The information encoded there is uniquely yours (unless you happen to have an identical twin). It is written in molecular code on a thread of deoxyribonucleic acid about 2 meters (2 yards) long. If all of the DNA in all of your cells were placed end to end, the resulting double helix would stretch from here to the Sun and back more than 60 times!

But even more astounding than that astronomical figure is the amount of information that is carried by that invisible 2-yard thread. It consists of approximately 1.2×10^9 triplet codons, each capable of encoding an amino acid, if there are no nonsense sequences. The same three-base DNA code could be adapted to represent the 26 letters of the English alphabet. This means that your DNA could encode 1.2×10^9 letters or 3.1×10^8 six-letter words. These words would fill 1300 volumes of 500 pages each. And you carry that library in the DNA of each of your cells! No wonder the Human Genome Project is a formidable task.

13.8   *The Sceptical Chymist*

Sometimes authors get carried away by their own hyperbole. It might be a good idea to check the correctness of the claim that the DNA in an adult human being would stretch from the Earth to the Sun over 60 times. You will need to know that the average distance between the Earth and the Sun is 93 million miles. The other necessary information is in the paragraphs above and conversion factors are in Appendix 1.

Once the base sequence of a segment of DNA is known, the code can be "read" and translated into the primary structure of the associated protein. Furthermore, knowledge of the sequence of bases in a piece of genetic material suggests the possibility of creating copies in the laboratory. The necessary procedures have been developed and used to synthesize strands of DNA. Of course these synthetic polynucleotides need not conform to naturally occurring genes, so the potential exists for creating new genetic material.

As successful as these *in vitro* investigations have been, the easiest and most efficient place to synthesize nucleic acids is in the cell, not laboratory glassware. Recombinant DNA techniques have enabled researchers to insert a segment of DNA from one organism into the DNA of a host organism—usually a strain of bacteria. Bacteria contain rings of DNA called **plasmids.** These rings can be removed and cut by the action of special enzymes. Meanwhile, segments of DNA are either prepared synthetically or isolated from other organisms. This foreign DNA is inserted and patched into the plasmid ring by other enzymes. The modified plasmids are then reintroduced into the bacteria. Once inside the "host," the synthetic mechanism of the bacteria takes over, making copies of the "guest" DNA.

The artificially introduced genetic material also directs the bacteria to synthesize the protein for which it codes. Thus, bacteria can be made to produce enzymes and hormones that belong to vastly different species. A specific example is provided by insulin, the hormone used to treat diabetes. Insulin is a small protein consisting of 51 amino acids. It is produced by the pancreas, and it influences many metabolic processes. Most familiar is its role in reducing the level of glucose in the blood by promoting the entry of that sugar into muscle and fat cells. People who suffer from diabetes have an insufficient supply of insulin and hence elevated levels of blood sugar. Left untreated, the disease can result in poor blood circulation, especially to the arms and legs, blindness, and early death. However, diabetes can be controlled by administering insulin either orally or by injection.

Until recently, all insulin used by diabetics was prepared from the pancreas glands of cows and pigs, collected in slaughter houses. But the insulin produced by cattle and hogs is not identical to human insulin. Bovine insulin differs from the human hormone in 3 out of 51 amino acids; porcine and human insulin differ in only one. These differences are slight, but sufficient to undermine the effectiveness of bovine and porcine insulin in some human diabetics. For many years, there seemed to be no hope of obtaining enough human insulin to meet the need. While insulin has been synthesized in the laboratory, the process is far too complex for industrial adaptation. However, the lowly bacterium, *E. coli*, has been making human insulin since about 1980. This unlikely bit of interspecies cooperation is a consequence of using recombinant DNA techniques to introduce the gene for human insulin into this common organism. Although the *E. coli* has no use for our insulin, it generates it in sufficient quantities to harvest, purify, and distribute to diabetics.

Similar methods in molecular engineering have been used to induce bacteria to produce human growth hormone, a protein that is being tested as a treatment for dwarfism. New, safer vaccines have also been made using bacteria as chemical factories. Traditionally, vaccines have been made from killed or weakened viruses. Once exposed to this foreign material, the body's immune system creates chemical defenses against future infection. But sometimes the virus particles are not completely inactivated, and disease, not immunity, results. To avoid this risk, the DNA encoding for a characteristic but noninfectious part of a virus—for example, its protein coat—can be introduced into plasmids. The bacteria will consequently produce this particular protein. The protein is then isolated, concentrated, and used as a vaccine, one that carries essentially no risk of infection.

13.9 ■ *Consider This*

"The Far Side" cartoons of Gary Larsen frequently focus on some of the quirkier aspects of science and technology. Try your hand at a cartoon done in the style of "The Far Side" addressing genetic engineering.

13.10 ■ *Consider This*

For centuries animal breeders, agricultural researchers, and observant farmers have been bringing about the genetic transformation of animals and plants through selective breeding. Many people who accept such activities without question are strongly opposed to accomplishing similar ends through direct manipulation of genetic material. Account for this difference in response.

■ *The Modern Prometheus(?)*

Nature is obviously an indispensable aid and ally in medicinal and biological chemistry. Much of our success has come from understanding and imitating natural processes. If we are as wise as we are intelligent, the chemistry of the future will use these processes to alter and improve nature. Molecular engineering has made it possible to create nucleic acids, proteins, enzymes, hormones, drugs, and other biologically important molecules that do not exist in nature. The new molecules might be designed to be more efficient catalysts than their naturally occurring counterparts, more effective and less toxic drugs for treating a wide range of diseases, modified hormones that actually work better than the original. It is not at all fanciful to imagine a whole range of enzymes, engineered to consume environmentally hazardous wastes that are impervious to naturally occurring enzymes. Thanks to our ability to manipulate matter, it is just a matter of time before acquired immune deficiency syndrome (AIDS) and at least some forms of cancer become as infrequent as polio or small pox.

Even more exciting is the possibility of eradicating certain genetic defects. Our growing knowledge of the human genome, coupled with our understanding of the chemistry of genetics, hold the promise of altering our inheritance. The molecular locus of some hereditary diseases is already known. For example, people who suffer from sickle-cell anemia have hemoglobin with a minor alteration in its primary structure. The amino acid valine replaces glutamic acid in two places in this large, oxygen-carrying molecule. This slight change is sufficient to cause the hemoglobin to polymerize or gel when the oxygen concentration is low. Because we know the genetic code for these two amino acids, we know the sort of error in base sequence that accounts for this substitution. And if we know the error, perhaps we can someday correct it by manipulating the genetic material.

The manipulation of our human heredity holds great promise for individuals and for our species. Prospects include the elimination of sickle-cell anemia, diabetes, hemophelia, phenylketonuria, and dozens of other serious hereditary traits. The next logical step would seem to be the creation of new organisms. Frost-resistant strains of strawberries have been developed in the laboratory, and a new species of tomato has been created that is tough enough for long-distance shipping and still tastes like a tomato. There has even been talk of corn plants genetically altered so they could "fix" nitrogen from the atmosphere, like soybeans. This "unnatural" species would thus flourish without the need of "artificial" fertilizers. Indeed, genetic engineering could do much to help solve world hunger and starvation.

But the risks and dangers of molecular biology are equally great. Bioengineering carries with it vast potential for good or evil. It is haunted by the specter of efforts to design a master race or to subjugate or eliminate "defectives" through genetic manipulation. Hence, there is an intentional irony in the title of this final section. Prometheus was the demigod who stole fire and the flame of learning from the gods and brought these incomparable gifts to humanity. "The Modern Prometheus" is the subtitle of *Frankenstein,* Mary Shelly's classic study of scientific knowledge run amok. How to use our ever-growing knowledge of the natural world will be one of the greatest challenges of the twenty-first century.

13.11 ■ *Consider This*

The genetically engineered tomato mentioned in the text has been created by removing a small piece of the tomato's DNA that causes over-ripening. Without this bit of genetic information, tomatoes can be picked when they have fully ripened on the vine and shipped to market without fear that they will over-ripen and turn to mush in transit. No new genetic information is introduced. Yet, some people are boycotting the sale of the transformed tomatoes. Devise a newspaper advertisement for a supermarket that plans to introduce these genetically engineered tomatoes. Or, if you oppose their sale, devise an advertisement arguing against their introduction.

13.12	*Consider This*
■	In 13.4 Consider This, you were asked to consider the value of a massive investment of public money in the Superconducting Super collider. If the SSC works as scientists hope, it should provide important information about the fundamental nature of matter itself. Another example of megabuck big science, the Human Genome Project, has as its aim the complete translation of human genetic information. In this international endeavor, the base sequence of all the DNA in a characteristic set of human genes will be determined. Suppose only one of these projects can be funded. Where would you put the money and why?

■ *Conclusion*

The last chapter of this book, as almost all those that preceded it, ends with a dilemma: how can we balance the great potential benefits of modern science and technology and the risks that seem inevitably to be part of the Faustian bargain that brought us knowledge? The authors have looked, with myopic professorial vision, into the cloudy crystal ball of the future. It is in the nature of science that we cannot confidently predict what new discoveries will be made by tomorrow's chemists. Nor can we know the applications of those discoveries. Such uncertainty is one of the delights of our discipline. A chemist must learn to live with ambiguity, indeed, to thrive on it.

But all citizens of this planet must at least develop a tolerance for ambiguity and a willingness to take risks. Life itself is a biological, intellectual, and emotional risk. Of course, we all seek to maximize benefits, but we must recognize that individual gain must sometimes be sacrificed for the benefit of society. We live in multiple contexts—the context of our families and friends, our towns and cities, our states, our countries, our planet. We have responsibilities to all. *You*, the readers of this book, will help create the context of the future. We wish you well.

■ *References and Resources*

"Biotechnology." Information Pamphlet. Washington: American Chemical Society, 1985.

Committee on Critical Technologies. *Critical Technologies: The Role of Chemistry and Chemical Engineering*. Washington: National Academy Press, 1992.

Committee on Materials Science and Engineering. *Materials Science and Engineering for the 1990s: Maintaining Competitiveness in the Age of Materials* (Summary). Washington: National Academy Press, 1989.

Farrauto, R. J.; Heck, R. M.; and Speronello, B. K. "Environmental Catalysts." *Chemical & Engineering News*, Sept. 2, 1992: 34–44.

Foner, S. and Orlando, T. P. "Superconductors: The Long Road Ahead." *Technology Review*, Feb./Mar. 1988: 36–47.

Good, M. L. (ed.). *Biotechnology and Materials Science: Chemistry for the Future*. Washington: American Chemical Society, 1988.

Pimentel, G. C. and Coonrod, J. A. *Opportunities in Chemistry: Today and Tomorrow*. Washington: National Academy Press, 1987.

Science **256**, May 8, 1992. Issue devoted to molecular advances in genetic disease.

Science **258**, Oct. 2, 1992. Issue devoted to the Human Genome Project.

Stinson, S. C. "Biotechnology Providing Springboard to New Functional Materials." *Chemical & Engineering News*, July 16, 1990: 26–32.

Thayer, A. M. "High-Temperature Superconductors on the Road to Applications." *Chemical & Engineering News*, Nov. 27, 1989: 9–24.

———. "Catalyst Suppliers Face Changing Industry." *Chemical & Engineering News*, March 9, 1992: 27–49.

■ *Exercises*

1. Convert the following temperatures from the Kelvin or absolute temperature scale to the Celsius scale.

   a. the boiling point of liquid nitrogen: 77 K

   b. the freezing point of water: 273 K

   c. typical room temperature: 298 K

2. Most of the MRI instruments currently in use are tuned to detect the hydrogen atoms in water molecules and therefore provide a way of examining the soft tissue in the body. Compare this with the operation of X-rays and the types of structures they are capable of imaging.

*3. Dispersion forces are present in all molecules and provide the major attractive forces in the absence of other stronger forces. Moreover, dispersion forces increase with increasing molecular size and mass. Therefore, the boiling points of a group of similar compounds should increase with increasing molar mass. Test this hypothesis by plotting the boiling points (bp) of the following compounds versus their molar masses (MM).

   H_2S (bp −61°C, MM 34 g)

   H_2Se (bp −47°C, MM 81 g)

   H_2Te (bp −2°C, MM 130 g)

   From your graph, predict the boiling point of H_2O, MM 18 g. Then account for the fact that the boiling point of H_2O is 100°C.

4. Some high temperature superconductors have been developed by replacing Y and/or Ba in the 1–2–3 compound with other elements. (Apparently the copper is required to retain this special behavior.) Consult a periodic table to identify elements that might be used as substitutes for Y and others that might substitute for Ba. Justify your choices.

5. Explain why catalysts can speed up spontaneous reactions or cause a particular spontaneous reaction, out of several, to be favored, but cannot cause a nonspontaneous reaction to occur.

6. Review the earlier chapters in this text and identify two specific instances where catalysts play significant roles in each of the following areas.

   a. energy production

   b. synthesis of materials

   c. environmental protection

7. Some of the elements that are most useful as catalysts are quite expensive. Consult a chemical catalog to determine the cost per gram of Ag, Au, Ir, Pd, Pt, Rh, and Ru.

8. To understand why catalysts are finely divided, calculate the surface area of a cube 10 mm on a side and the surface area when it is divided into 1000 cubes 1 mm on a side.

9. Carbon is one of the most common elements, and the procedure for making C_{60} is quite simple. One vaporizes carbon in an inert atmosphere and some of the vapors condense to form C_{60}. Nevertheless, fullerenes were not discovered until the 1980s. Speculate why.

10. Calculate the energy for a photon of red light (700 nm) and compare it with the energy of a photon with sufficient energy to split H_2O. (See Chapters 2 and 9.)

11. Suggest why photosynthesis developed in a way that uses two photons of red light rather than one photon of ultraviolet light to energize the decomposition of water.

12. The activity of most enzymes is dependent on the temperature and acidity of the solution in which they function. Account for this observation in terms of protein structure.

13. What sequence of DNA bases would match up with the series ATTGAC? What sequence would bond with the new series?

14. Radiation and certain chemicals can damage various parts of a cell. Why is damage to the nucleus particularly dangerous?

15. Make a list of some potentially beneficial outcomes of molecular engineering and a list of some potentially dangerous ones.

Appendix 1

Measure for Measure: Conversion Factors and Constants

Metric Prefixes

deci	(d)	1/10	= 10^{-1}		deka	(da)	10	=	10^1
centi	(c)	1/100	= 10^{-2}		hecto	(h)	100	=	10^2
milli	(m)	1/1000	= 10^{-3}		kilo	(k)	1000	=	10^3
micro	(μ)	$1/10^6$	= 10^{-6}		mega	(M)			10^6
nano	(n)	$1/10^9$	= 10^{-9}		giga	(G)			10^9

Length

1 centimeter (cm) = 0.394 inch (in)

1 meter (m) = 39.4 in = 3.24 feet (ft) = 1.08 yard (yd)

1 kilometer (km) = 0.621 miles (mi)

1 in = 2.54 cm = 0.0833 ft

1 ft = 30.5 cm = 0.305 m = 12 in

1 yd = 91.44 cm = 0.9144 m = 3 ft = 36 in

1 mi = 1.61 km

Volume

1 cubic centimeter (cm^3) = 1 milliliter (mL)

1 liter (L) = 1000 mL = 1000 cm^3 = 1.057 quarts (qt)

1 qt = 0.946 L

1 gallon (gal) = 4 qt = 3.78 L

Mass

1 gram (g) = 28.3 ounces (oz) = 0.00220 pound (lb)

1 kilogram (kg) = 1000 g = 2.20 lb

1 metric ton (mt) = 1000 kg = 2200 lb = 2.20 ton (t)

1 lb = 454 g = 0.454 kg

1 ton (t) = 2000 lb = 909 kg = 0.909 mt

Time

1 year (yr) = 365.24 day (d)

1 hr = 60 minute (min)

1 day = 24 hours (hr or h)

1 min = 60 seconds (s)

Energy

1 joule (J) = 0.239 calorie (cal)

1 cal = 4.184 joule (J)

1 kilocalorie (kcal) = 1 Dietary Calorie (Cal) = 4184 J = 4.184 kilojoule (kJ)

1 kilowatt hour (kWh) = 3,600,000 J = 3.60×10^6 J

Constants

Speed of light $(c) = 3.00 \times 10^8$ m/s

Planck's constant $(h) = 6.63 \times 10^{-34}$ J s

Avogadro's number $(N_A) = 6.02 \times 10^{23}$ mole^{-1}

Atomic mass unit $(u) = 1.66 \times 10^{-24}$ g

Appendix 2

The Power of Exponents

Scientific or exponential notation provides a compact and convenient way of writing very large and very small numbers. The idea is to make use of positive and negative powers of 10. Positive exponents are used to represent large numbers. The exponent, written as a superscript, indicates how many times 10 is multiplied by itself. For example:

$$10^1 = 10$$
$$10^2 = 10 \times 10 = 100$$
$$10^3 = 10 \times 10 \times 10 = 1000$$

Note that the positive exponent is equal to the number of zeroes between the 1 and the decimal point. Thus, 10^6 corresponds to 1 followed by 6 zeros or 1,000,000. This same rule applies to 10^0, which equals 1. One billion, 1,000,000,000, can be written as 10^9.

When 10 is raised to a negative exponent, the number is always less than 1. This is because a negative exponent implies a reciprocal, that is, 1 over 10 raised to the corresponding positive exponent. For example:

$$10^{-1} = 1/10^1 = 1/10 = 0.1$$
$$10^{-2} = 1/10^2 = 1/100 = 0.01$$
$$10^{-3} = 1/10^3 = 1/1000 = 0.001$$

It follows that the larger the negative exponent, the smaller the number. The negative exponent is always one more than the number of zeroes between the decimal point and the 1. Thus, 10^{-4} is equal to 0.0001. Conversely, 0.000001 in scientific notation is 10^{-6}.

Of course, most of the quantities and constants used in chemistry are not simple whole number powers of ten. For example, Avogadro's number is 6.02×10^{23} or 6.02 multiplied by a number equal to 1 followed by 23 zeroes. Written out, this corresponds to $6.02 \times 100,000,000,000,000,000,000,000$ or 602,000,000,000,000,000,000,000. Switching to very small numbers, a wavelength at which carbon dioxide absorbs infrared radiation is 4.257×10^{-6} m. This number is the same as 4.257×0.000001 or 0.000004257.

Your Turn

Express the following numbers in scientific notation.

a. 100,000 **b.** 250 **c.** 1234.56

d. 0.00001 **e.** 0.004 **f.** 0.0876

Express the following numbers in conventional decimal notation.

a. 1×10^7 **b.** 2.998×10^8 **c.** 15×10^6

d. 1×10^{-7} **e.** 3.336×10^{-9} **f.** 6.63×10^{-34}

Appendix 3

Clearing the Logjam

You may have encountered logarithms in mathematics courses but wondered if you would ever use them. In fact, logarithms (or "logs" for short) are extremely useful in many areas of science. The essential idea is that they make it much easier to deal with very large *ranges* of numbers, for example moving by powers of 10 from 0.0001 to 1,000,000.

It is certainly likely that you have met logarithmic scales without knowing it. The Richter scale for magnitudes of earthquakes is one example. On this scale, an earthquake of magnitude 6 is 10 times more powerful than one of magnitude 5. An earthquake of magnitude 8 would be 100 times more powerful than one of magnitude 6. Another example is the decibel scale. Each increase of 10 units represents a 10-fold increase in sound level. Thus, a normal conversation at 1 meter (dB 60) is 10 times louder than quiet music (dB 50). Loud music (dB 70) and extremely loud music (dB 80) are 10 times and 100 times as loud, respectively, as a normal conversation.

A simple exercise using a pocket calculator can be a good way to learn about logs. You will need a calculator that "does" logs and preferably has a "scientific notation" option. Start by finding the logarithm of 10. Simply enter 10 and press the "log" button. The answer should be 1. Next find the log of 100 and then the log of 1000. Write down the answers. What pattern do you see? (The pattern may be more obvious if you recall that 100 can be written as 10^2 and 1000 is the same as 10^3.) Predict the log of 10,000 and then check it out. Then try the log of 0.1 or 10^{-1} and log of 0.01 (10^{-2}). Predict the log of 0.0001 and check it out.

So far so good, but we have only been considering whole-number powers of 10. It would be helpful to be able to obtain the logarithm of any number. Once again, your handy little electronic wizard comes to the rescue. Try calculating the logs of 20 and 200, then 50 (5×10^1) and 500 (5×10^2). Predict the log of 5×10^3 or 5000. Now for something slightly more tricky: the log of 0.05. Try to make sense of the answer. Finally, try the log of 2473 and the log of 0.000404. In each case, does the answer seem to be in the right ballpark? If you see any other interesting relationships, you may want to experiment further.

pH is simply a special case of these logarithmic relationships. It is defined as the negative of the logarithm of H^+ molarity. In mathematical terms, pH = $-\log (M_{H+})$. If the hydrogen ion concentration, M_{H+}, is 0.0000582 mole/liter, what is the pH of the solution? (Remember to change the sign, in other words, multiply by -1.)

If we can convert hydrogen ion concentration into pH, how do we go in the reverse direction? Your calculator can do this for you if it has a button labelled "10^x." (Alternatively, it may use two buttons: first "Inv" and then "log.") To demonstrate the procedure, suppose you wish to find the number whose logarithm is 2.65. First make a prediction of the approximate value (remember the numbers with logs equal to 2 and 3). Then proceed as follows: enter 2.65 and hit 10^x (or follow whatever steps are appropriate for your calculator). The display should give the desired number. Now try to apply the same procedure to some solutions. If the pH of a solution is 4.3, what is its H^+ molarity? (To do this calculation, *first* change the sign of the pH from positive to negative.) Finally, test yourself by determining the H^+ concentrations in solutions of pH 3.3 and 10.7.

Appendix 4

Answers to Selected Exercises

Chapter 1

1. 33% oxygen, 50% nitrogen, 17% carbon dioxide
3. 9000 ppm
5. 0.10 atmosphere
7. a. carbon disulfide
8. a. NaF
11. a. i. $N_2 + O_2 \rightarrow 2\,NO$
    ii. $N_2 + 2\,O_2 \rightarrow 2\,NO_2$
12. a. 2×10^{25} molecules
14. b. 2.375×10^2
15. b. 7490

Chapter 2

1.
Element	No. Electrons	No. Outer Electrons	Element Resembled
Na	11	1	Li, K*
Al	13	3	B, Ga*
N*	7*	5	P, As*
F*	9*	7	Cl

*other answers possible

2. a. 6, 14   b. 53, 131

4. a. $\ddot{\underset{H\,\,\,H}{\overset{..}{S}}}$ b. $\ddot{O}::C::\ddot{O}$

5. a. 300×10^{-9} m $= 3.00 \times 10^{-7}$ m, ultraviolet b. 10×10^{-6} m $= 1.0 \times 10^{-5}$ m, infrared
6. a. 1.0×10^{15} s^{-1} b. 3.0×10^{13} s^{-1}
7. a. 6.6×10^{-19} J   b. 2.0×10^{-20} J
9. 3.28 m
16. a. 0.08   b. 0.01
18. 12% increase in skin cancer, assuming % increase in skin cancer $= 2 \times$ % decrease in O_3 concn.

Chapter 3

1. a. $2.2 \times 10^{-4}\%$ b. 11:59:59.81 pm (0.19 sec to midnight)
   c. 0.038 sec to midnight   d. 0.0037 sec to midnight
4. a. $H:C:::N:$ b. $\dot{\ddot{N}}::N::\ddot{O}:$ or $:N:::N:\ddot{O}:$
5. a. linear   b. linear
9. rocks: 80%, oceans: $5.6 \times 10^{-2}\%$, fossil fuels: $5.3 \times 10^{-3}\%$
10. 7.5 ppm
12. a. Cl, 17, 35   b. Cl, 17, 37
13. a. 1.995×10^{-23} g   b. 5.327×10^{-23} g   c. 9.277×10^{-23} g
15. a. 44.0 g   b. 18.0 g
16. a. 63.6% N, 36.4% O   b. 11.1% H, 88.9% O
17. a. 318 g N, 182 g O   b. 56 g H, 444 g O
19. a. 0.10 mole Na   b. 0.0020 mole Cu   c. 1.91 mole U

Chapter 4

1. a. 100 g $H_2O(l)$ at 25°C
3. a. −184 kJ, exothermic   b. +222 kJ, endothermic
6. a. $2 C_2H_6(g) + 7 O_2(g) \rightarrow 4 CO_2(g) + 6 H_2O (g)$, 1409 kJ/mole or 47.0 kJ/g
7. 155 kJ/mole of N—Br bonds
14. $CO(g) + H_2(g) + O_2(g) \rightarrow CO_2(g) + H_2O(g)$
    518 kJ are released when 1 mole CO and 1 mole H_2 are burned
15. a. 5.2×10^5 barrels of oil   b. 3.1×10^{12} kJ
17. a. 373 K   b. −269°C
18. a. I: 0.815   II: 0.664   III: 0.830

Chapter 5

1. 79.5 gal, 40.4 drums/week, 173 drums/month, 2110 drums/year
3. 4.3 L/day, 1600 L/year
6. 55.4 moles H_2O, 55.4 moles H_2, 1350 L H_2
7. liquid water: 3.34×10^{22} molecules, ice: 3.07×10^{22} molecules
10. a. $KC_2H_3O_2(s) + H_2O(l) \rightarrow K^+(aq) + C_2H_3O_2^-(aq)$
    b. $Ca(OH)_2(s) + H_2O(l) \rightarrow Ca^{2+}(aq) + OH^-(aq)$
11. a. 3200 cal released   c. 1.08×10^7 cal absorbed
13. 81,000 cal
14. a. 1,000,000 L/person
15. a. $Mg^{2+}(aq) + CO_3^{2-}(aq) \rightarrow MgCO_3(s)$
16. a. 1.52×10^{-2} mole Na^+ b. 7.6×10^{-3} mole Ca^{2+}, 0.304 g Ca^{2+}

Chapter 6

1. a. acid   b. base   c. neither
4. a. 0.10 M   b. 0.200 M
5. a. 0.50 M H^+, 0.50 M NO_3^- b. 0.15 M Ba^{2+}, 0.30 M OH^-
6. a. $M_{H+} = 4.0 \times 10^{-2}$, $M_{OH-} = 2.5 \times 10^{-13}$, acidic
   b. $M_{H+} = 1.0 \times 10^{-7}$, $M_{OH-} = 1.0 \times 10^{-7}$, neutral
7. a. 1.40   b. 7.00
8. a. $M_{H+} = 1.0 \times 10^{-11}$, $M_{OH-} = 1.0 \times 10^{-3}$, basic
   b. $M_{H+} = 3.2 \times 10^{-4}$, $M_{OH-} = 3.1 \times 10^{-11}$, acidic
   c. $M_{H+} = 1.0 \times 10^{-7}$, $M_{OH-} = 1.0 \times 10^{-7}$, neutral
11. 1.27×10^5 g or 279 lb $Ca(OH)_2$
12. $NaHCO_3 + H^+ \rightarrow Na^+ + H_2O + CO_2$ 84.0 g $NaHCO_3$
    $NaOH + H^+ \rightarrow Na^+ + H_2O$ 40.0 g $NaOH$
    $CaCO_3 + 2 H^+ \rightarrow Ca^{2+} + H_2O + CO_2$ 50.0 g $CaCO_3$
14. 1.56 tons $CaCO_3$

Chapter 7

1. 7.0×10^7 kg Ca^{2+}
3. 0.82 mole $NaHCO_3$/L, 5.50 mole NH_4Cl/L
5. $CaCO_3 + 2 HCl \rightarrow CaCl_2 + H_2O + CO_2$
   $Na_2CO_3 + 2 HCl \rightarrow 2 NaCl + H_2O + CO_2$
   $NaHCO_3 + HCl \rightarrow NaCl + H_2O + CO_2$
6. 63.1 g Na_2CO_3
8. a. 0.0125 mole Ca^{2+}/L   b. 1.75×10^9 moles Ca^{2+} c. 3.5×10^{12}g
9. a. $Pb^{2+}(aq) + CO_3^{2-}(aq) \rightarrow PbCO_3(s)$
   b. $Ba^{2+}(aq) + SO_4^{2-}(aq) \rightarrow BaSO_4(s)$

10. a. 430 ppm Na^+ by molecules or ions
13. a. Fe is oxidized to Fe^{2+}, Cu^{2+} is reduced to Cu
    b. H is oxidized to H^+ (or H_2O), Al^{3+} (or Al_2O_3) is reduced to Al
16. a. precipitation   b. oxidation-reduction

Chapter 8

1. a. 7 protons, 8 neutrons
3. 10.812, boron (B)
4. 5.6×10^{-13} kg
5. a. $^1_0n + ^{235}_{92}U \rightarrow ^{87}_{35}Br + ^{146}_{57}La + 3\,^1_0n$
7. a. to vaporize water: 5.73×10^{10} cal (assuming a starting temperature of 100°C) or 6.53×10^{10} (assuming a starting temperature of 25°C); to heat ocean water: 3.54×10^{10} cal
10. a. $^{131}_{53}I \rightarrow ^{131}_{54}Xe + ^0_{-1}e$
    b. $^{238}_{92}U \rightarrow ^{234}_{90}Th + ^4_2He$
12. a. mass reactants = 238.00018 g
    mass products = 237.99559 g
    mass change = -0.00459 g
    energy change = -4.13×10^{11} J, 4.13×10^{11} J released per mole of U-238
16. 1/8 (0.125), 1/32 (0.0313), 1/1024 (0.00098)
17. 6×10^{18} atoms, 12×10^{18} atoms, 6.3×10^{-4} g

Chapter 9

1. a. 3.7×10^{17} kJ, 1.4×10^{10} tons coal
    b. 1.1×10^{21} kJ   c. 3.4×10^{18} kJ
2. 1.76×10^8 cal
5. a. 2.2×10^{17} g $C_6H_{12}O_6$ b. 3.2×10^{17} g CO_2
7. -478 kJ/mole, -13.3 kJ/g
8. $2\,Li \rightarrow 2\,Li^+ + 2\,e^-$ anode (oxidation)
    $I_2 + 2\,e^- \rightarrow 2\,I^-$ cathode (reduction)
    $2\,Li + I_2 \rightarrow 2\,Li^+ + 2\,I^-$
10. a. 4.32×10^6 J   b. 0.031 gal   c. gas cost: 3.7 cents
13. Gardner, MA: 880 m², Washington, DC: 600 m², Conway, AR: 467 m²
15. a. $SiO_2 + C \rightarrow Si + CO_2$
17. 2.3×10^{-10} g, Ga: 2.0×10^{12} atoms, As: 1.8×10^{12} atoms
19. a. mass change = -0.00432 g, energy released = 3.89×10^{11} J

Chapter 10

1. Break 10 C=C bonds, form 19 C—C bonds (ignore ends of molecule) -554 kJ, exothermic, remove heat
2. 1430 monomer units
3. 3.24×10^{11} moles C_2H_4, 7.94×10^{12} L C_2H_4
6. a. 56.9% Cl   b. 7.74% H
9. 521,000 g/mole
10. polystyrene, 92.2% C
12. n CH_2=CH–CH=$CH_2 \rightarrow \text{---}(CH_2\text{–}CH\text{=}CH\text{–}CH_2)\text{---}_n$, addition
14. addition polymerization: reactive double bond(s)
    condensation polymerization: functional groups capable of reacting with functional groups on other molecules
15. a. Ala-Glu-Lys   Glu-Ala-Lys   Ala-Lys-Glu
       Lys-Glu-Ala   Lys-Ala-Glu   Glu-Lys-Ala
16. 6120 g (ignoring H_2O)   b. 5202 g (considering H_2O)

Chapter 11

1. 60–90 tablets

2. a. H
          ..
   H:C̈:C:::N
         H

   CH_3 carbon: 4 single bonds
   CN carbon: 1 single bond, 1 triple bond

5. 24.9 g salicylic acid

10. Note the following structural similarities: both molecules are built around a benzene ring with an attached amino ($-NH_2$) group. Opposite the $-NH_2$ is an S=O in sulfanilamide and a C=O in para-aminobenzoic acid. Moreover, the NH_2 attached to the S in sulfanilamide and the OH in para-aminobenzoic acid are roughly the same size. Thus, the overall structures of the two molecules are quite similar, so it is reasonable that they might both fit the same active site.

12. testosterone: $5.5 \times 10^{-7}\%$ or 55 ppm, estrone: $3.3 \times 10^{-7}\%$ or 33 ppm

14. new pocess: 42%, old process: 18%

18. Not all animals respond to a given drug in the same way that humans do; they may be more or less sensitive. Therefore, the choice of experimental animals is very important in drug testing.

Chapter 12

1. 6.8×10^{10} Cal, 34 million people

4. Because of the difference in the shape of the alpha and beta bonds, the cellulose molecule presumably does not fit the active site of the enzyme that break down starch into glucose units.

6.
$$
\begin{array}{l}
\quad\quad\quad\quad\quad\overset{O}{\overset{\|}{}}\;\overset{H}{\overset{|}{}} \\
CH_3(CH_2)_nC-O-C-H \\
\quad\quad\quad\quad\overset{O}{\overset{\|}{}}\quad| \\
CH_3(CH_2)_nC-O-C-H \quad + \quad 3\,NaOH \rightarrow \quad 3\,CH_3(CH_2)_nC-O^-Na^+ \\
\quad\quad\quad\quad\overset{O}{\overset{\|}{}}\quad| \\
CH_3(CH_2)_nC-O-C-H \\
\quad\quad\quad\quad\quad\quad| \\
\quad\quad\quad\quad\quad\quad H
\end{array}
$$

$$
+ \begin{array}{l}
\overset{H}{\overset{|}{}} \\
HO-C-H \\
\quad| \\
HO-C-H \\
\quad| \\
HO-C-H \\
\quad| \\
\quad H
\end{array}
$$

10. a. 6   b. 27

12. 13.6 kg, 6.8×10^{25} amino acids

14. 88 min studying, 16 min jogging (assumes body weight of 150 lb)

17. 3.1×10^{22} Na^+ ions, 3.1×10^{22} K^+ ions

18. 90.8 mg KI, 69.4 mg I^-

Chapter 13

1. a. $-196°C$

4. For Y: La, Sc, Zr, Hf, etc., for Ba: Sr, Ca

5. Catalysts cannot make a reaction take place under conditions where it is thermodynamically impossible.

8. $600\ mm^2$, $6000\ mm^2$

10. 2.84×10^{-19} compared to 7.62×10^{-19} J

13. TAACTG, ATTGAC

Glossary

A

acid a substance that releases hydrogen ions, H^+, usually in aqueous solution *153*

acid anhydride a compound, typically an oxide of a nonmetallic element, that reacts with water to generate an acid *163*

acid deposition the process by which acid is deposited through precipitation, fog, or airborne particles *153*

acid neutralizing capacity (ANC) the capacity of a lake to resist change in pH when acids are added to it *171*

activation energy the energy necessary to initiate a chemical reaction *101*

active site the region of an enzyme molecule where the catalytic activity occurs *307*

addition polymerization a process in which monomeric molecules combine to form a polymer without the elimination of any atoms *278*

aerosol a form of liquid in which the droplets are so small that they stay suspended in the air rather than settling *15*

allotropes two forms of the same element that differ in their molecular or crystal structure, and hence in their properties *29*

alpha particle a particle given off during radioactive decay and consisting of two protons and two neutrons; it has a mass of 4 amu and a charge of +2 units *218*

amino acid a compound containing a carboxylic acid group (–COOH), a basic amino group (–NH₂), and a characteristic identifying group; amino acids polymerize to form proteins *281, 345*

anabolic steroid a compound that promotes muscle growth but can have serious negative side-effects *316*

androgens male sex hormones *313*

anion a negatively charged ion *133*

anode the electrode at which oxidation occurs *128, 249*

aquifer a large natural underground reservoir *141*

atmospheric pressure the force with which the atmosphere presses down on a given surface area *6*

atom the smallest unit of an element that can exist as a stable independent entity *10*

atomic mass the mass of an atom expressed relative to a value of exactly 12 for carbon-12 *33*

atomic mass unit (amu) a unit used to express the mass of individual atoms and molecules, equal to $1.66 \times 10^{-24} g$ *73*

atomic number the number of protons in an atomic nucleus, equal to the number of electrons in an electrically neutral atom *30*

atomic weight (*see atomic mass*) *30*

Avogadro's number the number of objects in one mole, 6.02×10^{23} *73*

B

basal metabolism rate (BMR) the amount of food energy necessary to support basic body functions *354*

base a substance that releases hydroxide ions, OH^-, usually in aqueous solution *154*

beta particle an electron released during radioactive decay; it has a mass of 1/1838 amu and an electrical charge of –1 unit *217*

biochemistry the branch of chemistry that deals with the chemistry of living things *72*

biomass materials produced by biological processes *109*

biotechnology technology based on the manipulation and alteration of biological materials, especially genetic material *382*

bond energy the amount of energy that must be absorbed to break a specific chemical bond, usually expressed in kJ/mole of bonds *98*

breeder reactor a fission reactor that converts U-238 to fissionable Pu-239 while it generates energy *216*

buckminsterfullerenes (fullerenes) C_{60} and related compounds *284*

C

calorie the amount of heat necessary to raise the temperature of exactly one gram of water by one degree Celsius; 4.184 J *92*

Calorie the amount of heat necessary to raise the temperature of exactly one kilogram of water by one degree Celsius; 1000 cal; used in nutrition *92*

carbohydrate a compound containing carbon, hydrogen, and oxygen, the latter two in the same 2:1 atom ratio as found in water *335*

carbon cycle the cyclic process in which carbon and its compounds circulate through the animal, vegetable, and mineral kingdoms *71*

catalyst a chemical substance that participates in a chemical reaction and influences its speed without undergoing permanent change *45, 375*

cathode the electrode at which reduction occurs *128, 249*

cation a positively charged ion *133*

chemical element a substance that cannot be broken down into simpler stuff by any chemical means *9*

chemical equation a representation of a chemical reaction using chemical symbols and formulas *11*

chemical formula a representation of the elementary composition of a chemical compound *10*

chemical reaction

chemical reaction a process whereby substances described as reactants are transformed into different substances called products *11*

chlorofluorocarbon a compound composed of the elements chlorine, fluorine, and carbon *48*

coenzyme a substance, generally consisting of small molecules working in conjunction with an enzyme to enhance the enzyme's activity *357*

cold fusion the process by which deuterium nuclei purportedly combine at or near room temperature to yield larger nuclei and large quantities of energy *261*

compound a pure substance made up of two or more elements in a fixed, characteristic chemical combination and composition *9*

condensation reaction a process in which monomeric molecules combine to form polymers by the elimination of small molecules such as H_2O *280*

copolymer a polymer consisting of two or more different monomeric units *280*

covalent bond a chemical bond created when two atoms share electrons (usually an even number) *33*

covalent compound a compound consisting of molecules that are in turn made up of covalently bonded atoms; a molecular compound *136*

current the rate at which the electrons flow; measured in amperes *249*

D

daughter the isotope formed by the radioactive decay of the "parent" isotope *218*

density mass per unit volume; usually expressed in grams per cubic centimeter or grams per milliliter *131*

deoxyribonucleic acid (DNA) the compound constituting the genetic material of all living things *384*

desalination any process that removes ions from salty water, such as sea or brackish waters *147*

dispersion forces relatively weak interactions responsible for the attraction between molecules *372*

distillation a purification or separation process in which a solution is heated to the boiling point and the vapors are collected and condensed *105*

double bond a covalent bond consisting of two pairs of electrons shared between two atoms *34*

E

efficiency   the fraction of heat energy that is converted to work in a power plant *113*

electrochemical cell   a device that converts the energy released in a spontaneous chemical reaction into electrical energy *249*

electrode   an electrical conductor that serves as the site of chemical reaction in an electrochemical or electrolytic cell *128*

electrolysis   the electrical decomposition of a compound into its constituent elements *128*

electrolyte   a solution that conducts electricity or a compound that will conduct electricity when dissolved in water *133*

electrolytic cell   a device in which electrical energy is converted to chemical energy *249*

electromagnetic spectrum   the entire range of radiant energy, including x–, γ–, ultraviolet, visible, infrared, microwave, and radio radiation *36*

electron   a subatomic particle with a mass of 1/1838 amu and a charge of –1 unit that is of great importance in atomic structure and chemical reactivity *30*

electronegativity   a measure of the attraction of an atom for electrons *129*

endothermic   absorbing heat *98*

energy   the capacity to do work *92*

entropy   a measure of randomness in position or energy *115*

enzyme   a biochemical catalyst, a protein that influences the rate and direction of a chemical reaction *306, 376*

equilibrium   a condition in which the rate of a physical or chemical process is equaled by the rate of the reverse process and no net change is observed *140*

ester   a compound made by the reaction of a carboxylic acid and an alcohol *280*

estrogens   female sex hormones *313*

exothermic   releasing heat *97*

F

fat   a triglyceride; a compound made from fatty acids and glycerol, $C_3H_8O_3$ *337*

fatty acid   an acidic compound with a long hydrocarbon chain; a component of fats and oils *340*

first law of thermodynamics (law of conservation of energy)   energy is neither created nor destroyed; the energy of the universe is constant *93*

fission (nuclear)   a reaction in which a large atomic nucleus, such as uranium-235, splits when struck by a neutron to form two smaller fragments and release large quantities of energy *208*

free radical   an unstable chemical species with an unpaired electron *45*

frequency   in wave motion, the number of waves passing a fixed point in one second *36*

fuel cell   an electrochemical cell in which a fuel such as hydrogen is allowed to react with oxygen under controlled conditions and the energy is liberated as electricity *248*

functional groups   groupings of atoms that confer characteristic properties on the molecule and the compound *303*

fusion (nuclear)   a reaction in which the nuclei of light atoms combine to form heavier nuclei and release large quantities of energy *258*

G

gamma rays   high-frequency, high-energy electromagnetic radiation released during radioactive decay *36, 218*

greenhouse effect   the process by which atmospheric gases such as CO_2, CH_4, and H_2O trap and return a major portion of the heat (infrared radiation) radiated by the Earth *64*

H

half-life   the time required for the level of radioactivity to fall to one-half of its initial value *223*

heat   the form of energy that flows from a hotter to a colder body *92*

heat of combustion   the quantity of heat released when a fuel is burned; variously expressed in cal/g, cal/mole, J/g, or J/mole *97*

heat of fusion   the heat that must be absorbed to bring about melting or fusion of a solid; variously expressed in cal/g, cal/mole, J/g, or J/mole *140*

heat of vaporization   the heat that must be absorbed to change a liquid into its vapor; variously expressed in cal/g, cal/mole, J/g, or J/mole *140*

heterogeneous catalysis   a catalyzed reaction in which the reactants and the catalyst are in different physical states or phases *376*

homogeneous catalysis   a catalyzed reaction involving a single physical phase *377*

hormones   substances produced by the body's endocrine glands that can have a wide range of physiological functions, including serving as "chemical messengers" *305*

hydrocarbon   a compound of hydrogen and carbon *13*

hydrogen bond   a relatively weak electrostatic attraction between a hydrogen atom bearing a net positive charge and a nitrogen, oxygen, or fluorine atom bearing a net negative charge; hydrogen bonds exist between some molecules and within some molecules *130*

human genome   the totality of human genetic information *386*

I

inertial confinement   a process being tested as a way of inducing nuclear fusion *260*

infrared radiation   heat radiation; the region of the electromagnetic spectrum that is adjacent to the red end of the visible spectrum and is characterized by wavelengths longer than red light *36*

inorganic chemistry   the branch of chemistry that deals primarily with chemicals of a mineral origin *72*

ion exchange   a process in which ions are interchanged, usually between a solution and a solid *144*

ionic bond   a chemical bond created by the electrostatic attraction between oppositely charged ions *133*

ionic compound   a compound consisting of positively and negatively charged ions *133*

I (isomers)

isomers   different compounds with the same formula and the same number and elementary identity of atoms; isomers differ in molecular structure, that is, the way in which the constituent atoms are arranged *108, 301*

isotopes   two (or more) forms of the same element whose atoms differ in number of neutrons hence in atomic mass *33*

J

joule   a unit of energy corresponding to kg m²/s² *92*

K

Kelvin scale   the absolute temperature scale whose zero corresponds to –273°C *113*

L

Lewis structure   a representation of molecular structure based on the octet rule and using dots to represent electrons *33*

lipids   fats, oils, and related compounds *337*

lipoproteins   compounds consisting of lipid and protein portions *344*

M

macronutrients   the major classes of compounds required for nutrition; carbohydrates, fats, and proteins *327*

magnetic resonance imaging (MRI)   a procedure for medical diagnosis based on the absorption of radio-frequency energy by hydrogen nuclei in a magnetic field *371*

malnutrition   a condition caused by a diet lacking in the proper mix of nutrients, even though enough calories are eaten daily *327*

mass number   the sum of the number of protons and the number of neutrons in any atomic nucleus *33*

Meissner effect   the phenomenon in which a magnet floats over a superconductor *373*

mixture   a physical combination of two or more substances (elements or compounds) that may be present in variable amounts *9*

molar mass   the mass of one Avogadro number (one mole) of atoms, molecules, or whatever particles are specified; usually expressed in grams *75*

molarity (M)   the number of moles of solute present in one liter of solution *155*

mole   one Avogadro number of anything; 6.02×10^{23} atoms, molecules, electrons, etc. *74*

molecular compound   a compound consisting of molecules; a covalent compound *136*

molecular mass   the mass of a molecule expressed relative to a value of exactly 12 for carbon-12 *75*

molecular weight   (*see molecular mass*) *75*

molecule   a combination of a fixed number of atoms, held together by chemical bonds in a certain geometric arrangement *10*

monomer   a small molecule that combines with other monomers to yield a polymer *270*

N

n-type semiconductor   a material that will not normally conduct electricity well, but will do so under certain conditions through the movement of negatively charged electrons *255*

neutralization   the chemical reaction of an acid and a base *155*

neutron   a subatomic particle with a mass of 1 amu and no electrical charge *30*

nonelectrolyte   a substance that does not conduct electricity, either by itself or in solution *133*

nonspontaneous   a process that will not occur by itself, but only if energy is supplied from some external source *116*

nuclear magnetic resonance (NMR)   a physical technique based upon the absorption of energy of radio frequency by atomic nuclei in strong magnetic fields; used for determining molecular structure *370*

nucleus (atomic)   the center of an atom *30*

O

octet rule   a generalization that in most stable molecules, all atoms except hydrogen will share in eight outer electrons *34, 301*

optical isomers   molecules with the same chemical composition but different structures that interact differently with polarized light *308*

organic chemistry   the branch of chemistry that deals primarily with compounds of carbon *72, 300*

osmosis   the natural tendency for a solvent to move through a membrane from a region of higher solvent concentration to a region of lower solvent concentration *147*

oxidation   a process in which atom, ion, or molecule loses one or more electrons *133*

P

p-type semiconductor   a material that will not normally conduct electricity well, but will do so under certain conditions through the movement of positively charged "holes" *256*

parent   the isotope undergoing radioactive decay *218*

periodic table   an organization of the elements in order of increasing atomic number and grouped according to similar chemical properties and similar electron arrangements *30*

pH   the negative logarithm of the hydrogen ion concentration when the concentration is expressed as moles of H^+ ion per liter of solution; $pH = -\log(M_{H^+})$ *157*

phenylketonuria   a genetic disease in which the body cannot properly metabolize the amino acid phenylalanine *349*

photon   a "particle" of radiant energy *38*

photosynthesis   the process by which green plants use sunlight to power the conversion of carbon dioxide and water into sugars such as glucose plus oxygen *62*

photovoltaic cell   a device that converts radiant energy into electricity *254*

physical chemistry   the branch of chemistry that seeks to elucidate the structure of matter and discover the general principles governing its transformation *72*

Planck's constant   the proportionality constant relating the energy of a photon to the frequency of radiation; 6.63×10^{-34} joule second *38*

plasma   a gas-like form of matter consisting of electrons and positively charged ions *259*

plasmid   a ring of bacterial DNA used in molecular engineering *386*

polar covalent bond   a covalent bond in which the electrons are not equally shared, but displaced toward the more electronegative atom *129*

polymer   a substance consisting of long macromolecular chains *270*

polyunsaturated   having more than one carbon-carbon double bond per molecule; usually applied to fats and oils *339*

positron   a subatomic particle with the mass of an electron, but with a charge of +1 unit *258*

primary structure   the identity and sequence of the amino acids present in a protein molecule *382*

proteins   polymers made up of various amino acids as the monomeric units; essential components of the body and the diet *281*

proton   a subatomic particle with a mass of 1 amu and a charge of +1 unit *30*

Q

quantized   separated into discrete energy levels, as, for example, the electronic energy levels in an atom *38*

quantum mechanics   a branch of physical science that treats the structure of atoms and molecules by constructing theoretical models *39*

R

radiation absorbed dose (rad)   the absorption of 0.01 joule of radiant energy per kilogram of tissue *220*

radioactivity   the phenomenon in which certain unstable atomic nuclei emit radiation and thereby undergo structural transformation *217*

receptor site   a site on a cell or molecule where a hormone or other biologically active molecule can bind *307*

reduction   a process in which an atom, ion, or molecule gains one or more electrons *133*

roentgen equivalent man (rem)   a unit of radiation exposure equal to the number of rads absorbed multiplied by a constant characteristic of the type of radiation *220*

S

saturated   having only single carbon-carbon bonds; usually applied to fats and oils *339*

second law of thermodynamics   it is impossible to completely convert heat into work without making some other changes in the universe; heat will not of itself flow from a colder to a hotter body; the entropy of the universe is increasing *115*

secondary structure   helices, parallel chains, and other localized structural features in the overall structure of a protein molecule *383*

semiconductors   materials that do not normally conduct electricity well, but will do so under certain conditions *254*

single bond   a covalent bond consisting of one pair of electrons shared between two atoms *33*

solute   a component that dissolves in a solvent to form a solution *132*

solution   a homogenous mixture at the atomic, molecular, and/or ionic level, consisting of a solute (or solutes) dissolved in a solvent *132*

solvent   in a solution, the component present in the largest concentration, usually the liquid component in which the solute dissolves *132*

specific heat   the quantity of heat energy that must be absorbed in order to increase the temperature of one gram of a substance by one degree Celsius *138*

spontaneous   a process that can occur by itself, though it may be necessary to initiate the reaction *116*

steady state   a condition in which a dynamic system is in balance so that there is no net change in the concentration of at least some of the participants in the reactions *42*

steroids   a class of ubiquitous and diverse organic compounds that contain three six-membered carbon rings and one five-membered ring *311*

substrate   the species upon which the enzyme acts *307*

superconductor   a substance that can conduct electricity with essentially no resistance *369*

T

temperature   a property that determines the direction of heat flow; when two bodies are in contact, heat always flows from the object at the higher temperature to that at the lower temperature *92*

teratogen   a substance that gives rise to birth defects *319*

tertiary structure   the overall three-dimensional structure of a protein molecule *383*

tetrahedron   a four-cornered figure with four equal sides *66*

tokamak   a donut-shaped reaction vessel in which a plasma of light isotopes is contained by magnetic fields, with the object of inducing nuclear fusion *259*

transition metals   the 30 elements in the 3rd through 12th columns of the periodic table *377*

triple bond   a covalent bond consisting of three pairs of electrons shared between two atoms *35*

troposphere   the part of the atmosphere that lies directly on the surface of the Earth *6*

U

ultraviolet   the region of the electromagnetic spectrum that is adjacent to the violet end of the visible spectrum and is characterized by wavelengths shorter than violet light *36*

undernourishment   a condition in which the daily intake of food is insufficient to supply the body's energy requirements *327*

unsaturated   having at least one carbon-carbon double bond per molecule; often applied to fats and oils *339*

V

vitamins   organic compounds that serve a variety of functions essential to life, often by promoting metabolic processes *356*

voltage   a measure of difference in electrical potential *249*

W

wavelength   in wave motion, the distance between successive peaks *35*

work   work is done when movement occurs against a restraining force; equal to the force multiplied by the distance over which the motion occurs *92*

Z

zeolite   a claylike mineral made up of aluminum, silicon, and oxygen; often used as a water softener or catalyst *144*

Credits

Photographs

Chapter 1
Opener: © European Space Agency/SPL/Photo Researchers, Inc.; p. 2 (top): © European Space Agency/SPL/Photo Researchers, Inc.; p. 2 (bottom): © NASA/Peter Arnold, Inc.; p. 3: © Stephen J. Krasemann/Photo Researchers, Inc.; 1.3: © Philippe Plailly/SPL/Photo Researchers, Inc.; 1.4: © Will & Deni McIntyre/Photo Researchers, Inc.; 1.6: © Spencer Grant/Stock Boston

Chapter 2
Opener: NASA

Chapter 3
Opener: © Greg Vaughn/Tom Stack & Assoc.

Chapter 4
Opener: © Bob Daemmrich/The Image Works; 4.1: © Tom McHugh/Photo Researchers, Inc.; 4.11: © Werner H. Muller/Peter Arnold, Inc.; 4.13: Courtesy Truman Schwartz

Chapter 5
Opener: © Frans Lanting/Photo Researchers, Inc.; 5.6 A–B: © Tom Pantages; 5.13: © Lawrence Migdale/Photo Researchers, Inc.; 5.15: © Wm. C. Brown Communications, Inc./Bob Coyle, photographer

Chapter 6
Opener: © John Shaw/Tom Stack & Assoc.; 6.7: © Lawrence Migdale/Photo Researchers, Inc.; 6.9 A–B: Courtesy Truman Schwartz; 6.10: © Marshall Prescott/Unicorn Stock Photos

Chapter 7
Opener: U.S. Geological Survey, EROS Data Center; 7.2: © Ron Trinca Photography; 7.3: Bill Tillery; 7.5: © Mike Davis/Photolink; 7.6: Courtesy Truman Schwartz

Chapter 8
Opener: Courtesy Ray Dennis, Northern States Power; 8.1: Bettmann Newsphotos; 8.2: © Crandall/The Image Works; 8.6: © J. Sohm/The Image Works; 8.7: Reuters/Bettmann

Chapter 9
Opener: © Fabricius & Taylor/Stock Boston; 9.2 and 9.8: Courtesy Truman Schwartz

Chapter 10
Opener: © Ogust/The Image Works; p. 268: AP/Wide World Photos; 10.8: © Ben Blankenburg/Stock Boston

Chapter 11
Opener: © Kolvoord/The Image Works; 11.2 A–B: © Wm. C. Brown Communications, Inc./Bob Coyle, photographer

Chapter 12
Opener: © Joel Dexter/Unicorn Stock Photos; 12.6A: © Lund/Custom Medical Stock Photos; 12.6B: © Rosemary/Custom Medical Stock Photos; p. 348: © Leonard Lessin/Peter Arnold, Inc.; 12.11: © John Paul Kay/Peter Arnold, Inc.; 12.13: International Atomic Energy Agency, Public Information Products

Chapter 13
Opener: © Comstock; 13.2: © Will & Deni McIntyre/Photo Researchers, Inc.; 13.3: © Scott Camazine/Photo Researchers, Inc.; 13.5: Courtesy Edmund Scientific Company; 13.6: © Japan Airlines/Photo Researchers, Inc.; 13.8: Courtesy AC Rochester, General Motors; 13.11B: © Nelson Max/Peter Arnold, Inc.

Figures/Text Art

Chapter 1
1.1, 1.2, 1.5: Illustrious, Inc.

Chapter 2
2.1, 2.2, 2.3, 2.4, 2.5, 2.6, 2.7, 2.8, 2.9, 2.10, 2.11, 2.12, 2.13, 2.14, TA 2.10, TA 2.11: Illustrious, Inc.

Chapter 3
3.1, 3.2, 3.3, 3.4, 3.5, 3.6, 3.7, 3.8, TA 3.3, TA 3.5, TA 3.8, TA 3.11: Illustrious, Inc.

Chapter 4
4.2, 4.3, 4.4, 4.5, 4.6, 4.7, 4.8, 4.9, 4.10, 4.12, 4.14, 4.15: Illustrious, Inc.

Chapter 5
5.1, 5.2, 5.3, 5.4, 5.5, 5.7, 5.8, 5.9, 5.10, 5.11, 5.12, 5.14, 5.16, 5.17: Illustrious, Inc.

Chapter 6
6.1, 6.2, 6.3, 6.4, 6.5, 6.6, 6.8, 6.11: Illustrious, Inc.

Chapter 7
7.1, 7.4, 7.8: Illustrious, Inc.

Chapter 8
8.3, 8.4, 8.5, 8.8, 8.9, 8.10: Illustrious, Inc.

Chapter 9
9.1, 9.3, 9.4, 9.5, 9.6, 9.7, 9.9A, 9.9B, 9.10A, 9.10B, 9.11, 9.12, 9.13: Illustrious, Inc.

Chapter 10
10.1, 10.2, 10.3, 10.4, 10.5, 10.6, 10.7, 10.9A–B, 10.10, 10.11, 10.12, 10.13, 10.14, TA 10.2, TA 10.3, TA 10.4, TA 10.5, TA 10.6, TA 10.7, TA 10.8: Illustrious, Inc.

Chapter 11
11.1, 11.3, 11.4, 11.5, 11.6, 11.7, 11.8, 11.9, 11.10, 11.11, 11.12, 11.13, 11.14, 11.15, 11.16, TA 11.1A, TA 11.1B, TA 11.2, TA 11.3, TA 11.4, TA 11.5, TA 11.6, TA 11.7, TA 11.8: Illustrious, Inc.

Chapter 12
12.1, 12.2, 12.3, 12.4, 12.5, 12.7, 12.8, 12.9, 12.10, 12.14, 12.15, TA 12.2, TA 12.3, TA 12.4, TA 12.5, TA 12.6, TA 12.7, TA 12.8, TA 12.9, TA 12.10, TA 12.11, TA 12.14, TA 12.15, TA 12.16, TA 12.17, TA 12.18, TA 12.19, TA 12.20, TA 12.22, TA 12.25: Illustrious, Inc.

Chapter 13
13.1, 13.4, 13.7, 13.9, 13.10, 13.11A: Illustrious, Inc.

Quotations/Songs

Chapter 9
"Here Comes the Sun" by George Harrison. © 1969 HARRISONGS LTD. International Copyright Secured. All Rights Reserved.

Index